THE YEASTS
Volume 1

THE YEASTS

Edited by

ANTHONY H. ROSE

*School of Biological Sciences, Bath University,
Claverton Down, Bath, England.*

AND

J. S. HARRISON

*The Distillers Co. (Yeast) Ltd., Great Burgh, Epsom,
Surrey, England.*

Volume 1

BIOLOGY OF YEASTS

1969

ACADEMIC PRESS · LONDON and NEW YORK

ACADEMIC PRESS INC. (LONDON) LTD.
BERKELEY SQUARE HOUSE
BERKELEY SQUARE
LONDON, W1X 6BA

U.S. Edition published by

ACADEMIC PRESS INC.
111 FIFTH AVENUE
NEW YORK, NEW YORK 10003

Library of Congress Catalog Card Number : 72–85465

SBN: 12–596401–3

Made and printed in Great Britain by
William Clowes and Sons, Limited, London and Beccles

Contributors

DO CARMO-SOUSA, L., *Laboratory of Microbiology, Gulbenkian Institute of Science, Oeiras, Portugal.*

FOWELL, R. R., *The Distillers Co. (Yeast) Ltd., Great Burgh, Epsom, Surrey, England.*

GENTLES, J. C., *Department of Medical Mycology, University of Glasgow, Glasgow, Scotland.*

HARRISON, J. S., *Research Department, The Distillers Co. (Yeast) Ltd., Great Burgh, Epsom, Surrey, England.*

HAWTHORNE, D. C., *Department of Genetics, University of Washington, Seattle, Washington 98105, U.S.A.*

KREGER-VAN RIJ, N. J. W., *Laboratory of Bacteriology and Serology, State University of Groningen, Groningen, The Netherlands.*

LA TOUCHE, C. J., *Mycology Unit, The General Infirmary, Leeds, England.*

LAST, F. T., *Glasshouse Crops Research Institute, Littlehampton, Sussex, England.*

MATILE, PH., *Institut für allgemeine Botanik, Eidg. Technische Hochschule, Zurich, Switzerland.*

MOOR, H., *Institut für allgemeine Botanik, Eidg. Technische Hochschule, Zurich, Switzerland.*

MORTIMER, R. K., *Donner Laboratory, University of California, Berkeley, California 94720, U.S.A.*

PRICE, D., *Glasshouse Crops Research Institute, Littlehampton, Sussex, England.*

ROBINOW, C. F., *Department of Bacteriology and Immunology, University of Western Ontario, London, Ontario, Canada.*

ROSE, A. H., *School of Biological Sciences, Bath University of Technology, Bath, England.*

Preface

It is often said that yeasts have been more intimately associated than any other group of micro-organisms with the progress and well-being of mankind. Many would base this claim largely on the capacity of certain yeasts to bring about a rapid and effective alcoholic fermentation, an activity which has been exploited for centuries in the manufacture of alcoholic beverages and in the leavening of bread. Others may prefer to acknowledge the contribution which yeasts have made in the elucidation of the basic metabolic processes of living cells. Few, however, would deny that yeast species, although less numerous than those of other major groups of micro-organisms, are sufficiently important that information on them needs to be fully documented so that scholars can continue to exploit their characteristic qualities. Over the years, many excellent texts have been published that deal comprehensively with yeasts and their activities. The present set of three volumes aims to bring the literature on this subject up to date.

So much has been written on yeasts that it is now impossible to encompass our knowledge of their microbiology in one volume. Moreover, it is beyond the capacity of any one person to write such a review, and so recourse must be made to a team of specialists each reviewing his or her own field of interest.

The first of the present volumes deals with the biology of yeasts, the second with yeast physiology and biochemistry, and the third with yeast technology. Together, these volumes present an up-to-date account of knowledge on yeasts and their activities. At the same time, each volume is sufficiently complete in itself to be perused separately.

We are extremely grateful to the authors of the chapters, whose labours have made this effort possible. Their forbearance, often under conditions of pressure and always with a deadline looming ahead, is very greatly appreciated. Our hope is that readers of these volumes will consider the venture to have been worthwhile.

Anthony H. Rose
J. S. Harrison

July 1969

Contents

5. Yeasts Associated with Living Plants and their Environs . . 183

F. T. LAST and D. PRICE

6. Yeast Cytology 219

PH. MATILE, H. MOOR and C. F. ROBINOW

7. Sporulation and Hybridization of Yeasts 303

R. R. FOWELL

8. Yeast Genetics 386

ROBERT K. MORTIMER and DONALD C. HAWTHORNE

Contents of Volume 2

Physiology and Biochemistry of Yeasts

Contents of Volume 3

Yeast Technology

Abbreviations

The following abbreviations are used for names of yeast genera:

Aureobasidium	*A.*	*Lipomyces*	*L.*
Ashbya	*Ash.*	*Metschnikowia*	*M.*
Bullera	*B.*	*Nematospora*	*N.*
Brettanomyces	*Br.*	*Oospora*	*O.*
Candida	*C.*	*Pichia*	*Pi.*
Citeromyces	*Cit.*	*Pityrosporum*	*Pit.*
Coccidiascus	*Co.*	*Pullularia*	*Pull.*
Cryptococcus	*Cr.*	*Rhodotorula*	*Rh.*
Debaryomyces	*D.*	*Saccharomyces*	*Sacch.*
Endomycopsis	*E.*	*Saccharomycodes*	*S.*
Eremascus	*Erem.*	*Schizosaccharomyces*	*Schizosacch.*
Geotrichum	*G.*	*Schwanniomyces*	*Schw.*
Hansenula	*H.*	*Sporobolomyces*	*Sp.*
Hanseniaspora	*Ha.*	*Torulopsis*	*T.*
Itersonilia	*I.*	*Trichosporon*	*Trich.*
Kloeckera	*Kl.*	*Trigonopsis*	*Trig.*
Kluyveromyces	*Kluyv.*	*Zygosaccharomyces*	*Zygosacch.*

The abbreviations used for chemical compounds are those recommended by the *Biochemical Journal* (1967; **102**, 1). Enzymes are referred to by the trivial names recommended by the "Report of the Commission on Enzymes of the International Union of Biochemistry" (1961, Pergamon Press, Oxford). All temperatures recorded in this book are in degrees Centigrade.

Chapter 1

Introduction

A. H. Rose and J. S. Harrison
School of Biological Sciences, Bath
University of Technology, Bath, England, and
Research Department, The Distillers Co. (Yeast) Ltd.,
Great Burgh, Epsom, Surrey, England

Compared with other major groups of micro-organisms—algae, bacteria, fungi and protozoa—the yeasts are represented by comparatively few genera and species. According to a recent source (Chapter 2, p. 5), there are only about 350 species of yeasts, grouped into 39 genera, whereas the number of algal, bacterial, fungal and protozoal species are in the tens of thousands. The pioneer microbiologists quickly discovered that yeasts differ from algae mainly because they do not possess a photosynthetic capability, and from protozoa which, since they lack a rigid cell wall, have closer affinities with animal cells. Bacteria, in general, can be quickly distinguished from yeasts by their much smaller size. It is with the fungi that yeasts have by far the greatest affinity. Indeed, the most widely accepted definition of yeasts is that they are fungi whose usual and dominant form is unicellular. This definition does not state what is meant by "usual" or "dominant", and the problems which can arise in interpreting these terms are referred to by Kreger-van Rij (Chapter 2, p. 5). How is it then that yeasts have continued to be recognized as a group of organisms quite separate and distinct from other fungi? There are several reasons for this, but most stem from the advantages that the single-cell habit confers on yeasts. Single cells have a much higher metabolic rate, weight for weight, than mycelial or more highly organized fungi, largely because of the much greater surface area : volume ratio. Yeasts grow faster and reproduce more rapidly than mycelial fungi, and also bring about chemical changes much more quickly. The single-cell habit also allows yeasts to attain a much wider ecological distribution than mycelial forms, a distribution which resembles that normally associated with bacteria.

No other group of micro-organisms has been more intimately associated with the progress and well-being of the human race than the yeasts. Their contribution to Man's progress has been based very largely on the capacity of certain yeasts to bring about a rapid and efficient conversion of sugars into alcohol and carbon dioxide, and so to effect an alcoholic fermentation of sugary liquids such as grain extracts and milk. It has often been said that yeasts are the oldest of cultivated plants. This has been documented by a series of marvellously preserved models, exquisitely constructed, of a bakery and a brewery on a wealthy estate which were uncovered in excavations at Thebes, and known to date from the XIth Dynasty (about 2000 B.C.; Winlock, 1920). Microscopic examination of sediments contained in beer urns found in this and other tombs has revealed what appear almost certainly to be yeast cells. Norsemen certainly prepared alcoholic beverages by fermenting milk, probably by yeast action, a practice which is perpetuated today among certain nomadic tribes. It was from these early beginnings that the large industries that we know today developed—industries such as brewing, wine-making and the large-scale production of baker's yeast. These, and many other processes in which yeasts (notably strains of *Saccharomyces cerevisiae*) have contributed to the industrial and commercial progress of the past century, are described in Volume 3 of this treatise.

A second major contribution which yeasts have made to Man's progress has been in the elucidation of the basic biochemical and metabolic processes of living cells. Because of their use in the brewing and baking industries, strains of *Sacch. cerevisiae* have been available in bulk quantities for almost a century. It was hardly surprising, therefore, that much of the early work on the biochemistry of living cells was done with these organisms. In 1897, the Buchner brothers discovered that fermentation of sucrose could be accomplished by cell-free yeast juice, an observation that is often said to have given birth to the new science of biochemistry. Volume 2 in this treatise deals comprehensively with the physiology and biochemistry of yeasts.

The present volume is concerned with the biology of yeasts—their taxonomy, systematics, ecology, morphology and cytology, as well as with their genetics and the ways in which they form spores.

The Dutchman, Antonie van Leeuwenhoek, is credited with being the first man to have observed yeasts microscopically. In 1680, he sent descriptions and drawings of yeast cells to the Royal Society in London, some four years after he had made public his observations on bacteria (see Volume 3, Chapter 4). Not a little controversy has surrounded this claim of Leeuwenhoek to have been the first to see yeast cells, but some diligent researching by A. Chaston Chapman (1931) showed quite conclusively the justification for Leeuwenhoek's priority in this claim.

Leeuwenhoek was, unfortunately, not a scientist—he was a draper by profession—and it was left to the botanists of the day to lay the firm foundations of yeast microbiology. Although the famous Swedish botanist Linné (1707–1778) (Linneus) interested himself briefly in the underlying cause of fermentation, it was not until the first half of the nineteenth century that significant progress was made towards an understanding of the biology of yeasts, and through this to an appreciation of yeast physiology and biochemistry.

Cagniard-Latour in 1837 demonstrated that beer yeast consists of spherical bodies which are able to multiply, and which belong to the Vegetable Kingdom. In this, he received support from Schwann, who termed yeast "Zuckerpilz" or "sugar fungus" from which the name "Saccharomyces" originates, and later from Kutzing. However, this cellular or vitalistic theory of fermentation was vehemently and occasionally mockingly attacked by a trio of chemists—Liebig, Wohler and Berzelius. The euphoria which stemmed from the success with which the newly recognized branch of organic chemistry had succeeded in explaining hitherto complex and mysterious organic processes convinced this trio of chemists that chemical reactions, rather than the activities of living cells, could perfectly well explain the alcoholic fermentation of sugars.

It fell to Louis Pasteur finally to prove that fermentation is due to living cells—thereby winning the day for the vitalistic theory of fermentation. Pasteur continued to make masterly contributions to yeast microbiology and our understanding of fermentation in general, particularly in his "Études sur le Vin" (1866) and "Études sur la Bière" (1876) in which he clarified much concerning the effect of oxygen on alcoholic fermentation by yeast. The interested reader should turn to the Dubos (1950) biography for more detail on Pasteur's contributions, and to the case study made by Conant (1952).

After Pasteur, there followed a period of intense activity during which yeast taxonomy and systematics were rapidly embellished. Many famous microbiologists and botanists were associated with this era, including the great Danish student of yeasts, Emil Christian Hansen. A valuable work on yeast fermentations came from Alfred Jörgensen in 1886, and this was followed in 1912 by the first important treatise devoted entirely to yeasts which came from Alexandre Guilliermond working in Lyon. Guilliermond's book was the first to include keys for identifying yeasts. Advances in yeast taxonomy and systematics have in recent years been dealt with most comprehensively by workers in Delft in Holland. This group, which was inspired very largely by A. J. Kluyver, has produced a series of authoritative texts on yeast taxonomy (Stelling Dekker, 1931 ; Lodder, 1934 ; Diddens and Lodder, 1942) the latest of which is about to be published (Lodder, 1970).

As the science of microbiology has grown at an ever-increasing rate, so other aspects of the biology of yeasts have been studied in depth. These studies have been reviewed extensively over the years in several texts, including that by Ingram (1955), and those edited by Roman (1957), Cook (1958) and Reiff *et al.* (1960, 1962). The present volume and its companions are the latest to review the literature on yeasts comprehensively and, in so doing, we hope that they do justice to the work of those microbiologists who, over the past two centuries, have devoted their lives to the study of some of man's smallest and most faithful servants.

References

Chapman, A. C. (1931). *J. Inst. Brew.* **37**, 433–436.

Conant, J. B. (1952). "Pasteur's Study of Fermentation", Case 6. Harvard Case Histories in Experimental Science. 57 pp. Harvard University Press, Cambridge, Massachusetts.

Cook, A. H., ed. (1958). "The Chemistry and Biology of Yeasts". 763 pp. Academic Press, New York.

Diddens, H. A. and Lodder, J. (1942). "Die Hefesammlung des Centraalbureau voor Schimmelcultures. II. Die anaskosporogenen Hefen". 511 pp. North-Holland Publ. Co., Amsterdam.

Dubos, R. J. (1950). "Louis Pasteur, Free Lance of Science". Little Brown & Co., Boston, Massachusetts.

Guilliermond, A. (1912). "Les Levures". *In* "Encyclopédie Scientifique", (A. A. Toulousse, ed.), Octave Doin, Paris. See "The Yeasts". Translated and revised by F. W. Tanner, 1920. Wiley, New York.

Ingram, M. (1955). "An Introduction to the Biology of Yeasts". 273 pp. Pitman, London.

Jörgensen, A. (1886). "Die Mikroorganismen der Gärungsindustrie". See "Microorganisms and Fermentation". Translated and edited by A. Hansen, 1948. 550 pp. Griffin & Co. Ltd., London.

Lodder, J. (1934). "Die Hefesammlung des Centraalbureau voor Schimmelcultures. II. Die anaskosporogenen Hefen". 256 pp. North-Holland Publ. Co., Amsterdam.

Lodder, J., ed. (1970). "The Yeasts". North-Holland Publ. Co., Amsterdam.

Lodder, J. and Kreger-van Rij, N. J. W. (1952). "The Yeasts, a Taxonomic Study". 713 pp. North-Holland Publ. Co., Amsterdam.

Pasteur, L. (1866). "Études sur le Vin". 264 pp. Imprimeurs Impérials, Paris.

Pasteur, L. (1876). "Études sur la Bière". 383 pp. Gauthier-Villars, Paris.

Reiff, F., Kautzmann, R., Lüers, H. and Lindemann, M. (1960). "Die Hefen. I. Die Hefen in der Wissenschaft". 1024 pp. Verlag Hans Carl, Nürnburg.

Reiff, F., Kautzmann, R., Lüers, H. and Lindemann, M. (1962). "Die Hefen. II. Technologie der Hefen". 983 pp. Verlag Hans Carl, Nürnburg.

Roman, W., ed. (1957). "Yeasts". 273 pp. W. Junk, The Hague.

Stelling-Dekker, N. M. (1931). "Die Hefesammlung des Centraalbureau voor Schimmelcultures. I. Die sporogenen Hefen". 587 pp. North-Holland Publ. Co., Amsterdam.

Winlock, C. E. (1920). *Bull. met. Mus. Art* **12**, 128–149.

Chapter 2

Taxonomy and Systematics of Yeasts

N. J. W. KREGER-VAN RIJ

Laboratory of Bacteriology and Serology
State University of Groningen, Groningen, The Netherlands

I. Introduction

The group of micro-organisms known as the "yeasts" is by traditional agreement limited to fungi in which the unicellular form is predominant. Vegetative reproduction is usually by budding. This group does not constitute a taxonomic unity, although it comprises subdivisions of narrowly related species. The diversity of the yeasts is illustrated by the fact that 39 genera and some 350 species are recognized. The features by which the species are distinguished are for the smaller part morphological, and for the greater part physiological.

Of the two topics mentioned in the title of this chapter, taxonomy deals with the principles of classification, and systematics with the results of the application of these principles. In practice, it appears to be difficult to keep these subjects clearly separated. The principles are often best explained and illustrated by their application. Moreover, before discussing the principles and practice of classification, the special features of the yeasts used for that purpose, i.e. the criteria for classification, should be mentioned. In the selection of the criteria, we are also guided by taxonomic principles.

Notwithstanding these difficulties, I have divided the subject matter of this chapter into three sections. In the first section, the criteria used for differentiation and classification will be discussed, in which special attention is given to criteria for the standard description of species. In the section on taxonomy, the phylogenetic theories of Guilliermond and of Wickerham are introduced, both of which have influenced and governed the classification of yeasts. The classification of the higher taxa by different authors is not discussed, but is summarized in a table. The principles of classification of genera and species are considered in more detail. In the last section, a survey of the classification is given, with emphasis on the delimitation and subdivision of the genera.

Yeast taxonomy and systematics are still subject to a process of development by the description of many new species and the introduction of various new features for distinction. An account of the recent state of development is given in "The Yeasts", edited by Lodder (1970), a comprehensive study of yeast classification to which thirteen authors have contributed. This monograph contains detailed descriptions of the species and the methods for identification, including tables and keys. As far as possible, the present chapter follows the classification of the species as given by the authors of this work.

II. Criteria Used in the Classification of Yeasts

A. CHOICE OF CRITERIA

Features which are different in the various yeasts may be used as criteria in classification. What is generally indicated as a feature is the response of the yeast to certain prescribed test conditions: a change in the test conditions may bring about a different response. This holds both for morphological and physiological features. Since differences are important in systematics, it is necessary to compare the responses under identical conditions. This has led to the development of certain standardized tests which have become more or less generally accepted. However, the attachment of certain laboratories to their own methods, with which they are thoroughly acquainted, has meant that, even in the standard tests, slight variations occur. These are not considered to be of great importance but, in the event of doubt when comparing a yeast strain with the standard description of a certain species, it is advisable to use identical methods.

The important details of a test method are the condition and amount of the inoculum, the composition of the medium, the apparatus used, the time and temperature of incubation, and the method of evaluating the results of the test.

For use in systematics, it is desirable that a property of a yeast be constant under standard conditions. However, several properties of yeasts show variations. Yeasts are examined as cultures on certain media, which means that selection is started from the moment of isolation. Keeping yeasts for a prolonged period on these media extends the selection and may lead to a fairly constant product which differs, however, from the original organism. The use of lyophilized cultures may considerably diminish the risk of a change in properties, although selection is not entirely excluded.

B. VARIABILITY OF CRITERIA

Some features used in systematics which may be variable are discussed in the following section of the chapter.

Firstly, cultures of many yeasts are known to change from the smooth (S) to the rough (R) colony form. This change is not only apparent from the aspect and consistency of the streak cultures, but also involves the shape and association of cells as well as pellicle formation. The glistening, smooth and soft culture changes to a mat, rough, tough form; the round to short-oval cells growing singly or in pairs change into more elongate ones in tree-like formations; the absence of pseudomycelium gives way to a well-developed pseudomycelium; also the rough strains may form a dry, mat pellicle. In mycelial yeasts there is a tendency to produce more hyphae and fewer budding cells, and also fewer arthrospores.

Wickerham (1964b) recognizes these changes as mutation from fermentative to oxidative forms in which many genes are involved, with transitional stages. He studied the phenomena in a strain of *Hansenula petersonii*, and proposed the theory that the genes in question are not individually mutated, but controlled by other genes which regulate metabolism in accordance with the environment. The changes in metabolism involve a decreased fermentative capacity, the utilization of other metabolites instead of sugars, and the production of hydrophobic instead of hydrophilic molecular structures on the outer cell wall. Thus the environment strongly influences the S–R variation. For instance, the smooth form of *H. petersonii* appeared to be stable on agar containing glucose, peptone, yeast extract and malt extract; on malt-agar the colony form changed from smooth to rough. This change in colony form on malt-agar is observed in many yeasts. It is conspicuous in the so-called *guilliermondii* group which includes *Pichia guilliermondii*, *Pi. scolyti*, *Pi. rhodanensis* and related species, none of which ferments maltose, or only weakly. The S–R variation especially affects the morphological characteristics used in classification, such as pseudomycelium formation, and this has to be reckoned with.

A second variation in yeast strains kept in a collection is the loss of sporulating ability. Freshly isolated strains very often sporulate abundantly on various media. As they are maintained for a longer period on a special medium, sporulating conditions seem to become more restricted, or asporogenous forms get the upper hand, until it is difficult to find any spores at all when the usual induction methods are used.

It is well known that spores often yield cultures which readily sporulate again. Beijerinck (1898) used this phenomenon to restore sporulation in old cultures and to isolate new sporulating strains by dry-heat treatment. Wickerham and Burton's (1954) heat treatment in liquid medium may have the same result.

A change in the colony form from smooth to rough may result in a loss of sporulating ability under the usual conditions. Wickerham (1964b) observed the reverse in a strain of *Hansenula petersonii* where spores were found in the rough but not in the smooth form. It may be expected that a change in the physiology of the yeast strain requires a change in the conditions needed for sporulation.

Thirdly, in some yeast species, the cells in culture gradually change from haploid to diploid. This results from a diploidization not followed by sporulation, and outgrowing of the haploid cells by diploid ones. Wickerham and Burton (1962) mention this for cultures of the homothallic species *Hansenula mrakii* and *H. californica* in which use of laboratory media suppressed sporulation. This change may also be expected in yeasts which are both homo- and heterothallic, and in which diploidized unisexual cells sporulate with difficulty or not at all, for instance in *Pichia membranaefaciens*. Kreger-van Rij (1964) found that old diploid strains of this species, which did not sporulate any more, could still be mated with the appropriate mating type after which sporulation followed. Freshly isolated mating-type strains of this species which did not sporulate alone easily changed from haploid to diploid on malt-agar.

Finally, the influence of the culture medium on the yeast has been apparent in a number of species, and includes changes in fermentation and assimilation of some sugars especially in members of the genus

Saccharomyces. Saccharomyces capensis acquired by mutation the ability to ferment maltose (Roberts and van der Walt, 1960). On malt-agar, the originally maltose-negative cells were readily outgrown by the maltose-positive ones. Another type of variation which is also influenced by the presence of maltose in the medium is the acquisition of the ability to ferment sucrose involving α-glucosidase. This type of change was found in *Sacch. italicus* and *Sacch. steineri* (Peynaud and Domercq, 1956), and in a number of yeast species, all producing melibiase, studied by Scheda and Yarrow (1966). Changes in the cultures are known to be very slow. The last authors nevertheless assume that a mutation is involved, but that the mutation rate is low; the ability to ferment sucrose has no selective value on malt-agar.

It seems possible that a failure to find a sharp distinction between species fermenting maltotriose and those not fermenting it, in strains kept on malt-agar, is also due to mutation or to adaptation to this sugar. Gilliland (1956) has suggested the use of maltotriose fermentation as a criterion for differentiating between certain *Saccharomyces* species. True brewer's yeasts which have been in contact with starch and wort are expected to be positive. Kudrjawzew (1960) also used this property to distinguish between *Sacch. cerevisiae* and *Sacch. vini*, and between *Sacch. carlsbergensis* and *Sacch. uvarum*. Most of the strains of *Sacch. vini* and *Sacch. uvarum*, which are unable to ferment maltotriose, had been isolated from wine and other fruit juices.

In the genus *Saccharomyces*, which comprises groups of narrowly related species which may differ in a single feature depending on a single gene, transitions from one species to another are possible. From an ecological point of view, the acceptance of a species in its original form may be justifiable.

C. CRITERIA USED IN THE STANDARD DESCRIPTION

The procedures for establishing the criteria used for the description of yeast species have for the greater part been described by Wickerham (1951) and by Lodder and Kreger-van Rij (1952). For details the reader is referred to these works. The literature on other methods will be given in the following text.

1. *Vegetative Reproduction*

The manner of vegetative reproduction, including the shape and size of the cells, is studied in standard liquid and on standard solid media, the latter on slants, on slides or in Dalmau plates. There are two means of vegetative reproduction in yeasts: by the formation of buds, or of hyphae with cross-walls.

Buds may have a narrow or a broad base. They occur multilaterally on the cell (Fig. 1) or are confined to special regions; for instance, chiefly on the shoulders of elongate cells or exclusively at one or both poles (Figs. 2 and 3). The latter mode of budding may lead to the formation of the so-called multiple scars, i.e. concentric ridges of bud scars (see p. 57). Budding cells which remain attached to each other are referred to as pseudomycelium (Figs. 4, 5 and 6). Very often, there is a differentiation among them of cells which grow out to a greater length, the pseudomycelium cells, and others which remain short and are referred to as blastospores or blastoconidia. Blastospores arise terminally on the pseudomycelium cells; on the other hand, blastoconidia arise often on small protuberances, but not terminally.

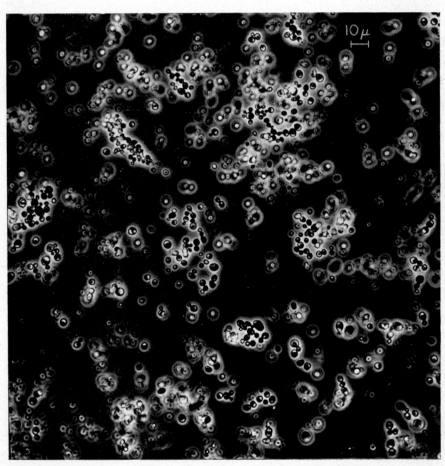

FIG. 1. Photomicrograph (phase contrast) showing multilateral budding in *Debaryomyces hansenii*.

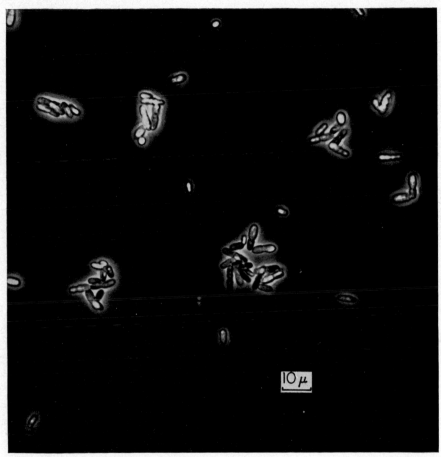

FIG. 2. Photomicrograph (phase contrast) showing monopolar budding in *Pityrosporum ovale*.

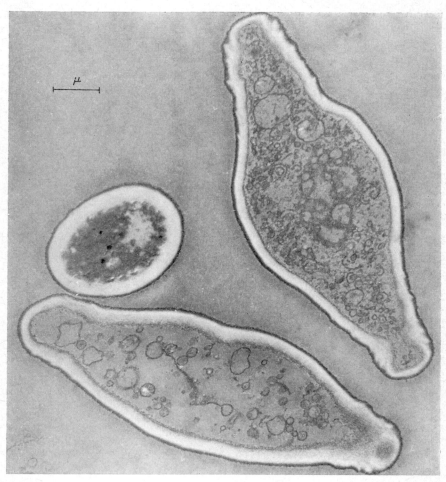

FIG. 3. Electron micrograph of a thin section through cells of *Hanseniaspora osmophila* showing bipolar budding with multiple scars. The organisms were fixed with osmic acid and uranyl acetate.

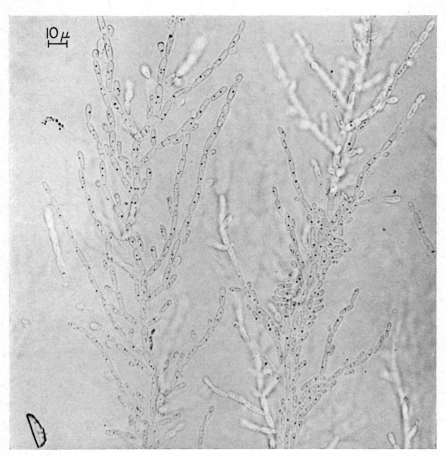

FIG. 4. Photomicrograph of *Pichia membranaefaciens* showing little differentiation of cells.

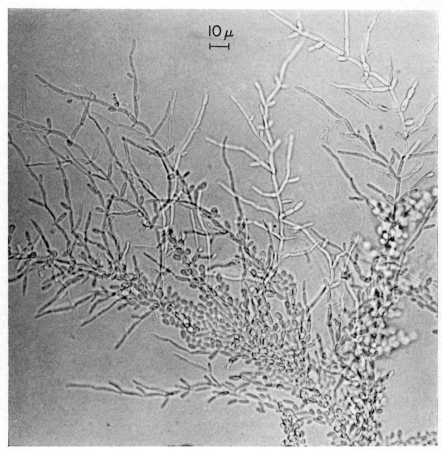

FIG. 5. Photomicrograph showing well developed pseudomycelium in *Candida parapsilosis*.

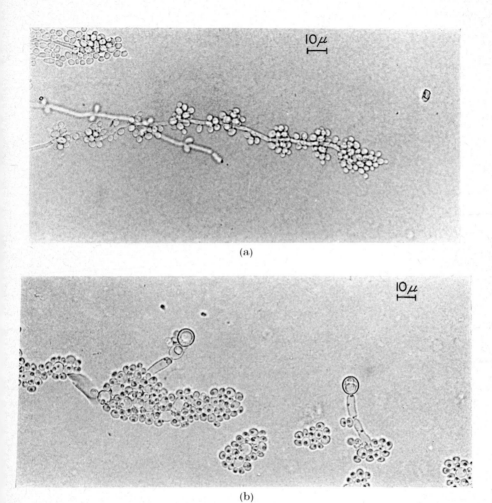

(a)

(b)

Fig. 6. Photomicrographs of *Candida albicans* showing (a) pseudomycelium, and (b) chlamydospores.

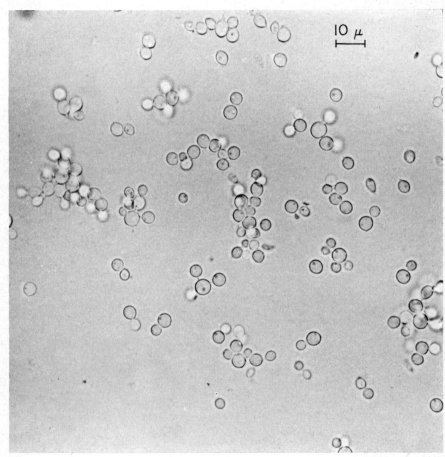

FIG. 7. Photomicrograph showing bud formation on small stalks in
Sterigmatomyces halophilus.

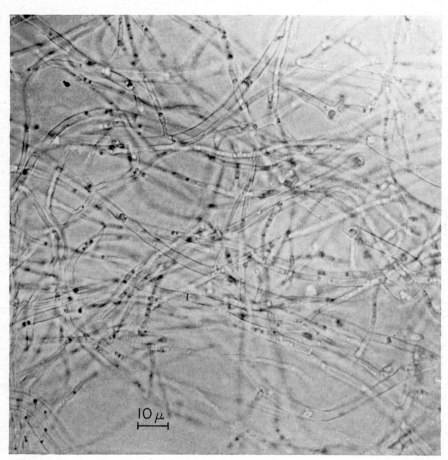

FIG. 8. Photomicrograph showing true mycelium in *Endomycopsis burtonii*.

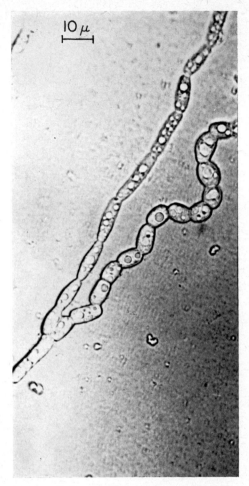

FIG. 9. Photomicrograph showing arthrospores in *Trichosporon cutaneum*.

One special type of cell includes those formed on stalks by budding. This is typical of members of the genus *Sterigmatomyces*, in which the cells are not forcefully discharged (Fig. 7). Ballistospores also arise on stalks, but are discharged by a water-drop mechanism.

True hyphae, which may form side branches, grow at the tip. Cross-walls are formed without constriction (Fig. 8). The hyphae may be split up at the cross-walls to form arthrospores (Fig. 9). Cross-wall formation in single cells followed by fission is typical of the genus *Schizosaccharomyces*.

Electron microscope studies have been made of bud formation and cross-wall formation in a few yeast species. It appears that there is a similarity between the broad cross-walls of *Schizosaccharomyces pombe* (Streiblová *et al.*, 1966) and the narrow ones in budding yeasts such as *Saccharomyces cerevisiae* (Agar and Douglas, 1955; Hashimoto *et al.*, 1959) and *Rhodotorula glutinis* (Marchant and Smith, 1967). On the other hand, Marchant and Smith (1968) reported differences in the mode of bud separation between *Saccharomyces cerevisiae* and *Rhodotorula glutinis*.

The distinction between budding and fission has always played an important part in yeast taxonomy. At present, however, there is a tendency to depreciate the importance of this distinction. It may be worthwhile to reserve judgement until a closer, more extensive, examination of the ultrastructure of budding and fission has established their resemblances and differences.

2. Sexual Reproduction

The main features of sexual reproduction in ascogenous yeasts are the mode of ascus formation and the shape of the spores. The shape of the ascus, the number of spores per ascus, the colour of the spores, their possible easy liberation from the ascus and their mode of germination may also be typical criteria.

Ascospore formation plays an important role in yeast systematics. However, knowledge of the factors controlling it is very restricted. Miller and coworkers (Miller and Hoffmann-Ostenhof, 1964) have contributed to a better insight into the sporulation of *Saccharomyces cerevisiae*. A number of different sporulation media are used. They are inoculated with the yeast in a well-nourished condition, the pre-sporulation medium being of importance especially when the sporulation medium does not support growth. Various temperatures of incubation may be tried, generally lower than the optimum temperature for growth. Sporulation is often confined to a restricted region of the streak culture.

On the sporulation media, the ascus may be formed by a direct change from a vegetative cell, or sporulation may be preceded by conjugation either of two single cells or of a mother cell and a bud (Figs. 10 and 11). Conjugation of mycelial cells is termed *anastomosis*. A type of conjugation which may also occur in mycelial yeasts is gametangiogamy, a conjugation of two cells which act as gametangia and which are outgrowths of the hyphae. The hyphal cells remain haploid and are separated by cross-walls from the fused gametangia which form the ascus. If spores are formed in a culture from a single spore or from a haploid cell, the yeast is homothallic; if a mixture of cells of opposite mating type is required for sporulation, the yeast is heterothallic. Several yeast species are known to be both homothallic

2

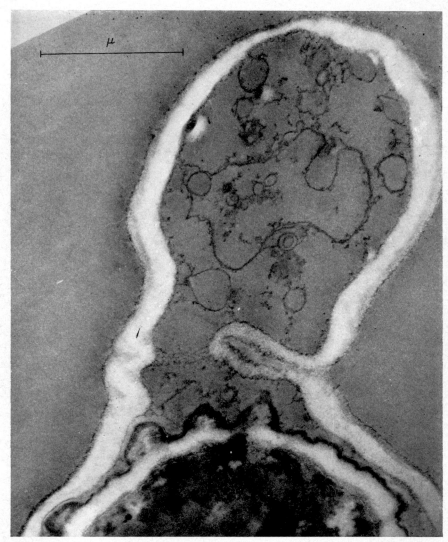

Fig. 10. Electron micrograph of a section through an ascus of *Debaryomyces phaffii* showing conjugation between mother cell and bud. The cells were fixed with osmic acid and uranyl acetate.

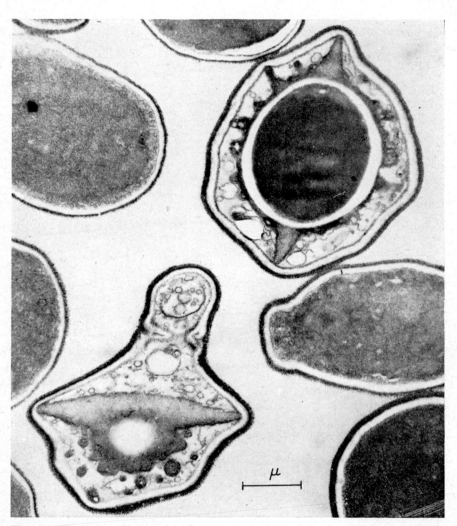

FIG. 11. Electron micrograph of a section through asci of *Schwanniomyces alluvius,* one of which shows conjugation between mother cell and bud. The asci were fixed with osmic acid and uranyl acetate.

and heterothallic. In these yeasts, the culture from a single haploid cell sporulates after diploidization, but it also represents one of the mating types. Very often, but not always, mixtures of cells of opposite type sporulate more readily than cells of a single type. Cells of opposite mating-type may agglutinate when mixed, as for instance in the species *Hansenula wingei* (Wickerham, 1956).

The course of diploidization of the nucleus has been studied in a number of yeast species, but is still unknown in many others.

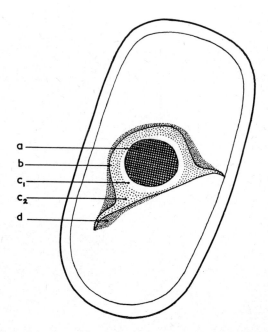

FIG. 12. Schematic drawing of a section of an ascus with one hat-shaped spore: a, indicates the protoplast; b, a thin dark outer layer of the spore wall; c_1, a light zone of the inner layer; c_2, a darker zone of the inner layer; d, osmiophilic material surrounding the spore.

Variation in the shape of the ascospores of yeasts provides a good criterion for differentiation. Round, oval, reniform, cylindrical and needle-shaped spores occur. A ledge in the middle or on one side of round or oval spores gives them a saturn- or a hat- or helmet-shaped appearance. Wartiness of the spore wall is another typical feature revealed by light microscopy. Electron microscope studies of sections of spores undertaken for taxonomic purposes by Kawakami (1960), Kawakami *et al.* (1961), Kreger-van Rij (1964, 1966a), and Besson (1966) show more detail and also afford an insight into the internal structure of the spores

(Fig. 12). Fixation with osmic acid or with potassium permanganate reveals, in thin sections, the protoplast with the plasma membrane surrounded by the spore wall. In the wall, layers of different electron-density are visible. In the simplest form there is a dark outer layer and a lighter inner layer. Variations in the thickness and shape of these layers occur; for instance, a dark warty outer layer (Figs. 13 and 14), or a light inner layer forming warts (Figs. 15 and 16). The inner layer is not always of even electron-density; dark and light zones may be distinguished, the latter as the inner zone, the former between this and the outer layer. A dark line occasionally separates these zones. The outer zone may be very dark, and thus not distinguishable from the dark outer layer. Observations on the germination of hat-shaped spores show that the light inner zone becomes the wall of the vegetative cell, while the rest of the spore wall, including the ledge which belongs

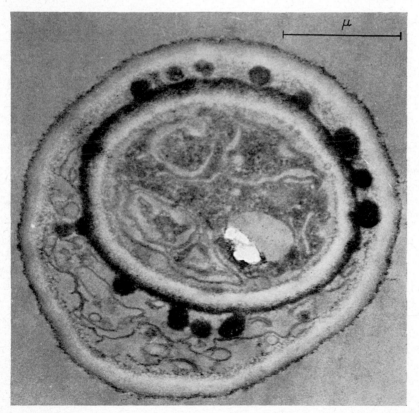

FIG. 13. Electron micrograph of a section through an ascus of *Pichia terricola*. The spore has a warty dark outer layer. The spores were fixed with osmic acid and uranyl acetate.

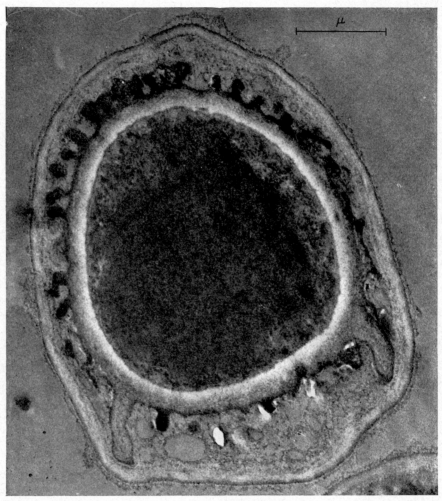

FIG. 14. Electron micrograph of a section through an ascus of *Pichia fluxuum*. The spore has a ledge and a warty dark outer layer. The asci were fixed with osmic acid and uranyl acetate.

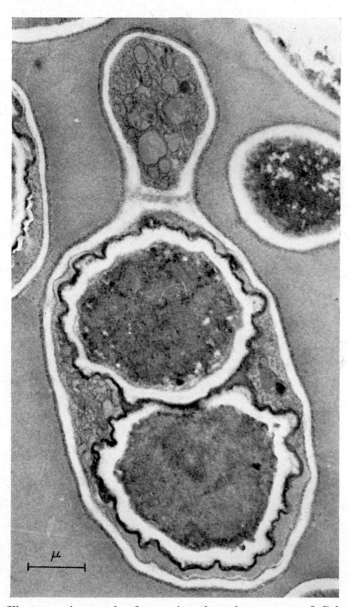

FIG. 15. Electron micrograph of a section through an ascus of *Debaryomyces vanriji*. The spore wall has warts formed by the light inner layer. The asci were fixed with osmic acid and uranyl acetate.

to the darker zone of the inner layer, vanishes (Figs. 17 and 18). Removal of the outer layers proceeds either by stretching and a gradual disintegration, or by rupturing as the cells swell. In the latter case, the empty shell generally remains attached to the cell. Observation by light microscopy led to the conclusion that the spore had two membranes. Schiönning (1903) based the genus *Saccharomycopsis* on this feature. At present, it seems that this distinction does not depend on the presence of one or two layers in the spore wall, but rather on the composition of these layers and the mode of germination. The cell

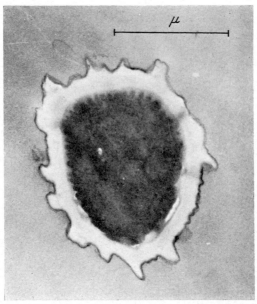

FIG. 16. Electron micrograph of a section through an ascospore of *Schizosaccharomyces malidevorans*. The warts on the spore wall are formed by the light inner layer. The spores were fixed with osmic acid and uranyl acetate.

which has arisen from the spore by swelling and shedding its outer layers may reproduce by forming one or more buds or a mycelial hypha. Very little is yet known about the micromorphology of spore germination in various species.

Electron micrographs of preparations fixed with osmic acid show that yeast spores are surrounded in the ascus by an osmiophilic material (Figs. 12d and 19). When liberated, this material adheres to the spores and is probably the cause of their agglutination.

In the genera *Sporidiobolus*, *Rhodosporidium*, and *Leucosporidium* which are heterobasidiomycetes, the first stage in sexual reproduction is

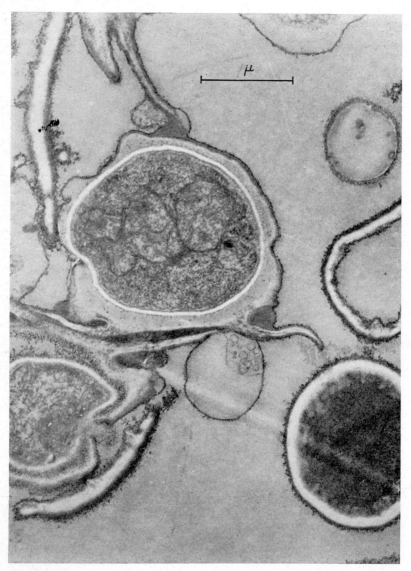

FIG. 17. Electron micrograph of a section through a hat-shaped spore of *Pichia pijperi*. The different layers shown in the schematic drawing (Fig. 12) are clearly visible. The spores were fixed with osmic acid and uranyl acetate.

2*

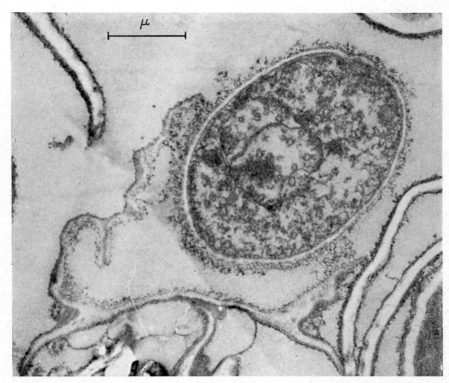

FIG. 18. Electron micrograph of a section through a germinating spore of
Pichia pijperi.

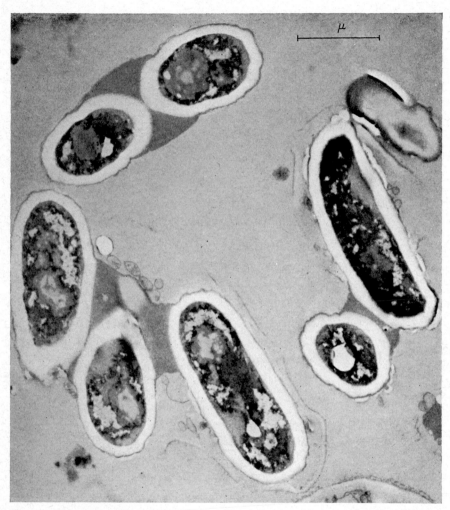

FIG. 19. Electron micrograph of a section through free ascospores of *Kluy-veromyces osmophilus*. The spores are stuck together with osmiophilic material. They were fixed with osmic acid and uranyl acetate.

the formation of dikaryotic cells. In *Rhodosporidium* and *Leucosporidium* this may be the result of a mating reaction of two yeast cells (Banno, 1963, 1967; Fell *et al.*, 1969). From the dikaryotic cell, hyphae with clamp connections develop eventually leading to the formation of chlamydospores (Fig. 20). According to Banno (1967), diploidization and meiosis

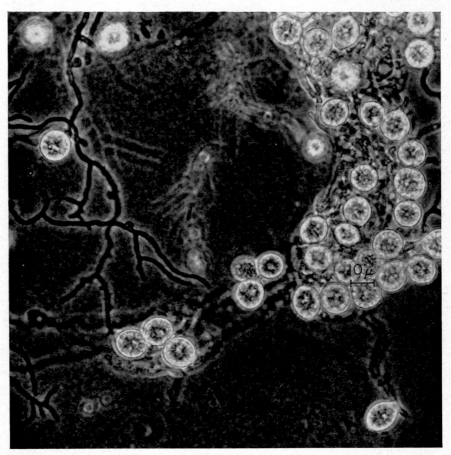

FIG. 20. Photomicrograph (phase contrast) of *Sporidiobolus ruinenii* showing true mycelium with clamp connections and chlamydospores. Reproduced by courtesy of Dr. J. Ruinen.

occur in the chlamydospores which on germination produce a promycelium, from which haploid yeast cells, the sporidia, are formed by budding.

Vegetative endospores were observed by do Carmo-Sousa (1966) in strains of *Trichosporon* and *Oosporidium* species. They may also occur in *Endomycopsis fibuligera* (Kreger-van Rij, 1964).

3. *Physiological Criteria*

The principal physiological criteria used in the classification of yeasts are fermentation and assimilation.

The standard description of species includes the capacity to ferment certain sugars tested in Durham tubes for a fixed period of time. Characterization of the fermentation as vigorous, good, slow or weak depends on the rate of fermentation, and is observed as the quantity of gas in the insert tube checked every two days over a period of two weeks. The usual sugars tested are glucose, galactose, sucrose, maltose, lactose and raffinose. Additional sugars, such as trehalose and cellobiose, and the polysaccharides inulin and starch may also be tried. In fermentation of the trisaccharide raffinose, different enzymes are involved. Complete fermentation of this sugar may depend either on the presence of invertase and melibiase, or of melibiase and α-glucosidase.

The value of the assimilation of carbon compounds as a character for the distinction of yeast species has been emphasized by Wickerham (1951) and Wickerham and Burton (1948). In the standard description of species, the results of tests with 30–40 compounds are recorded. These compounds have been chosen for their differentiating value, although detailed knowledge about the biochemical background of their utilization is often lacking. Permeability of the cell for the compound and presence of the required enzyme system both determine whether this compound is utilized. Barnett (1968a) has given a survey of the biochemistry of the utilization of the carbon compounds used in the standard description. This author (Barnett, 1968b) also made a study of the use of polyols by yeasts.

Wickerham's list of carbon compounds includes D-glucose, D-galactose, L-sorbose, maltose, sucrose, cellobiose, trehalose, lactose, melibiose, raffinose, melezitose, inulin, soluble starch, D-xylose, L-arabinose, D-arabinose, D-ribose, L-rhamnose, D-glucosamine hydrochloride, ethanol, glycerol, *i*-erythritol, adonitol, dulcitol, D-mannitol, D-sorbitol, α-methyl-D-glucoside, salicin, potassium D-gluconate, calcium 2-keto-D-gluconate, potassium 5-keto-D-gluconate, potassium sodium saccharate, pyruvic acid, DL-lactic acid, succinic acid, citric acid, ethyl acetoacetate, and *i*-inositol.

Komagata *et al.* (1964) studied the utilization of hydrocarbons by a great number of yeast species. It appeared that *n*-paraffins containing more than nine carbon atoms were easily attacked by a restricted number of species, most of them belonging to the genus *Candida*. Scheda and Bos (1966) also investigated the assimilation of hydrocarbons with a view to their use in taxonomy.

The differentiating value of the assimilation of carbon compounds varies. The assimilation of a number of them, for instance erythritol and the β-glucosides, is considered to be sufficient for differentiating between species in some genera. Insufficient knowledge of the taxonomic value to be attached to the utilization of each compound leads to a preference for using more than one compound for the distinction of species. However, Barnett (1966, 1968a) has shown that enzymic links exist between the utilization of some compounds. For instance, a single enzyme is probably responsible for hydrolysing the β-glucosides, cellobiose, arbutin and salicin, and the splitting and assimilation of these compounds represent one characteristic instead of three.

The method for testing assimilating ability is of great importance, since by the presence or absence of the opportunity for adaptation, very different results may be obtained. Wickerham and Burton (1948) have prescribed a test in a chemically-defined liquid medium with static cultures examined after three weeks. To obtain quick preliminary results, it is also possible to use auxanographic tests which take two or three days, and for which the positive results agree with those obtained using the liquid medium test. If necessary, the inactive compounds may be tested afterwards using the liquid medium.

Assimilation of β-glucosides may be quickly and conveniently examined by growing the yeast on a solid medium containing arbutin, splitting of which gives a brown colour with iron salts present.

Assimilation of nitrogenous compounds is tested in the same way as that of the carbon compounds, namely in liquid medium. Potassium nitrate is the most important test compound because its utilization is highly valued for the differentiation of yeasts. Other compounds which may be tried are sodium nitrite (Wickerham, 1957; van der Walt, 1963), ethylamine hydrochloride (van der Walt, 1962) and creatine and creatinine (Kreger-van Rij and Staib, 1963). Utilization of nitrogenous compounds may also be tested with the auxanographic method which generally gives the same results for nitrate. For nitrite, it is preferred to the liquid medium test. Creatine and creatinine give distinctive results in the auxanogram for *Debaryomyces hansenii* and *Cryptococcus neoformans*; *D. hansenii* assimilates creatine and *Cr. neoformans* creatinine.

Apart from these fermentation and assimilation tests, the standard description includes a number of criteria which are of less general importance but which may be typical of certain genera and species and are, therefore, very useful for identifications. They are: starch formation on solid or in static or shaken liquid medium; growth in a vitamin-free medium; vitamin requirements (van Uden and do Carmo-Sousa, 1956; van Uden and Farinha, 1958); growth on 50% (w/w) glucose-yeast extract agar; growth on 10% (w/v) NaCl + 5% (w/v) glucose in Wickerham's (1951) yeast-nitrogen base medium;

NaCl tolerance; growth at 37°; maximum temperature for growth (van Uden and do Carmo-Sousa, 1956; van Uden and Farinha, 1958); acid production on chalk-agar; production of esters; urea hydrolysis (Seeliger, 1956); gelatin liquefaction; actidione resistance (van der Walt and van Kerken, 1961a); splitting of fat. Some of these criteria will be briefly discussed.

Aschner *et al.* (1945) found that certain yeast species under suitable conditions produce compounds in the cell wall, the capsule or in the medium which give a blue colour with iodine solution. This formation of starch, as it is usually indicated, requires media with pH values below 5. Later on, other species were found to give this reaction after growth in a medium with a higher pH value. Production of starch by a yeast often correlates with the ability to assimilate inositol. The characteristic of starch formation plays an important role in the differentiation of the anascosporogenous yeasts. It occurs in the genera *Bullera*, *Cryptococcus*, *Leucosporidium*, *Candida*, *Trichosporon* and *Oosporidium*.

Growth in a vitamin-free medium, a test devised by Wickerham (1951), is restricted to relatively few yeast species and may, therefore, be useful for the identification of a yeast strain. For this test, and also for establishing the vitamin requirements of a yeast, the conditions have to be strictly standardized if the results are used for comparison.

Wickerham (1951) considered the ability to grow at 37° of importance as an indication that the yeast might grow in warm-blooded animals. This temperature, and also the maximum temperature for growth, are considered to have taxonomic value for the differentiation of certain species.

Acid production is typical of the genera *Dekkera* and *Brettanomyces*, although it is also found in other genera.

The image which the standard description gives of a yeast species is aimed mainly at classification purposes, that is, it is given in criteria which differentiate it from other species. Within a species, variations in characteristics may occur among the strains, some of which are of practical importance, such as the amounts of ethanol produced.

D. OTHER CRITERIA

Some features are typical of a limited group of yeasts and have not been standardized for general differentiation. Nevertheless, they have an influence in taxonomy. These are, for instance, the negative Pasteur effect typical of most *Brettanomyces* species (Wikén *et al.*, 1961), and the ability to form respiratory-deficient strains known as *petite-colonie* mutants, a property found in most diploid *Saccharomyces* species (Bulder, 1963).

Apart from the properties mentioned above, there are others, not established in routine tests, which are of importance in yeast taxonomy because they are considered to reveal either relationship or the absence of it. To these properties belong the chemical composition of the cell wall and of extracellular compounds, the antigenic structure of the cell, and the guanine plus cytosine (GC) contents of DNA.

The composition of the cell wall has been studied by chemical, physical and enzymic analysis, the physical examination comprising chromatography and Röntgen analysis. These studies have not been systematically carried out from a taxonomic point of view; the results are fragmentary and, for a large part, concern only *Saccharomyces cerevisiae*. Phaff (1963) reviewed the literature on the cell wall, and since then several new publications have appeared.

The components of the yeast cell-wall are glucans, mannans, chitin, proteins and lipids. Qualitative and quantitative differences in these components were found to exist among the yeast species. For instance, chitin appeared to be present in unusually high quantities in *Endomycopsis capsularis* but could not be detected in *Schizosaccharomyces* spp. (Roelofsen and Hoette, 1951; Kreger, 1954). Two types of mannan were found to exist, the first one, from *Sacch. cerevisiae* and several other yeast species giving an insoluble copper complex (Garzuly-Janke, 1940), and the second, from *Rhodotorula aurantiaca* and *Sporobolomyces roseus* (Crook and Johnston, 1962) and from *Nadsonia elongata* (Dyke, 1964) not giving an insoluble copper complex. Spencer and Gorin (1968) determined in apiculate yeasts the proton magnetic resonance spectra of mannans and galactomannans which gave an insoluble copper complex, and found distinct differences between the various species. They suggested that these spectra might be useful in classification. Tanaka and Phaff (1967), using bacterial enzymes as indicators, found different glucans in *Sacch. cerevisiae* and *Sacch. (Lodderomyces) elongisporus*. *Schizosaccharomyces* spp. also contained a glucan different from that in *Sacch. cerevisiae* (Kreger, 1954).

Extracellular compounds, either present in the capsule or excreted into the medium, include polysaccharides and lipids. Phosphomannans of varying composition are produced by *Hansenula* species. Wickerham and Burton (1962) believe that the ratio of mannose to phosphate in these compounds correlates with the evolutionary stage of the species; the most highly phosphorylated mannans, which are very mucous, are formed by the most primitive species.

In 1952 Wickerham directed attention to the formation of extracellular starch under special conditions in members of the genera *Cryptococcus*, *Lipomyces* and by the ascomycete *Taphrina*, and speculated on a possible relationship between these genera on this basis. On the other hand, Slodki *et al.* (1966) found similar extracellular heteropolysaccharides formed by a *Cryptococcus* sp. and the basidiomycete *Tremella*, and suggested a relationship on that basis.

Production of extracellular lipids by yeasts, reviewed by Stodola *et al.* (1967), has not yet been used in yeast systematics, although the authors point out that it may be a useful criterion. It is typical of some *Cryptococcus* and *Rhodotorula* species (Ruinen and Deinema, 1964).

Tsuchiya and coworkers (1965) have made an analysis, using the slide agglutination method, of the antigenic structure of yeast species of several genera. Both thermostable and thermolabile antigens were found. The authors obtained results which may be useful for the systematics of yeasts, especially for the differentiation of genera. Their data partly corroborate the existing generic differentiation. For instance, species in the genus *Hansenula*, which seems to be very homogeneous, appeared to be rather uniform in antigenic structure. On the other hand, failure to find mutual antigens in species indicating the absence of a direct relationship, a situation which was found for *Schizosaccharomyces pombe*, *Candida albicans* and *Cryptococcus neoformans*, confirms the very separate taxonomic positions of these species. These obvious correlations between antigenic structure and other taxonomic criteria make it worth considering a relationship between different genera when a striking resemblance in the antigenic structure exists as, for instance, between *Debaryomyces hansenii*, *Citeromyces matritensis* and *Schwanniomyces occidentalis*. The genera *Debaryomyces*, *Citeromyces* and *Schwanniomyces*, although easily distinguishable, have also some common features. However, resemblances in antigenic structure may be found which do not correlate at all with other features of these yeasts. In such cases, it seems unjustifiable to accept a taxonomic relationship exclusively on a serological basis, as proposed by Tsuchiya *et al.* (1965) for *Schizosaccharomyces pombe*, *Candida lipolytica* and *Torulopsis bacillaris*. Moreover, the very few antigens known for these species make a narrow basis for this proposition.

Biguet *et al.* (1965) made an electrophoretic analysis of the antigens of several yeast species. In contrast with Tsuchiya's method, which demonstrates the antigens of the cell wall which are generally insoluble, the electrophoretic analysis reveals the presence of soluble antigens of the whole cell which are precipitated by antibodies. Biguet *et al.* (1965) point to the similarity of results obtained with these two complementary methods. They found a great resemblance in the immunoelectrophoretograms of strains presumed to be the perfect and the imperfect forms of species, for instance, *Hansenula anomala* and *Candida pelliculosa*.

Storck (1966) pointed to the value which the nucleotide composition of fungal DNA may have for taxonomy. This composition is expressed as the percentage of guanine plus cytosine residues (GC content) in DNA. Other authors cited by Storck (1966) determined the GC contents of DNAs from several fungi. Storck (1966) examined 41 strains of different species of zygomycetes, ascomycetes, basidiomycetes and deuteromycetes. The 14 ascomycetes varied in GC content between 38 and 54%, and the Hemiascomycetoideae, among which were six yeast species, between 39 and 45%. The latter range was similar to that of the zygomycetes (38–48%). The nine basidiomycetes had higher percentages, namely 44–63%. Among these were *Sporobolomyces roseus* with 50%, and *Sp. salmonicolor* with 63%. *Rhodotorula mucilaginosa* resembled the latter species in having a GC content of 61–63%. The six yeast species belonging to the genera *Debaryomyces*, *Pichia*, *Saccharomyces*, *Kluyveromyces* and *Schizosaccharomyces* did not show remarkable differences. Although these data are as yet very sparse, they seem to be promising in showing similarities and differences which correlate with other important characteristics. They may be especially helpful for the classification of the larger groups in the yeasts among the other fungi.

E. IDENTIFICATION OF YEASTS

The extensive list of criteria used in the standard description may well discourage anyone who wants to identify a yeast. However, not all criteria have the same distinctive value. This partly depends on the genus to which the yeast belongs. Furthermore, the procedures which lead to determination of genera and species, such as the use of keys and tables, indicate which criteria are of first importance to arrive at in an identification. Some of these criteria are typical of a few or even of one known species. Of course, to be quite certain of the identity of a given yeast with a described species, it may be necessary to establish also the other criteria in the standard description. Those which vary within one species are of little value. For the introduction of a new species, publication of the full standard description is desirable.

The first step in the identification is generally the establishment of the genus, since most keys to species assume this knowledge.

For the ascosporogenous yeasts, characteristics of sexual reproduction are of great importance, and induction of spore formation is, therefore, the first step necessary for identification. Of equal importance is the mode of vegetative reproduction. Nitrate assimilation, fermentation, and pellicle formation are some of the criteria which lead to a restricted group of genera or to one genus. Among the yeasts forming ballistospores and in the asporogenous yeasts, most genera are recognized by features of the vegetative reproduction.

Keys and tables of genera will indicate which criteria are required for recognizing species. These criteria are generally fermentation and assimilation reactions with a number of carbon compounds.

III. Taxonomy

A. PHYLOGENETIC THEORIES

In the construction of classification systems for yeasts, phylogenetic relationships have been assumed. Phylogeny bears least upon one of the smallest taxa of the system, the species, and most upon the higher taxa. Whereas the classification into families, subfamilies and tribes has primarily theoretical value, the classification into genera is of practical importance because of the generic name used in the binomials of the species. It is, therefore, in the delimitation of the genus that phylogenetic considerations are in practice of first importance.

Any phylogenetic system is based on a hypothesis in which primitive and more developed properties are recognized, and in which one or more other properties are considered to be relatively stable and to constitute the binding elements of the species on a developmental line. The taxonomist, in proposing a system of classification, evaluates more or less consciously the properties used for differentiation. This evaluation, however, is far from absolute. It is often, rightly, determined by the objects that are being classified (Donk, 1964). For instance, nitrate assimilation in the genus *Hansenula*, where it is one of the features distinguishing the genus from related genera, is valued more highly than in the genus *Cryptococcus*, where it separates species. Moreover, the evaluation of a feature may change as it is better understood; and, finally, the appreciation of different authors may vary.

History is an important fact in systematics. The present systems of yeast classification are the products of a gradual development. Changes in classification are generally the cause of changes in nomenclature, which are confusing for the non-taxonomist. Therefore, prudence is desirable in this respect; it may be preferable to maintain a classification which is far from satisfactory so long as the alternative is not better.

The classification system of the ascosporogenous yeasts given by Lodder (1970) is rooted in earlier systems of Hansen (1904), Klöcker (1907, 1924), Guilliermond (1909) and Stelling-Dekker (1931). Lodder *et al.* (1958) have given a review of the development of these systems to which the reader is referred for more information. A few points will be mentioned here.

Although Hansen and Klöcker were already aware of different levels of relationship between the yeasts, Guilliermond (1909) was the first to render account of his phylogenetic views by describing the principles. He assumed, in the fungi which produce the same type of primitive ascus, an evolutionary development from mycelial to budding species, and observed stages of it in the genera *Eremascus, Endomyces* (= *Endomycopsis* Stelling-Dekker) and *Saccharomyces*. Another line connected *Eremascus* via *Endomyces* with *Schizosaccharomyces*. Strains of *Eremascus* form only true mycelium, those of *Endomycopsis* true mycelium with budding cells, and strains of *Saccharomyces* only budding cells. In the second line, strains of *Endomyces* produce true mycelium that falls apart into arthrospores, and those of *Schizosaccharomyces* form single cells by fission. From this point of view, vegetative reproduction, especially budding or fission, is of primary importance.

Wickerham (1951, 1970) and Wickerham and Burton (1962) proposed a phylogenetic theory with a view different from that of Guilliermond. They considered the shape of the spores a more important character than the characteristics of budding or fission. Wickerham made a very intensive study of the genus *Hansenula*, which is characterized by the shape of the spores and by nitrate assimilation. According to his theory, which he derived from observations on *Hansenula*, the fundamental evolutionary trend is a progressive change from haploid to diploid species. This goes together with a decreased dependence on external vitamins, the ability to ferment an increased number of sugars, and the capacity to form more hyphae. The last feature is in sharp contrast with Guilliermond's theory. Wickerham's evolutionary lines in the genus *Hansenula* connect in the first place species with the same ecological background, such as isolates from conifers or free-living yeasts. In the second place, Wickerham assumes an evolutionary distinction between homothallic and heterothallic species. Even if they resemble each other closely in other respects, such as *Hansenula canadensis* and *H. wingei*, they are placed on different phylogenetic lines, the former homothallic, the latter heterothallic.

B. DIFFERENTIATION OF THE HIGHER TAXA

In the present systems of classification two of the characteristics mentioned above, namely the mode of vegetative reproduction and the shape of the spores, are of importance for the distinction of the higher taxa. There are a few other characters which are considered to be of prime importance as indications of relationships between yeasts, but these are not yet routinely established. They include the composition of the cell wall, which plays an important part in determining the antigenic structure, and the GC contents of the DNA of the yeasts. Both of these properties have been discussed in a preceding section (p. 34).

To date, a general and detailed classification of yeasts has been given by only a few authors, namely Kudrjawzew (1960), Windisch (1960), Novák and Zsolt (1961), and Lodder (1970). Kudrjawzew (1960) leaves

TABLE I. *Some Systems of Yeast Classification*

Author	Order	Family	Subfamily
Kudrjawzew (1960)	Saccharomycetales	Saccharomycetaceae	
		Schizosaccharomycetaceae	
		Saccharomycodaceae	
Windisch (1960)	Endomycetales	Endomycetaceae	
		Saccharomycetaceae	
	(Fungi Imperfecti)	Cryptococcaceae	
		Sporobolomycetaceae	
Novák and Zsolt (1961)	Endomycetales	Lipomycetaceae	
		Schizosaccharomycetaceae	
		Saccharomycetaceae	Prosaccharomycetoideae
			Multisporoideae
			Levigatosporoideae
			Verrucosporoideae
			Prohansenuloideae
			Pichioideae
			Hansenuloideae
		Hansenulaceae	
		Fabosporaceae	
		Nematosporaceae	
		Sporobolomycetaceae	
		Cryptococcaceae	Trichosporoideae
			Procandidoideae
			Candidoideae
			Torulopsoideae
			Cryptococcoideae
Lodder (1970)	Endomycetales	Endomycetaceae	
		Saccharomycetaceae	Schizosaccharomycoideae
			Nadsonioideae
			Saccharomycoideae
			Lipomycetoideae
		Spermophthoraceae	

out of his system most of the asporogenous genera. Lodder (1970) gives a classification of the ascosporogenous yeasts only. Differences between the primary divisions in these systems exist in the classification of certain groups either as a family, or as a subfamily, in a further splitting up of groups at a higher or lower level, and in the classification of the genus *Schizosaccharomyces*. The four systems up to and including the subfamilies are given in Table I. They will not be discussed here in greater detail. In view of the practical importance of the delimitation of genera and species, a few remarks will be made on these taxa.

C. DIFFERENTIATION OF THE GENERA

Characterization of the ascosporogenous genera is mainly based on the shape of the spores. In addition, the shape of the ascus and the mode of vegetative reproduction play an important role. This has led to the recognition of a number of small genera which are morphologically very homogeneous and in which the species differ in only a few physiological properties. Of the 22 sporogenous genera, no less than 10 genera comprise only one species, and six others include between two and five species. Of the larger genera, *Hansenula*, *Kluyveromyces* and *Debaryomyces* are rather homogeneous. In *Pichia* and *Saccharomyces*, distinct groups may be recognized, whereas the genus *Endomycopsis* is very heterogeneous.

The three genera in which ballistospores are formed differ in the presence or absence of clamp connections, and in the shape of the ballistospores. The genera *Sporidiobolus* and *Bullera* each comprise two or three species; *Sporobolomyces* includes nine species.

In the differentiation of the 14 genera of asporogenous yeasts, the mode of vegetative reproduction is of first importance. A few physiological criteria, such as fermentative ability and the formation of starch, are also used for distinction. Seven genera, each with between one and four species, are easily recognized by these means. The asporogenous genera comprise several species, sporogenous forms of which are also known. Thus, the genera *Dekkera* and *Hanseniaspora* are the perfect counterparts of *Brettanomyces* and *Kloeckera* respectively. In the same way, *Rhodotorula* and *Rhodosporidium* are probably forms of *Sporidiobolus*. For a number of genera, such as *Trigonopsis* and *Schizoblastosporion*, no perfect forms are as yet known. On the other hand, the genera *Torulopsis* and *Candida* comprise imperfect species of several ascosporogenous genera. This means that the genus *Candida* especially is very heterogeneous.

D. DIFFERENTIATION OF THE SPECIES

The subdivision of genera into species is determined by the typical characteristics of each genus. In morphologically homogeneous genera, physiological properties differentiate the species, whereas in morphologically heterogeneous genera both morphology and physiology play a role.

The delimitation of a species depends on the variations allowed in its criteria. Some species are very uniform in all properties, others show wide variations. The characteristics of the strains which are considered to constitute the species determine these variations.

Interfertility is not a general criterion for the differentiation of species. It occurs among species of the genus *Kluyveromyces* and among those of *Saccharomyces*; it has not yet been found among the species of *Hansenula*.

Several authors consider ecology important for taxonomy. This seems to be quite acceptable, since conditions of growth are operative in selection of the species. However, although many data are available about the yeast species found in all kinds of environments, knowledge about the conditions in nature which make the complete life cycle of a particular yeast species possible is very scarce. Many isolations are probably made of yeasts in a special stage, such as spores from soil, or dried, capsulated cells from the air. The easy transportability of yeasts by insects may also give a biased picture of their habitat.

The application of numerical taxonomy for the differentiation of yeast species has been described by Poncet (1967) for the genus *Pichia*. The author studied 30 single strains of different species, and compared the results with those obtained with the classical methods by Kreger-van Rij (1964) and Boidin *et al.* (1965b). From this comparison, it appeared that a certain range of 'distances' between two species was critical either for their separation or for synonymy. An independent decision in these cases, based entirely on the numerical method, was not made. It should be noted that the representation of a species by a single strain is not correct, unless the description of the species is based on one strain only. Variations in the features which are characteristic of the species are thus not considered. Of the features used by Poncet (1967), many are known to vary within one species. Limited value only can therefore be given to the conclusions arrived at in this study.

IV. Systematics

In the following section of this chapter, a survey of the classification of yeasts will be given with the emphasis on a more detailed discussion of the genera.

As mentioned before, classification of the higher taxa is based on hypotheses which are all the more speculative since many data considered to be useful are scarce or lacking. The concepts of various

authors about this classification are not identical. I do not feel com-
pelled to add another proposition. Since, however, it is necessary to give
a general view of the classification of yeasts, a certain scheme will be
followed.

Three subdivisions of yeasts are recognized:
    A. Ascosporogenous yeasts
    B. Ballistosporogenous yeasts
    C. Yeasts forming no ascospores or ballistospores

A. ASCOSPOROGENOUS YEASTS

The ascosporogenous yeasts are classified by Gäumann (1964) in the
order Endomycetales, in the subclass Prototunicatae, which are primitive ascomycetes with an undifferentiated ascus wall.

In the Endomycetales, we recognize two yeast families: the Saccharomycetaceae and the Spermophthoraceae (Table II). The latter
have needle-shaped spores; the spores of the former are of various other
shapes.

In the Saccharomycetaceae, several groups may be recognized, often
classified as subfamilies. The first group comprises the single genus
Schizosaccharomyces in which vegetative reproduction is exclusively by
fission. In the second group are included all ascosporogenous genera
which show multilateral budding and in which the ascus and the
ascospores do not have the peculiar shape typical of the genus *Lipomyces* (see p. 53) (Fig. 21). These genera are: *Endomycopsis, Hansenula,
Pachysolen, Citeromyces, Dekkera, Pichia, Debaryomyces, Schwanniomyces, Wingea, Kluyveromyces, Saccharomyces, Saccharomycopsis* and
Lodderomyces. The third group consists of the genus *Lipomyces* with

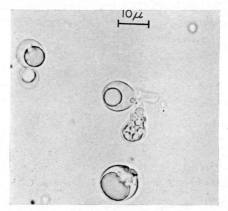

FIG. 21. Photomicrograph of a sac-like ascus attached to a cell of *Lipomyces
starkeyi*. The cells contain large oil globules.

TABLE II. *Characteristics of the Yeast Genera in the Families Saccharomycetaceae and Spermophthoraceae*

SACCHAROMYCETACEAE

I

Schizosaccharomyces
Fission cells (true mycelium and arthrospores).
Ascospores round, oval or reniform, smooth or warty; 4–8 per ascus.
Fermentation positive.

II

Endomycopsis
Budding cells, true mycelium (arthrospores).
Ascospores round, oval, hat-, saturn- or sickle-shaped.
Fermentation positive or negative.
Nitrate assimilation positive or negative.

Hansenula
Budding cells (true mycelium).
Ascospores hat- or saturn-shaped, round or hemispherical.
Fermentation positive or negative.
Nitrate assimilation positive.

Pachysolen
Budding cells.
Ascus consists of a cell with a tube, both thick-walled.
Ascospores hat-shaped.
Fermentation positive.
Nitrate assimilation positive.

Citeromyces
Budding cells.
Ascospores round, warty.
Fermentation positive.
Nitrate assimilation positive.

Dekkera
Budding cells.
Ascospores hat-shaped.
Fermentation positive.
Nitrate assimilation positive or negative.
Short-lived on malt agar.

Pichia
Budding cells (some true mycelium).
Ascospores round, hat- or saturn-shaped, smooth or warty.
Fermentation positive or negative.

Debaryomyces
Budding cells.
Conjugation between mother cell and bud.
Ascospores round or oval, warty.
Fermentation positive or negative.

Schwanniomyces
Budding cells.
Conjugation between mother cell and bud.
Ascospores round or oval with an equatorial ledge, warty.
Fermentation positive.

Wingea
Budding cells.
Conjugation between mother cell and bud.
Ascospores oblate ellipsoidal.
Fermentation positive.

Kluyveromyces
Budding cells.
Ascospores reniform, crescentiform, round or oval; easily liberated; 1–16 per ascus.
Fermentation positive.

Saccharomyces
Budding cells.
Ascospores round or oval, smooth or warty.
Fermentation positive.

Saccharomycopsis
Budding cells.
Ascospores oval.
Fermentation positive, weak or slow.
No growth on malt agar.

Lodderomyces
Budding cells.
Ascospores oval.
Fermentation positive but slow.

III

Lipomyces
Budding cells, capsulated.
Cells with protuberances may be formed.
"Active buds" may conjugate or directly change into asci.
Ascospores oval, amber-coloured, smooth, warty or with ridges; 1–16 per ascus.
Fermentation negative.

IV

Nadsonia
Budding cells, bipolar on a broad base.
Conjugation between mother cell and bud.
Ascospores round, warty, brown.
Fermentation positive.

Saccharomycodes
Budding cells, bipolar on a broad base.
Generally conjugation of germinating spores in the ascus.
Ascospores round with a ledge; 2–4 per ascus.
Fermentation positive.

Hanseniaspora
Budding cells, bipolar on a broad base.
Ascospores round, hat-, helmet-, or saturn-shaped, smooth or warty.
Fermentation positive.

Wickerhamia
Budding cells, bipolar on a broad base.
Ascospores cap-shaped, 1–16 per ascus.
Fermentation positive.

SPERMOPHTHORACEAE

Nematospora
Budding cells, true mycelium.
Asci large, cylindrical.
Ascospores spindle-shaped, 4–8 per ascus.
Fermentation positive.

Metschnikowia
Budding cells.
Asci large, cylindrical or clavate.
Ascospores needle-shaped, 1–2 per ascus.
Fermentation positive.

Coccidiascus
Budding cells.
Asci large, banana-shaped.
Ascospores spindle-shaped, 8 per ascus.

the typical shape of ascus and ascospores mentioned above. The fourth group comprises genera in which budding is bipolar, namely *Nadsonia*, *Saccharomycodes*, *Hanseniaspora* and *Wickerhamia*.

In the Spermophthoraceae, the yeast genera with needle-shaped spores are *Nematospora*, *Metschnikowia* and *Coccidiascus*.

There are some ascosporogenous fungi which are not considered to be yeasts, but which are probably related to yeasts. Gäumann (1964) also classifies these in the Endomycetales. These genera include, for instance, *Dipodascus* in the Dipodascaceae, *Eremascus* and *Endomyces* in the Endomycetaceae, and *Ashbya* in the Spermophthoraceae. The genus *Taphrina*, in which budding yeast cells occur, is classified in another order, namely the Taphrinales. Of these genera, only *Endomyces* will be very briefly discussed.

The genus *Endomyces* was first described by Reess in 1870. Stelling-Dekker (1931) confined it to organisms which form hyphae that split up into arthrospores but do not form budding cells. The two species included by her in the genus were *Endomyces decipiens* and *Endomyces magnusii*. The first species produces hat-shaped spores; no conjugation was observed to precede ascus formation. For *Endomyces magnusii*, Guilliermond (1937) described conjugation of gametangia leading to the formation of an ascus with four oval spores. Van der Walt (1959b) described a third species, *Endomyces reessii*. It also has gametangiogamy and produces four oval thick-walled spores per ascus. From this it appears that the genus *Endomyces* is only homogeneous from the point of view of vegetative reproduction.

In the genus *Schizosaccharomyces*, vegetative reproduction is exclusively by fission, either of single cells or of mycelium splitting up into arthrospores. The absence of budding distinguishes this genus from all other yeasts. The cell wall of *Schizosaccharomyces* is also exceptional in that chitin and mannan are lacking and another type of glucan, the so-called S-glucan, is present in addition to the usual yeast glucan (Kreger, 1967). The ascus in *Schizosaccharomyces* is formed after somatogamous conjugation. The spores are round, oval or reniform. They may have distinct warts (Fig. 16, p. 26). The spores of most species give a positive starch test with iodine solution. Fermentation is vigorous.

The place which *Schizosaccharomyces* takes in a classification of yeasts is still a point of consideration. Guilliermond (1937) classified it in close connection with *Endomyces* which it resembles in the mode of vegetative reproduction. Klöcker (1924), on the other hand, suggested that it may be closer to the budding yeasts (Saccharomycetaceae) than to the yeasts forming hyphae (Endomycetaceae). Both points of view still have their adherents.

The four *Schizosaccharomyces* species recognized by Slooff (1970) are differentiated by the shape of the spores and the number of spores formed per ascus, by the presence or absence of mycelium, and the fermentation of a few sugars. Slooff (1970) found that all species formed starch when tested in the liquid medium formulated by Wickerham (1952).

Strains of this genus have been isolated from grapes, wine, fruit wine, sugar and molasses. *Schizosaccharomyces japonicus* and its variety (*Schizosacch. japonicus* var. *versatilis*) are naturally-occurring respiratory-deficient yeasts (Bulder, 1963.)

Stelling-Dekker (1931) created the genus *Endomycopsis* for those ascosporogenous yeasts which form both true mycelium and budding cells. Guilliermond (1909) considered these species as forms intermediate between the entirely mycelial organisms of the genus *Eremascus* and the exclusively budding yeasts like *Saccharomyces*. Kreger-van Rij (1964, 1970) provisionally retained the genus *Endomycopsis*, although aware of the fact that it is heterogeneous in several respects, for instance, in the mode of ascus formation and in the shape of the spores. She was, however, of the opinion that species in this genus showed some features which would sharply distinguish them when they were placed in other genera. Some of these features require a closer examination. In the first place, very little is known about the micromorphology of budding and fission in *Endomycopsis* species. Secondly, information about the transition from haplophase to diplophase in various species is very incomplete. The formation of asci exclusively on the mycelial hyphae, which is typical of most *Endomycopsis* species, is part of this problem.

Some authors, who do not favour a differentiation based on the presence or absence of abundant true mycelium which formally distinguishes *Endomycopsis* from other yeast genera, have already removed *Endomycopsis* species to the genera *Hansenula*, *Pichia*, *Schwanniomyces* and *Guilliermondella*.

Wickerham (1970) pointed to the resemblance between the aerial ascophores of *Cephaloascus fragrans* and the hyphae that bear asci in *E. platypodis* which he believed to be indicative of a relationship. However, there are also distinct differences between the two species. Schippers-Lammertse and Heyting (1962) described how ascophores of *Cephaloascus fragrans* resulted from a conjugation between a hyphal cell and a gametangium formed by the neighbouring cell, or from anastomosis between two hyphal cells. The ascophore bears exclusively asci; it is generally encrusted with brown particles. In *E. platypodis* (Baker and Kreger-van Rij, 1964), gametangiogamy was not observed; the so-called ascophores in this species are not encrusted and they do not bear asci exclusively.

Kreger-van Rij (1970) accepted 10 species in the genus *Endomycopsis* which are differentiated by the shape of the spores, the number of spores per ascus, nitrate assimilation, and fermentation and assimilation of a number of carbon compounds.

The species *E. monospora*, *E. bispora* and *E. platypodis*, which have several features in common, are all associated with bark beetles. The latter two species are nitrate-positive. Strains of *E. fibuligera*, which ferments starch, have been isolated from starchy material such as chalky bread and chinese yeast (raji). The strains of *E. vini* came from wine. This species has round to oval spores with a broad ledge in the middle and a second ledge at one side of this (Fig. 22). The second ledge may not appear to be very distinct in the light microscope, but it is clearly visible in electron micrographs. Its electron density in osmic acid-fixed preparations appeared to be different from that of the equatorial ledge. As far as known, *E. burtonii* is the only heterothallic species in the genus *Endomycopsis*. It is the only species forming arthrospores. Wickerham

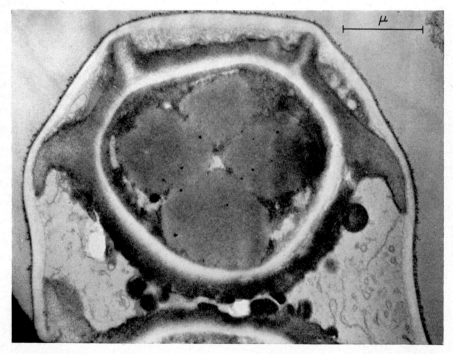

Fig. 22. Electron micrograph of a section through an ascospore of *Endomycopsis vini*. The spore has two ledges, one dark equatorial ledge and one smaller subequatorial ledge which is less electron-dense. The spores were fixed with osmic acid and uranyl acetate.

and Burton (1952), who named the species *E. chodatii*, detected its heterothallism. Cells of several species described as asporogenous, such as *Sporotrichum anglicum* and *Trichosporon behrendii*, appeared to conjugate and to sporulate when mixed with the appropriate mating type of *E. burtonii*.

Species of the three genera, *Hansenula*, *Pachysolen* and *Citeromyces*, are all able to assimilate nitrate. In the next genus, *Dekkera*, this ability varies among the strains.

The genus *Hansenula* has been extensively studied by Wickerham (1951, 1970) and Wickerham and Burton (1962), whose publications contain an abundance of data. The ascospores produced by *Hansenula* spp. are round, hemispherical, hat- or saturn-shaped; they are generally easily liberated from the ascus. Wickerham includes in this genus species forming true mycelium as well as pseudomycelium. The species are homothallic or heterothallic; they have haploid, or a mixture of haploid and diploid, or diploid cells. The capacity for fermentation varies among the species.

The genus *Hansenula* differs from the genus *Pachysolen* by the shape of the ascus, which is not rod-shaped, and from the genus *Dekkera* by having long-lived cultures. On the base of phosphomannans excreted by a number of *Hansenula* species, Wickerham and Burton (1962) proposed that these species are related to *Pachysolen* spp. and to some *Pichia* species. In the nitrate-positive genus *Citeromyces*, spores are of a different shape, namely round and warty.

Wickerham (1970) placed the 25 species which he accepted in *Hansenula* into five phylogenetic lines, two heterothallic and three homothallic. The first principle underlying the arrangement on these lines is a development from haploid to diploid species. Secondly, an evolvement of habitat is taken into consideration. The species considered to be primitive have been isolated from the frass of coniferous trees. In one of the heterothallic and one of the homothallic lines, the species are considered to have evolved to the free-living state, which means that they have been isolated from soil or fruit. This evolution coincides with the acquisition of greater fermentative powers, an increase in the ability to synthesize vitamins, and a change from mucoid to butyrous and mat and dry colonies. In the other heterothallic and the second homothallic lines, the species are considered to have evolved to a greater dependence upon coniferous trees. This goes together with a decreased physiological activity. On the third homothallic line, the habitat of the species, as far as known, progresses from coniferous to broad-leaved trees.

Wickerham's proposed evolutionary lines have the great advantage of giving a clear survey of the genus, eventually with possible relations to other genera. On the other hand, the lines may give a very simplified picture. For a number of species which differ in a few physiological properties, such as *H. fabianii*, *H. subpelliculosa*, *H. anomala* and *H. ciferrii*, a single line of development seems acceptable. For other more divergent species, the assumed phylogenetic relationships are more speculative, the more so if the number of available strains is small and, therefore, knowledge about the habitat is scarce.

Boidin and Adzet (1957) have described the genus *Pachysolen* with the single species *Pachysolen tannophilus*, isolated from a tan liquor of vegetable origin. The hat-shaped spores of this yeast are formed in an ascus of peculiar shape. It is a cell with a tube of variable length as an outgrowth. The walls of the cell and the tube are very thick, with the exception of the tip of the tube where the protoplast containing the spores is situated. These spores are easily liberated. The species ferments glucose.

Wickerham and Burton (1962) pointed to the relationship between *Pachysolen* and *Hansenula* which, apart from the shape of the spores and the ability to assimilate nitrate, appears in the similarity of the phosphomannans produced by *Pachysolen tannophilus*, *Hansenula capsulata* and *H. holstii*.

The genus *Citeromyces* described by Santa Maria (1957) is represented by the single species *Citeromyces matritensis*. Cells of strains of opposite mating-type of this species agglutinate when mixed (Wickerham, 1958) and, after conjugation, produce asci with one or two round warty spores. However, diploid strains of one mating type may also sporulate alone. The non-sporulating form of this species has been described as *Torulopsis globosa*. *Citeromyces matritensis* has been isolated from sugar and from liquids with a high sugar content. It ferments glucose and sucrose vigorously.

The genus *Dekkera*, described by van der Walt in 1964, comprises two species, strains of which were originally classified in the asporogenous genus *Brettanomyces*. In 1960, van der Walt and van Kerken observed in these strains hat- and helmet-shaped spores which were easily liberated from the ascus. Sporulation was promoted by the addition of vitamins to the medium. The species are homothallic. They have the typical physiological features of the genus *Brettanomyces* which will be discussed below. The yeasts reproduce by multilateral budding; they do not form mycelium. Nitrate assimilation varies among the strains of both species, which differ in their ability to assimilate β-glucosides. Strains of *Dekkera bruxellensis* have been isolated from lambic and stout beers, those of *Dekkera intermedia* from beer, wine and 'tea beer fungus'.

The remaining genera of ascosporogenous yeasts in the second group are all nitrate-negative. The genus *Pichia* has recently been studied by Boidin *et al.* (1964, 1965a, b; Pignal and Boidin, 1965) and by Kreger-van Rij (1964, 1966a, 1970). It shares several morphological and physiological features with the genus *Hansenula*, such as the shape of the spores, formation of pseudomycelium and a pellicle, and rate of fermentation; it differs from the genus *Hansenula* in that the species do not assimilate nitrate. The greater number of *Pichia* species have ascospores with a ledge which gives them a hat- or saturn-shaped appearance. The species with round spores without a ledge resemble in this respect species of *Saccharomyces*, *Kluyveromyces* and *Debaryomyces*, but in other respects they are distinguishable from these genera. In *Pi. membranaefaciens* and *Pi. ohmeri*, spores with and without a ledge occur. In *Pi. fluxuum* the ledge is very narrow and camouflaged by the rough spore wall; it is, however, distinctly visible in electron micrographs (Fig. 14, p. 24). The warts on the spores in this species and also in *Pi. terricola* (Fig. 13, p. 23), when examined in ultrathin sections under the electron microscope, appear to be of a different nature than those of *Debaryomyces* spores in that they are formed by the outer layer of the spore wall. In most species, the spores are easily liberated from the ascus. Half of the *Pichia* species are haploid and homothallic, showing conjugation between mother cell and bud before ascus formation. They include species that form spores with and without a ledge. A number of mostly haploid species, which are all heterothallic, are included in the *guilliermondii* group. In some of these species, true hyphae may be observed to a limited extent which seems to relate them to the genus *Endomycopsis*. Homothallism and heterothallism were observed in *Pi. membranaefaciens* and *Pi. anyophorae*. It is conceivable that other homothallic species may also appear to have mating types. Most *Pichia* species form pseudomycelium. Boidin *et al.* (1965b) also accepted in the genus *Pichia* species that produce abundant true mycelium, species which Kreger-van Rij (1970) retained in the genus *Endomycopsis*. Although *Pichia* spp., in contrast with *Hansenula* spp., do not assimilate nitrate, in some species nitrite-positive strains occur, as for instance in *Pi. pinus*. This species seems to be narrowly related to the genus *Hansenula* where a similar species, *H. glucozyma*, has been described by Wickerham (1970). Fermentation ability varies among *Pichia* species; it is seldom vigorous. Pellicle formation is found in many of the species. Kreger-van Rij (1970) accepts 35 species in the genus *Pichia*, species that are distinguishable by the shape of the spores, assimilation and fermentation reactions, growth at 37° and vitamin requirements. Sources of *Pichia* strains include bark beetles, *Drosophila*, exudates of trees, tanning liquid and fruit.

The genus *Debaryomyces* is characterized by warty ascospores; the warts are formed by the inner layer of the spore wall (Fig. 15, p. 25). Conjugation between mother cell and bud precedes ascus formation (Fig. 10, p. 20). Fermentation may be good, but is generally slow and certainly not vigorous. The rate of fermentation is one of the features which distinguishes *Debaryomyces* spp. from members of the so-called *Torulaspora* group. The species of this latter group, which is retained in the genus *Saccharomyces*, have a vigorous fermentation. In most of them, the spore wall is also found to be warty (Kreger-van Rij, 1970). Warty spores also occur in some *Pichia* species but, in contrast with *Debaryomyces*, these species are diploid and the warts are formed by the outer layer of the spore wall.

Kreger-van Rij (1970) recognizes eight species in *Debaryomyces*. There are two groups among them, each with rather similar species distinguishable only by a few fermentation and assimilation reactions. Species in the first group have small round cells and a fermentation capacity which, if present, is only weak; species in the second group have bigger cells and a fermentation capacity which, if present, is good.

Debaryomyces hansenii, a representative of the first group, has many synonyms indicating that it is frequently encountered. Most of the strains assimilate creatine when tested by the auxanographic method (Kreger-van Rij and Staib, 1963), and several assimilate nitrite (Wickerham, 1957). *Debaryomyces castelli*, *D. cantarellii* and *D. phaffii* of the second group are very similar. They have fermentative ability and generally produce one spore per ascus. The fourth species of this group, *D. vanriji*, does not ferment any sugars and it forms four spores per ascus.

The genus *Schwanniomyces* is characterized by the shape of the spores, which are round to oval with a warty wall and a ledge in the middle. The warts and the ledge are electron-dense when fixed with osmic acid or potassium permanganate. This dark layer bursts on germination, and the light inner layer becomes the cell wall. Ascus formation is preceded by conjugation between mother cell and bud (Fig. 11, p. 21). Ferreira and Phaff (1959) studied nuclear behaviour with the aid of stained preparations. From this, they concluded that meiosis occurred in the bud, the so-called meiosis bud. They found that yeast autolysate (1:10)-agar, supplemented with 2% glucose, was a very good sporulation medium for the strain of *Schw. alluvius* which they examined.

The four species which Phaff (1970) recognizes in *Schwanniomyces* ferment glucose readily. They are morphologically very similar, and can be differentiated on the assimilation of a few carbon compounds. All strains of this genus have been isolated from soil.

The genus *Wingea* was described by van der Walt (1967). The author had previously (van der Walt, 1959a) classified the single species, *Wingea robertsii*, in the genus *Pichia*, but the unusual shape of the spores induced him to create a new genus for it. The spores are oblate ellipsoidal in shape. In sections, the outer layer of the wall appears to be broadened at the poles. Single spores have a light brown colour, and an abundantly sporulating culture is dark chocolate brown. Conjugation between mother cell and bud precedes ascus formation. Free spores were never observed in the cultures. The species ferments glucose, sucrose, maltose and raffinose, and assimilates most of the carbon compounds tested. The strains studied by van der Walt all came from larval feeds of *Xylocopa caffra*, the South African carpenter bee.

The genus *Kluyveromyces* (van der Walt, 1956, 1965) is characterized by ascospores which are reniform, crescentiform, round or oval, and which are easily liberated from the ascus. The number of spores per ascus is in most species between one and four but up to 16 and more in *Kluyv. polysporus* and *Kluyv. africanus*. Fermentation is vigorous. Several species produce a red pigment, closely related or identical with pulcherrimin (Wickerham and Burton, 1956a).

Wickerham (1955) indicated species of the present genus *Kluyveromyces* as *Dekkeromyces* species, but he never validated this name. Kudrjawzew (1960) had assigned these species to the genus *Fabospora*. Boidin *et al.* (1962) classified the species of this genus in *Guilliermondella*, on the basis of a resemblance between their spores and those of *Guilliermondella selenospora*. Kreger van Rij and van der Walt (1963) showed, however, that this resemblance is superficial, and that the shape of the spores is fundamentally different.

Wickerham and Burton (1956a, b) found that, in the genus *Dekkeromyces*, most if not all species hybridized with another. They were homothallic with the exception of *Sacch. lactis* (*Kluyv. lactis*) which is heterothallic. *Kluyveromyces fragilis* was believed to be a natural hybrid between *Kluyv. lactis* and *Zygosacch. ashbyi* (*Kluyv. marxianus*). The genus includes some 20 species which are differentiated on the number of spores per ascus, the shape of the spores, and a number of fermentation and assimilation reactions. The lactose-fermenting species, *Kluyv. marxianus*, *Kluyv. lactis*, *Kluyv. fragilis* and *Kluyv. sociasi*, have been frequently isolated from milk and milk products. *Kluyveromyces marxianus* and *Kluyv. fragilis* have also been isolated from human sources, including sputum and faeces. Sources of isolations of the other *Kluyveromyces* species are *Drosophila* and soil. *Kluyveromyces osmophilus* (Fig. 19, p. 29), isolated from sugar, is an osmophilic yeast which has a much higher growth rate on media with a high concentration of sugar than on those containing a low concentration. It cannot be adapted to

3

give a high rate of sugar fermentation in a medium with a low sugar concentration (Kreger-van Rij, 1966b).

In the genus *Saccharomyces*, the ascospores are round or oval, warty in a number of species but smooth in most others. In contrast with the spores of *Kluyveromyces* spp. they are not easily liberated from the ascus. On malt extract, a dry pellicle is not formed. All species have a vigorous fermentation.

Different groups may be recognized in the genus *Saccharomyces*. In the first group, indicated as *Saccharomyces sensu stricto* which comprises most species, there is no conjugation immediately preceding ascus formation. The species are diploid, or possibly of a higher ploidy. Homothallism and heterothallism were observed in them. Bulder (1963) found that the species of *Saccharomyces s. s.* were petite-positive, that is, respiratory-deficient mutants could easily be obtained from them. This is in contrast with the species in the two following groups. Well-known species, such as *Sacch. cerevisiae* and *Sacch. uvarum*, belong to the first group.

The second group is that designated by the former genus *Zygosaccharomyces*. In members of this group, conjugation immediately precedes ascus formation, either between mother cell and bud, or between different single cells. The first and the second groups are distinct in that transitions from haploid to diploid species or *vice versa* occur very infrequently. The second group contains several osmophilic species, such as *Sacch. rouxii*. Wickerham and Burton (1960) isolated mating types from this species, although it is also known to be homothallic.

The third group comprises the former genus *Torulaspora* and the species *Debaryomyces globosus*. The species in this group are haploid. They generally form protuberances on the cells under special conditions. Conjugation occurs either via the protuberances between different single cells, or directly between mother cell and bud. Diploidization without apparent conjugation, with only protuberances present on the cells, has also been described. Most species of this group have warty spores (Kreger-van Rij, 1970). The warts may be very indistinct and only visible by electron microscopy. Almost all *Torulaspora* strains can grow without the addition of vitamins, which is a conspicuous feature in the genus *Saccharomyces*.

Two *Saccharomyces* species, namely *Sacch. montanus* and *Sacch. kluyveri*, take a special position in the genus. They are the only species which split β-glucosides. *Saccharomyces montanus* is haploid or diploid and homothallic. *Saccharomyces kluyveri* may occur in various degrees of ploidy (Wickerham, 1958); it is homothallic as well as heterothallic. Cells of opposite mating-type agglutinate when mixed.

Van der Walt (1962) has tested the ability to utilize ethylamine as a sole source of nitrogen in a number of *Saccharomyces* species. It appeared that the species of the *Saccharomyces s. s.* group did not assimilate this compound, but that several *Zygosaccharomyces* species were positive.

The species in the three groups are mainly differentiated on the fermentation and assimilation of a few carbon compounds. The narrow relationship between several species of *Saccharomyces s. s.* is shown by their interfertility. Differences in a single enzyme between species mean that mutations from one species to another are possible. Scheda and Yarrow (1966) described mutants which had acquired the ability to ferment galactose, sucrose or maltose.

The genus *Saccharomycopsis* comprises the single species *Saccharomycopsis guttulata* which has its habitat in the stomach of the rabbit. Shifrine and Phaff (1958a, 1959) made an extensive study of this species, which cannot be grown under the usual conditions. It has large cylindrical budding cells and forms oval to cylindrical ascospores. Upon germination, the outer layer of the spore bursts. *Saccharomycopsis guttulata* has some outstanding physiological characteristics. It grows at temperatures between 35° and 40°, in a pH range of 2–6·5, and it requires amino acids for growth. Richle and Scholer (1961) found that good growth was possible on a solid medium provided the concentration of carbon dioxide in the atmosphere was increased to about 10% (v/v). The yeast also grows well in a liquid medium containing 2% glucose, 1% proteose peptone and vitamins, at pH 4·5. Glucose, sucrose and raffinose are weakly or slowly fermented.

Phaff (1970) points out the typical physiological characteristics that distinguish the genus *Saccharomycopsis* from *Saccharomyces*. Moreover, Shifrine and Phaff (1958b) found *Saccharomycopsis guttulata* to have an unusual cell-wall composition, with a high protein and a low mannan content in comparison with *Sacch. cerevisiae*.

Lodderomyces is a genus devised by van der Walt (1966) for the species *Lodderomyces elongisporus*, originally described by Recca and Mrak (1952) as *Saccharomyces elongisporus*. A typical feature of the genus is the occurrence of asci containing one, or occasionally two, big oval spores. Slow fermentation distinguishes the genus from *Saccharomyces*. The species *Lodderomyces elongisporus* is homothallic. Scheda and Bos (1966) found that its two available strains assimilated hydrocarbons. Van der Walt (1966) suggested that *Lodderomyces elongisporus* is the perfect form of *Candida parapsilosis*.

The genus *Lipomyces* has outstanding characteristics in the formation of oval amber-coloured spores in an ascus which arises in a peculiar way. The spores are liberated from the ascus; the spore wall is smooth, warty or has ridges. Starkey (1946) first described the asci as buds growing out to sac-like shapes containing between four and 16 or even more spores (Fig. 21, p. 42). The vegetative cells often contain a large lipid globule. Under suitable conditions, the cells have a capsule which gives the culture a mucous appearance. Fermentative ability is absent. The starch reaction, typical of the genus *Cryptococcus* the species of which also form mucous cultures, is very weak in *Lipomyces*. Growth is rather slow.

In 1966, Bab'eva and Meavad discovered in isolates related to *Lipomyces* species another type of ascus formation. They found that the vegetative cells produced protuberances by which conjugation took place, often with buds still attached to the mother cell. The spores are formed in one of the conjugating cells, namely in that with the pro-tuberance. The authors considered the species which had this type of conjugation to belong to a new genus, *Zygolipomyces*. Krasilnikov *et al.* (1967) described two species in this genus, both forming oval amber-coloured spores with longitudinal ridges on the wall. Slooff (1970) has observed, in the species originally described as *Lipomyces*, conjugation between 'active buds' and vegetative cells. Active buds have abnormal shapes; they may also change directly, without apparent conjugation, into asci as described by Starkey (1946). Slooff (1970) did not find principal differences in ascus formation between *Lipomyces* and *Zygolipomyces*, and discarded the latter genus.

For distinguishing species the use of different properties has been proposed. Lodder and Kreger-van Rij (1952) used the assimilation of lactose, which was not accepted by Connell *et al.* (1954). Slodki and Wickerham (1966) found a difference in the composition of the capsular polysaccharides in various strains and used this for differentiation. Nieuwdorp *et al.* (1969) observed, in *Lipomyces*, strains with smooth and warty spores when observed in ultrathin sections with the electron microscope. A distinction on the basis of the spore wall—smooth, warty, or with ridges, characteristics which are not always distinct in light microscopy but very clear in electron micrographs—seems acceptable. All *Lipomyces* strains have been isolated from soil.

The next group of ascosporogenous yeasts is characterized by bipolar budding on a broad base. Stelling-Dekker (1931) classified this group in the tribe of the Nadsonieae. Lodder and Kreger-van Rij (1952) followed this classification. Lodder (1970) places these yeasts in a subfamily, Nadsonioideae. Kudrjawzew (1960) also recognized this group as a family of the Saccharomycodaceae.

Streiblová *et al.* (1964) found that bipolar budding produced a special kind of scar on the cells of yeasts in the genera *Saccharomycodes*, *Hanseniaspora* and *Nadsonia* which are in this group. These so-called multiple scars are made up of a series of concentric ridges on the cell wall resulting from buds being formed at the same site. I made the same observations in electron micrographs of *Hanseniaspora* (Fig. 3, p. 12) and of *Wickerhamia* the fourth genus of this group, but not in *Nadsonia* (N. J. W. Kreger-van Rij, unpublished observations). In members of the last genus, older scars were rather indistinct in sections, perhaps because, when fixed with osmic acid or permanganate, the cell wall does not have a distinct dark outer region as in most other yeasts. Although more than one bud was formed at a pole, the ridges of the scars were not all concentrically arranged and a more complicated pattern of budding seems probable. Under the light microscope, more buds were observed at one pole, and these were still attached to the cell so obviously they could not be formed at the same site. This corroborates the original description of *Guilliermondia* (= *Nadsonia*) *fulvescens* by Nadson and Konokotina (1912). The buds are often formed on a small protuberance at each pole.

The four genera of this group, described by Phaff (1970), all differ markedly in the shape of the ascospores; diploidization and ascus formation, in so far as it has been described for three of them, also have divergent features. On the basis of these characteristics, it is questionable whether they constitute a natural group.

The genus *Saccharomycodes*, with the single species *S. ludwigii*, has round spores with a narrow ledge, usually with four spores in each ascus (Kreger-van Rij, 1969; Fig. 23). Upon germination, they conjugate in pairs while still in the ascus, and thus give diploid vegetative cells. Winge and Laustsen (1939) found that this yeast is a heterozygote which, by conjugation of genetically different spores, retains this character. *Saccharomycodes ludwigii* ferments glucose and sucrose. Strains of this species have been isolated from exudate of an oak, from grape must, and from wines (Ribéreau-Gayon and Peynaud, 1960).

In the genus *Hanseniaspora*, the spores are round to hemispherical with a ledge which makes them saturn-, helmet- or hat-shaped, or alternatively round without a ledge but with a distinctly warty wall. Kreger-van Rij and Ahearn (1968) made a study of the morphology of the spores of *Ha. uvarum*, *Ha. valbyensis* and *Ha. guilliermondii*. They observed a ledge on the spores of *Ha. uvarum* which had not been previously described, and which by light microscopy is not always distinctly visible. Miller and Phaff (1958) postulated a diploidization during germination of a single spore based on nuclear stainings of the spores of two *Hanseniaspora* species.

The species of the genus show a vigorous fermentation. They all split β-glucosides. Phaff (1970) mentions their absolute requirement for inositol and pantothenate. The species are differentiated by the shape of the spores, the number of spores in each ascus, their liberation from the ascus, and by the fermentation and assimilation of maltose. Most strains have been isolated from soft fruit, such as grapes, from soil, and *Drosophila*.

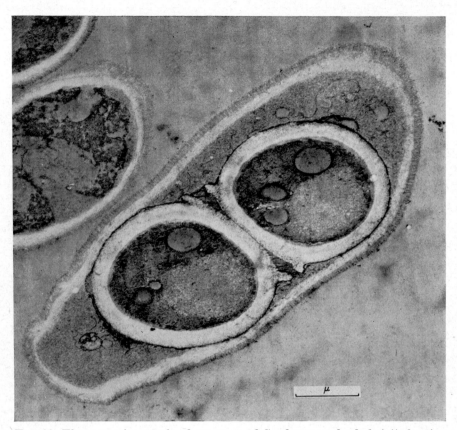

FIG. 23. Electron micrograph of an ascus of *Saccharomycodes ludwigii* showing two of the four spores lying in a pair.

Yeasts in the genus *Nadsonia* have a very peculiar mode of ascus formation which has been described by Nadson and Konokotina (1912, 1926). A conjugation takes place between mother cell and bud; the nucleus of the daughter cell moves into the mother cell and forms a dikaryon. Thereafter, the contents of these cells move into a second bud formed at the opposite end of the mother cell. In this bud, nuclear fusion and meiosis result in the formation of one, and occasionally two or four spores. The ascus may be detached from the mother cell. Skovsted (1943) studied mutation in a strain of *Nadsonia richteri*

(= *Nadsonia elongata*). He found various cell shapes and giant colony forms in the mutant. The ascospores are big and round with thin, rather long warts formed by the outer layer of the spore wall (Fig. 24). Single spores are yellowish-brown, and a culture containing many spores is dark brown.

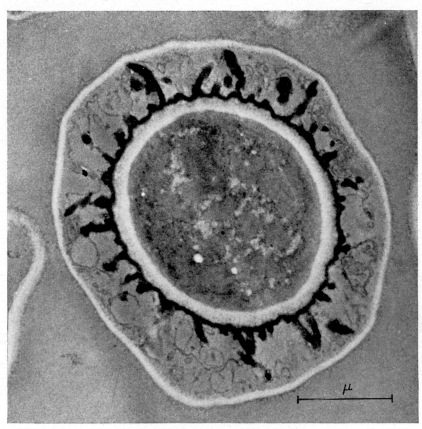

FIG. 24. Electron micrograph of a section through an ascus of *Nadsonia elongata*. The asci were fixed with osmic acid and uranyl acetate.

The two species of the genus, *Nadsonia fulvescens* and *Nadsonia elongata*, both have fermentative ability. They differ in the fermentation and assimilation of a number of carbon compounds, in the maximum temperature for growth, and in the ability to grow without the addition of vitamins. Strains of both species have been isolated from exudates of trees.

The genus *Wickerhamia* is characterized by cap-like spores (Fig. 25). Electron micrographs of sections of the spores show that the ellipsoidal spore body has a ledge attached to the bottom. The ledge has an inverted edge and is short at one side of the spore and very broad at the opposite side. The latter is the peak of the cap, and forms part of the spore wall. Usually one spore is present in the oval ascus but, according to Soneda

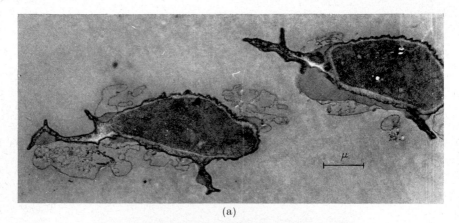

(a)

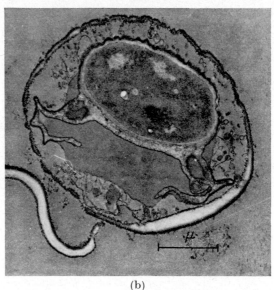

(b)

FIG. 25. Electron micrographs of sections in two directions through ascospores of *Wickerhamia fluorescens*. The spores were fixed with osmic acid and uranyl acetate.

(1960) who first described the genus, up to 16 may be formed in one ascus. The ascus ruptures in the middle, resulting in two half egg shell-like parts. It is not yet known where diploidization occurs in the life cycle. The single species, *Wickerhamia fluorescens*, ferments glucose, galactose and sucrose. It may excrete riboflavin into the medium, hence its name. The only strain available was isolated from dung of a wild squirrel in Japan.

Yeasts in the genera *Nematospora*, *Metschnikowia* and *Coccidiascus* are all characterized by needle- or spindle-shaped spores. They have been considered to be related on this basis, and were classified together in a subfamily, the Nematosporoideae (Stelling-Dekker, 1931), or in a family, the Spermophthoraceae (Gäumann, 1964). In the latter family were also classified *Ashbya*, a genus members of which form mostly true mycelium and only few yeast cells, and *Spermophthora* reported to produce both sporangiospores and ascospores. In *Ashbya* and *Spermophthora*, plurinucleate stages occur and, therefore, they have not been included in the yeasts.

The genus *Nematospora* comprises one species, *N. coryli*, which is a pathogen for fruit, and is reported to be transferred and introduced

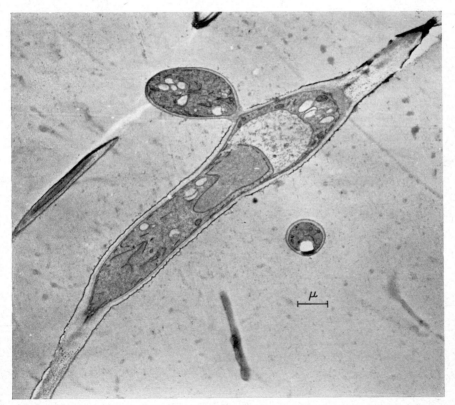

FIG. 26. Electron micrograph of a section through an ascospore of *Nematospora coryli* which has formed one bud. The section is not through the whole length of the spore. The spores were fixed with potassium permanganate.

3*

into the plant by insects. It has been isolated from various fruits, such as hazelnuts, tomatoes, cotton bolls and citrus fruits.

Nematospora coryli forms budding cells as well as true mycelium. Manuel (1938) observed conjugation preceding ascus formation. The asci are cylindrical and much larger than the budding cells. They generally contain eight spores. The spores are very long and pointed at both ends, one end relatively short and the other very thinly drawn out. The latter structure is often indicated as a whip attached to the spore, a description which is deceptive. The protoplast lies in the middle and here, on germination, buds are formed (Fig. 26). It is pointed towards the long end of the spore and flattened at the shorter point which may give the impression of a septum. Electron micrographs confirmed these light-microscope observations. The spore wall which forms the short point appears homogeneous and slightly refractive in young spores, but in older spores shows parts which stain blue with Sudan Black B, indicating the presence of lipids (N. J. W. Kreger-van Rij, unpublished observations). The spores are easily liberated from the ascus. *Nematospora coryli*, in the description of do Carmo-Sousa (1970), ferments glucose, sucrose, maltose and raffinose.

The genus *Metschnikowia*, of which Miller and van Uden (1970) give a detailed description, is characterized by needle-shaped spores, pointed at both ends, and occurring singly or in pairs in a large elongate ascus. The spores are thinner than those of *Nematospora*. Two spores lying closely together in the ascus may be mistaken for a single one. The species are heterothallic (Miller *et al.*, 1967; Wickerham, 1964a). The formation of pseudomycelium is rudimentary; true mycelium is lacking. Fermentative ability, if present, is restricted to glucose and occasionally galactose.

The genus *Metschnikowia* has a long history to which important data have recently been added. Metschnikoff (1884) described the first species, *Monospora bicuspidata*, present as a parasite in *Daphnia magna*. A second species, parasitic in the brine shrimp *Artemia salina*, was described by Kamienski (1899) who gave the genus its present name. These yeasts had not then been grown in pure culture. Van Uden and Castelo-Branco (1961) were the first to obtain pure cultures of similar organisms isolated from sea water, fish intestines and kelp. A second important development was Wickerham's finding of mating types in *Candida pulcherrima*, *Candida reukaufii* and a third species, all of which he transferred to the new genus *Chlamydozyma* (Wickerham, 1964a). This genus was characterized by the formation of chlamydospores. Mating reactions were also observed between one of the *Chlamydozyma* species and a *Metschnikowia* species. These did not yield ascospores, but showed the relationship between these genera. The important final step in this history was made by Pitt and Miller (1968) who discovered conditions which induced sporulation in the *Chlamydozyma* species; these are now classified as *Metschnikowia* species.

Of the five species accepted by Miller and van Uden (1970) in this genus, *M. bicuspidata*, *M. krissii* and *M. zobellii* have been isolated from brine shrimp, sea water, fish intestines and kelp. *Metschnikowia pulcherrima* has been isolated from fruits and flowers. This species forms big round oil-containing chlamydospores, the so-called 'pulcherrima cells'; these may turn into asci. Several strains produce pulcherrimin, a red pigment. *Metschnikowia reukaufii* closely resembles *M. pulcherrima*, but forms oval to cylindrical chlamydospores which may turn into clavate asci. Strains of this species frequently occur in nectar of flowers.

Coccidiascus legeri, the single species of the genus, has never been isolated. Chatton (1913) observed this yeast in cells of the intestine of *Drosophila funebris*. He mentioned in his description of genus and species that ascus formation is probably preceded by conjugation. Two connected cells, each with a nucleus, were observed to change into a banana-shaped ascus containing eight fusiform spores, each screwed in a helix. The yeast has budding cells and no mycelium.

B. BALLISTOSPOROGENOUS YEASTS

The ballistosporogenous yeasts comprise the genera *Sporidiobolus*, *Sporobolomyces* and *Bullera*, which are classified in the family of the Sporobolomycetaceae (Table III). Derx (1948) also included in this family the genera *Itersonilia* and *Tilletiopsis*, both forming ballistospores, but no budding cells. The Sporobolomycetaceae were considered to belong to the basidiomycetes because of the ballistospores which resemble basidiospores in that they are shot off by a drop-excretion mechanism. This conception was not generally accepted because a sexual cycle had not been observed.

The description of the genus *Sporidiobolus* by Nyland (1948, 1949) has thrown new light on this question. Apart from budding cells producing ballistospores in the same way as *Sporobolomyces*, *Sporidiobolus* spp. form a dikaryotic mycelium with clamp connections, typical of basidiomycetes. Laffin and Cutter (1959), as the result of a study of the life cycle of *Sporidiobolus johnsonii*, proposed a scheme for the alternation of the haploid and diploid stages in this species, and thus confirmed its basidiomycetous nature. It is noteworthy that, in this scheme, the ballistospores are not basidiospores but asexual cells.

In the species of the genera *Sporobolomyces* and *Bullera*, no sexual phase has yet been found. The combination of these genera with *Sporidiobolus* in the Sporobolomycetaceae is based on the character of

TABLE III. *Main Characteristics of the Genera of Ballistosporogenous Yeasts with the Family Sporobolomycetaceae, and of Yeasts that do not Form Ascospores or Ballistospores*

SPOROBOLOMYCETACEAE

Sporidiobolus
Budding cells, true mycelium with clamp connections, chlamydospores, sporidia.
Ballistospores asymmetrical.
Fermentation negative.
Nitrate positive.

Sporobolomyces
Budding cells, (true mycelium).
Ballistospores asymmetrical.
Fermentation negative.
Nitrate assimilation positive or negative.

Bullera
Budding cells.
Ballistospores symmetrical.
Fermentation negative.
Nitrate assimilation positive or negative.

———

Rhodosporidium
Budding cells, true mycelium with or without clamp connections, chlamydospores, sporidia.
Orange or pink cultures.
Fermentation negative.
Nitrate assimilation positive.

Leucosporidium
Budding cells, (pseudomycelium), true mycelium with or without clamp connections, chlamydospores, sporidia.
Hyaline cultures.
Fermentation positive or negative.
Nitrate assimilation positive or negative.

———

CRYPTOCOCCACEAE

Rhodotorula
Budding cells.
Red or yellow cultures.
Fermentation negative.
Nitrate assimilation positive or negative.
Starch not produced.
Inositol not assimilated.

Cryptococcus
Budding cells.
Hyaline, red or black cultures.
Fermentation negative.
Nitrate assimilation positive or negative.
Starch produced.
Inositol assimilated.

Torulopsis
Budding cells.
Fermentation positive or negative.
Nitrate assimilation positive or negative.
Starch not produced.
Inositol not assimilated.

Candida
Budding cells, pseudomycelium, (true mycelium, chlamydospores).
Fermentation positive or negative.
Nitrate assimilation positive or negative.
Starch may or may not be produced.
Inositol may or may not be assimilated.

Trigonopsis
Budding cells, oval or triangular.
Fermentation negative.

Schizoblastosporion
Budding cells, bipolar on a broad base.
Fermentation negative.

Kloeckera
Budding cells, bipolar on a broad base.
Fermentation positive.

Pityrosporum
Budding cells, monopolar on a broad base.
Fermentation negative.

Brettanomyces
Budding cells.
Fermentation positive.
Nitrate assimilation positive or negative.
Short-lived on malt-agar.

Trichosporon
Budding cells, true mycelium and arthrospores.
Fermentation positive or negative.
Nitrate assimilation positive or negative.
Starch may or may not be produced.
Inositol may or may not be assimilated.

Oosporidium
Budding cells on a broad base, true mycelium.
Fermentation negative.
Nitrate assimilation positive.
Starch produced.
Inositol not assimilated.

Sterigmatomyces
Budding cells, formed on sterigmata.
Fermentation negative.
Nitrate assimilation positive or negative.

ballistospore formation. Laffin and Cutter (1959) agreed with this combination, but Nyland (1949) objected to the combination of perfect and imperfect forms in one family.

Another genus closely resembling *Sporidiobolus*, with a sexual stage but not forming ballistospores, is *Rhodosporidium*. Banno (1967) described the species *Rhodosporidium toruloides* in which a dikaryotic mycelium is formed after conjugation of yeast cells of opposite mating-type of the species *Rhodotorula glutinis*. Meiosis is considered to take place upon germination of the chlamydospores, yielding a haploid phase of yeast cells. In contrast with this, Laffin and Cutter (1959) considered the yeast phase in *Sporidiobolus* to be diploid. Notwithstanding this difference, it is very probable that *Sporidiobolus* and *Rhodosporidium* are closely related genera, and species of the genera *Rhodotorula*, *Rhodosporidium*, *Sporobolomyces* and perhaps also *Itersonilia* may well represent stages of *Sporidiobolus*. About the conditions which induce these stages very little is known. Banno (1967) classified the genus *Rhodosporidium* in the order of the Ustilaginales in the Basidiomycetes. This is in accordance with Gäumann's suggestion (1964) that *Sporobolomyces* might comprise species representing vegetative stages of species belonging to the Ustilaginales.

Very few data are available about a possibly analogous relationship between species of *Cryptococcus*, *Bullera* and *Tremella*. Slodki *et al.* (1966) found similar extracellular heteropolysaccharides in *Cryptococcus laurentii* and *Tremella* species; *Cr. laurentii*, except for the fact that it does not form ballistospores, closely resembles *Bullera alba*. The genera *Rhodosporidium*, *Leucosporidium*, *Rhodotorula* and *Cryptococcus*, although probably related to the genera forming ballistospores, will be discussed under yeasts that do not form ascospores or ballistospores.

Nyland (1948, 1949) described the genus *Sporidiobolus* which is characterized by the formation of budding yeast cells, and also mycelium with clamp connections and brown thick-walled chlamydospores. Asymmetrical ballistospores are produced on sterigmata arising on yeast cells or on the mycelium. The formation of mycelium with clamp connections and chlamydospores distinguishes this genus from *Sporobolomyces*.

Nyland (1948, 1949) and Laffin and Cutter (1959) studied the life cycle of the species *Sporidiobolus johnsonii*. They found that karyogamy took place in the chlamydospores which were formed on dikaryotic mycelium. The chlamydospores germinated with a short germ-tube from which yeast cells arose by budding. Laffin and Cutter (1959), on the basis of the shape of the survival curve after exposure of the cells to ultraviolet radiation and on the results of mating experiments with monokaryotic 'operative' cells, came to the conclusion that the yeast cells were diploid. The data on meiosis were less clear. Nyland (1949) did not observe it at all, but Laffin and Cutter (1959), observing meiotic-like figures in yeast cells, presumed that it took place in these cells prior to the appearance of mycelium. They never obtained haploid yeast cells.

The genus *Sporidiobolus* comprises two species, *Sporidiobolus johnsonii*, originally isolated from a pustule of *Phragmidium rubi-idaei* on a leaf of a *Rubus idaeus* in Washington, and first described by Nyland, and *Sporidiobolus ruinenii* (Fig. 20, p. 30), described by Phaff (1970). Strains of the latter species were isolated by Ruinen (1963) from the leaves of *Malpighia coccigera* in the Bogor Botanical Garden in Indonesia. Both species assimilate nitrate. They have no fermentative ability. According to Phaff (1970), they differ in the assimilation of a few carbon compounds, and in the ability to grow at 37°. The strains may occur in the mucous form or in the non-mucous form. On malt-agar, the cultures are pink to red, and occasionally dark brown when many chlamydospores are present. Ruinen (1963) observed external lipid production in the cultures of *Sporidiobolus ruinenii*.

Kluyver and van Niel (1924–1925) first described the genus *Sporobolomyces* for yeasts which form kidney- or sickle-shaped spores on sterigmata. The spores are discharged by a drop-excretion mechanism. Apart from the formation of these so-called ballistospores, the yeasts reproduce by budding and by the formation of mycelium. No dikaryotic mycelium with clamp connections is formed as in *Sporidiobolus*. Sterigmata may arise both on budding cells and on mycelial hyphae. *Sporobolomyces* spp. differ from *Bullera* spp. which are also characterized by the formation of ballistospores, and in the shape of the latter which are asymmetrical in *Sporobolomyces* but symmetrical in *Bullera*. In contrast with *Bullera* spp., most *Sporobolomyces* species are pink to red, although one, *Sp. singularis*, is hyaline.

Phaff (1970) recognizes nine species of *Sporobolomyces*. They are differentiated on the assimilation of nitrate, on the presence or absence of true mycelium, and on the assimilation of a few carbon compounds. Derx (1930) found that fresh isolates of *Sporobolomyces* species were often mucous, and changed into the non-mucous form when kept in culture. He isolated several strains from leaves infected with smut. Other sources of isolation were the atmosphere, fruit, soil, grass and sea water.

The genus *Bullera* is characterized by multilateral budding and by the formation of symmetrical ballistospores. In the latter feature, they differ from *Sporobolomyces* in which the ballistospores are asymmetrical. The cultures of *Bullera* are hyaline or yellowish. The strains may occur in the mucous or in the dry form.

Strains of two of the three *Bullera* species recognized by Phaff (1970) are still available. The single strain of the third species, *B. grandispora*, is no longer available. *Bullera alba* is nitrate-negative and assimilates inositol; *B. tsugae* is nitrate-positive and does not assimilate inositol. Strains of *B. alba* have been isolated from plants and from the air. The single strain of *B. tsugae*, described by Phaff and do Carmo-Sousa (1962), has been isolated from frass.

C. YEASTS THAT DO NOT FORM ASCOSPORES OR BALLISTOSPORES

Yeasts which do not form ascospores or ballistospores are classified in the family of the Cryptococcaceae (Gäumann, 1964) with the exception of the genera *Rhodosporidium* and *Leucosporidium*. These genera are known to form sporidia, which are considered to be basidiospores. The genera *Rhodosporidium, Leucosporidium, Rhodotorula* and *Crytococcus* are probably related, and the latter two may represent stages of Basidiomycetes. The possibility is not excluded that, in *Candida* and *Trichosporon*, species may occur which also belong to this group. In the genera not belonging to this group, and also in *Candida*, species occur which are certainly imperfect forms of ascosporogenous yeasts, and a few genera, e.g. *Kloeckera*, have their counterpart in a sporogenous genus. But, generally, the data for relationships between perfect and imperfect forms are still too scarce to make a more natural classification of the asporogenous yeasts possible. At present, criteria of vegetative reproduction chiefly determine the delimitation of the genera. These genera are: *Rhodosporidium, Leucosporidium, Rhodotorula, Cryptococcus, Torulopsis, Candida, Trigonopsis, Kloeckera, Schizoblastosporion, Pityrosporum, Brettanomyces, Trichosporon, Oosporidium* and *Sterigmatomyces*.

In 1967, Banno described the genus *Rhodosporidium* with the single species *Rhodosporidium toruloides*. This species is the perfect form of a number of strains of *Rhodotorula glutinis*. It has budding cells, and mycelium with clamp connections and chlamydospores. Banno found that some strains of *Rh. glutinis* represented opposite mating types. When mixed, the monokaryotic cells conjugated and developed a dikaryotic mycelium with clamp connections. On the mycelium, pear-shaped chlamydospores were formed in which karyogamy occurred and which, when mature, had a thick brown wall. On germination of the chlamydospores with a promycelium consisting of 2–4 cells, reduction division took place. Each cell of the promycelium formed one yeast-like cell indicated as a sporidium. After budding they produced the yeast phase. Banno (1967) compared this type of germination of the chlamydospores or resting cells with that found in the Ustilaginales. The sporidia can be considered as basidiospores.

A small percentage of the chlamydospores yielded upon germination diploid yeast cells which produced a dikaryotic mycelium. Banno (1967) presumed that, in these, meiosis failed during germination of the resting spore, but that a somatic reduction of the diploid nucleus might occur in the yeast cells leading to a dikaryotic mycelium. This site of meiosis seems to agree with that found in *Sporidiobolus* by Laffin and Cutter (1959).

Banno (1967) tried to obtain mating reactions among several *Rhodotorula* and *Sporobolomyces* species, but he had success in only five of the 17 strains of *Rh. glutinis*; among these strains, the number of conjugations was very small under the conditions used. Two of the conjugating *Rh. glutinis* strains had been isolated from wood pulp and from the air respectively. Banno (1967) isolated two other *Rhodosporidium* strains in their perfect stage from the air.

As indicated by the name, the genus *Rhodotorula* originally comprised red yeasts. The red colour is due to the presence of carotenoid pigments. Lodder (1934) included yeasts with yellow carotenoid pigments in the genus. Lodder and Kreger-van Rij (1952) retained the characteristic of the distinct red or yellow colour of the cultures for *Rhodotorula*, thus distinguishing it from *Cryptococcus*, a very similar genus. Species of the latter might be yellowish or pinkish, and might also contain carotenoid pigments. This not very exact distinction between the genera has led several authors to propose other ways of classification. Nakayama *et al.* (1954) suggested using the production of starch as a means of differentiating between *Rhodotorula* and *Cryptococcus*. Hasegawa *et al.* (1960) transferred all *Cryptococcus* species in which carotenoid pigments could be detected to the genus *Rhodotorula*, and subdivided the genus into two subgenera, *Rubrotorula* for red yeasts, and *Flavotorula* for yellow yeasts. In the genus *Cryptococcus*, only *Cr. neoformans* was retained. Since the colour of the culture may vary with the composition of the medium, Hasegawa *et al.* (1960) used a special medium when comparing strains, namely potato-yeast extract broth.

Phaff and Spencer (1966) proposed a differentiation between *Rhodotorula* and *Cryptococcus* based on the utilization of inositol, which is absent in the former and positive in the latter. Phaff and Ahearn (1970) have defined the genus *Rhodotorula* in accordance with these characters. Inability to assimilate inositol and to produce starch are mentioned in their definition. Fermentative ability is also absent. The authors recognize nine species, differing in the assimilation of nitrate and of a number of carbon compounds. Generally, the number of carbon compounds assimilated by the various species is great, but variation between the strains of one species is also considerable.

Ahearn (1964) who made a study of morphological and physiological properties of *Rhodotorula* species, isolated new strains which produce mycelium with clamp connections and chlamydospores. These strains, which he identified as *Rh. glutinis*, are now probably better placed in *Rhodosporidium*. *Rhodotorula* strains have been isolated from various sources, including leaves, flowers, the atmosphere, soil and marine sources.

The genus *Cryptococcus* as defined by Phaff and Fell (1970) has multilateral budding, no pseudomycelium or true mycelium, and does not ferment, while all species assimilate inositol. Most species have capsulated cells and produce mucous cultures. Starch production is generally found only at low pH values; in two species that produce red cultures, starch is also formed at higher pH values. Most species give hyaline cultures, but *Cr. macerans*, *Cr. infirmo-miniata* and *Cr. hungaricus* are red; *Cr. laurentii* may be orange while *Cr. flavus* is yellow and *Cr. ater* dark brown or black in old cultures. The assimilation of nitrate varies among the species. Seeliger (1956) found that all *Cryptococcus* species which he examined gave a positive urease test. *Rhodotorula* species were also positive, but *Torulopsis* species were negative.

The genus *Cryptococcus* shows a strong resemblance with the genus *Rhodotorula*. A former distinction based on the colour of streak cultures appeared to be unsatisfactory. Phaff and Spencer (1966) proposed using the assimilation of inositol for differentiation; this character is positive in *Cryptococcus* and absent in *Rhodotorula*. This feature correlated with the presence of certain cell-wall components found in a number of species of both genera, namely the presence of an acidic heteropolysaccharide containing D-xylose, D-mannose and D-gluconic acid in *Cryptococcus* and a new type of mannan in *Rhodotorula*. Another genus which has several features in common with *Cryptococcus* is *Torulopsis* but, in the latter, the species are either capable of fermenting or if not are unable to assimilate inositol and do not produce starch.

Phaff and Fell (1970) accept 17 species in the genus *Cryptococcus*; these are differentiated on the colour of the culture, nitrate assimilation, starch formation, ability to grow at 37°, and the assimilation of a number of carbon compounds. The pathogenic species, *Cr. neoformans*, is one of the few *Cryptococcus* species that grows at 37°. It is also one of the few which assimilates creatinine in the auxanographic test (Staib, 1963). This author (Staib, 1963) also found that the presence of ground thistle seed (*Guizotia abyssinica*) in the medium gave a dark brown colour in the culture, a reaction which is specific for *Cr. neoformans*. Strains of *Cr. neoformans* have been isolated from material of human origin, from mastitis of a cow, and from bird manure (Staib, 1962).

The genera *Torulopsis* and *Candida* have very wide boundaries, and comprise all species that show multilateral budding and which, by the absence of a specific character, could not be placed in any of the other asporogenous genera.

Torulopsis differs from *Candida* by the inability to form pseudomycelium, from *Cryptococcus* by the absence of starch production, and from *Rhodotorula* by the absence of visible carotenoid pigments. Generally, cultures on malt-agar are not short-lived as in *Brettanomyces*. *Torulopsis lactis-condensi* and *T. pintolopesii* are exceptions, but these species are respiratory-deficient (Bulder, 1963). In some species, a very rudimentary pseudomycelium may be formed. The formation of pseudomycelium is known to vary depending on the smooth or rough form of the yeast. It is not excluded that, among the five species which van Uden and Vidal Leiria (1970) have transferred from the genus *Torulopsis* to the genus *Candida*, this kind of variation had taken place. The species *T. pinus*, *T. ingeniosa* and *T. molischiana* all produce capsules, but no starch is formed and they do not assimilate inositol.

Van Uden and Vidal-Leiria (1970) include 36 species in the genus *Torulopsis*. Fourteen of these species assimilate nitrate, among them being *T. globosa* the imperfect form of *Citeromyces matritensis*, and *T. molischiana* the asporogenous form of *Hansenula capsulata*. Of the other nitrate-positive species, no perfect forms are known. Five nitrate-negative species have a corresponding sporogenous form, namely *T. bovina*—*Sacch. telluris*, *T. mogii*—*Sacch. rouxii*, *T. holmii*—*Sacch. exiguus*, *T. sphaerica*—*Kluyveromyces lactis*, and *T. candida*—*Debaryomyces hansenii* and *D. marama*.

Most *Torulopsis* species ferment one or more sugars vigorously. Many species are represented by very few strains. Consequently, the data about their habitats are scarce. Many isolates of *T. candida* are available, several of them of human origin. Most strains of *T. glabrata* and of *T. inconspicua* are also of human origin. Strains of the species *T. haemulonii*, *T. torresii* and *T. maris* have all been isolated from sea water. The sodium chloride tolerance of the first two species is relatively high, namely 16–20% (w/v) (van Uden and Vidal-Leiria, 1970). *Torulopsis halonitratophila*, isolated from soy mash and described by Onishi (1960), is according to this author obligately halophilic at 30°. Several strains of *T. stellata* (syn. *T. bacillaris*) have been isolated from over-ripe grapes used for the production of *vins blancs liquoreux*. This species has an absolute requirement for inositol (Ribéreau-Gayon and Peynaud, 1960).

The genus *Candida*, in the diagnosis of van Uden and Buckley (1970), comprises species that show multilateral budding, or bipolar budding but not on a broad base. Pseudomycelium is formed by all species. True mycelium may occur, but no arthrospores are produced. Polar budding was observed in *C. aquatica*; it resulted in cells assuming a star-like arrangement (Jones and Slooff, 1966). Pseudomycelium varies from well-developed to primitive. Well developed pseudomycelium is differentiated into pseudomycelial cells and blastospores in a special configuration; primitive pseudomycelium is undifferentiated. Van Uden and Buckley (1970) have pointed to the unsatisfactory and artificial results obtained with the use of this characteristic for the distinction of the genera *Candida* and *Torulopsis*. In the first place, it may vary within the strain (see p. 7). Secondly, in ascosporogenous genera like *Hansenula*, it varies among the species, so that asporogenous forms of these species are classified in different genera. Nevertheless, Van Uden and Buckley (1970) decided against a combination of *Candida* and *Torulopsis* in one genus because of the confusion it might cause by the many changes of names involved. Species producing true mycelium include *C. lipolytica* and *C. ciferrii*. In old strains of *C. albicans*, true mycelium may also be found. This species, and also the narrowly related species *C. stellatoidea*, form chlamydospores under special conditions (Fig. 6, p. 15). This feature is so typical that it is frequently used for the identification of *C. albicans*. This species, which is important from a medical point of view because it may be a pathogen, is described in a monograph by Winner and Hurley (1964) who also mention other identification methods.

Starch is produced at low pH values by *C. humicola*. In this respect, and also in the assimilation of inositol, this species resembles *Cryptococcus* species. It produces true mycelium as well as pseudomycelium.

A number of *Candida* species also form starch at higher pH values, on malt-agar or on glucose-peptone agar. These species are nitrate-positive, and have fermentative ability; also, they do not grow at 25°, but at a lower temperature such as 15°. Di Menna (1966) isolated *C. frigida*, *C. gelida* and *C. nivalis* from Antarctic soil; Komagata and Nakase (1965) isolated *C. curiosa*, which is considered to be synonymous with *C. nivalis*, from frozen food. *Candida aquatica*, which was isolated from water scums, also belongs to this group.

Van Uden and Buckley (1970) recognize some 80 species in the genus *Candida*. The large size of the genus does not facilitate its surveyability and accessibility. The elaborate key leading to the recognition of species includes almost all of the properties of the standard description. It might be convenient if a subdivision could be made into more or less homogeneous groups, preferably corresponding with the sporogenous genera. The genus comprises the imperfect forms of several genera, for instance *Hansenula*, *Kluyveromyces* and *Pichia*.

Some 20 *Candida* species are nitrate-positive. Amongst them are asporogenous *Hansenula* species, as *C. utilis* and *C. silvicola*, the starch-positive species mentioned above, and a group of capsulated starch-negative yeasts, isolated from leaves of tropical plants and described by Ruinen (1963). To this group belong *C. foliorum, C. javanica* and *C. diffluens*, all of which produce extracellular lipids. *Candida muscorum* described by di Menna (1958) also resembles yeasts in this group. *Candida macedoniensis* and *C. pseudotropicalis* are the imperfect forms of *Kluyveromyces marxianus* and *Kluyv. fragilis* respectively.

A number of *Candida* species, such as *C. rhagii, C. lusitaniae, C. oregonensis* and *C. tenuis*, resemble *Pichia* species of the *guilliermondii* group. It is probable that they are also heterothallic. Mating reactions in the species of this group may be scarce, as in *P. guilliermondii*. Many strains which differ from this species only by the absence of a mating reaction are maintained in the genus *Candida* as *C. guilliermondii*. Apparently, optimal conditions for this reaction are not yet known for all species.

The assimilation reactions constitute an important distinguishing property in *Candida*. The uncertain value of these reactions and the great variability among the strains often make it difficult to delimit a species. This may lead to excessive lumping as for instance in *C. sake* as described by van Uden and Buckley (1970).

The genus *Trigonopsis* is characterized by the occurrence of triangular cells. This shape originates from the formation of buds on slight protuberances at different sites on the cell. The buds are ellipsoidal. Cells which have formed more than three buds and, accordingly, have more bulges and are of a more complicated shape, also occur. The single species, *Trig. variabilis*, does not assimilate nitrate and has no fermentative ability. Strains of it have been isolated from beer and grapes.

The genus *Kloeckera* is characterized by bipolar budding with the formation of multiple scars, and by its fermentative ability. Nitrate is not assimilated. Young cells are oval, older ones lemon-shaped. Phaff (1970) recognizes four *Kloeckera* species, two of which are asporogenous forms of *Hanseniaspora* species. *Kloeckera apiculata* may comprise asporogenous strains of *H. valbyensis, H. guilliermondii* and *H. uvarum*; *Kloeckera corticis* is the imperfect form of *H. osmophila*. The four *Kloeckera* species differ in the assimilation of sucrose and maltose. Like *Hanseniaspora* spp., they have an absolute requirement for inositol and pantothenate and they split β-glucosides.

The genus *Schizoblastosporion* resembles *Kloeckera* in that the species in it multiply by bipolar budding on a broad base but, in contrast with *Kloeckera*, it has no fermentative ability. Nitrate is not assimilated. Nearly all strains of the single species *Schizoblastosporion starkeyi-henricii* have been isolated from soil, namely in the U.S.A., Norway, Denmark and New Zealand.

The genus *Pityrosporum* is characterized by monopolar budding on a broad base, a feature not encountered in any other yeast genus. Mother cell and bud, after separation, form buds at the site of a bud scar and birth scar respectively. The cells have been described as bottle-shaped.

Of the three species accepted by Slooff (1970) in the genus, only *Pit. canis* grows on malt-agar; *Pit. ovale* (Fig. 2, p. 11) and *Pit. orbiculare* require lipids to be added to this medium. They all grow well on Littman Oxgall agar at 37° (Martin-Scott, 1952). Barfatani *et al.* (1964) studied the ultrastructure of *Pit. orbiculare*, and Swift and Dunbar (1965) that of *Pit. ovale* and *Pit. canis*. These authors found a relatively thick cell wall with a corrugated inner surface closely adhering to the plasma membrane. Barfatani *et al.* (1964) refer to two separate membranes of the cell wall of which the inner membrane exhibits indentations.

Pityrosporum canis has been isolated from infected dog's ears, and *Pit. ovale* and *Pit. orbiculare* from the human skin, the former from scalp scales, the latter from normal skin as well as from lesions of tinea versicolor. In tinea versicolor, *Pit. orbiculare* is frequently associated with a fungus, *Malassezia furfur*, which has not yet been obtained in pure culture (see p. 163). Keddie and Shadomy (1963) suggested that these organisms are identical, the yeast being the conidial stage of the fungus. They consider as evidence for this identity a similarity in conidia formation of the fungus from phialides and bud formation in the yeast, as well as common antigens in the fungus and the yeast. Moreover, Keddie (1966), studying the ultrastructure of *Malassezia furfur* in excized skin, found a dented construction in the cell wall similar to that mentioned above for the *Pityrosporum* species. Conclusive evidence for the identity of both forms might be obtained from the finding of conditions which would make growth of the fungus in pure culture possible, conditions which should turn the yeast into the mycelial form.

The genus *Brettanomyces* is primarily characterized by physiological properties. The yeasts of this genus grow slowly on malt-agar; they are short lived on this medium. A vigorous production of acetic acid is typical of most species of *Brettanomyces*. Cultures in malt extract or on malt-agar have a typical odour. All species are fermentative. Custers (1940) found that oxygen stimulated fermentation by *Br. claussenii*, a phenomenon which he named a negative Pasteur effect. Wikén *et al.* (1961) demonstrated this effect in all the six *Brettanomyces* species studied. Nitrate assimilation varies among the species.

Cells of *Brettanomyces* spp. multiply by budding; they may be arranged as pseudomycelium. Pointed, so-called ogive, cells frequently occur. Long non-septate hyphae, described as blastese, are formed by the species *Br. anomalus* and *Br. custersianus*. The sporulating forms of *Br. bruxellensis* and *Br. intermedius* have been classified in the genus

Dekkera. Most *Brettanomyces* strains have been isolated from beer and wine.

In the genus *Trichosporon* budding cells, often arranged as pseudo-mycelium, and true septate mycelium which breaks up into arthrospores are formed (Fig. 9, p. 18). The presence of arthrospores distinguishes the genus both from *Candida* and *Oosporidium*. Do Carmo-Sousa (1966) observed asexual endospores formed by internal budding in strains of *Trich. cutaneum, Trich. capitatum* and *Trich. fermentans*. In *Trich. cutaneum* they were also formed by protoplasmic cleavage.

Do Carmo-Sousa (1970) accepts seven species in *Trichosporon*. *Trichosporon pullulans* is the only species which assimilates nitrate. A few species, among which is *Trich. fermentans*, show a slow or weak fermentation of glucose, occasionally also of galactose. *Trichosporon pullulans* and strains of *Trich. cutaneum* give a positive starch test; this reaction is doubtful in *Trich. inkin*. *Trichosporon cutaneum* provided most of the *Trichosporon* strains studied. It has a long list of synonyms. The assimilation reactions show a wide variation with uniformity only for lactose, xylose and ethanol, all of which are positive, whereas inulin is negative. It cannot be excluded that, within these wide limits, different species are included, e.g. *Trich. multisporum*, but apparently it is not yet possible to distinguish them with the present methods. The origin of the strains of *Trich. cutaneum* also varies considerably, but many strains have been isolated from white piedra and some from skin lesions.

Haskins and Spencer (1967) have described an organism in a new genus *Trichosporonoides*, as *Trichosporonoides oedocephalis*. This genus resembles *Trichosporon* in every respect with the exception of the production of conidiospores on the swollen tip of aseptate sporophores. The conidia have a dark brown wall. The species ferments glucose, galactose, sucrose and maltose. It assimilates nitrate and erythritol. It has been isolated from a honey comb. The authors pointed to the resemblance of the conidiosphores of this species with those of the ascomycete *Oedocephalum*, and of the phycomycete *Cunninghamella*.

The genus *Oosporidium*, originally described by Stautz (1931), has been accepted and newly diagnosed by do Carmo-Sousa (1970) to include the single species *Oosporidium margaritiferum*. Its cells reproduce by budding on a broad base and by septum formation in the longer cells. In contrast with *Trichosporon*, arthrospores are not formed in *Oosporidium*. The cells are usually arranged in chains. Asexual endo-spores are formed by protoplasmic cleavage. *Oosporidium margariti-ferum* grows very slowly on malt-agar on which it produces a pink or yellow pigment. It assimilates nitrate, is non-fermentative and gives a weak starch reaction. Two strains of this species came from slime flux.

The genus *Sterigmatomyces*, described by Fell (1966, 1970), is exceptional among the yeasts in that the cells reproduce exclusively by forming buds on sterigmata. Unlike ballistospores, these buds when mature are not forcefully discharged, but are disjointed at a septum in the mid-region of the sterigma. The cells are round or oval; true hyphae are not formed. The sterigmata vary in length from 1.5 to 26 μ, each bearing a single cell. More than one sterigma may be formed by one cell. The two species accepted by Fell (1970), *Sterigmatomyces halophilus* (Fig. 7, p. 16) and *Sterigmatomyces indicus*, differ in the assimilation of nitrate. They have no fermentative ability. It appeared that the addition of 3% sodium chloride to the culture medium stimulated growth. Strains of both species have been isolated from sea water.

V. Acknowledgements

In presenting a survey of yeast systematics in this chapter, I have profited considerably by perusing the new edition of "The Yeasts", edited by J. Lodder (North-Holland Publishing Co.). I am greatly indebted to the authors of this book, namely Drs. D. G. Ahearn, H. Buckley, L. do Carmo-Sousa, J. W. Fell, J. Lodder, M. W. Miller, H. J. Phaff, W. Ch. Slooff, N. van Uden, M. Vidal-Leiria, and L. J. Wickerham who put the manuscripts of their contributions at my disposal, and to the North-Holland Publishing Co. for giving me permission to refer to this edition which is to be published in 1970. I acknowledge the hospitality of the Laboratory of Ultrastructural Biology of the State University of Groningen which made preparation of the electron micrographs possible. I am very grateful to Dr. J. Lodder and Dr. C. J. E. A. Bulder for reading the typescript. My thanks are also due to Dr. J. J. Wachters who made most of the light micrographs, and to Mr. L. Hoekstra for drawing Fig. 12.

References

Agar, H. D. and Douglas, H. C. (1955). *J. Bact.* **70**, 427–434.
Ahearn, D. G. (1964). Ph.D. Thesis: University of Miami.
Aschner, M., Mager, J. and Leibowitz, J. (1945). *Nature, Lond.* **156**, 295.
Bab'eva, I. P. and Meavad, K. (1966). *Mikrobiologiya* **35**, 824–828.
Baker, J. M. and Kreger-van Rij, N. J. W. (1964). *Antonie van Leeuwenhoek* **30**, 433–441.
Banno, I. (1963). *J. gen. appl. Microbiol., Tokyo* **9**, 249–251.
Banno, I. (1967). *J. gen. appl. Microbiol., Tokyo* **13**, 167–196.
Barfatani, M., Munn, R. J. and Schjeide, O. A. (1964). *J. invest. Derm.* **43**, 231–234.
Barnett, J. A. (1966). *Nature, Lond.* **210**, 565–568.

Barnett, J. A. (1968a). *In* "The Fungi" (G. C. Ainsworth and A. S. Sussman, eds.) Vol. III, pp. 557–595. Academic Press, New York.

Barnett, J. A. (1968b). *J. gen. Microbiol.* **52**, 131–159.

Beijerinck, M. W. (1898). *Zentbl. Bakt. ParasitKde (Abt II)* **4**, 657–663, 721–730.

Besson, M. (1966). *Bull. Soc. mycol. Fr.* **82**, 489–503.

Biguet, J., Tran Van Ky, P., Andrieu, S. and Fruit, J. (1965). *Mycopath. Mycol. appl.* **26**, 241–256.

Boidin, J., Abadie, F., Jacob, J. L. and Pignal, M. C. (1962). *Bull. Soc. mycol. Fr.* **78**, 155–203.

Boidin, J., Abadie, F. and Lehodey, Y. (1965a). *Bull. Soc. mycol. Fr.* **81**, 5–23.

Boidin, J. and Adzet, J. M. (1957). *Bull. Soc. mycol. Fr.* **73**, 331–342.

Boidin, J., Pignal, M. C. and Besson, M. (1965b). *Bull. Soc. mycol. Fr.* **81**, 566–606.

Boidin, J., Pignal, M. C., Lehodey, Y., Vey, A. and Abadie, F. (1964). *Bull. Soc. mycol. Fr.* **80**, 396–438.

Bulder, C. J. E. A. (1963). Thesis: University of Delft.

Carmo-Sousa, L. do (1966). Proc. 2nd Internat. Symp. on Yeasts, Bratislava, pp. 87–92.

Carmo-Sousa, L. do (1970). *In* "The Yeasts" (J. Lodder, ed.) North-Holland Publishing Co., Amsterdam.

Chatton, E. (1913). *C. r. Séanc. Soc. Biol.* **65** II, 117–120.

Connell, G. H., Skinner, C. E. and Hurd, R. C. (1954). *Mycologia* **46**, 12–16.

Crook, E. M. and Johnston, I. R. (1962). *Biochem. J.* **83**, 325–331.

Custers, M. Th. J. (1940). Thesis: University of Delft.

Derx, H. G. (1930). *Annls mycol.* **28**, 1–23.

Derx, H. G. (1948). *Bull. bot. Gdns. Buitenz. Ser. III* **17**, 465–472.

Donk, M. A. (1964). *Persoonia* **3**, 199–324.

Dyke, K. G. H. (1964). *Biochim. biophys. Acta* **82**, 374–384.

Fell, J. W. (1966). *Antonie van Leeuwenhoek* **32**, 99–104.

Fell, J. W. (1970). *In* "The Yeasts" (J. Lodder, ed.), North-Holland Publishing Co., Amsterdam.

Fell, J. A. and Phaff, H. J. (1970). *In* "The Yeasts" (J. Lodder, ed.). North-Holland Publishing Co., Amsterdam.

Fell, J. A., Statzell, A., Hunter, I. L. and Phaff, H. J. (1969). In press.

Ferreira, J. D. and Phaff, H. J. (1959). *J. Bact.* **78**, 352–361.

Garzuly-Janke, R. (1940). *Zentbl. Bakt. ParasitKde (Abt II)* **102**, 361–365.

Gäumann, E. (1964). "Die Pilze". Birkhäuser Verlag, Basel.

Gilliland, R. B. (1956). *C. r. Trav. Lab. Carlsberg Sér. physiol.* **26**, 139–148.

Guilliermond, A. (1909). *Revue gén. Bot.* **21**, 353–391, 401–419.

Guilliermond, A. (1937). "La sexualité, le cycle de développement, la phylogénie et la classification des levures". Masson & Cie, Paris.

Hansen, E. C. (1904). *Zentbl. Bakt. ParasitKde (Abt II)* **12**, 529–538.

Hasegawa, T., Banno, I. and Yamauchi, S. (1960). *J. gen. appl. Microbiol., Tokyo* **5**, 200–216, **6**, 196–215.

Hashimoto, T., Conti, S. F. and Naylor, H. B. (1959). *J. Bact.* **77**, 344–354.

Haskins, R. H. and Spencer, J. F. T. (1967). *Can. J. Bot.* **45**, 515–520.

Jones, E. B. G. and Slooff, W. Ch. (1966). *Antonie van Leeuwenhoek* **32**, 223–228.

Kamienski, Th. (1899). *Trudy̆ imp. S-peterb. Obshch. Estest.* **30**, 363–364.

Kawakami, N. (1960). *Mem. Fac. Engng Hiroshima Univ.* **1**, 207–237.

Kawakami, N., Nehira, T. and Kodama, K. (1961). *Mem. Fac. Engng Hiroshima Univ.* **1**, 407–414.

Keddie, F. M. (1966). *Sabouraudia* **5**, 134–137.

Keddie, F. and Shadomy, S. (1963). *Sabouraudia* **3**, 21–25.

Klöcker, A. (1907). *In* "Handbuch der technischen Mykologie" (F. Lafar, ed.) Vol. 4, pp. 168–191. Fischer, Jena.

Klöcker, A. (1924). "Die Gärungsorganismen". Urban & Schwarzenberg, Berlin.
Kluyver, A. J. and van Niel, C. B. (1924–1925). *Zentbl. Bakt. ParasitKde* (*Abt II*) **63**, 1–20.
Komagata, K. and Nakase, T. (1965). *J. gen. appl. Microbiol.*, *Tokyo* **11**, 255–267.
Komagata, K., Nakase, T. and Katsuya, N. (1964). *J. gen. appl. Microbiol.*, *Tokyo* **10**, 313–321.
Krasilnikov, N. A., Bab'eva, I. P. and Meavad, K. (1967). *Mikrobiologiya* **36**, 923–931.
Kreger, D. R. (1954). *Biochim. biophys. Acta* **13**, 1–9.
Kreger, D. R. (1967). Symposium on Yeast Protoplasts. Jena, 1965, pp. 81–88. Akademie Verlag, Berlin.
Kreger-van Rij, N. J. W. (1964). Thesis : University of Leiden.
Kreger-van Rij, N. J. W. (1966a). Proc. 2nd Internat. Symposium on Yeasts, Bratislava, pp. 51–58.
Kreger-van Rij, N. J. W. (1966b). *Mycopath. Mycol. appl.* **29**, 137–141.
Kreger-van Rij, N. J. W. (1969). *Can. J. Microbiol.* in press.
Kreger-van Rij, N. J. W. (1970). *In* "The Yeasts" (J. Lodder, ed.), North-Holland Publishing Co., Amsterdam.
Kreger-van Rij, N. J. W. and Ahearn, D. G. (1968). *Mycologia* **60**, 604–612.
Kreger-van Rij, N. J. W. and Staib, F. (1963). *Arch. Mikrobiol.* **45**, 115–118.
Kreger-van Rij, N. J. W. and Walt, J. P. van der (1963). *Nature, Lond.* **199**, 1012–1013.
Kudrjawzew, W. I. (1960). "Die Systematik der Hefen". Akademie Verlag, Berlin.
Laffin, R. J. and Cutter, V. M. (1959). *J. Elisha Mitchell scient. Soc.* **75**, 89–96, 97–100.
Lodder, J. (1934). "Die anaskosporogenen Hefen, I. Hälfte". *Verh. K. Akad. Wet. Sect. II*, **32**, 1–256.
Lodder, J. (1970). *In* "The Yeasts" (J. Lodder, ed.), North-Holland Publishing Co., Amsterdam.
Lodder, J. and Kreger-van Rij, N. J. W. (1952). "The Yeasts, a Taxonomic Study", 713 pp. North-Holland Publishing Co., Amsterdam.
Lodder, J., Slooff, W. Ch. and Kreger-van Rij, N. J. W. (1958). *In* "The Chemistry and Biology of Yeasts" (A. H. Cook, ed.), pp. 1–62. Academic Press, New York.
Manuel, J. (1938). *C. r. Séanc. Soc. Biol.* **207**, 1241–1243.
Marchant, R. and Smith, D. G. (1967). *Arch. Mikrobiol.* **58**, 248–256.
Marchant, R. and Smith, D. G. (1968). *J. gen. Microbiol.* **53**, 163–169.
Martin-Scott, I. (1952). *Br. J. Derm.* **64**, 257–273.
di Menna, M. E. (1958). *J. gen. Microbiol.* **18**, 269–272.
di Menna, M. E. (1966). *Antonie van Leeuwenhoek* **32**, 25–28.
Metschnikoff, E. (1884). *Virchows Arch. path. Anat. Physiol.* **96**, 177–195.
Miller, J. J. and Hoffmann-Ostenhof, O. (1964). *Z. allg. Mikrobiol.* **4**, 273–294.
Miller, M. W., Barker, E. R. and Pitt, J. I. (1967). *J. Bact.* **94**, 258–259.
Miller, M. W. and Phaff, H. J. (1958). *Mycopath. Mycol. appl.* **10**, 113–141.
Miller, M. W. and van Uden, N. (1970). *In* "The Yeasts", (J. Lodder, ed.). North-Holland Publishing Co., Amsterdam.
Nadson, G. A. and Konokotina, A. G. (1912). *Wschr. Brau.* **29**, 309–313, 332–336.
Nadson, G. A. and Konokotina, A. G. (1926). *Annls Sci. nat.* (*Bot.*) **8**, 165–182. (reviewed in *Bull. Inst. Pasteur* (1926). **14**, 667–668).
Nakayama, T., Mackinney, G. and Phaff, H. J. (1954). *Antonie van Leeuwenhoek* **20**, 217–228.

Nieuwdorp, P. J., Bos, P. and Slooff, W. Ch. (1969). *Antonie van Leeuwenhoek* in press.

Novák, E. K. and Zsolt, J. (1961). *Acta bot. hung.* **7**, 93–145.

Nyland, G. (1948). *Mycologia* **40**, 478–481.

Nyland, G. (1949). *Mycologia* **41**, 686–701.

Onishi, H. (1960). *Bull. agric. Chem. Soc. Japan* **24**, 226–230.

Peynaud, E. and Domercq, S. (1956). *Annls Inst. Pasteur* **91**, 574–580.

Phaff, H. J. (1963). *A. Rev. Microbiol.* **17**, 15–30.

Phaff, H. J. (1970). *In* "The Yeasts" (J. Lodder, ed.), North-Holland Publishing Co., Amsterdam.

Phaff, H. J. and Ahearn, D. G. (1970). *In* "The Yeasts" (J. Lodder, ed.), North-Holland Publishing Co., Amsterdam.

Phaff, H. J. and Carmo-Sousa, L. do (1962). *Antonie van Leeuwenhoek* **28**, 193–207.

Phaff, H. J. and Fell, J. W. (1970). *In* "The Yeasts" (J. Lodder, ed.), North-Holland Publishing Co., Amsterdam.

Phaff, H. J. and Spencer, J. F. T. (1966). Proc. 2nd Internat. Symposium on Yeasts, Bratislava, p. 59–65.

Pignal, M. C. and Boidin, J. (1965). *Bull. Soc. mycol. Fr.* **81**, 197–226.

Pitt, J. I. and Miller, M. W. (1968). *Mycologia* **60**, 663–685.

Poncet, S. (1967). *Antonie van Leeuwenhoek* **33**, 345–358.

Recca, J. and Mrak, E. M. (1952). *Fd. Technol.*, *Champaign* **6**, 450–454.

Reess, M. (1870). "Botanische Untersuchungen über die Alkoholgärungspilze", Leipzig.

Ribéreau-Gayon, J. and Peynaud, E. (1960). "Traité d'oenologie I". Béranger, Paris.

Richle, R. and Scholer, H. J. (1961). *Pathologia Microbiol.* **24**, 783–793.

Roberts, C. and Walt, J. P. van der (1960). *C. r. Trav. Lab. Carlsberg* **32**, 19–34.

Roelofsen, P. A. and Hoette, I. (1951). *Antonie van Leeuwenhoek* **17**, 297–313.

Ruinen, J. (1963). *Antonie van Leeuwenhoek* **29**, 425–438.

Ruinen, J. and Deinema, M. H. (1964). *Antonie van Leeuwenhoek* **30**, 377–384.

Santa Maria, J. (1957). *Boln Inst. nac. Invest. agron.*, Madr. No. 37, 269–276.

Scheda, R. and Bos, P. (1966). *Nature, Lond.* **211**, 660.

Scheda, R. and Yarrow, D. (1966). *Arch. Mikrobiol.* **55**, 209–225.

Schiönning, H. (1903). *C. r. Trav. Lab. Carlsberg* **6**, 103–125.

Schippers-Lammertse, A. F. and Heyting, C. (1962). *Antonie van Leeuwenhoek* **28**, 5–16.

Seeliger, H. P. R. (1956). *J. Bact.* **72**, 127–131.

Shifrine, M. and Phaff, H. J. (1958a). *Antonie van Leeuwenhoek* **24**, 193–209.

Shifrine, M. and Phaff, H. J. (1958b). *Antonie van Leeuwenhoek* **24**, 274–280.

Shifrine, M. and Phaff, H. J. (1959). *Mycologia* **51**, 318–328.

Skovsted, A. (1943). *C. r. Trav. Lab. Carlsberg Sér. Physiol.* **23**, 409–453.

Slodki, M. E. and Wickerham, L. J. (1966). *J. gen. Microbiol.* **42**, 381–385.

Slodki, M. E., Wickerham, L. J. and Bandoni, R. J. (1966). *Can. J. Microbiol.* **12**, 489–494.

Slooff, W. Ch. (1970). *In* "The Yeasts" (J. Lodder, ed.), North-Holland Publishing Co., Amsterdam.

Soneda, M. (1960). *Nagaoa* **7**, 9–13.

Spencer, J. F. T. and Gorin, P. A. J. (1968). *J. Bact.* **96**, 180–183.

Staib, F. (1962). *Zentbl. Bakt. ParasitKde* (*Abt I*) **186**, 274–275

Staib, F. (1963). *Zentbl. Bakt. ParasitKde* (*Abt I*) **191**, 429–432.

2. TAXONOMY AND SYSTEMATICS OF YEASTS

Starkey, R. L. (1946). *J. Bact.* **51**, 33–50.

Stautz, W. (1931). *Phytopath. Z.* **3**, 163–229.

Stelling-Dekker, N. M. (1931). "Die sporogenen Hefen". *Verh. K. Acad. Wet. Sect. II*, **28**, 1–547.

Stodola, F. H., Deinema, M. H. and Spencer, J. F. T. (1967). *Bact. Rev.* **31**, 194–213.

Storck, R. (1966). *J. Bact.* **91**, 227–230.

Streiblová, E., Beran, K. and Pokorný, V. (1964). *J. Bact.* **88**, 1104–1111.

Streiblová, E., Málek, I. and Beran, K. (1966). *J. Bact.* **91**, 428–435.

Swift, J. A. and Dunbar, S. F. (1965). *Nature, Lond.* **206**, 1174–1175.

Tanaka, H. and Phaff, H. J. (1967). Symposium on yeast protoplasts. Jena, 1965, pp. 113–129. Akademie Verlag, Berlin.

Tsuchiya, T., Fukazawa, Y. and Kawakita, S. (1965). *Mycopath. Mycol. appl.* **26**, 1–15.

Uden, N. van and Buckley, H. (1970). *In* "The Yeasts" (J. Lodder, ed.), North-Holland Publishing Co., Amsterdam.

Uden, N. van and Carmo-Sousa, L. do (1956). *Port. Acta biol.* **4**, 7–17.

Uden, N. van and Castelo-Branco, R. (1961). *J. gen. Microbiol.* **26**, 141–148.

Uden, N. van and Farinha, M. (1958). *Port. Acta biol.* **6**, 161–178.

Uden, N. van and Vidal-Leiria, M. (1970). *In* "The Yeasts" (J. Lodder, ed.), North-Holland Publishing Co., Amsterdam.

Walt, J. P. van der (1956). *Antonie van Leeuwenhoek* **22**, 265–272.

Walt, J. P. van der (1959a). *Antonie van Leeuwenhoek* **25**, 337–348.

Walt, J. P. van der (1959b). *Antonie van Leeuwenhoek* **25**, 458–464.

Walt, J. P. van der (1962). *Antonie van Leeuwenhoek* **28**, 91–96.

Walt, J. P. van der (1963). *Antonie van Leeuwenhoek* **29**, 52–56.

Walt, J. P. van der (1964). *Antonie van Leeuwenhoek* **30**, 273–280.

Walt, J. P. van der (1965). *Antonie van Leeuwenhoek* **31**, 341–348.

Walt, J. P. van der (1966). *Antonie van Leeuwenhoek* **32**, 1–5.

Walt, J. P. van der (1967). *Antonie van Leeuwenhoek* **33**, 97–99.

Walt, J. P. van der and Kerken, A. E. van (1960). *Antonie van Leeuwenhoek* **26**, 292–296.

Walt, J. P. van der and Kerken, A. E. van (1961a). *Antonie van Leeuwenhoek* **27**, 81–90.

Walt, J. P. van der and Kerken, A. E. van (1961b). *Antonie van Leeuwenhoek* **27**, 206–212.

Wickerham, L. J. (1951). *Tech. Bull. U. S. Dep. Agric.* No. 1029, pp. 1–56.

Wickerham, L. J. (1952). *A. Rev. Microbiol.* **6**, 317–332.

Wickerham, L. J. (1955). *Nature, Lond.* **176**, 22.

Wickerham, L. J. (1956). *C. r. Trav. Lab. Carlsberg Sér. physiol.* **26**, 423–442.

Wickerham, L. J. (1957). *J. Bact.* **74**, 832–833.

Wickerham, L. J. (1958)., *Science N.Y.*, **128**, 1504–1505.

Wickerham, L. J. (1964a). *Mycologia* **56**, 253–266.

Wickerham, L. J. (1964b). *Mycologia* **56**, 398–414.

Wickerham, L. J. (1970). *In* "The Yeasts" (J. Lodder, ed.), North-Holland Publishing Co., Amsterdam.

Wickerham, L. J. and Burton, K. A. (1948). *J. Bact.* **56**, 363–371.

Wickerham, L. J. and Burton, K. A. (1952). *J. Bact.* **63**, 449–451.

Wickerham, L. J. and Burton, K. A. (1954). *J. Bact.* **67**, 303–308.

Wickerham, L. J. and Burton, K. A. (1956a). *J. Bact.* **71**, 290–295.

Wickerham, L. J. and Burton, K. A. (1956b). *J. Bact.* **71**, 296–302.

Wickerham, L. J. and Burton, K. A. (1960). *J. Bact.* **80**, 492–495.

Wickerham, L. J. and Burton, K. A. (1962). *Bact. Rev.* **26**, 382–397.

Wiken, T., Scheffers, W. A. and Verhaar, A. J. M. (1961). *Antonie van Leeuwen-hoek* **27**, 401–433.

Windisch, S. (1960). *In* "Die Hefen. I. Die Hefen in der Wissenschaft (F. Reiff, R. Kautzmann, H. Lüers and M. Lindemann, eds.), pp. 23–178. Verlag Hans Carl, Nürnberg.

Winge, Ö. and Laustsen, O. (1939). *C. r. Trav. Lab. Carlsberg Sér. physiol.* **22**, 357–370.

Winner, H. I. and Hurley, R. (1964). "Candida albicans". J. & A. Churchill Ltd., London.

Chapter 3

Distribution of Yeasts in Nature

LÍDIA DO CARMO-SOUSA

Laboratory of Microbiology, Gulbenkian Institute of Science,
Oeiras, Portugal

I. Introduction

Yeast populations of several hundred species are continuously building up and dying off in terrestrial as well as in aquatic environments. They play their part in the dynamics of the biological and chemical turnover in soil, plants, animals and water, where they are active as competitors for nutrients, antagonists or symbiotic associates or as victims of the behaviour of their neighbours. The atmosphere appears to be a medium of dispersal rather than a biotope for them.

Among the physicochemical factors that affect the ecology of yeasts, the most important appear to be the energy sources, nutrients, temperature, pH value and water.

Being devoid of photosynthetic power, the yeasts depend strictly on the presence of organic carbon as an energy and carbon source. Simple sugars, such as glucose, fructose and mannose, are assimilated by all yeast species so far studied. Other monosaccharides and their polyol and acid derivatives, as well as oligosaccharides and even some polysaccharides, are utilized by different yeast species according to characteristic patterns. Also, hydrocarbons are a suitable carbon and energy source for many yeasts. Species of *Cryptococcus*, *Rhodotorula*, *Candida* and *Torulopsis*, which are able to utilize a great variety of different carbon compounds, are most frequently encountered in substrata such as water masses, soil and green leaves, where simple carbon compounds are present in very low concentrations. Species of *Saccharomyces* and *Pichia*, which assimilate few carbon compounds, are abundant in fruit juices, sugary plant exudates and other materials rich in simple sugars.

Ammonium nitrogen is suitable as a nitrogen source for any strain of yeast. In addition, many species are able to utilize nitrate nitrogen; only a few will live on nitrite nitrogen. The concentration of nitrite, as well as of other nitrogen compounds such as urea, must be relatively low otherwise toxic effects are exerted.

Many yeasts are dependent on an external supply of one or more vitamins of the B complex for growth. It appears that, at least in some cases, this dependency can vary with factors such as temperature (Kreger-van Rij, 1958).

Growth, sporulation and survival are intimately related to ambient temperatures. Psychrophiles are common in bodies of cold water and in terrestrial polar areas. Psychrophobic species such as *Saccharomycopsis guttulata*, *Saccharomyces telluster* (*Candida bovina*), *Candida slooffii* and *Torulopsis pintolopesii*, which are able to grow only within a narrow range of temperatures with 20–28° as a lower limit and 42–45° as an upper limit, are adapted to life in the digestive tract of warm-blooded animals. *Saccharomycopsis guttulata* and *Saccharomyces telluster* survive outside the animal body in the form of ascospores.

Surface water run-off, rainfall and percolation through soil are factors that alter yeast population densities as well as species distributions in niches. Drought and desiccation also affect survival. It is interesting to verify that, in environments like soil, the phyllosphere and the atmosphere which are more exposed to conditions of low water content, the most frequent yeast species are those that can produce extracellular polysaccharide-protein complexes. Such macromolecular complexes may act as protective coats on the yeast cells.

Competition for nutrients is probably the single most important factor in yeast ecology. It is possible that yeasts also act as antagonists of other organisms (e.g. bacteria) through the induction of environmental pH changes. As is the case with many bacteria, organic acids are excreted as end products of the metabolic activities of many yeasts. But, unlike most bacteria, yeasts grow well at relatively low pH values.

A few yeast species, for example *Candida albicans* and *Cryptococcus neoformans*, may become pathogenic for warm-blooded animals (see Chapter 4, p. 107). Yeasts may be inhibited or suppressed by antibiotics and lytic enzymes that are produced by other organisms. Some animals may act as predators of yeasts. Drosophila flies are an example. Through the industry of man, yeasts have found a wide range of beneficial applications in the biosphere (see Volume 3 of this treatise). Endosymbiotic associations of yeasts with insects will be discussed further.

Most work on yeast distribution has been devoted to determining the yeast species present in particular types of substrata. As a consequence of this qualitative approach, numerous yeast species have been discovered and described. By 1952 Lodder and Kreger-van Rij could recognize 165 valid species in their taxonomic system. In Lodder's forthcoming treatise (Lodder, 1970), about 350 valid species will be accepted. Such an increase in the number of known taxa reflects to some extent the intensification of ecological work in the yeast domain in the last 15 years. However, not much of the effort spent in this direction has been relevant to substrate-yeast species relationships, mainly because inappropriate techniques have been used.

The significance of the results of qualitative and quantitative analyses of the yeast flora in any natural substrate depends on the adequacy of the methods employed with respect to sampling, culturing and yeast identification. In general, sampling with sterile precautions is recommended; the time lapse between sampling and culture processing should be decreased to a minimum, and cooling should be maintained during transportation; the conditions of isolation should preferably be selective for the species that are active in the ecological niche under study; standard techniques of yeast identification recommended by yeast taxonomists should be followed.

Enrichment techniques that make use of liquid media for isolating yeasts from nature are prone to distort the pattern of the relative frequencies of yeast species distributions. Di Menna (1957) showed that the growth of yeasts that occasionally occur in soils may be selectively favoured, while that of significant representatives of the yeast flora of the sample may be masked if the latter do not thrive as easily in the culture medium as the former. A comparison of the results obtained by van Uden et al. (1958) and van Uden and Carmo-Sousa (1962c) for yeasts in the digestive tract of swine leads to similar conclusions.

Culturing on the surface of agar-containing medium is preferable to using incorporation methods since, in pour-plates, it is difficult to judge differences between the colonies and to isolate them. Inocula are better taken from suspensions of known weights of samples followed by serial dilutions in sterile water or saline.

Yeasts isolated from a given substrate may sometimes be survivors of contaminations even if present in relatively large numbers. In this connection, the findings of Yoneyama (1956) in soil contaminated with tree exudates are illustrative.

To establish which species are active in a particular substrate, methods which differ from case to case may have to be adopted. Repeated sampling of the same site at adequate time intervals, followed by a quantitative and qualitative analysis of the samples, may be a suitable approach in many cases. Phaff *et al.* (1964) studied the yeast flora of a slime flux of *Ulmus carpinifolia* along these lines.

The literature up to 1956 on yeasts isolated from natural substrates, namely leaves, flowers, sweet fruit, grain, root crops, fleshy fungi, exudates of trees, insects, dung and soil was reviewed by Lund (1958). Phaff *et al.* (1966) discussed some aspects of the association of yeasts with plants (leaves, flowers, tree exudates, plant pathogenic yeasts), animals (warm-blooded animals and insects), soils and water. An extensive review on marine yeasts was published recently (van Uden and Fell, 1968). The present chapter refers to data published mainly after 1956 and up to 1967. Pathogenic yeasts and yeasts in industrial products are not discussed. Aquatic and terrestrial environments are considered separately, the latter being subdivided into four groups, namely soils, plants, animals and atmosphere. Within each group, sampling and isolation techniques as well as results of qualitative and quantitative studies are discussed in relation to several types of substrates. Quantitative estimates of yeast populations in different geographical areas, expressed as total counts and as relative frequencies, are reviewed. Finally, attention is drawn to particular characteristics of yeasts and their respective substrates whenever they appear relevant to ecological relationships. As far as taxonomic nomenclature is concerned it was decided to use the names adopted in the original papers. The reader interested in the synonymy of any of the names may consult Lodder (1970).

II. Terrestrial Environments

A. SOIL

Yeasts have been found in soils of widely different texture, chemical composition, humidity and pH value at various geographical locations and diverse climatic conditions, in bare soils as well as in soils that support a natural vegetation (forests, bush, grasslands) or are cultivated by man (orchards, vineyards, cropfields, gardens).

The best technique for sampling soils for yeast surveys appears to be that used by di Menna (1957). Sterile containers are thrust into the sides of freshly dug pits at a depth of 2–10 cm where yeast populations are more abundant. Some workers have used enrichment techniques in liquid media for qualitative yeast surveys (Table I). As discussed in the introduction such techniques are generally unsuitable for ecological surveys.

Di Menna (1957) found that, among the media she used, a solid medium composed of glucose (4%), peptone (1%) and agar (2%; pH 4) permitted the highest yeast counts (up to 2×10^5 viable units per gram of soil). Zambrano and Casas-Campillo (1959) preferred a glucose-peptone-agar medium (pH 7) to which Rose Bengal and streptomycin were added.

Di Menna (1966) incubated agar plates at 4° for 4–5 weeks when studying Antarctic soils while, for studying New Zealand soils (see Table II), agar plates were incubated at room temperature for 4–6 days. Zambrano and Casas-Campillo (1959) incubated plates at 28° for one week for studying Mexican soils from tropical and subtropical areas.

Quantitative estimates of yeast populations in soils have been reported as total counts of viable units per unit weight of soil (dry or wet) and as relative frequencies of yeast species. These frequencies have been determined by picking up at random a large number of yeast colonies from a given set of isolation plates, identifying them, and expressing the number of isolates for each yeast species as a percentage of the total number of isolates.

For some yeast species, the soil may be only a reservoir where they can survive protected against desiccation and drought until dispersed by animals, growing plants and wind to suitable substrates. Yoneyama (1956, 1957b) in Japan has found that soils under pine trees and various species of *Quercus* harboured yeasts associated with such trees, namely *Saccharomyces cerevisiae* var. *tetrasporus*, *Sacch. pinimellis* and *Schizosaccharomyces versatilis*. For one site, he counted 250–1,500 viable units of *Sacch. cerevisiae* var. *tetrasporus* per gram of soil. Soil collected some centimetres away from the area subjected to contamination was devoid of the yeast species referred to.

Some evidence has accumulated that soil may also offer true ecological niches for yeasts. Although found in lower numbers as compared to other micro-organisms in soil, yeasts are able to build up significant populations in a highly competitive environment (Miller and Webb, 1954). The number of viable units per gram of soil may be of the order of thousands. Di Menna (see Table II) concluded from qualitative and quantitative surveys of the yeast flora of New Zealand soils that yeast populations varied qualitatively from place to place with soil type and vegetation but not with season, while the density of yeast populations was different from place to place and also varied with season. She observed that *Candida curvata* was more frequent in forest soils while species of *Cryptococcus* dominated in grassland soils.

4

TABLE I. *Yeasts Isolated from Soil by Enrichment Techniques*

Country	Type of soil	No. of sites	Percentage of soils containing yeasts	Dominant yeast species	References
Spain	Associated with vine and other plants producing sugary fruits	8	100	*Saccharomyces ellipsoideus* *Torulaspora rosei* *Cryptococcus albidus*	Capriotti (1958)
Sweden	Mostly clay cultivated with different plants	16	75	*Debaryomyces castellii* *Torulaspora nilssoni* *Candida vanriji*	Capriotti (1959)
Norway	Associated with *Festuca, Juniperus, Salix* and *Sedum* species	2	100	*Schizoblastosporion starkeyi-henricii* *Candida humicola* *Cryptococcus laurentii*	Roberts (1960)
	Associated with *Betula* and *Vaccinium* species				
U.S.A. (Florida)	Sandy beach and coastal area with mangrove vegetation	10	90	*Torulaspora delbrueckii* *Hansenula anomala* *Torulopsis glabrata* *Candida tropicalis*	Capriotti (1962)
Finland	Mostly sandy-argilous cultivated with different plants	32	66	*Rhodotorula mucilaginosa*	Capriotti (1963)
Italy	Soil associated with bats in caves at 5°	1	100	*Saccharomyces* (several species)	Martini (1963)
	Greenhouse at 22° from vessels with different plants	8	100	*Saccharomyces ellipsoideus* *Pichia fermentans* *Zygosaccharomyces mellis*	Capriotti and Rainieri (1964)

TABLE I contd. Yeasts Isolated from Soil by Enrichment Techniques

Country	Type of soil	No. of sites	Percentage of soils containing yeasts	Dominant yeast species	References
Bahamas	Cave associated with bats	1	100	*Cryptococcus* (several species) *Rhodotorula pilimane*	Orpurt (1964)
U.S.A. (Pennsylvania, New Jersey, South Carolina, Georgia, Florida)	Associated with vines, gardens, meadows and shrubs	26	36	*Pichia fermentans* *Hansenula anomala*	Capriotti (1967)
U.S.A. (Alaska)	In forest and associated with potatoes and cabbage	12	25	*Saccharomyces ellipsoideus* *Hansenula* (3 species) *Rhodotorula glutinis*	Capriotti (1967)

TABLE II. *Yeasts Isolated from New Zealand Soils*

Type of soil	No. of areas	No. of samples	Percentage of soils containing yeasts	No. viable units/g soil	Yeast species of higher percentage relative frequency	References
Silt loams under pasture	1	35	100	$6 \times 10^3 – 2 \times 10^5$	*Cryptococcus albidus* *Cryptococcus terreus* *Candida curvata* *Schizoblastosporium starkeyi-henricii*	di Menna (1957)
Tussock grasslands	3	27	100	—	*Cryptococcus albidus* *Cryptococcus terreus* *Candida humicola* *Candida curvata*	di Menna (1958)
Tussock grasslands	6	55	100	$5 \times 10^2 – 3.8 \times 10^4$	*Cryptococcus albidus* *Cryptococcus diffluens* *Cryptococcus terreus* *Candida curvata* *Trichosporon cutaneum*	di Menna (1960a)
Silt loams under pasture	4	65	100	$1 \times 10^3 – 8 \times 10^4$	*Cryptococcus albidus* *Cryptococcus laurentii* *Cryptococcus terreus* *Candida curvata* *Trichosporon cutaneum* *Torulopsis ingeniosa* *	di Menna (1960b)
Silt loams under forest	2	20	100	$1 \times 10^3 – 3 \times 10^4$	*Candida curvata* *Trichosporon cutaneum* * *Hansenula californica* * *Hansenula mrakii* *	di Menna (1960b)

* Abundant in one area only.

TABLE III. *Yeasts Isolated from Soils from Polar Areas*

Type of soil	No. of samples	Percentage of soils containing yeasts	No. viable units/g soil	Yeast species of higher percentage relative frequency	References
Antarctica	138	43	5 to more than 10^4	*Candida scottii* *Cryptococcus* sp. *Rhodotorula* sp.	di Menna (1960c, 1966)
East Greenland	8	100	2×10^2–$5\cdot6 \times 10^4$	*Candida scottii* *Candida gelida* *Cryptococcus* sp. *Rhodotorula* sp.	di Menna (1966)

Di Menna (1960c, 1966) studied the yeast flora associated with 138 soil samples from Antarctica (Table III). Yeasts were detected in only 60 of these samples, 48 of which had less than 10^3 viable units per gram; 14 had 10^3–10^4 viable units per gram and six had 10^4 or more. A few samples gave only five viable units per gram. *Candida scottii* was the most frequent species, followed by species of *Cryptococcus* and *Rhodotorula* (mainly *Cr. laurentii*, *Cr. albidus* and *Rh. mucilaginosa*). Most strains of *Candida scottii* were obligate psychrophiles. Other obligate psychrophiles present in some samples were *Candida nivalis*, *C. gelida* and *C. frigida*. Di Menna (1966) pointed out that these psychrophiles, as well as species of *Cryptococcus* and *Rhodotorula* which are facultative psychrophiles, are capable of building up populations characteristic of polar areas. Other yeast genera represented in Antarctic soils were *Debaryomyces*, *Torulopsis* and *Trichosporon*.

Di Menna (1966) observed that the occurrence of yeast populations in Antarctic soils was concomitant with the presence of other plants (including mosses and algae). There was no apparent correlation between the qualitative or quantitative distribution of yeasts and differences in latitude, fauna or pH value of the soil. She also suggested the possibility of finding obligate osmophiles with suitable techniques in Antarctic soils which have a high content of soluble salts.

The yeast species identified by Soneda (1961) in 'soily' materials from Antarctica were also found by di Menna (1960c, 1966) in her surveys. Di Menna (1966) studied eight soil samples from East Greenland (Table III). All samples contained yeasts in numbers which varied from 2×10^2 to 56×10^3 viable units per gram. The dominant species were *Cryptococcus albidus*, *Cr. laurentii*, *Candida gelida*, *C. scottii* and *Rhodotorula glutinis*. This last species was also found by Capriotti (1967) in Alaska.

Reddy and Knowles (1965) reported on the fungal flora of raw humus from a boreal forest under black spruce in Canada. Yeasts attained 26–64% of the total number of fungal colonies on plates inoculated with soil washings, and 7–28% on plates inoculated with washed soil particles. These results were obtained by incubating plates for one week at 30°. Species identification was not reported. Zambrano and Casas-Campillo (1959) studied 17 soils from tropical and subtropical areas in Mexico. Ten of these soils had yeast populations ranging from 10^3 to 10^5 viable units per gram of dry weight. Yeasts were identified in nine genera. Species of *Saccharomyces* were the most frequent. Capriotti (1962) concluded from his qualitative studies on European and American soils that yeasts are more frequent in soils of warmer areas such as Italy, Spain and Florida than in soils of cold areas such as Holland, Sweden and Finland.

From the available data it appears that the most frequent yeast species in soils, particularly uncultivated soils, are capable of producing extracellular slime. This property confers on them the ability to resist desiccation and drought. They are usually able to utilize a large number of different carbon compounds so that they more successfully survive precarious nutritive conditions.

The role of yeasts in the organic cycle of soil appears to involve mainly the utilization of products resulting from a primary attack on vegetable matter carried out by other organisms. For instance, soil yeasts cannot degrade cellulose. Most of them, however, assimilate cellobiose (di Menna, 1959a) which results from the breakdown of cellulose. There is some evidence that antibiotics produced by bacteria and streptomycetes may affect the distribution of yeast populations in soil (di Menna, 1962). Additional references on the yeast flora of soils are given by Capriotti (1959).

B. PLANTS

During the last ten years, most work on yeasts associated with plant materials has been concerned with exudates from tree trunks and with the phyllosphere. Other materials such as decaying wood, flowers and fruit, moss and mushrooms have occasionally been studied.

1. *Exudates from Tree Trunks*

The work by Phaff *et al.* (1964) in California on the yeast flora of a single slime flux from *Ulmus carpinifolia* is a good example of a thorough ecological yeast survey. Quantitative and qualitative modifications of the yeast populations in the flux were followed over a one-year period. Samples were collected periodically with aseptic precautions and well mixed. Serial dilutions from known weights of the samples were plated onto 5% malt agar (pH 3·7) and incubated at room temperature for 7–10 days. Yeast colonies were observed under a light microscope. Counts were made of the colonies having the same morphological type. The microscopic shape of the cells in the colonies of each type was checked. A few colonies of each type were selected for isolation and identification. Populations of *Pichia pastoris* attained 18×10^3–79×10^4 viable units per gram dry weight, the highest counts being observed in March during blooming and seed development but before leaves appeared. *Trichosporon penicillatum* attained 4×10^2–17×10^4 viable units per gram dry weight with its highest counts in April and May. Other yeasts occasionally present were probably contaminants brought in by insects and dust.

A number of other papers on yeasts present in slime fluxes in California have been published. Shehata *et al.* (1955) isolated species of

Hansenula, Pichia, Candida and *Trichosporon* from *Quercus kelloggii*. In another series of 64 slime-flux samples from *Quercus kelloggii*, the most frequent species were *Debaryomyces fluxorum, Hansenula mrakii, Pichia pastori* and *Pichia silvestris*. In 46 slime-flux samples from *Abies concolor*, the most frequent species were *Debaryomyces fluxorum* and *Pichia silvestris* (Carson *et al.*, 1956; Phaff and Knapp, 1956). Miller *et al.* (1962) found that species of *Hansenula, Pichia* and *Saccharomyces* were the most frequent ones in slime fluxes from ten trees classified in six different genera.

Kobayashi (1953) reported on the isolation of species of *Saccharomyces, Debaryomyces, Endomycopsis, Hanseniaspora, Sporobolomyces* and *Trichosporon* from the exudates of four different genera of trees in Japan. Yoneyama (1955, 1956, 1957a, b), also in Japan, has shown that sweet-sap exudates from pine protuberances (pine honey) are ecological niches for *Torulopsis candida, Schizosaccharomyces versatilis, Saccharomyces cerevisiae* var. *tetrasporus* and *Saccharomyces pinimellis*. The same yeast species, except *Torulopsis candida*, were found to be commonly associated with the barks of *Quercus* in Japanese forests (Yoneyama, 1957b). Yeasts associated with other tree exudates in California belonged to the genera *Cryptococcus, Rhodotorula* and *Torulopsis* (Shehata *et al.*, 1955; Miller *et al.*, 1962). Three species of *Candida* were isolated from a tree gum collected in a moist forest in Cameroon (Boidin *et al.*, 1963). Further references on yeasts associated with trunk exudates are given by Kobayashi (1953).

2. *Phyllosphere*

The most important work on yeasts associated with leaves has been conducted by di Menna (Table IV) in New Zealand. Fresh samples of leaves, either mown mechanically or clipped with sterile scissors, were collected into sterile polythene bags. Known weights of samples were suspended in sterile tap water (suspension in 0·1% peptone-water did not alter the results) for 15 minutes with eventual shaking by hand. Serial dilutions were plated onto a glucose (4%)–peptone (1%)–agar (2%) medium (pH 4) and incubated at room temperature for 4–6 days. The most frequent yeast species (Table IV) are all able to produce extracellular slime, whether organized in a capsule or not, and assimilate a large number of different carbon compounds. As already stated for the yeasts most frequently isolated from soils, these properties are related respectively with the ability of resisting desiccation and drought and of surviving in poor nutritive environments. Yeast populations were most abundant in April and decreased with autumn rainfall. Di Menna (1959b) observed that yeasts form microscopic colonies at the edges of the epidermal cells of the grass leaves and are easily washed

TABLE IV. *Yeasts Isolated from Plant Materials in New Zealand*

Material	No. of areas	No. of samples	Total populations	Most frequent species	References
Leaves of pasture grass and herbs	1	8	—	*Torulopsis aerea* *Cryptococcus laurentii*	di Menna (1957)
Roots of grasses and herbs	1	5	—	*Cryptococcus terreus* *Schizoblastosporion starkeyi-henricii* *Candida curvata*	di Menna (1957)
Freshly collected leaves of pasture plants	4	24	31×10^3–1×10^8	*Cryptococcus laurentii* *Rhodotorula graminis* *Torulopsis ingeniosa*	di Menna (1959b)
Leaves of pasture plants kept frozen for up to 5 months after collection	3	6	6×10^5–33×10^7	*Cryptococcus laurentii* *Rhodotorula marina* *Rhodotorula graminis*	di Menna (1959b)
Freshly fallen leaves and twigs from forest trees	1	2	—	*Cryptococcus laurentii* *Candida humicola*	di Menna (1960b)
Decomposing litter and humus	1	2	—	*Candida humicola* *Trichosporon cutaneum*	di Menna (1960b)
Leaves of tussocks and pasture plants	1	2	—	*Cryptococcus laurentii*	di Menna (1960a)
Roots of tussocks	1	1	—	*Candida humicola*	di Menna (1960a)
Roots of pasture plants	1	1	—	*Cryptococcus terreus*	di Menna (1960a)

out. The qualitative yeast pattern also varied with season but not with locality. Yeast species on the grass leaves are different from those associated with grass roots. The latter are similar to those in local soil. Qualitative changes were also correlated with the ageing of plants, probably due to modifications of the cell exudates of the leaves which serve as nutrients for the yeasts.

Ruinen (1963) reported on a qualitative survey of the yeasts associated with mature leaves of tropical foliage in Indonesia, Surinam and the Ivory Coast. She identified 22 species belonging to the genera *Hansenula*, *Sporobolomyces*, *Candida*, *Cryptococcus* and *Rhodotorula*. Species of the last two genera were the most frequent. Verona and Rambelli (1962), in Italy, found apiculate yeasts (*Hanseniaspora* and *Kloeckera* spp.) and species of *Lipomyces*, *Saccharomyces*, *Rhodotorula* and *Trichosporon* in leaf litter of eucalyptus, chestnut and oak trees. Additional data on yeasts from the phyllosphere are given in Chapter 5 (p. 183) of this volume.

3. *Other Plant Materials*

a. Decaying wood. Yeasts found in decaying wood in California belonged to the genera *Hansenula*, *Pichia*, *Saccharomyces* and *Candida* (Shehata *et al.*, 1955; Phaff and Knapp, 1956) and from similar material from Cameroon to *Saccharomyces* (Boidin *et al.*, 1963).

b. Flowers. Yeasts associated with flowers (Miller *et al.*, 1962; Boidin *et al.*, 1963; Capriotti and Rainieri, 1964) are similar to those reported by earlier workers (Lund, 1958). Capriotti and Rainieri (1964) concluded from qualitative analyses that included an enrichment technique, that the yeast flora of open flowers differs from that of the buds.

c. Fruit. Sasaki and Yoshida (1959) collected apples, grapes, cherries and strawberries aseptically in 16 different areas in Japan, and isolated yeasts present on the surfaces of the fruits by rubbing these on the surface of agar plates. *Torulopsis candida* was the most widespread yeast. Furthermore, *Sporobolomyces roseus* and *Rhodotorula glutinis* were frequent on apples. The most frequent yeast species on apples during the ripening season in Canada (Williams *et al.*, 1956) were *Candida malicola* and *Rhodotorula glutinis* var. *rubescens* in a total population that varied from 17×10^2 to 199×10^2 viable units per gram of apple. Yeast numbers were lower in summer than in spring and autumn. Yeast populations on cider apples in England varied from 12 to 95,200 viable units per gram, *Candida pulcherrima* being the dominant species (Bowen and Beech, 1964).

Batista *et al.* (1961d) isolated yeasts associated with the skin, pulp and seeds of 66 different kinds of fruit in Brazil. The most frequent species were *Candida guilliermondii*, *C. parapsilosis* and *Trichosporon cutaneum*.

Miller and Phaff (1962) found that pollinated Calimyrna figs support the growth of a specific microflora in their internal tissues which includes *Candida guilliermondii* var. *carpophila*. This yeast does not cause spoilage. In mature figs, spoilage is usually caused by apiculate yeasts (*Hanseniaspora* and *Kloeckera* spp.) and *Torulopsis stellata*. Boidin *et al.* (1963) isolated species of *Pichia*, *Candida*, *Kloeckera* and *Torulopsis* from fruit collected in Cameroon.

d. *Moss*. Di Menna (1960c) found that the dominant yeast species in a sample of moss from Antarctica were *Cryptococcus laurentii*, *Cr. albidus* and *Rhodotorula minuta* in a total population of 75×10^3 viable units per gram.

e. *Mushrooms*. Ramirez-Gomes (1957) analysed the yeast flora of mushrooms in France. Pieces of 134 mushrooms (43 genera, 107 species) were taken aseptically and incubated for a few days for enrichment of the yeast population. The washings of each piece were plated onto malt-agar supplemented with 0·25% sodium propionate. *Sporobolomyces albidus*, *Candida curvata*, *Saccharomyces cerevisiae*, *Torulopsis inconspicua* and *Candida anomala* were the most frequent species in 112 positive samples. *Candida humicola* was found to be associated with *Clavaria* and *Pleurotus* (Phaff and Knapp, 1956) in California and with *Podoscypha* (Boidin *et al.*, 1963) in Cameroon.

C. ANIMALS

Studies on the ecology of yeasts in relation to animal-hosts that live exclusively or mainly in terrestrial environments have been limited to mammals, birds and insects.

1. *Mammals and Birds*

Yeasts associated with humans are discussed in Chapter 4 (p. 107) of this volume. The distribution of yeast species in the digestive tract of other warm-blooded animals has been surveyed by a number of authors.

Easy and effective techniques of sampling *post-mortem* consist in making aseptic incisions in portions of the digestive tract through which contents may be squeezed out (Parle, 1957), or taken by means of sterile instruments (spoons, syringes, pipettes). Clarke and di Menna (1961) collected samples from the bovine rumen *in vivo* through fistulas kept under aseptic conditions or with sterile oesophageal tubes. A culture medium composed of glucose (2%), peptone (1%), yeast extract (0·5%) and agar (2%) with antibiotics has proved useful for qualitative-quantitative studies (van Uden and Carmo-Sousa, 1962c).

Incubation of isolation cultures is best carried out at 37° since many transients which have their maximum temperatures for growth below

that of the animal body will be eliminated in this way. Parle (1957)
analysed materials from the stomach, small intestine and large intestine
of various mammals in New Zealand. Van Uden and Carmo-Sousa
(1957, 1962a, b, c) and van Uden *et al.* (1958) surveyed the caecal contents
of domestic and wild free-living mammals in Portugal and Portuguese
East Africa. Yeasts in the crops of turkeys in the U.S.A. were studied by
Manfre *et al.* (1958).

Batista *et al.* (1961a, b, c) in Brazil and Saëz (1959, 1960a, b, 1963)
in France reported on yeasts isolated from faeces of domestic and
wild mammals and birds in captivity. Saëz (1963) observed that the
offspring of wild mammals in captivity harboured no yeasts in their
digestive tracts if they were less than 24 hours old. Species of *Candida*
were found in animals one to ten days old. Further references on qualitative surveys of yeasts in the digestive tract of warm-blooded animals
may be found in the review by Hurley (1967).

Clarke and di Menna (1961) stated that yeast populations attained
low numbers in the rumen of cows in New Zealand. The dominant yeast
species apparently were not associated with the feed materials. Van
Uden and Carmo-Sousa (1962c) showed that yeast populations are
relatively low in the small intestine of pigs, but increase greatly in the
caecum and rectum. In these lower parts of the tract, the highest
count was 9×10^6 viable units per gram (wet weight). Apparently,
yeast populations were higher in animals fed on a grain diet than in
animals fed a green vegetable diet. Soneda (1959), in Japan, reported
that yeast populations in the faeces of wild animals in captivity were
of the order of 10^2-10^6 viable units per gram in six out of ten carnivores,
$4 \times 10^4-4 \times 10^6$ in nine out of ten omnivores, and $5 \times 10^2-2 \times 10^6$
in a group of herbivores.

Yeasts that occur in the alimentary canal of warm-blooded animals
may be classified as obligatory saprophytes, facultative saprophytes or
simple transients (van Uden, 1960, 1963). Yeasts in the first group
(Table V) have seldom been found outside the animal body. Most of
them have a narrow temperature range for growth, are dependent on a
number of growth factors and are capable of growth at pH values between 1·0 and 2·0. *Saccharomycopsis guttulata* is most frequent in rabbits
and chinchillas; *Torulopsis pintolopesii* is frequent in mice, rats and
guinea pigs; *Candida slooffii* in pigs and bush-pigs; *Saccharomyces
tellustris* and its imperfect stage (*Candida bovina*) in pigs and turkeys;
Candida albicans and *Torulopsis glabrata* in humans and several other
warm-blooded animals (monkeys, pigs, hedgehogs, wart hogs, opossum,
sheep, fowl, sparrows); and *Candida stellatoidea* in humans. Nothing is
known about advantages or disadvantages that hosts acquire from these
obligate saprophytes.

TABLE V. *Yeasts Isolated from the Digestive Tracts of Warm-Blooded Animals*

Designation	Ecological group — Characteristics	Yeast species	Animal hosts
Obligatory saprophytes	Normal habitat inside the animal body	*Saccharomycopsis guttulata*	Rabbits; chinchillas
		Torulopsis pintolopesii	Mice; rats; guinea pigs
		Candida slooffii	Pigs; bush pigs
		Saccharomyces tellustris (Candida bovina)	Pigs; turkeys
		Candida stellatoidea	Humans
		Candida albicans	Humans and several other warm-blooded animals
		Torulopsis glabrata	Humans and several other warm-blooded animals
Facultative saprophytes	Normal habitat inside and outside the animal body	*Candida tropicalis*	Humans; cattle; pigs
		Candida krusei	Humans; cattle; horses; pigs
		Candida parapsilosis	Humans; horses
		Candida guilliermondii	Humans; baboons; horses
		Trichosporon cutaneum	Horses; cattle
		Pichia membranaefaciens	Pigs
Transients	Normal habitat outside the animal body		

The group of facultative saprophytes (Table V) comprises yeasts that are able to build up populations in the digestive tracts of warm-blooded animals as well as in other natural substrates. *Candida tropicalis* is frequent in humans, cattle and pigs; *Candida krusei* in humans, cattle, horses and pigs; *Candida parapsilosis* in humans and horses; *Candida guilliermondii* in humans, baboons and horses; *Trichosporon cutaneum* in horses and cattle (Parle, 1957; van Uden and Carmo-Sousa, 1957; van Uden *et al.*, 1958; Clarke and di Menna, 1961). The yeast species mentioned in this group may, in some instances, become pathogenic for their hosts (see Chapter 4, p. 107).

Yeast species that are ingested with food or other material, but are unable to grow inside the animal body, are included in the group of transients. Some of them will be destroyed by the digestive processes at one or another section of the alimentary canal; many will survive and be redistributed on other substrates.

The ecological classification of a given yeast species in the above groups may depend on the nature of the host species being considered. Quantitative techniques may reveal, for example, that a yeast which occurs as an obligate saprophyte in one host species is a transient when occurring in other host species.

Gustafson (1959) reviewed the literature on the ecology of *Pityrosporon* and reported that he had observed yeasts morphologically identifiable with this genus in smears of ear-wax from pigs, cows, horses and roe, and in skin scrapings from elks. Cultures of *Pit. ovale* were recovered by him from less than 50% of the samples that showed positive smears. *Pityrosporon canis* was identified in a few samples of pig ear-wax. No cultures were obtained from roe. Smears and cultures were negative for ear-wax samples from sheep and elks. He used sterile wooden spatulas for collecting samples, which were smeared on wort-agar plates overlayered with a thin layer of sterile olive oil containing penicillin.

2. *Insects*

It is well known that insects are vectors in the dispersal of yeasts, as well as of other micro-organisms. Yeasts, on the other hand, play an important role as nutrient suppliers for many insects. The degree of dependency of the insects on yeasts for food varies from a relatively random choice, as is the case with *Drosophila* spp., to very strict types of association as with the intracellular symbiosis in Cerambycidae (Table VI). General reviews on symbiotic associations in insects have appeared from Richards and Brooks (1958) and Koch (1960, 1963).

Studies on yeasts associated with insects require the latter to be collected alive, anaesthetized (e.g. with ether) and dissected aseptically.

TABLE VI. *Yeasts Associated with Insects*[*]

Insect	Yeast species	Type of association	Geographical area	References
Drosophila species (flies)	Apiculate including *Pichia*	Exosymbiosis	U.S.A.	Camargo and Phaff (1957)
Scarabeidae (coprophagous beetles)	*Trichosporon cutaneum*	Exosymbiosis	Italy	Malan and Gandini (1966)
Blastophaga psenes (fig wasp)	*Candida guilliermondii* var. *carpophila*	Exosymbiosis	U.S.A.	Phaff and Miller (1961)
Platypus cylindrus (ambrosia bark-beetle)	*Endomycopsis platypodis Candida* species	Exosymbiosis	England	Baker (1963) Baker and Kreger-van Rij (1964)
Gnathotrichus materiarius (ambrosia bark-beetle)	*Endomycopsis fasciculata*	Exosymbiosis	U.S.A.	Batra (1963)
Cerambycidae (bark beetles)	*Candida tenuis Candida rhagii Candida parapsilosis* var. *intermedia*	Intracellular symbiosis	Germany	Jurzitza (1959); Jurzitza et al. (1960)
Cerambycidae (bark beetles)	*Candida tenuis Candida parapsilosis* var. *intermedia*	Intracellular symbiosis	Chile	Grinbergs (1962)

* Only the most relevant of the reviewed data are listed.

The appropriate organs may be directly streaked on agar plates. Jurzitza (1959), when studying the intracellular symbionts of Cerambycidae, observed mycetomata in Ringer solution under the microscope and prepared single-cell cultures of yeast cells present in them.

A number of authors have used malt-agar as an isolation culture medium. *Drosophila* flies feed on substrata that are frequently rich in yeast populations, and yeasts are readily digested in their digestive tract. Camargo and Phaff (1957) found that *Hanseniaspora uvarum* and *Pichia kluyveri* were the dominant yeast species in the intestinal tracts of *Drosophila* spp. feeding on tomatoes in tomato fields in California. The same yeast species were the most frequent in fermenting tomatoes picked in the same fields.

The yeast flora of nests and larvae of coprophagous beetles belonging to the Scarabeidae was analysed by Malan and Gandini (1966) in Italy. *Trichosporon cutaneum* was by far the most frequent species in the nest walls and in the guts of the larvae. The same yeast species was also found in the dung of sheep and cattle that was used by the beetles as nest-building material. This species has some cellulolytic activity.

A most interesting study of a symbiotic association of *Candida guilliermondii* var. *carpophila* with the fig wasp, *Blastophaga psenes*, was reported by Phaff and Miller (1961) in California. By following the insect life-cycle, these authors showed that the yeast is inoculated by the adults into the figs. It is frequently found in the wasp-containing flower ovaries where it may act as a nutrient source for the insect (see also Volume 3, Chapter 9).

Trichosporon sericeum and other yeasts belonging to the genera *Candida*, *Torulopsis* and *Saccharomyces* were isolated from termites collected in Cameroon (Boidin *et al.*, 1963). Some species of termites are 'ambrosia-growers' and cultivate special 'fungus-gardens' for food supply. Whether the above yeasts were termite-symbionts grown in such 'gardens' was not mentioned.

Yeasts are also associated with 'ambrosia-growers' that attack the bark of various trees (Callaham and Shifrine, 1960; Baker, 1963). *Endomycopsis* sp. and *Candida* sp. were isolated in England from the bodies of *Platypus cylindrus* and from the tunnels made by the beetles in the barks of *Quercus* trees (Baker, 1963). *Endomycopsis* sp. was later described as *E. platypodis* by Baker and Kreger-van Rij (1964). Both yeasts were seeded by the adults in the tunnels and utilized as food by the larvae. Batra (1963) described *Endomycopsis fasciculata* as a symbiont of the ambrosia beetle, *Gnathotricus materiarius*, that infested *Pinus strobus* in the U.S.A. The yeast served as food for the larvae, and possibly also for the adults.

Miller *et al.* (1962) isolated yeasts from insect frass and from unidentified beetles and larvae found in the bark of species of *Quercus, Abies, Pinus, Cercocarpus* and *Populus*. Most cultures belonged to the genus *Candida*. Species of *Cryptococcus, Rhodotorula, Torulopsis, Hansenula, Pichia* and *Saccharomyces* were also identified.

Species of Cerambycidae that attack the bark of conifers and dead deciduous trees harbour intracellular symbionts in the intersegmental sacs of the adult females. The eggs are externally infected therefrom. The symbionts pass to the intestinal tract of the larvae, are trapped into diverticula and form mycetomata. Jurzitza (1959) and Jurzitza *et al.* (1960) conducted investigations on this type of association in Germany. They concluded that *Candida tenuis, C. rhagii* and *C. parapsilosis* var. *intermedia* were specific symbionts for species of Cerambycidae belonging to the genera *Rhagium, Harpium, Leptura* and *Gaurotes*. *Candida tenuis* and *C. parapsilosis* var. *intermedia* were also the most frequent yeast species isolated by Grinbergs (1962) from the intestinal tract of Cerambycidae in Chile. However, mycetomata were not observed by him. The intracellular symbionts of the Cerambycidae excrete vitamins of the B-complex and amino acids into the culture media (Jurzitza, 1959). They appear to be sources of such compounds for the insects.

Intracellular symbionts isolated from intestinal mycetomata of several species of Anobiidae have been classified by some authors in the yeast domain. More detailed studies, however, indicate that such organisms may rather be related to the Taphrinales (Kühlwein and Jurzitza, 1961; Jurzitza, 1964; van der Walt, 1961). From a morphological point of view it is not unlikely that the organism isolated by Grinbergs (1962) from the digestive tract of a larva of Anobiidae in Chile and identified by him as *Trichosporon* sp. also belongs to the same group.

Grinbergs (1962) reported on *Candida parapsilosis* var. *intermedia, Candida* sp. and *Hansenula* sp. isolated from the digestive tract of a larva of Curculionidae. Wistreich *et al.* (1960) and Moore and Wistreich (1961) stated that yeasts were apparently absent from the midgut of larvae of two species of *Tenebrio*. Other yeast species found in association with insects, probably through the feeding habits of the latter, are: *Endomycopsis wickerhamii* isolated from the larval gut and frass of Cossidae in South African Cycadales (van der Walt, 1959); *Endomycopsis scolyti* isolated from the frass and body of *Scolytus* parasite in *Abies* and *Pseudotsuga* in California (Phaff and Yoneyama, 1961); *Sporobolomyces singularis, Bullera tsugae, Cryptococcus skinneri* and *Candida oregonensis* isolated from frass presumably produced by *Scolytus tsugae* in the bark of *Tsuga heterophylla* in California (Phaff and Carmo-Sousa, 1962); *Hansenula holstii* from frass in coniferous trees, U.S.A. (Wickerham, 1960); *Sporobolomyces roseus* (Phaff and Knapp, 1956) and *Candida shehatae* (Buckley and van Uden, 1967) from unidentified wood-destroying insects; *Candida berthetii* and *Candida melinii* from the frass of xylophagous larvae (Boidin *et al.*, 1963); *Pichia stipitis* from larvae of *Cetonia* sp., *Doreus parallelopipedus* and *Laphria* sp. (Pignal, 1967); *Torulopsis apicola* from the intestinal tract of honey bees (Hajsig, 1958). Lavie (1954) reported that *Kloeckera apiculata* var. *apis* and *Torulopsis apis* act as antagonists of parasitic mites in the respiratory canals of honey bees.

D. ATMOSPHERE

The literature concerning the distribution of yeasts in air was reviewed by Gregory (1961). The scant information available is based mainly on the study of air samples taken at several heights from ground level up to about 3,000 metres. Yeasts have been found in numbers that are usually low, but are relatively higher at ground levels than in the upper air. They apparently originate from the vegetation layer above the soil surface. Except for the ballistospores of *Sporobolomyces* and *Bullera* spp., which are actively discharged into the air, the other yeasts in the atmosphere are probably carried by dust particles conveyed by mechanical disturbances and movements of air masses. For a review of the air sampling techniques, the monograph by Gregory (1961) should be consulted.

Di Menna (1955) found that indoor and outdoor air in one city in New Zealand had yeast numbers that averaged one viable unit per two ft^3. Species of *Cryptococcus*, *Rhodotorula*, *Sporobolomyces* and *Debaryomyces* were dominant. Species of *Cryptococcus* and *Rhodotorula* were also the most frequent in outdoor air in Budapest (Vöros-Felkai, 1966, 1967), and at ground levels in Texas, U.S.A. (Al-Doory, 1967). Adams (1964) concluded that airborne yeasts at some fruit and vegetable sites in Canada belonged mainly to the genera *Torulopsis*, *Cryptococcus* and *Kloeckera*. Yeast populations attained 20% of the total fungal flora.

Yeasts occupied third place among the most frequent groups of fungi in the atmosphere over a city in Kansas (U.S.A.) at a height of 150 ft (Kramer and Pady, 1960, cited by Adams, 1964). In Texas (U.S.A.) they represented 5·5% of the total fungal viable units at ground level (Al-Doory, 1967). Apparently there are no seasonal variations on the incidence of yeast species at ground levels (di Menna, 1955; Vöros-Felkai, 1966, 1967).

A series of papers on micro-organisms of the upper atmosphere has been published. Instrumentation for isokinetic sampling of air at high altitudes has been developed (Timmons *et al.*, 1966). Fulton (1966a, b) reported that yeasts were present in air samples taken at altitudes

TABLE VII. *Yeasts Isolated from Sea Water*

No. of stations	Depth range of casts (m)	No. of samples	Sampling technique	Percentage of samples containing yeasts	Depth range of positive samples (m)	Viable yeasts range	Viable yeasts mean	Relative frequency of species	References
					Black Sea				
21	2,000	174	Niskin biosampler	48	0–2,000	0–150	—	*Candida diddensii* 22% *Rhodotorula rubra* 18% *Debaryomyces hansenii* 18% *Rhodotorula glutinis* 16% *Cryptococcus laurentii* 8% *Cryptococcus albidus* 6% *Candida guilliermondii* 6% Other species 16%	Meyers *et al.* (1967b)
					North Sea				
12	1	>72	ZoBell microbiological sampler	99	1	<10–>3,000	—	*Debaryomyces hansenii* *Candida* *Rhodotorula* *Hanseniaspora*	Meyers *et al.* (1967a)

between 152 and 3,127 metres above the mean terrain level in Texas (U.S.A.). Yeasts represented 3% of the total fungal population.

III. Aquatic Environments

In a comprehensive review of the literature on the ecology of marine yeasts, van Uden and Fell (1968) discussed the distribution of yeast populations in oceans and seas, estuaries and inland waters, sediments, weeds, algae, invertebrates, fishes, marine mammals and birds. Conditions for suitable sampling and isolation of marine yeasts were critically examined. Additional work on yeasts associated with aquatic environments (not cited in the review of van Uden and Fell) is discussed below and summarized in Table VII.

Meyers *et al.* (1967b) noted that yeast populations in the Black Sea were denser in the upper 1,000 metres of water. At greater depths, only 25% of the samples yielded yeasts. Their findings appeared to be correlated with the absence of equalizing vertical currents, decreased oxygen tensions and high concentrations of hydrogen sulphide at the lower levels of the Black Sea. Meyers *et al.* (1967a) registered an increase of population densities of *Debaryomyces hansenii* in the North Sea during summer. The increases were often co-incident with blooms of *Noctiluca miliaris*.

Hedrick and Soyugenc (1967) reported on yeasts in water and sediment of Lake Ontario in North America. Yeast populations attained 10 viable units per 100 ml at a depth of one metre, 130 units at mid-depth and 460 units near the bottom. The sediment contained 46 viable units per 100 ml. The most frequent species were *Candida guilliermondii* and *Rhodotorula mucilaginosa*. The variation of population densities with depth appeared to be related to variations in the concentrations of organic nitrogen and nitrate nitrogen.

Unidentified yeasts have been isolated from air above the Atlantic, Pacific and Arctic Oceans. Though in low numbers they were present at the North Pole at a height of 3,000 metres (Gregory, 1961). Van Uden and Castelo-Branco (1963) did not obtain any yeast growth on agar plates exposed for 5, 10 and 20 minutes on a pier on the Pacific Coast in California, U.S.A.

References

Adams, A. M. (1964). *Can. J. Microbiol.* **10**, 641–646.
Al-Doory, Y. (1967). *Mycopath. Mycol. appl.* **32**, 313–318.
Baker, J. M. (1963). *In* "Symbiotic Associations" (P. S. Nutman and B. Mosse, eds.), pp. 232–265. University Press, Cambridge.
Baker, J. M. and Kreger-van Rij, N. J. W. (1964). *Antonie van Leeuwenhoek* **30**, 433–441.

Batista, A. C., Fischman, O., Vasconcelos, C. T. de and Rocha, I. G. da (1961a). *Publções Inst. Micol. Recife* No. 327, 3–27.

Batista, A. C., Vasconcelos, C. T. de, Fischman, O. and Silva, J. O. da (1961b). *Publções Inst. Micol. Recife* No. 326, 3–16.

Batista, A. C., Vasconcelos, C. T. de, Fischman, O. and Staib, F. (1961c). *Publções Inst. Micol. Recife* No. 325, 3–27.

Batista, A. C., Vasconcelos, C. T. de, Lima, J. A. de and Shome, S. K. (1961d). *Publções Inst. Micol. Recife* No. 329, 3–21.

Batra, L. R. (1963). *Am. J. Bot.* **50**, 481–487.

Boidin, J., Pignal, M. C., Mermier, F. and Arpin, M. (1963). *Cahiers de la Maboké* **1**, 86–100.

Bowen, J. F. and Beech, F. W. (1964). *J. appl. Bact.* **27**, 333–341.

Buckley, H. R. and Uden, N. van (1967). *Mycopath. Mycol. appl.* **32**, 297–301.

Callaham, R. Z. and Shifrine, M. (1960). *J. Forens. Sci.* **6**, 146–154.

Camargo, R. de and Phaff, H. J. (1957). *Fd Res.* **22**, 367–372.

Capriotti, A. (1958). *Revta Ciencia Apl.* **12**, No. 61.

Capriotti, A. (1959). *K. LantbrHögsk. Annlr.* **25**, 185–220.

Capriotti, A. (1962). *Arch. Mikrobiol.* **41**, 142–146.

Capriotti, A. (1963). *Annali Fac. Agr. Univ. Perugia* **18**, 45–60.

Capriotti, A. (1967). *Arch. Mikrobiol.* **57**, 406–413.

Capriotti, A. and Rainieri, L. (1964). *Arch. Mikrobiol.* **48**, 325–331.

Carson, H. L., Knapp, E. P. and Phaff, H. J. (1956). *Ecology* **37**, 538–544.

Clarke, R. T. J. and di Menna, M. E. (1961). *J. gen. Microbiol.* **25**, 113–117.

di Menna, M. E. (1955). *Trans. Br. mycol. Soc.* **38**, 119–129.

di Menna, M. E. (1957). *J. gen. Microbiol.* **17**, 678–688.

di Menna, M. E. (1958). *N.Z. Jl agric. Res.* **1**, 939–942.

di Menna, M. E. (1959a). *J. gen. Microbiol.* **20**, 13–23.

di Menna, M. E. (1959b). *N.Z. Jl agric. Res.* **2**, 394–405.

di Menna, M. E. (1960a). *N.Z. Jl agric. Res.* **3**, 207–213.

di Menna, M. E. (1960b). *N.Z. Jl agric. Res.* **3**, 623–632.

di Menna, M. E. (1960c). *J. gen. Microbiol.* **23**, 295–300.

di Menna, M. E. (1962). *J. gen. Microbiol.* **27**, 249–257.

di Menna, M. E. (1966). *Antonie van Leeuwenhoek* **32**, 29–38.

Fulton, J. D. (1966a). *Appl. Microbiol.* **14**, 237–240.

Fulton, J. D. (1966b). *Appl. Microbiol.* **14**, 245–250.

Gregory, P. H. (1961). "The Microbiology of the Atmosphere". Leonard Hill [Books] Ltd., London.

Grinbergs, J. (1962). *Arch. Mikrobiol.* **41**, 51–78.

Gustafson, B. A. (1959). *Acta path. microbiol. scand.* **48**, 51–55.

Hajsig, M. (1958). *Antonie van Leeuwenhoek* **24**, 18–22.

Hedrick, L. R. and Soyugenc, M. (1967). *Proc. 10th Conf. Great Lakes Research.*

Hurley, R. (1967). *Rev. med. vet. Mycol.* **6**, 159–176.

Jurzitza, G. (1959). *Arch. Mikrobiol.* **33**, 305–332.

Jurzitza, G. (1964). *Arch. Mikrobiol.* **49**, 331–340.

Jurzitza, G., Kühlwein, H. and Kreger-van Rij, N. J. W. (1960). *Arch. Mikrobiol.* **36**, 229–243.

Kobayashi, Y. (1953). *Bull. natn. Sci. Mus., Tokyo* **33**, 31–46.

Koch, A. (1960). *A. Rev. Microbiol.* **14**, 121–140.

Koch, A. (1963). *In* "Recent Progresses in Microbiology, VIII, Montreal, 1962", pp. 150–161. Univ. Toronto Press, Canada.

Kramer, C. L. and Pady, S. M. (1960). *Trans. Kans. Acad. Sci.* **63**, 53–60.

104 LÍDIA DO CARMO-SOUSA

Kreger-van Rij, N. J. W. (1958). *Antonie van Leeuwenhoek* **24**, 137–144.
Kühlwein, H. and Jurzitza, G. (1961). *Arch. Mikrobiol.* **40**, 247–260.
Lavie, P. (1954). *C. R. hebd. Séanc. Acad. Sci., Paris* **238**, 947–949.
Lodder, J., ed. (1970). "The Yeasts", 2nd ed. North Holland Publ. Co., Amsterdam.
Lodder, J. and Kreger-van Rij (1952). "The Yeasts. A Taxonomic Study". 713 pp. North Holland Publ. Co., Amsterdam.
Lund, A. (1958). *In* "The Chemistry and Biology of Yeasts" (A. H. Cook, ed.), pp. 63–91. Academic Press, New York.
Malan, C. E. and Gandini, A. (1966). *Centro Entomol. Alpina Forestale, Consiglio Nazionale delle Ricerche*, Pub. No. 99.
Manfre, A. S., Wheeler, H. O., Feldman, G. L., Rigdon, R. H., Ferguson, T. M. and Couch, J. R. (1958). *Am. J. vet. Res.* **19**, 689–695.
Martini, A. (1963). *Arch. Mikrobiol.* **45**, 111–114.
Meyers, S. P., Ahearn, D. G., Gunkel, W. and Roth, F. J. Jr. (1967a). *Mar. Biol.* **1**, 118–123.
Meyers, S. P., Ahearn, D. G. and Roth, F. J. Jr. (1967b). *Bull. mar. Sci.* **17**, 576–596.
Miller, M. W. and Phaff, H. J. (1962). *Appl. Microbiol.* **10**, 394–400.
Miller, M. W., Phaff, H. J. and Snyder, H. E. (1962). *Mycopath. Mycol. appl.* **16**, 1–18.
Miller, J. J. and Webb, N. S. (1954). *Soil Sci.* **77**, 197–204.
Moore, J. and Wistreich, G. A. (1961). *J. Insect Path.* **3**, 399–402.
Orpurt, P. A. (1964). *Can. J. Bot.* **42**, 1629–1633.
Parle, J. N. (1957). *J. gen. Microbiol.* **17**, 363–367.
Phaff, H. J. and Carmo-Sousa, L. do (1962). *Antonie van Leeuwenhoek* **28**, 193–207.
Phaff, H. J. and Knapp, E. P. (1956). *Antonie van Leeuwenhoek* **22**, 117–130.
Phaff, H. J. and Miller, M. W. (1961). *J. Insect Path.* **3**, 233–243.
Phaff, H. J., Miller, M. W. and Mrak, E. M. (1966). "The Life of Yeasts". 186 pp. Harvard Univ. Press, Cambridge, Mass.
Phaff, H. J. and Yoneyama, M. (1961). *Antonie van Leeuwenhoek* **27**, 196–202.
Phaff, H. J., Yoneyama, M. and Carmo-Sousa, L. do (1964). *Riv. Patol. veg., Padova* **4**, 485–497.
Pignal, M. C. (1967). *Bull. mens. Soc. linn. Lyon* **36**, 163–168.
Ramirez-Gomez, C. (1957). *Microbiologia esp.* **10**, 215–247.
Reddy, T. K. R. and Knowles, R. (1965). *Can. J. Microbiol.* **11**, 837–843.
Richards, A. G. and Brooks, M. A. (1958). *A. Rev. Ent.* **3**, 37–56.
Roberts, C. (1960). *C. r. Trav. Lab. Carlsberg* **32**, 75–88.
Ruinen, J. (1963). *Antonie van Leeuwenhoek* **29**, 425–438.
Saëz, H. (1959). *Revue Mycol.* **24**, 426–433.
Saez, H. (1960a). *Recl. Méd. vét. exot. Éc. Alfort* **136**, 567–573.
Saëz, H. (1960b). *Cah. Méd. Vét.* **29**, 1–9.
Saëz, H. (1963). *Revue Mycol.* **28**, 52–61.
Sasaki, Y. and Yoshida, T. (1959). *J. Fac. Agric. Hokkaido Univ.* **51**, 194–220.
Shehata, A. M. el Tabey, Mrak, E. M. and Phaff, H. J. (1955). *Mycologia* **47**, 799–811.
Soneda, M. (1959). *Nagaoa* **6**, 1–24.
Soneda, M. (1961). "On Some Yeasts from the Antarctic Region". *Seto mar. Biol. Lab.*
Timmons, D. E., Fulton, J. D. and Mitchell, R. B. (1966). *Appl. Microbiol.* **14**, 229–231.

Uden, N. van (1960). *Trans. N.Y. Acad. Sci.* **89**, 59–68.

Uden, N. van (1963). *In* "Recent Progress in Microbiology VIII, Montreal, 1962", pp. 635–643. Univ. Toronto Press, Canada.

Uden, N. van and Carmo-Sousa, L. do (1957). *J. gen. Microbiol.* **16**, 385–395.

Uden, N. van and Carmo-Sousa, L. do (1962a). *Antonie van Leeuwenhoek* **28**, 73–77.

Uden, N. van and Carmo-Sousa, L. do (1962b). *Sabouraudia* **2**, 8–11.

Uden, N. van and Carmo-Sousa, L. do (1962c). *J. gen. Microbiol.* **27**, 35–40.

Uden, N. van, Carmo-Sousa, L. do and Farinha, M. (1958). *J. gen. Microbiol.* **19**, 435–445.

Uden, N. van and Castelo-Branco, R. (1963). *Limnol. Oceanogr.* **8**, 323–329.

Uden, N. van and Fell, J. (1968). *In* "Advances in Microbiology of the Sea" (M. R. Droop and E. J. Ferguson Wood, eds.), pp. 167–201. Academic Press, New York.

Verona, O. and Rambelli, A. (1962). *Annali. Fac. Agr. Univ. Pisa* **23**, 37–46.

Vöros-Felkai, G. (1966). *Acta microbiol. hung.* **13**, 53–58.

Vöros-Felkai, G. (1967). *Acta microbiol. hung.* **14**, 305–308.

Walt, J. P. van der (1959). *Antonie van Leeuwenhoek* **25**, 344–348.

Walt, J. P. van der (1961). *Antonie van Leeuwenhoek* **27**, 362–365.

Wickerham, L. J. (1960). *Mycologia* **52**, 171–183.

Williams, A. J., Wallace, R. H. and Clarke, D. S. (1956). *Can. J. Microbiol.* **2**, 645–648.

Wistreich, G. A., Moore, J. and Chao, J. (1960). *J. Insect Path.* **2**, 320–326.

Yoneyama, M. (1955). *Bot. Mag., Tokyo* **68**, 341–346.

Yoneyama, M. (1956). *J. Sci. Hiroshima Univ. Ser. B, div. 3*, **7**, 91–102.

Yoneyama, M. (1957a). *Bot. Mag., Tokyo* **70**, 92–96.

Yoneyama, M. (1957b). *J. Sci. Hiroshima Univ. Ser. B, div. 2*, **8**, 19–38.

Zambrano, G. and Casas-Campillo, C. (1959). *Revue Latino-Am. Microbiol.* **2**, 77–88.

Chapter 4

Yeasts as Human and Animal Pathogens

J. C. GENTLES AND C. J. LA TOUCHE

Department of Medical Mycology, University of Glasgow, Glasgow, Scotland and Mycology Unit, The General Infirmary, Leeds, England

I. Introduction

Diseases of man and animals caused by fungi are known as mycotic infections or mycoses. These range from mild, chronic infections to acute conditions which may affect only the superficial keratinized or mucosal parts of the body or involve the viscera and circulatory fluids. Yeasts may be responsible for any form of mycosis, and a single species may, depending on various factors, cause superficial or serious systemic disease or both.

Among the first fungi to be studied and recognized in their role as causal agents of diseases in man was the yeast now known as *Candida albicans*. In 1839, Langenbeck showed that the disease called thrush was caused by this fungus, and in 1842 a detailed description was published by Gruby. Although descriptions of certain of the ringworm fungi were also published about this time, progress in medical mycology was slow and spasmodic for almost a century. This neglect resulted in a considerable duplication of reports and records of fungal diseases which led to a multiplicity of names for these and their causal agents. Important species of pathogenic yeasts such as *Candida albicans* and *Cryptococcus neoformans* were particularly affected in this respect, so that each of these species is now encumbered with a lengthy and tedious synonymy. Lodder and Kreger-van Rij (1952) list 87 synonyms for *C. albicans* and 39 for *Cr. neoformans*.

In the United Kingdom, the Medical Research Council (M.R.C.), in an effort to attain some uniformity in nomenclature of pathogenic fungi, prepared a memorandum, "Nomenclature of Fungi Pathogenic for Man and Animals", in which the names used comply with the rules laid down in the *International Code of Botanical Nomenclature*. The names of fungi recommended by the M.R.C. Memorandum 3rd edition (1967) are used throughout this text. With regard to names of diseases, the situation is more complex and there are many aspects still under dispute. However, the growing tendency for linking disease names with those of the causal fungi (e.g. aspergillosis, cryptococcosis) rather than with the type of lesion produced (e.g. tinea, ringworm) or the part of the body affected (e.g. otomycosis, madura foot) will in time lead to a more specific designation of the mycoses.

The number of fungal pathogens of man and animals is small compared with that represented by bacteria and viruses, and in many cases their distribution is geographically restricted. Nevertheless, in the endemic regions, the incidence of certain mycoses has been shown to be considerable. Coccidioidomycosis, caused by *Coccidioides immitis* which lives naturally in the soil of certain regions, has been shown to affect

50% of the population in certain parts of Mexico (Gonzáles-Ochoa, 1967) and this mycosis was diagnosed in 58% of dogs within one year after their introduction to an endemic area in Tucson, Arizona (Converse *et al.*, 1967).

Pathogenic yeasts and certain other fungal pathogens, such as the ringworm fungi, have a very wide distribution geographically, and the incidence of disease caused by them is correspondingly high.

Customs engendered by modern civilization contribute in large measure to the incidence in man; for instance, ringworm of the feet and nails may occur in more than 50% of some sections of the community which frequently use communal bathing facilities (Gentles and Holmes, 1957).

In surveys of the incidence of fungal disease in man and animals and in records of fungi isolated in mycological laboratories doing routine diagnostic work, yeasts are prominent and sometimes predominate. In the laboratories of medical mycology of Glasgow and Leeds, these fungi formed 16–20% of all pathogenic fungi isolated during one year (1967). Of these isolates, 70–80% were confirmed either as the cause of the disease or as an exacerbating factor in its aetiology.

Unlike many other pathogenic fungi, which occur naturally as saprophytes in soil or similar habitats, the natural habitat of most pathogenic yeasts is the human or animal body where they exist frequently as commensals. This characteristic introduces a major problem in the diagnosis of disease caused by them, so that their presence, even in overt lesions, requires prolonged and careful evaluation from an aetiological aspect. Their presence, even as commensals, is nevertheless extremely important from an epidemiological point of view, since progress from the role as harmless commensal to virulent pathogen may be induced by a number of factors.

The pathogenic yeasts do not form a homogeneous or even a naturally related group, in so far as they are represented by several genera which have distinct morphological characteristics. Like the vast majority of other pathogenic fungi, however, they are all members of the Fungi Imperfecti. The two most important genera are *Candida* and *Cryptococcus*; of lesser importance are *Torulopsis*, *Trichosporon*, *Rhodotorula* and *Pityrosporum*. Certain species of all these genera are capable of producing clinically defined disease in a human or animal host both spontaneously and experimentally.

In a chapter such as this, it is pertinent to consider the questions: what is a yeast and what is a pathogen? The first is adequately defined elsewhere and it is sufficient in the present context that we should state that we have not included accounts of such fungi as *Sporothrix schenckii* and *Histoplasma capsulatum*, the saprophytic states of which under natural conditions are mycelial, and which exist as yeasts only in the parasitic state in the animal and human body or on special cultural media incubated at 37°. The second question, that concerning pathogenicity, is less easy to explain or define, hence a certain arbitrary selection of species is necessary. The characteristics of fungi capable of producing disease, the nature of the infections they cause, and their frequency and distribution, all vary considerably. Certain fungi which are not infrequent as parasites of man and animals are incapable of growth at body temperature (37°) *in vitro*. Others, the thermophilic fungi, are able to grow well at this temperature and above, but the majority in this group are not known as animal pathogens. Clearly, the ability to develop at 37° may be an important criterion of pathogenicity when associated with other factors, but is not an absolute requirement and does not indicate that any particular species is likely to be pathogenic. Among other aspects to be considered are the effect of a given fungus on the host and the incidence of a particular mycosis. Is *Malassezia furfur*, the aetiological agent of the very common, mild and superficial condition, pityriasis versicolor, any less a pathogen than *Cryptococcus neoformans* or *Candida albicans*, which are respectively the aetiological agents of the rather rarely occurring but more serious cryptococcosis of the central nervous system and iatrogenic systemic candidiasis? Should a fungus be considered a pathogen because it has been possible to induce disease by means of administering massive doses to experimental animals whose resistance has been lowered by immunosuppressive drugs prior to inoculation? In view of the ever-increasing use of such drugs, especially in prolonged treatment with broad-spectrum antibiotics, and the established connection between fungal infection and physiological deficiencies or underlying infectious disease, the answer must be in the affirmative. Moreover, in recent years, fungi have become increasingly of importance because of the complications which may result from their invasion of tissues following certain surgical procedures; for example, organ transplant, open-heart surgery, and corneal grafts, in which administration of immunosuppressive drugs is routine procedure. Such fungi may be established pathogens or truly opportunistic in that they are normally harmless saprophytes and capable of pathogenicity only because the host mechanism which normally inhibits their development has been suppressed. Belonging to this category are the yeasts which are common inhabitants of the animal body and are therefore of particular significance. It is hoped that the examples which have been selected for this chapter will be sufficient to emphasize the importance of this group in human and animal disease, but that, at the same time, it will be appreciated that there is a need for caution in interpreting the significance of yeasts isolated from disease processes in an animal body. As already indicated, well established pathogenic species may be present in the body in the role of harmless commensals, while common and usually harmless saprophytes may become pathogens of sinister significance.

II. The Genus Candida

At least seven species of *Candida* have been found associated with pathogenic manifestations in man. These are, in their order of pathogenicity according to Stanley and Hurley (1967): 1. *C. albicans* (Robin) Berkhout; 2. *C. tropicalis* (Cast.) Berkhout; 3. *C. stellatoidea* (Jones et Martin) Langeron et Guerra; 4. *C. pseudotropicalis* (Cast.) Basgal; 5. *C. parapsilosis* (Ashf.) Langeron et Talice; 6. *C. guilliermondii* (Cast.) Langeron et Guerra; and 7. *C. krusei* (Cast.) Berkhout.

A. CANDIDA ALBICANS

By far the most pathogenic, as well as the most common species of *Candida* occurring in (or associated with) man, is *C. albicans*. For this reason it has been the species most studied and its relationship to man has received most attention from pathologists and microbiologists. Much of our knowledge of the genus *Candida* is therefore centred around it. However, since the pathogenicity and biology of other species of *Candida* are becoming increasingly known and appear to follow along

TABLE I. *Sites of Involvement in Superficial Candidiasis*

Cutaneous	Mucosal
INTERTRIGAL	DIGESTIVE TRACT
Interdigital clefts (hands and feet)	Mouth
Genito-crural folds (also genitalia)	Pharynx
Perianal skin	Oesophagus
Inframammary folds	Stomach
Axillae	Intestines
OTHER CUTANEOUS SITES	GENITAL TRACT
Nail folds and nails	Vagina
Oral commissures	URINARY TRACT
External auditory meatus	Urethra
Scalp	Bladder
Cornea	Ureters
GENERALIZED CUTANEOUS	RESPIRATORY
Extension from primary site (in infants)	Naso-lachrymal duct
Chronic diffuse candidiasis	Nasal passages
	Paranasal sinuses
	Nasopharynx, larynx
	Trachea, bronchi

TABLE II. *Sites of Involvement in Systemic Candidiasis*

CENTRAL NERVOUS SYSTEM Brain Meninges	DIGESTIVE SYSTEM Oesophagus Stomach Intestines
CIRCULATORY SYSTEM Heart valves Myocardium Blood vessels	URINARY SYSTEM Kidneys (including pelvis)
RESPIRATORY SYSTEM Lungs Bronchial tree	OTHER ORGANS Liver Spleen Pancreas

TABLE III. *Factors Predisposing to Infection by* Candida albicans (*and Other Pathogenic Species of* Candida)

| HORMONAL DISTURBANCES AND OTHER IDIOPATHIC STATES
Diabetes
Hypoparathyroidism
Hypoadrenocorticism
Carcinoma
Leukaemia
Pernicious anaemia
Aplastic anaemia
Agranulocytosis
Bronchiectasis
Malformation of the urinary tract
Ulceration of the digestive tract
Debility
Malabsorption
Malnutrition
Moribund state

PRE-EMINENTLY RECEPTIVE STATES
Pregnancy
Infancy and old age
Carbohydrate-rich diet
Maceration of skin
Skin surface contact with carbohydrates

DRUG THERAPY
Antibiotics
Corticosteroids
Contraceptive drugs | INFECTIOUS DISEASE
Tuberculosis
Chronic bronchitis
Influenza
Typhoid and other enteric infections
Bacterial endocarditis

SURGERY
Open heart operations
Bowel resections
Colostomy
Tooth extractions
Eye operations (corneal grafts)
Ear operations (skin grafts)

ACCIDENTAL INTRODUCTION OF CANDIDA BY INTRAVENOUS INJECTION OR INDWELLING URINARY CATHETERS
Blood transfusions
Glucose saline drips and other supportive fluids
Drugs, especially in addiction

ACCIDENTAL TRAUMA
Eye injury
Burns |

general lines the pattern exhibited by *C. albicans*, the dissertation on this species which follows applies in general to them also. Particular aspects relating to each species will receive individual treatment under the relevant headings.

Besides occurring commonly as a commensal in the mouth, intestines, and vagina, *C. albicans* is frequently found associated with a variety of lesions of the cutaneous and mucosal surfaces (Table I). It is capable of deep-seated infection, involving individual viscera or the bloodstream, whereby it may become disseminated to a number of viscera and is then often fatal (Table II). Its importance as a pathogen derives in part from its high incidence as a commensal, and the rapidity with which it can spread from the sites of its commensal activity when the natural resistance of the host to infection has been altered by predisposing factors. Among these factors, in addition to pre-eminently receptive but natural states such as infancy, old age and pregnancy, are debilitation, underlying hormonal disturbance, neoplastic disease and severe or chronic infection (Zimmerman, 1955; Table III). To these may be added the influence of corticosteroid and antibiotic therapy, surgical intervention, the risk of directly introducing the fungus into the bloodstream by injection during administration of supportive fluids or drugs (Scholer, 1963a), and finally the introduction of the fungus into the tissues by accidental trauma.

1. *Incidence of* Candida albicans *in Man*

The literature on *C. albicans* in its relation to man is voluminous and includes many reports of its incidence in normal subjects as well as in those suffering from disease. All agree that the incidence of this species is relatively high in normal subjects, especially so during the neonatal period, during old age and during pregnancy. Furthermore, a definite correlation (81·5%) was established by Harris *et al.* (1958) between the incidence of *C. albicans* in the maternal vagina at term and its incidence in the mouth of her newborn child. It is generally accepted that during pregnancy the glycogen content of vaginal mucosal cells is considerably increased, and this is thought to encourage the growth of yeasts. It has also been shown by Cruickshank (1934) that lactic-acid bacilli make use of this glycogen, and in converting it into lactic acid provide an acid environment which again favours the growth of yeasts such as *Candida*.

According to Taschdjian and Kozinn (1957), the presence of *C. albicans* in infants' mouths on the third or fourth day after birth resulted in clinical thrush during the neonatal period in 98% of the infants so affected. The percentage incidence of *C. albicans* in the principal sites of its commensal activity, the mouth, vagina and intestines, according

TABLE IV. *Incidence of Candida albicans in the Oral Cavity*

Authority	Category	Incidence (%)	Authority	Category	Incidence (%)
Marples and di Menna (1952)	Young adults	50·5	Harris et al. (1958)	Newborn infants	4·0
Rahim (1964)	Unselected adults	17·16	Rahim (1964)	Newborn infants	7·5
Schaulow et al. (1967)	Non-pregnant females	31·0	Schaulow et al. (1967)	Newborn infants	5·7
Smits et al. (1966)	Patients on admission	28·0	Somerville (1964)	Newborn infants	11·5
Young et al. (1951)	Young adults	46·0	Taschdjian and Kozinn (1957)	Newborn infants	3·77
Borowski et al. (1963)	Pregnant females	33·1	Harris et al. (1958)	Newborn infants from mothers infected with *Candida albicans*	28·0
Rahim (1964)	Pregnant females	9·0	Schaulow et al. (1967)		46·0
Schaulow et al. (1967)	Pregnant females	6·6	Somerville (1964)		31·0
Somerville (1964)	Pregnant females	46·4			

to various authors is shown in Tables IV, V and VI. For obvious reasons there is much variation in such data, if only because of the different methods employed by different investigators to obtain their results. Thus, Marples and di Menna (1952) showed quite clearly that, in taking samples from the mouth, the percentage incidence of samples positive for *Candida* was much lower if swabs were used than if mouth washings were collected and plated out for culture. Young *et al.* (1951) also used this method in their survey of the yeast flora in the mouths of young adults. Our own experience tends to confirm the inadequacy of swabs

TABLE V. *Incidence of* Candida albicans *in the Vagina*

Authority	Non-Pregnant Females (%)	Authority	Pregnant Females (%)
Schaulow *et al.* (1967)	12·7	Bret and Coupe (1958)	20·0
Mizuno (1961)	16·1	Borowski *et al.* (1963)	23·6
Mackenzie (1961)	4·6	Harris *et al.* (1958)	17·6
		Johnson and Mayne (1948)	37·0
		Negroni (1934)	31·0
		Rahim (1964)	33·3
		Schaulow *et al.* (1967)	26·1
		Somerville (1964)	39·4

TABLE VI. *Incidence of* Candida albicans *in Stools of Normal Subjects*

Authority	Category	Incidence (%)
Benham and Hopkins (1933)	Unspecified	18·0
Brabander *et al.* (1957)	Young adults	30·65
Marples and di Menna (1952)	Children	30·8
Somerville (1964)	Unspecified	13·5
Taschdjian and Kozinn (1957)	Unspecified	10·0

for this purpose in the case of adults. Marples and di Menna (1952) also found that dentures tended to increase the incidence of *C. albicans* in the mouth. Thus, in a sample of 99 subjects, 80 had natural teeth and of these 45% carried *C. albicans*, while of the 19 who wore dentures 68% carried *C. albicans*. Results obtained from our own investigations largely agree with these findings. The reason for this state of affairs is of course not far to seek, since food trapped between the dental plate and the oral mucosa provides a pabulum for *C. albicans* (Fig. 1) as well as for other micro-organisms; and unless scrupulous cleanliness is observed, the resulting growth between meals or during the night in cases where dentures are not removed or cleaned before bedtime, can be very considerable. The probable role of *C. albicans* in "denture stomatitis" was discussed by Cawson (1966) who isolated *Candida* species from the affected mucosa of 94% of patients presenting with this form of stomatitis (Cawson, 1963).

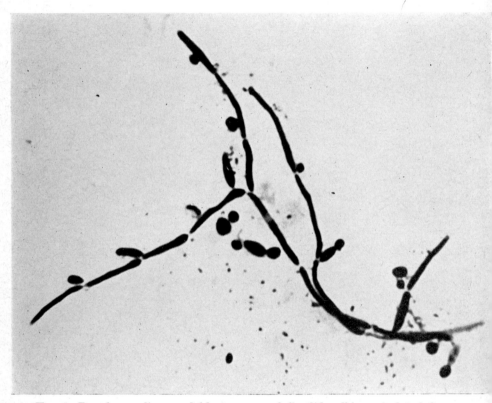

FIG. 1. Pseudomycelium and blastospores of *Candida albicans* isolated from a denture scraping. Gram-stained. Magnification, × 1,000.

The incidence of *C. albicans* in the faeces of normal subjects of all ages is also high, and indicates that the intestinal tract is a common habitat for this species (Table VI, p. 115). According to numerous reports, its presence in the intestines becomes clinically significant in a relatively high proportion of patients during the prolonged administration of certain broad-spectrum antibiotics (Table VII). Cormane and Goslings (1963) suggested that there is competition between bacteria and *Candida* species in the intestinal tract for available carbohydrate, particularly glucose, and that, when the numbers of bacteria are significantly decreased by administration of antibiotics, these yeasts, which are not suppressed by them, are thereby enabled to multiply unhindered by the bacteria.

TABLE VII. *Effect of Antibiotics on the Incidence of* Candida albicans

Antibiotic	Duration of treatment (days)	Incidence (%) in (a) mouth (b) sputum	Incidence (%) in rectal swabs
Penicillin	5	(a) 55	15
Tetracycline	5	(a) 78	25
None	—	(a) 24	7
(after Smits *et al.*, 1966)			
Tetracycline	3–5	(b) 47	55
(after Anderson, 1958)			

Candida albicans is seldom recovered from the healthy skin. Drouhet (1960) has repeatedly maintained this opinion, and stated that he was unable to demonstrate it in any of 2,000 samples. However, Marples and Somerville (1966) recorded it on the normal skin of 1·4% students aged 17–22 years and from as many as 27% of subjects aged 60 years and upwards.

2. *Pathological Aspects*

a. General considerations. Usually no consistent morphological pattern is discernible in the parasitic growth of *C. albicans* in lesions, whether these are superficial or deep seated. The two growth phases, budding cells (blastospores) and mycelium (or pseudomycelium), known respectively as yeast phase and mycelial phase, vary widely in their relative preponderance irrespective of the site in which they occur. Some authors, including Gresham and Whittle (1961), Taschdjian and Kozinn (1961) and Rogers (1966), have insisted that the development of the mycelial phase *in vivo* indicates a change from the commensal or saprophytic to the parasitic habit. While not disputing this contention, it is probably fair to state that the morphological pattern observed at any given time at any given site is determined by host factors prevailing at that time at a local tissue level or at a constitutional level. These factors are as yet not clearly defined. Louria *et al.* (1963), in a study of the pathogenesis of *C. albicans* infections in mice, could find no evidence that invasiveness could be correlated specifically with either the mycelial or the yeast phase; transformation into the mycelial phase was followed by progressive infection only in the kidney. Haley (1965) observed that *C. albicans* had developed hyphae in acid urine obtained from patients, and later demonstrated by experiments with rabbit urine contained in excised and washed rabbit bladders that *C. albicans* developed hyphae in acid urine in these bladders but not in the alkaline urine. She concluded from these results that the development of hyphae was a response to the influence of the bladder tissue as well as to the acid pH value of the urine in which the fungus was growing.

In tissues and body secretions, the yeast phase consists of sub-globose

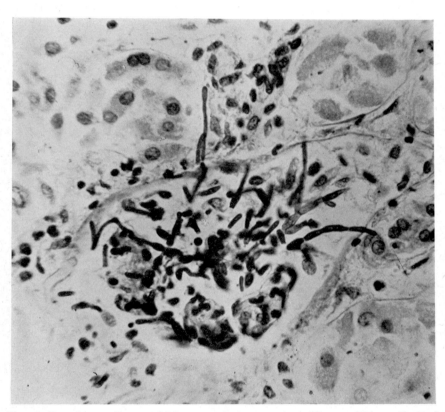

Fig. 2. Pseudomycelium and blastospores in a section of human kidney. Stained with periodic acid-Schiff reagent and haematoxylin. Magnification, × 500.

to oval cells of varying length and breadth, many of them bearing buds in different stages of development. These buds are situated terminally or subterminally. Pseudomycelium, which develops from blastospores by a process of elongation, consists of elongated, straight or curved cells of varying length joined end-to-end (Fig. 2). In cases where true mycelium is formed it does not differ from the septate mycelium of other fungi. Blastospores are often found attached to the cells of both pseudo-mycelium and true mycelium of *C. albicans*, and when this occurs it enables one to distinguish such mycelia from those of other pathogenic fungi such as dermatophytes or aspergilli. The number of these attached blastospores varies considerably in different preparations for a number of reasons which must necessarily include mechanical disturbance occurring during processing of the material. Occasionally, inflated intercalary cells are seen along the length of the pseudomycelium and large globose cells resembling, at least in size, the chlamydospores seen in certain cultures. Such forms are probably expressions of degenerate growth and have little significance. They have been noted by Blyth (1959) and Hurley (1966) in deep-seated lesions in experimentally infected animals, and by the authors (unpublished observations, 1968) in the human kidney in a case of haemorrhagic measles.

b. Superficial candidiasis. Cutaneous and mucosal infections by *C. albicans* occur far more frequently than visceral or systemic infections.

(i). *Mouth.* When fissures at the corners of the mouth (angular cheilosis; Fig. 3) are complicated by *Candida* infection, their surface is coated with a white, more or less viscous, deposit consisting of dead epithelial debris

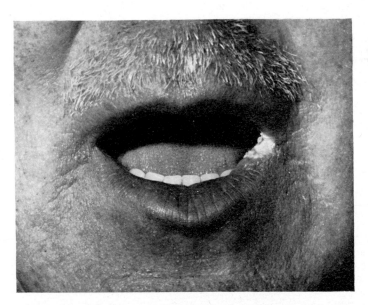

FIG. 3. Angular cheilosis complicated by infection with *Candida albicans*. Reproduced by courtesy of the General Infirmary at Leeds, England.

and yeast elements. This condition is usually associated with the wearing of dentures. Acute *Candida* infection of the oral mucous membranes, known as thrush, may involve all parts of the mouth including the tongue, but the commonest and best known manifestation is a lesion on the palate. When fully developed, it is characterized by the appearance of multiple white to cream-coloured curd-like spots or patches on a more or less inflamed background (Fig. 4). In neglected

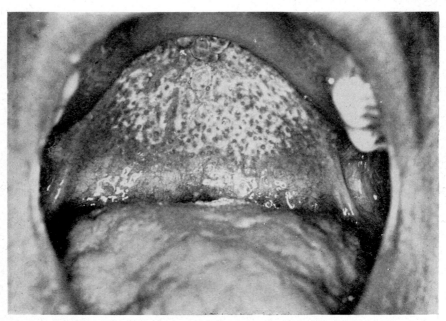

FIG. 4. Oral thrush due to *Candida albicans*. Reproduced by courtesy of the General Infirmary at Leeds, England

cases, these may coalesce to form membraneous plaques. Microscopically, the white material is seen to consist of pseudomycelium and blastospores of *Candida* mixed with epithelial debris, leucocytes and bacteria and also particles of food (Fig. 5). If left untreated, thrush may spread to the pharynx and even to the oesophagus.

Ludlam and Henderson (1942) reported 20 cases of thrush oesophagitis, occurring mostly in premature infants. *Candida* infection of the gastric and intestinal mucosal layers has also been reported by Beemer *et al.* (1954) and others. The gross features of oesophageal thrush have been described by Winner and Hurley (1964) in the following terms: "the mucous membrane may be covered by a thick, shaggy, cream-coloured confluent membrane for some considerable part of its length, for example, the whole of the oesophagus may be so covered".

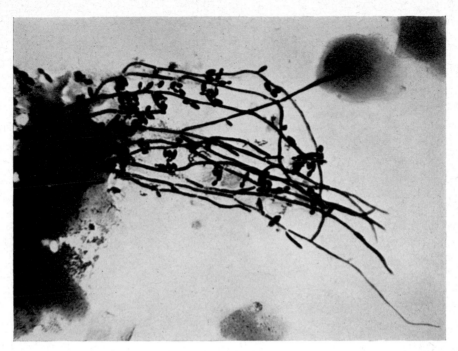

FIG. 5. *Candida albicans* in a direct preparation from a case of oral thrush. Gram-stained. Magnification, × 500.

(ii). *External auditory canal.* Mycotic infection of the external ear canal due to *Candida* is usually a mixed infection in which bacteria participate (e.g. *Staphyloccocus* spp., diphtheroids, coliforms, *Micrococcus* spp.). The species of *Candida* most frequently encountered are *C. albicans* and *C. parapsilosis*. Species of *Aspergillus* (*A. niger*, *A. fumigatus*) are also occasionally associated with yeasts in such infections (Smyth, 1962; Gregson and La Touche, 1961; La Touche, 1966; Powell *et al.*, 1962; Durcan *et al.*, 1968).

(iii). *Eye.* Although mycotic infections of the eye were considered to be rare in the past, they are becoming increasingly recognized, usually in the form of corneal ulcers. In addition to many common saprophytic fungi *C. albicans* and other *Candida* species have been recorded as the causal agents (Mendelblatt, 1953; Sykes, 1946; Mitsui and Hanabusa, 1955; Manchester and Georg, 1959). Infection of the cornea due to *Candida* can also develop as an extension of *Candida* granuloma and certain other forms of cutaneous candidiasis, or may be associated with oral lesions. Mycotic infection of the conjunctiva due to *C. albicans*, first recorded by Norton (1927), also occurs.

(iv). *Finger clefts and toe clefts.* Maceration of the finger clefts, especially in women engaged in household activities which include much washing,

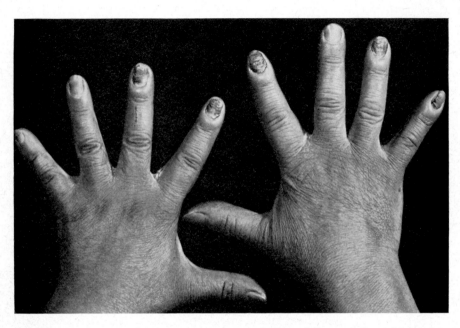

FIG. 6. Chronic paronychia and maceration of finger clefts complicated by infection with *Candida albicans*. Reproduced by courtesy of the General Infirmary at Leeds, England.

is frequently complicated by cutaneous candidiasis (Fig. 6). It occurs also occasionally in men, especially bartenders, and also in confectioners and fruit canners (Schwartz *et al.*, 1951). Cutaneous candidiasis also complicates macerated skin in the toe clefts.

(v). *Nail folds and nails.* Chronic paronychia, a chronic inflammatory condition affecting the nail folds, is due usually to a mixed infection by *Candida* and bacteria. Usually the proximal nail fold is affected first, and becomes swollen and painful. Inflammatory exudate, which collects, may be extruded from it by pressure and often contains *Candida* elements. The nail plate may also become involved and, although *Candida* develops in it usually only to a limited extent, the nail does not grow normally and is usually characterized by a somewhat crinkly surface traversed by one or more transverse grooves (Fig. 6). A brownish or greenish discolouration commonly occurs along the lateral borders, and the nail may also be somewhat eroded. *Candida* elements are often collected in quantity from the grooves between the lateral folds and the nail plate.

(vi). *Submammary folds, axillae, genitocrural folds.* Infection of the skin folds occurs relatively frequently, especially in obese women. The characteristic lesion is a uniform patch of erythema with satellite spots on

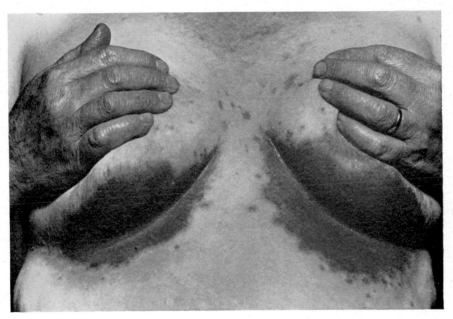

Fig. 7. Skin of submammary folds infected with *Candida albicans*. Reproduced by courtesy of the General Infirmary at Leeds, England.

the periphery. It extends from the depth of the fold over the areas of skin in contact (Fig. 7). The affected skin is moist, especially in the deeper parts of the fold where it is white and macerated. *Candida albicans* is more abundant in this part of the lesion than at its periphery. In both men and women, the genitocrural folds or peri-anal skin may be affected in a similar manner. In children, *Candida* infection of the napkin area, extending occasionally beyond it, occurs not infrequently

5*

in infants who develop thrush during the neonatal period. The rash which results from candidiasis in this area is sometimes not easy to distinguish from napkin rashes due to other causes, but the presence of blastospores and mycelial growth in microscopic preparations made from the lesion establishes the diagnosis.

c. Systemic candidiasis. Most of the viscera are susceptible to involvement in systemic infections due to *C. albicans*, particularly the kidneys, less frequently the heart, liver, central nervous system, lungs, spleen and deeper tissues of the digestive tract, and still less frequently the pancreas and thyroid. The sites of visceral involvement are listed in Table II (p. 112).

(i). *Respiratory system.* Although species of *Candida*, especially *C. albicans*, are frequently found in sputum samples and bronchial secretions, there are very few authentic records in the literature of primary infection of the bronchial mucosa or of the lungs. Invasion of the bronchial wall does occur under certain circumstances, as for instance in the presence of ulceration of the bronchial mucosa when, according to Plummer (1966), there is usually also severe underlying pathology. Primary invasion of the lung parenchyma by *Candida* in fatal cases of pneumonia has been difficult to prove, because of the possibility that such spread might be a post-mortem event (Plummer, 1966). The problem of pulmonary candidiasis is discussed fully by Winner and Hurley (1964) who state that, to be reasonably certain that *Candida* recovered from sputum originated in the lungs, it is necessary to collect the specimen in a sterile container and to refrigerate it until cultured or to culture within an hour or two; also that the fungus should be grown frequently or invariably from the sputum and not from the mouth or throat or postnasal swabs; or it should be grown from a bronchial cast, plug or membrane from a bronchiectatic cavity or pulmonary abscess.

Involvement of the lungs in cases of septicaemic or disseminated candidiasis is well established and well documented (Symmers, 1966; Bendel and Race, 1961; Debré *et al.*, 1955; Gausewitz *et al.*, 1951). In such cases, *Candida* has been demonstrated in micro-abscesses or granulomata in the lung parenchyma or in the lumen of blood vessels. Recently, an interesting account of a case of systemic candidiasis due to *C. albicans* was published by Berge and Kaplan (1967). This was a fatal case with involvement of the lungs and liver. In these organs, numerous granulomata were present containing asteroid bodies at the centre of which were blastospores of *C. albicans*. These asteroid bodies, the first to be recorded in association with infection by *Candida* in man, were surrounded by a wide eosinophilic zone.

(ii). *Circulatory system*. Of particular interest, because of recent developments in heart surgery, are the ever-increasing reports in the literature of infection of the heart valves by species of *Candida*. Utz (1966) comments on reports of 41 patients having suffered from this form of candidiasis. Of this number, 14 had developed the infection after cardiac surgery, nine had been treated for long periods intravenously for bacterial endocarditis, and six were addicted to drugs. The remainder (unclassified) had been receiving antibiotic therapy. The species concerned were: *C. albicans*, 20; *C. parapsilosis*, 9; *C. guilliermondii*, 5; *C. tropicalis*, 1; *C. stellatoidea*, 1; *C. krusei*, 1; and unidentified species of *Candida*, 4. The result of infection of the heart valves by *Candida* is the development of large vegetations consisting of dense mycelial growth and blastospores. In this group of cases enumerated by Utz (1966), there was a fatality rate of 88%. Autopsy showed that, in some cases, fragments of considerable size had broken off from the vegetations on the

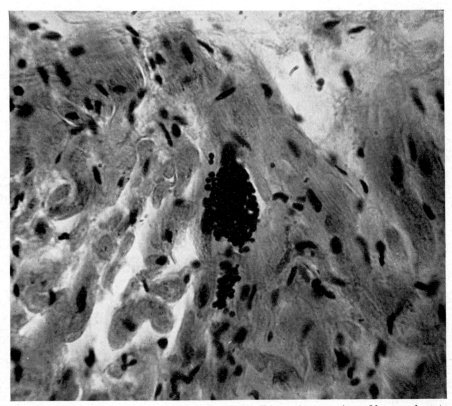

FIG. 8. Patch of yeast-phase cells of *Candida albicans* in a section of human heart muscle. Stained with periodic acid-Schiff reagent and haematoxylin. Magnification, × 500.

heart valves and had formed emboli in blood vessels in other parts of the body.

In cases of *Candida* septicaemia, it is not unusual to find, in the course of histological examination of necropsy material, that the heart muscle is involved in the infection. Scattered micro-abscesses containing varying amounts of the fungus may be found, sometimes with the mycelial phase predominant and sometimes the yeast phase. Occasionally, little groups of yeast-phase cells are found in parts of the heart muscle where no tissue response is observable (Fig. 8).

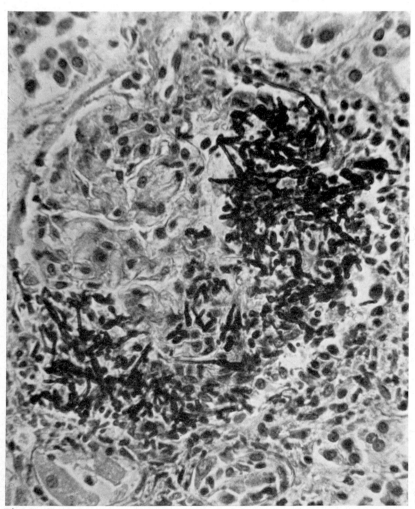

FIG. 9. *Candida albicans* in human kidney glomerulus with associated inflammatory exudate. Stained with periodic acid-Schiff reagent and haematoxylin. Magnification, × 500.

(iii). *Urinary system.* In most accounts of disseminated candidiasis or candida septicaemia, involvement of the kidneys is reported. Lehner (1964) noted this in 40 of 45 necropsy reports. Small abscesses are found in both the cortex and medulla (Symmers, 1966; Hurley, 1964; Lhener, 1964). Yeast-phase and mycelial-phase are found together (Fig. 9), or occasionally patches of yeast-phase cells only (Fig. 10). The lesions are found mainly in the glomeruli and tubules, but may also occur in the interstitial tissue. The fungus grows profusely in the glomeruli, eventually obliterating them. It also grows well in the tubules, completely filling parts of them. From these sites it extends into the surrounding

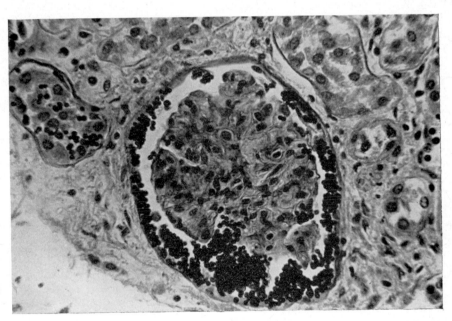

Fig. 10. Yeast-phase only of *Candida albicans* in part of a section of human kidney. Stained with periodic acid-Schiff reagent and haematoxylin. Magnification, × 500.

interstitial tissue. There is usually intense tissue reaction to the presence of the fungus, but sometimes very little reaction occurs and both states may be observed occasionally in the same section of kidney. The exudate surrounding the fungus is made up mainly of polymorphonuclear leucocytes, but there is also a varying admixture of round cells in some cases (Fig. 9).

(iv). *Central nervous system*. Utz (1966) enumerated 27 cases of meningitis due to infection by *Candida* which had been reported in the literature. In 26 of these, *C. albicans* was the aetiological agent; in the remaining case the species was not identified. The overall fatality rate in this group was 41%. Roessmann and Friede (1967) published an account of *Candida* infection in the brains of two infants, both of whom had operations to correct abnormalities of the bowel. Both had oral thrush in life, and *Candida* was found post-mortem in the lungs and kidneys as well as the brain. Brain lesions varied from frank, acute abscesses to granulomatous lesions and perivascular cuffing with sparse lymphocytic infiltrates. In both cases *Candida* was cultured, *C. albicans* being identified in the first case, and was presumably the species isolated in the second case. There is little doubt that, in both cases, dissemination occurred from the gut.

3. *Incidence of* Candida albicans *in Animals*

Candida albicans has been found in a wide range of animals, both as commensal and pathogen. It has been isolated mainly from the digestive tract. Van Uden (1960) recorded it in horses, swine, sheep, goats, African bush pigs, African wart hogs, baboons and African pied crows. Ainsworth and Austwick (1959) listed among animals suffering from candidiasis due to this species: fowls, turkeys, ducks, geese, guinea fowl and pigeon, as well as cattle, pigs, dogs, hedgehogs, guinea pigs, various other rodents and monkey. According to Austwick *et al.* (1966), outbreaks of candidiasis occur frequently in turkeys (Blaxland and Fincham, 1950), artificially reared partridges (Keymer and Austwick, 1961), and piglets (Osborne *et al.*, 1960). *Candida albicans* appears to be widely distributed among birds, especially cage birds. In Rostock, Deutsche Demokratik Republik, Kaben and Preuss (1967a, b) isolated it as a commensal from the droppings of a variety of birds kept at the zoo, as 20·4% of total yeasts represented by about 53 species (including 18 species of *Candida*). Evans (1968) isolated it from 26·5% of 137 samples of pigeon droppings, also as a commensal. It has, however, been reported as a cause of candidiasis in cage birds. Vallée *et al.* (1964) isolated it from Java sparrows (*Munia striata domestica*), linnets (*Carduelis cannabina*), and Chinese quail (*Excalfactoria*) suffering from candidiasis; and according to Austwick *et al.* (1966) parrots and budgerigars may become similarly infected.

4. *Experimental Infection in Animals*

Mice and rabbits are the animals most frequently used for demonstrating experimentally the pathogenicity of *Candida* species (Winner and Hurley, 1964), and for studies on the pathogenesis of experimentally induced *Candida* infections, both in previously healthy animals (Louria *et al.*, 1963) and in animals modified by various means, for example, by treatment with various antibiotics (Blyth, 1966) or by previous bacterial infection, X-ray dosage or alloxan poisoning to induce a diabetic state (Hurley, 1966). Such experiments show generally a consistent affinity by *Candida* for the kidney, and to a less degree the heart and brain. But, in the modified host, a more widespread involvement of the viscera and a greater proliferation of the fungus in the lesions take place. These experiments also show that the severity of the pathological process is roughly proportional to the amount of inoculum administered.

5. *Incidence of* Candida albicans *in the Environment*

Reports of extraneous sources of *C. albicans* are rare. The yeast has been isolated by Gitter and Austwick (1959) from the drinking water, the atmosphere and the litter of a pen occupied by piglets, many of which were suffering from candidiasis, and by Keymer and Austwick (1961) from grit and grass inside and outside pens in which young partridges, also suffering from candidiasis, were kept. Rogers and Beneke (1964), in Brazil, found the yeast in the soil of caves and chicken runs, in sand near a dwelling house, and in soil of a park flowerbed. Van Uden *et al.* (1956) isolated it from flowers of gorse (*Ulex* sp.) and leaves of the common myrtle (*Myrtus communis*) in Portugal. Clayton and Noble (1966) recovered it, together with other yeasts, from pillows and blankets in hospital wards in the London area, and Jennison (1966) made isolates from floors, cubicles, radiators, windows, and window-sills in a nursery of a maternity hospital where many of the patients were suffering from candidiasis. From the nature and rarity of such reports, it is clear that extraneous sources of *C. albicans* are of little significance in the epidemiology of candidiasis, except in the case of the immediate environment of humans or animals suffering from overt candidiasis.

6. *Diagnosis of Infection by* Candida albicans

Specimens from suspected infections of cutaneous or mucosal sites are best examined microscopically by mounting in 10–20% (w/v) potassium hydroxide to clear and soften the tissue. Particular attention should be paid to the presence of blastospores, pseudomycelium or mycelium, and when present a rough quantitative estimate of their proportions should be made. The significance and assessment of these findings depend very much on the experience of the investigator and the circumstances attending the case. The ubiquity of *C. albicans* as a commensal in mucosal sites should always be borne in mind; and unless the patient being investigated has been previously treated with anti-

fungal drugs or is suffering from some underlying disease, little significance should be attached to the presence of a few blastospores or a
little pseudomycelial growth. In any case, culture is necessary for
diagnostic confirmation and for specific identification; and again, having regard to the possible effect of previous therapy if any, the number
of colonies obtained can serve as a rough estimate of the degree of involvement of the fungus in the aetiology of the lesion or diseased
state. Specimens of sputum, bronchial secretions, vaginal secretions or
discharge or pus, and centrifuged deposits of urine may also be investigated by these methods. In the case of thick, sticky sputum, treatment
of the specimen with a preparation of pancreatic secretion very much
facilitates microscopical examination and does not affect the viability
of the fungus. In the case of specimens of cerebrospinal fluid, mounting
in finely divided Indian ink or nigrosin permits differentiation between
pus cells and yeast cells.

Biopsy material, when sufficient has been taken for culture, should be
fixed in a solution made by mixing 10 ml formalin and 90 ml M-NaCl,
embedded and sectioned. Special fungal stains such as periodic acid-
Schiff, Gomori's methenamine silver or Gridley stain, should be used.

TABLE VIII. *Fermentation and Assimilation Reactions of* Candida *species*

Yeast	Fermentation* (alternative or occasional reactions are shown in brackets)					Assimilation				
	Glucose	Galactose	Sucrose	Maltose	Lactose	Glucose	Galactose	Sucrose	Maltose	Lactose
Candida albicans	AG	A (AG)	— (A)	AG	—	+	+	+	+	—
Candida guilliermondii	AG	A (AG)	AG	—	—	+	+	+	+	—
Candida krusei	AG	—	—	—	—	+	—	—	—	—
Candida parapsilosis	AG	A (AG)	A	— (A)	—	+	+	+	+	—
Candida pseudotropicalis	AG	AG (A)	AG	—	AG	+	+	+	—	+
Candida stellatoidea	AG	—	—	AG	—	+	+	—	+	—
Candida tropicalis	AG	AG	AG	AG	—	+	+	+	+	—

* A indicates acid production; G, gas production,

Candida species grow well and are usually isolated easily on common culture media such as malt-extract or glucose-peptone agar. It is advisable, however, to inoculate several slopes or Petri dishes and to incubate at 25–28° as well as 37°. Moreover, at least some of the culture media should be supplemented with antibacterial agents and, if mould contamination is likely to be troublesome, the antifungal antibiotic actidione may also be incorporated. Certain pathogenic yeasts are sensitive to actidione, however, and this must always be borne in mind. Following isolation, the cultures should be streaked out to purify.

Pure cultures should be transferred to specialized media and used to determine the fermentation and assimilation pattern (Table VIII) for species identification.

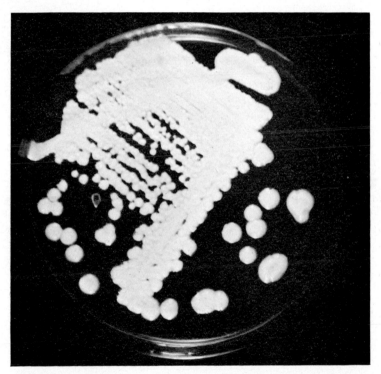

FIG. 11. Growth of *Candida albicans* on glucose-peptone agar.

7. *Mycological Aspects*

After one week at 37° on malt-extract or glucose-peptone agar, colonies of *Candida albicans* are white to cream coloured, opaque, smooth,

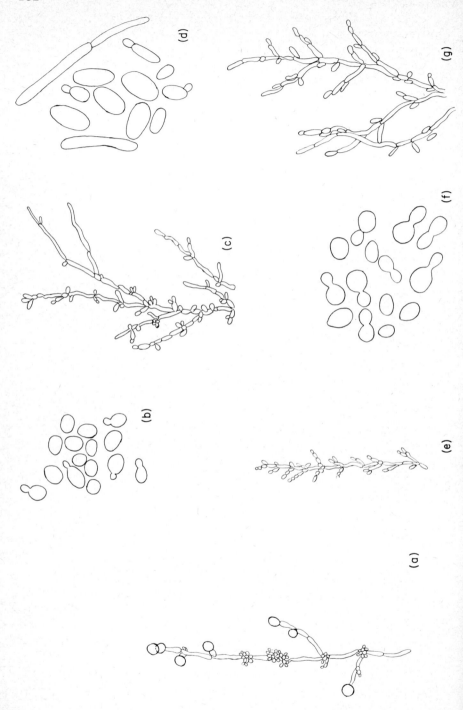

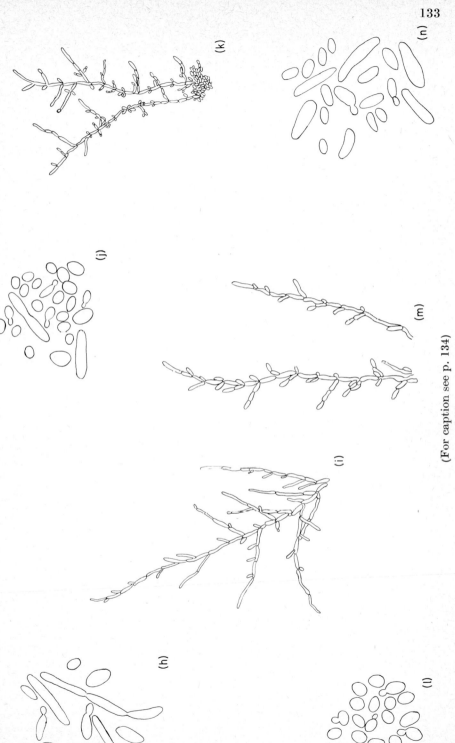

(k)

(n)

(j)

(m)

(i)

(h)

(l)

(For caption see p. 134)

firm and glistening and appreciably raised (Fig. 11). Occasionally their surface is rough. In consistency they are paste-like. These primary colonies are made up, almost uniformly, of blastospores which are sub-globose to oval, sometimes elongate, and measure 3·0–6·5 μ × 3·5–12·5 μ, and subculture to corn-meal, rice-infusion or potato-carrot (with added bile) agar at 27° is necessary to induce mycelial growth (Fig. 12). This usually occurs within 24 h, especially if the inoculum is cut into the agar in the petri dish or on the slide (in slide culture). The characteristic, large (7–17 μ) thick-walled, round cells (so-called chlamydospores; Fig. 13; Table IX) are usually also formed within this period, but some strains require a longer period to develop them and a few fail to do so. Their development is enhanced by the addition of Tween 80 to corn-meal agar or rice-infusion agar, and sometimes the sudden transference from the incubator to the lower temperature of the laboratory bench provides the stimulus necessary for their production in reluctant strains. Rapid diagnosis of C. albicans may also be made by inoculating serum or egg white and incubating at 37°. Within a few hours, short "germ tubes" are formed giving the cells a drum-stick form (Buckley and van Uden, 1963; Fig. 14). The assimilation and fermentation reactions of C. albicans are given in Table VIII (p. 130).

FIG. 12. Line drawings showing the morphology of vegetative forms of Candida species. (a) Candida albicans: slide culture on potato-carrot extract-agar. Magnification, × 400. (b) Candida albicans: cells after 7 days growth on malt extract-agar. Magnification, × 2,000. (c) Candida tropicalis: slide culture on potato-carrot extract-agar. Magnification, × 450. (d) Candida tropicalis: cells after 7 days growth on malt extract-agar. Magnification, × 2,000. (e) Candida stellatoidea: slide culture on potato-carrot extract-agar. Magnification, × 400. (f) Candida stellatoidea: cells after 7 days growth on malt extract-agar. Magnification, × 2,000. (g) Candida pseudotropicalis: slide culture on potato-carrot extract-agar. Magnification, × 800. (h) Candida pseudotropicalis: cells after 7 days growth on malt extract-agar. Magnification, × 2,000. (i) Candida parapsilosis: slide culture on potato-carrot extract-agar. Magnification, × 550. (j) Candida parapsilosis: cells after 7 days growth on malt extract-agar. Magnification, × 2,000. (k) Candida guilliermondii: slide culture on potato-carrot extract-agar. Magnification, × 450. (l) Candida guilliermondii: cells after 7 days growth on malt extract-agar. Magnification, × 2,000. (m) Candida krusei: slide culture on potato-carrot extract-agar. Magnification, × 550. (n) Candida krusei: cells after 7 days growth on malt extract-agar. Magnification, × 2,000.

TABLE IX. *Morphological Characteristics of Candida Species*

Yeast	Morphology in	
	Liquid medium (3% malt extract)	Slide culture, or cut-streak inoculum on corn-meal or rice-infusion agar
Candida albicans	Cells globose to oval, occasionally elongate (3·6·5 × 3·5–12·5 μ)	Blastospores in dense bunches at septa along abundant, well formed pseudomycelium; chlamydospores formed
Candida tropicalis	Cells short-oval with rounded ends (5–9 × 6–12 μ)	Abundant pseudomycelium; also true mycelium; well developed verticils of blastospores
Candida parapsilosis	Cells round-oval or pseudomycelial (2·5–5 × 4–8 μ)	Well developed pseudomycelium with giant cells; poorly developed verticils of few blastospores
Candida stellatoidea	Cells oval, occasionally with one end pointed; occasionally elongate (2·5–6 × 5–12 μ)	Well developed pseudomycelium; blastospores in rounded clumps and in short chains; chlamydospores in short chains
Candida pseudotropicalis	Cells short-oval or oval (3·5 × 4·5–9 μ)	Fine pseudomycelium; verticils of blastospores in chains irregularly developed
Candida guilliermondii	Cells subglobose or short-oval (2–5 × 3–6 μ) Occasionally giant cells (5·2 × 7 μ)	Pseudomycelium well developed and in part aerial; blastospores abundant in chains
Candida krusei	Cells large cylindrical, occasionally short-oval, usually within range 2·5–5·5 × 7·5–21·5 μ	Pseudomycelium well developed, but disintegrates easily; blastospores in verticils in chains

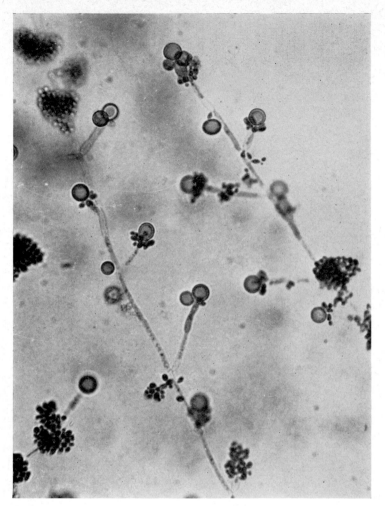

FIG. 13. Chlamydospores of *Candida albicans* grown on corn meal-agar. The preparation was stained with lactophenol cotton blue. Magnification, × 300.

B. CANDIDA TROPICALIS

1. *Incidence and Distribution*

Candida tropicalis is especially associated with vulvo-vaginitis. Hurley (1966) found it to compose 38%, Mizuno (1961) 5·9%, Singh and Sharma (1963) 3% and Smith *et al.* (1963) 4·4% of yeast isolates from this condition. Its occurrence in deep-seated lesions and in disseminated forms of candidiasis has been reported by a number of authors, and it has been also recorded from unusual sites such as the

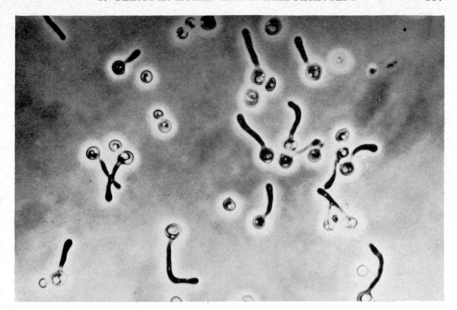

FIG. 14. Germ tubes formed by *Candida albicans* grown in egg albumin. Phase-contrast; magnification, × 1,000. Reproduced by permission of N. van Uden and Helen Buckley, and the editors of *Sabouraudia*.

petrous bone in a case of otogenic osteitis (Fendel and Pietsch, 1965) and in the eye (Graf, 1963).

The isolation of this species from the vegetations occurring on heart valves (Utz, 1966; Conn *et al.*, 1960) in cases of endocarditis, and from the blood and viscera by Symmers (1966) and others in cases of septicaemia, leaves little doubt as to its importance as a pathogen for man.

A number of reports in the literature list this species as occurring in the digestive tract of animals, among which are those of Kawakita and van Uden (1965) who found it in gulls and terns, and van Uden (1960) who found it in the caecal contents of African bush pigs (*Potamochoerus choeropotamus*). It has been recorded also by Austwick *et al.* (1966) and by Schulte and Scholz (1962) in foetuses in cases of bovine mycotic abortion, and by Barbesier (1960) and Scholer (1963b) in cases of bovine mastitis.

Austwick *et al.* (1966) have isolated this species from animal feeding stuffs, especially brewer's grains, in the environment of cattle suffering from mastitis due to yeasts. According to Lodder and Kreger-van Rij (1952) some of the strains examined by them were derived from rotten pineapples and from so-called "tea-fungus".

2. Cultural Characteristics

The gross cultural features of *C. tropicalis* are somewhat variable. Colonies are usually soft, cream coloured and glistening, becoming tough and folded with time, and they may have a hairy appearance. Cells are short-oval with rounded ends, sometimes globose, and measure 5–9 μ × 6–12 μ. On corn meal-agar or rice infusion-agar and in slide culture, abundant pseudomycelium bearing well developed verticils of blastospores is produced. The fermentation and assimilation reactions are given in Table VIII (p. 130).

C. CANDIDA STELLATOIDEA

1. Incidence and Distribution

Candida stellatoidea has been isolated from a number of human sources but most commonly from the human vagina. Jones and Martin (1938) obtained it frequently from this source, and Mizuno (1961) found it in 11·7% of cases of vulvo-vaginitis as well as in 7·9% of subjects in whom the vagina was symptomless. Hurley (1966) also cultured it in cases of vulvo-vaginitis. Kaul *et al.* (1960) found it as a cause of thrush in infants, most of whom were artificially fed, and Jansson (1962) obtained it as 2% of yeasts isolated from sputum. Guze and Haley (1958) recovered *C. stellatoidea* from two patients suffering from pyelonephritis, and Jamshidi *et al.* (1963) reported it as the aetiological agent of endccarditis following cardiac surgery. Its pathogenicity has been demonstrated in inoculated animals by Hurley and Winner (1964) and Hasenclever and Mitchell (1961).

2. Cultural Characteristics

Candida stellatoidea closely resembles *C. albicans* in its general habit, but especially in that it is the only other species of *Candida* which develops chlamydospores. However, it can be distinguished from *C. albicans* by its gross colony appearance and by the fact that it tends to develop chlamydospores in chains. It differs also in that it fails to produce acid with sucrose and, according to Conant (1940), it has a slower growth rate than *C. albicans* on blood-agar. When injected into rabbits intracutaneously, *C. stellatoidea* fails to produce abscesses, a feature of *C. albicans*, and when inoculated intravenously it causes a peculiar head twist whereas *C. albicans* kills the animal within five days.

Colonies are cream coloured and roughly star-shaped. Cells are oval, occasionally with one end pointed. Some are elongate. They measure 2·5–6 μ × 5–12 μ. Pseudomycelial formation is well developed, and the yeast bears blastospores in rounded clumps, as in *C. albicans*, and also in short chains. Chlamydospores tend to form in chains rather than singly or in clumps as in *C. albicans*, and may show a supporting cell. Fermentation and assimilation reactions are given in Table VIII (p. 130).

D. CANDIDA PSEUDOTROPICALIS

1. *Incidence and Distribution*

Candida pseudotropicalis has been isolated only rarely from cutaneous lesions, but it has been quite frequently cultured from the respiratory tract in patients suffering from respiratory diseases of various types (Behounkova *et al.*, 1965; Herbeuval *et al.*, 1955). It has also been recovered in cases of septicaemia (Beuthe, 1955; Skobel *et al.*, 1955; Symmers, 1964; Vince, 1959; Wegman, 1954). Its pathogenicity was demonstrated by Hurley and Winner (1964) in mice and rabbits when these were experimentally inoculated.

Candida pseudotropicalis has also been isolated from animals. Thus, Scholer (1963b) isolated it from milk from a cow with mastitis. Inert sources of this species mentioned by Lodder and Kreger-van Rij (1952) include samples of cream and an experimental milk and whey plant.

2. *Cultural Characteristics*

Colonies of *C. pseudotropicalis* are cream coloured, soft and rather flat with a rather sinuous margin. Cells are short-oval measuring 3–6 μ × 4·5–9 μ, occasionally more elongate and up to 15–18 μ. In slide culture on corn meal-agar or other suitable solid media, pseudomycelium is well developed but rather delicate in nature and bears irregularly formed verticils of blastospores in chains. The fermentation and assimilation reactions are given in Table VIII (p. 130).

E. CANDIDA PARAPSILOSIS

1. *Incidence and Distribution*

Candida parapsilosis occurs commonly on healthy human skin as well as in cutaneous, mucosal and deep-seated lesions. Schirren (1963) found that it formed 11·6% of yeast species isolated from healthy human skin; Schönborn (1967) obtained 654 isolates from skin and nail material investigated during 1964–65, and Hansen (1963) isolated it in 18·5% of lesions of the toe clefts which did not yield dermatophytes. It has been isolated also by a number of investigators, including ourselves, from nails and nail folds in cases of chronic paronychia (Sturde, 1956; Kruspl, 1963). It occurs also in the external auditory meatus (C. J. La Touche, unpublished observations) and has been reported as being involved in the aetiology of corneal ulcers (Manchester and Georg, 1959; Suie and Havener, 1963) and in infections of the conjunctiva (Ainley and Smith, 1965). It has also been isolated from a number of

cases of septicaemia and endocarditis (Wilson, 1961; Roberts and
Rabson, 1962; Andriole *et al.*, 1962; Scholer *et al.*, 1963; Utz, 1966). As
regards its occurrence in animals, Bisping *et al.* (1964) isolated it from
a case of abortion in a cow, as did Smith (1967). Van Uden and do
Carmo-Sousa (1962a) found that it formed 19·2% of the yeast flora in
the caecum of baboons. Hajsig *et al.* (1964) showed that *C. parapsilosis*
could provoke endometritis in heifers. Lodder and Kreger-van Rij
(1952), who examined 27 strains, state that 22 of these were of human
origin. Among the remainder, one was from a Dutch brewery, and
another from the so-called "tea-fungus" which is a mixture of bacteria
and yeasts used in the preparation of tea-beer in Eastern Europe and
the Far East. Batista *et al.* (1961d, f) isolated it in Recife, Brazil, from
the faeces of horses and mules, and from fruit in Manaus, Brazil.

2. Cultural Characteristics

Colonies of *C. parapsilosis* are cream coloured and usually smooth,
but occasional strains are rough or wrinkled. These may also originate
as sectors of smooth colonies (Schönborn, 1967). Cells are oval or somewhat pseudomycelial, and measure 2·5–5·0 μ × 4–8 μ. Pseudomycelium
formation is abundant and is distinctive because of the formation of
wide, elongate and often curved giant cells, giving a chain-like appearance. Blastospores are formed, but in small numbers in poorly developed
verticils. The assimilation and fermentation reactions of *C. parapsilosis*
are given in Table VIII (p. 130).

F. CANDIDA GUILLIERMONDII

1. Incidence and Distribution

Strains of *C. guilliermondii* cultured by Schirren (1963) from healthy
skin amounted to 3·5% of a collection of isolates representing 12 species
of yeasts. Kruspl (1963) recovered it from finger nails in cases of chronic
paronychia where it formed 1·3% of yeasts, and Whittle and Gresham
(1963) isolated the yeast from finger-nail folds which they had previously
occluded in order to induce maceration. Böhme (1965) obtained a
number of isolates from hair, skin and nails. The organism is not often
listed among species recovered in cases of vulvo-vaginitis, but Kearns
and Gray (1963) isolated it from two patients suffering from this condition. It is apparently also infrequent in the mouths of newborn infants,
although Kaul *et al.* (1960) cultured 11 strains in a survey of 113 infants
within a few days after birth. While *C. guilliermondii* has not often been
recorded from human faeces, Pan and Pan (1964) isolated it on six
occasions from stools in a random survey of 152 children in Taipei
(Formosa). Serious deep-seated infections with *C. guilliermondii* have

been reported by Utz (1966) who found it in five cases of endocarditis, and by Symmers (1966) who isolated it on one occasion from this condition. Utz (1966) also cultured it from the blood in four cases of *Candida* septicaemia. According to Batista *et al.* (1961a, b) this species is frequently present in the faeces of animals such as horses, mules, cattle and poultry.

2. Cultural Characteristics

Colonies of *C. guilliermondii* are cream coloured, smooth and glistening, with rare strains wrinkled and dull. Cells are small, 2–5 μ × 3–6 μ, sub-globose or short-oval. Occasional giant cells, up to 7 μ long, may be formed. On corn meal-agar or rice infusion-agar and in slide culture, pseudomycelial development is abundant and shows irregular verticils of blastospores in chains or in wreath-like formation. The pseudomycelium sometimes projects above the surface of the colony. The fermentation and assimilation reactions of *C. guilliermondii* are given in Table VIII (p. 130).

G. CANDIDA KRUSEI

1. Incidence and Distribution

Candida krusei has been isolated in the frequency of 8% of yeasts from the respiratory tract (Behounkova *et al.*, 1965) and, according to results obtained by a number of investigators (Petru, 1965; Mizuno, 1961; Dutt-Choudhuri and Dutt, 1961; Singh and Sharma, 1963; Hurley and Morris, 1964), it would seem to be a relatively frequent constituent of the vaginal yeast flora whether vaginitis is present or not. It has also been isolated from about 7% of newborn infants suffering from oral thrush by Kaul *et al.* (1960) and from the faeces of adults (Silveira and Correia, 1960), sick children (Dhom *et al.*, 1964) and a random sample of healthy children (Pan and Pan, 1964). It is a rare cause of keratomycosis (Hoffmann, 1963) and has been mentioned on a few occasions as being associated with deep-seated infection in humans; viz. candida endocarditis (Andriole *et al.*, 1962; Wolfe and Henderson, 1951) and septicaemia (Garin *et al.*, 1965).

Candida krusei has been isolated from various parts of the alimentary tract of many species of animal, notably horses, cattle, pigs and poultry by Batista *et al.* (1961a, b, c), from diseased fowl by Prophet (1963), from 60 of 100 pigs by Mehnert and Koch (1963), and from free-living baboons by van Uden and do Carmo-Sousa (1962a). Austwick *et al.* (1966) cultured *C. krusei* from material obtained from scouring calves, from the milk of cows suffering from mastitis, and from a bovine foetus in a case of mycotic abortion. Its pathogenicity was demonstrated in mice by Winner and Hurley (1964). There are few records of *C. krusei* having been cultivated from inert sources, but Lodder and Kreger-van Rij (1952) studied 23 strains isolated from a variety of materials including compressed yeast, beer, palm wines, salted cucumbers, apple, fermenting cocoa and paprica.

2. *Cultural Characteristics*

Colonies of *C. krusei* are cream to yellowish in colour and usually smooth, flat and dull; occasional strains are wrinkled. The majority of cells are large and cylindrical, $2 \cdot 5 – 5 \cdot 5$ μ $\times$ $7 \cdot 5 – 21 \cdot 5$ μ, but a few short-oval cells are usually also present. Pseudomycelium is formed in abundance and bears well formed verticils of blastospores in chains. The fermentation and assimilation reactions of *C. krusei* are given in Table VIII (p. 130).

H. CANDIDA SPECIES OF DOUBTFUL PATHOGENICITY

A number of other species of *Candida* have been found closely associated with man and animals. Their incidence is generally low and their pathogenicity, while proven in some cases, is of doubtful significance. Among them are the following:

Candida slooffii van Uden and do Carmo-Sousa
Candida zeylanoides (Cast.) Langeron and Guerra
Candida viswanathii Viswanathan and Randhawa

Candida slooffii was first isolated by van Uden and do Carmo-Sousa (1957) from material contained in the caecum of horses, and later by van Uden (1960) from the digestive tract in 50% of domestic pigs as well as in African bush pigs (*Potamochoerus choeropotamus*). Smith (1967)

TABLE X. *Fermentation and Assimilation Reactions of Three Candida Species of Doubtful Pathogenicity*

Yeast	Fermentation*					Assimilation				
	Glucose	Galactose	Sucrose	Maltose	Lactose	Glucose	Galactose	Sucrose	Maltose	Lactose
Candida slooffii	AG	—	—	—	—	+	—	—	—	—
Candida zeylanoides	—	—	—	—	—	+	—	—	—	—
Candida viswanathii	AG	AG	AG	AG	—	+	+	+	+	—

* A indicates acid production; G, gas production

found it in the hyperplasic mucosal lesions of the *pars oesophagea* of 35 otherwise healthy piglets. *Candida zeylanoides* was cultured from cutaneous sources by Drouhet (1960) and by Mahnke *et al.* (1961) from the lower respiratory tract of human cadavers.

Candida viswanathii was isolated by Viswanathan and Randhawa (1959) from cerebrospinal fluid in a fatal case of meningitis and again later from sputum by Sandhu and Randhawa (1962). The fermentative and assimilative reactions of these three species are shown in Table X.

III. Cryptococcus neoformans

This yeast is the sole causative agent of cryptococcosis, a disease which has been recorded in both animals and man from almost every country in the world. Infections may occur in any part of the body, but involvement of the central nervous system with subacute or chronic meningitis is the form most often diagnosed. Skin lesions may result from dissemination of the disease or from direct inoculation of the fungus, although this is rare. It is now accepted that, as for other serious mycoses such as histoplasmosis and coccidioidomycosis, infection follows inhalation of the fungus and establishment of primary foci in the lungs, with only a small proportion of such infections progressing to the secondary disseminated stage. Since pulmonary infections are usually asymptomatic and transitory, and since *Cryptococcus neoformans* infection, past or present, cannot be satisfactorily diagnosed by serological procedures such as skin testing or complement fixation, the evidence supporting this view is at present largely circumstantial. Emmons (1954, 1955) showed clearly that the natural habitat of *Cr. neoformans* is bird excreta, where it is frequently present in high concentrations, and that exposure of animals and man to this fungus is therefore not uncommon. Many patients with cryptococcal meningitis give a history of a recent respiratory infection (Terplan, 1948; Durant *et al.*, 1960; Haugen and Baker, 1954).

Linden and Steffen (1954) report a case of pulmonary cryptococcosis in a patient hospitalized for pemphigus vulgaris, and which would have probably passed unrecognized had not the patient been under close medical observation. In 1960, Emmons reported on pulmonary infections which had occurred in a number of workmen while cleaning an old tower heavily contaminated with pigeon droppings. At the time of the outbreak, it was believed that another fungal pathogen, *Histoplasma capsulatum*, known to be closely associated with animal (mainly bats) excreta, was responsible but, in retrospect with the knowledge that *Cr. neoformans* is frequently present in high concentrations in bird droppings, Emmons believes that the pulmonary infections of the workers were caused by *Cr. neoformans*. Transmission of infection from man to man does not occur.

The first known case of cryptococcosis occurred in Germany in 1894, and the causal fungus was isolated from a "sarcoma-like" lesion of the tibia (Busse, 1894, 1895, 1896; Buschke, 1895). The disease progressed and, on the death of the patient, multiple lesions were found in lungs, spleen, kidney, bones and skin. Some 40 years later, the identity of the isolate which was obtained was confirmed as *Cr. neoformans* by Benham (1935). During the two decades following the Busse–Buschke case, a number of other cases of cryptococcosis in man were reported; the organism was also isolated from a lesion in the lung of a horse (Frothingham, 1902). The first report of central nervous system involvement was made in 1905 (von Hansemann, 1905) and the first *ante mortem* diagnosis of the disease in this form (meningitis) was made by Versé (1914).

A. PREDISPOSING FACTORS IN CRYPTOCOCCOSIS

The onset of cryptococcosis cannot always be related to abnormal states resulting from other diseases or particular forms of therapy. Nevertheless, because of the ubiquitous nature of the fungus and the low incidence of clinical disease, it seems clear that establishment of recognizable infection may be related to some lowering of host resistance. Various diseases, such as Hodgkin's disease, leukaemia, sarcoidosis, rheumatoid arthritis and diabetes, have been suggested as predisposing to cryptococcosis. Whether these diseases are primarily responsible for lowered resistance or whether this results from the effects of certain drugs in use for their treatment is difficult to establish. Treatment for these diseases frequently involves corticosteroids, and it is well established that these compounds significantly decrease resistance to infections by other fungi such as *Candida* and *Aspergillus* species. The diagnosis of pulmonary cryptococcal infection in persons without symptoms of pulmonary disease suggests that primary pulmonary infections are not uncommon, and that it is the disseminated form for which a lowering of resistance in the host is perhaps a prerequisite. This is the form which usually manifests itself in disease of the central nervous system.

B. INCIDENCE OF CRYPTOCOCCOSIS

Not all cases of cryptococcosis are published, nor indeed, are all cases diagnosed as such. It is difficult, therefore, to indicate the true incidence of the disease. Littman and Zimmerman (1956), in a most interesting and informative account of this aspect of cryptococcosis, emphasized the gaps in our knowledge and pointed out that the greatest incidence of diagnosed cases of cryptococcosis occurs in those hospitals which have a recognized interest in medical mycology; and that reliable data on the frequency of minor, inapparent pulmonary infections due to *Cr. neoformans* are lacking. Nevertheless they show that the number of cases reported by other authors up to 1955 adds up to a total of more than 300, and state that *Cr. neoformans* is the most frequent cause of mycotic meningitis in man. Ajello (1967) reports that the annual number of deaths attributable to *Cr. neoformans* in the United States since 1952 averages 66, and that in the 12-year period 1952–63, the total of fatalities was 788. Littman and Schneierson (1959) postulate that, if the ratio of primary to secondary disseminated infections is the same for cryptococcosis as for coccidioidomycosis, then 5,000–15,000 cases of subclinical or clinical pulmonary cryptococcosis will occur every year in New York City alone.

The incidence of the disease in animals is not known, but there are numerous reports of its occurrence. Ainsworth and Austwick (1959) cite references to infections with *Cr. neoformans* in the horse, dog, and cat and to outbreaks of mastitis in herds of cattle involving 106 of 235 animals in one instance and 50 of 280 animals in another. Infections in 51 goats are recorded by Sutmöller and Poelma (1957) and in a high proportion of koalas examined by Bollinger and Finckh (1962). Ajello (1967) cites references to cryptococcosis in 13 species of animal. If one considers that, as for man, many cases of the disease will go undiagnosed, the incidence in animals and the range of host species must be accepted as greater than this.

C. PATHOLOGICAL ASPECTS

Unlike many fungi pathogenic for man and animals, *Cr. neoformans* does not undergo radical changes in morphology when parasitic in animal tissues. It is present as spherical to ovoid cells of variable size, but usually in the region of 5–10 μ diameter, surrounded by the characteristic mucinous capsule which, with the majority of staining procedures, fails to stain and appears as a halo around the cell. For this reason, the mucicarmine and alcian blue stains are useful, and may serve when necessary to assist differentiation of *Cr. neoformans* in tissue from other pathogenic yeasts (Fig. 15). The characteristic cryptococcal lesion (Baker and Haugen, 1955) consists of an aggregate of encapsulated cells (Fig. 16) intermixed with a reticulum of lightly formed connective tissue which enlarges to compress surrounding tissues. In the central nervous system, the subarachnoid space is often distended by gelatinous material and the brain substance frequently contains gelatinous cysts (Hildick–Smith *et al.*, 1964). These cysts are sometimes centred around a blood vessel (Fig. 17). Such accumulations of encapsulated cells are also found in infected lungs and cause considerable distension of alveoli

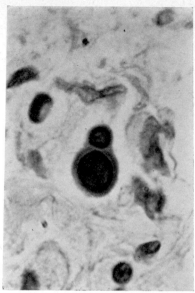

FIG. 15. *Cryptococcus neoformans* in human lung tissue. The section shows budding cell with the capsule stained by alcian blue. Magnification, × 2,000.

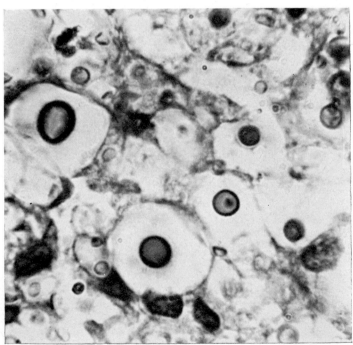

FIG. 16. *Cryptococcus neoformans* growing in human meninges. The section was stained with periodic acid-Schiff reagent. Magnification, × 2,000.

(Littman and Zimmerman, 1956). In cryptococcal mastitis in cattle, *Cr. neoformans* cells are present in abundance in the granulomata which develop in the mammary glands and in the udders (Innes *et al.*, 1952; Littman and Zimmerman, 1956). The capsule size is usually increased on transfer from *in vitro* to *in vivo* (Neill *et al.*, 1950; Bergman, 1961) and may be up to five times the diameter of the cell (Littman and Zimmerman, 1956). There is no doubt that the capsular material plays

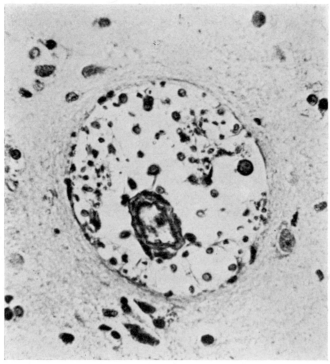

Fig. 17. *Cryptococcus neoformans* in perivascular cysts in human brain. The section was stained with periodic acid-Schiff reagent. Magnification, × 500.

an important role in the relationship between this pathogen and its hosts. All strains of *Cr. neoformans* are pathogenic, although the virulence, as measured by the number of viable cells required to establish infection or cause death of experimental animals within a fixed period of time, may vary. It has been reported by Drouhet *et al.* (1950, 1961) and Grose *et al.* (1968) that virulence increases with capsule size, although this has not been the experience of other workers who have investigated this aspect (Emmons, 1952; Kao and Schwarz, 1957; Littman and Tsubura, 1959; Hasenclever and Mitchell, 1960; Bergman, 1961, 1965).

6

However, Hasenclever and Emmons (1963) and Evans (1969) found that, in general, the virulence of isolates from bird droppings is lower than those obtained from cryptococcal infections in man and animals, and dropping isolates usually have a much smaller capsule (Kao and Schwarz, 1957; Emmons, 1962; Evans, 1969) than those isolated from disease processes.

The capsule is also important, according to Drouhet *et al.* (1950, 1961), in relation to the tissue reaction which is involved. While there is no histological reaction which is peculiar to infection by fungi and which may be used to distinguish a mycotic lesion from other inflammatory diseases (Symmers, 1968), cryptococcal infections are peculiar in that the tissue reaction is frequently minimal or absent. Drouhet *et al.* (1950, 1961) have found in both experimental animals and man, that if the yeast is encapsulated there is generally no tissue reaction whereas decapsulated *Cr. neoformans* cells are accompanied by host cellular reactions.

Cryptococcal infections also differ immunologically from other important mycoses. There is no simple diagnostic test based on dermal sensitivity such as is available for coccidioidomycosis and histoplasmosis, and the conventional serological methods of demonstrating antibody fail or give equivocal results (Seeliger, 1964). More complex procedures, such as that combining the complement-fixation and fluorescent-antibody techniques (Walter and Atchinson, 1966) and latex-particle agglutination (Bloomfield *et al.*, 1963; Gordon and Vedder 1966), show promise for the future, but are still largely in the experimental stages and require further evaluation. Confirmation of a diagnosis of cryptococcosis, at present, is therefore dependent on mycological procedures, viz. the recognition or culture of *Cr. neoformans* from biopsy or necropsy material, or, for instance, from cerebrospinal fluid where the yeast may usually be found if the central nervous system is involved.

D. EPIDEMIOLOGICAL ASPECTS

Prior to the discovery by Emmons (1954, 1955) of the association of *Cr. neoformans* with bird droppings, the yeast was isolated as a saprophyte from natural sources on very few occasions, the best known instance being that of Sanfelice (1894) from peach juice. On the other hand, there were several reports of its isolation from sources closely associated with humans and animals or their excreta or secretions (Klein, 1901; Carter and Young, 1950). For some time, it was therefore believed that cryptococcosis was probably of endogenous origin. Emmons's work has clearly established the important role of bird droppings in relation to the distribution of *Cr. neoformans* and the exogenous

nature of infections. Ajello (1967) believes that the isolations by Sanfelice, Klein, and Carter and Young, can probably be explained by airborne contamination. *Cryptococcus neoformans* has now been isolated
from the excreta of a number of common bird species in many different
countries with widely differing climatic conditions. Pigeons, which are
present in high numbers and in close association with man in many
large cities, have been mainly studied and most frequently implicated.
A considerable proportion of their old nests and accumulations of droppings at roosting sites contain *Cr. neoformans*, as is illustrated by the
findings of Emmons (1955) that 57%, Kao and Schwarz (1957) 38%,
and Emmons (1960) 69%, of samples contained this yeast. Captive and
domestic birds may also be implicated as shown by the isolation of *Cr.
neoformans* from the droppings of canaries (Staib, 1961), show birds
(Bergman, 1963), chickens (Ajello, 1958), and pigeon lofts (Swatek *et al.*,

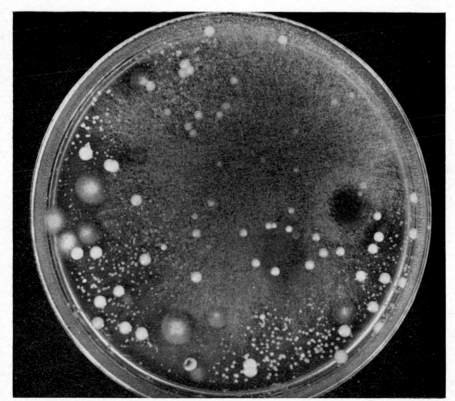

FIG. 18. A Petri dish showing numerous colonies of *Cryptococcus neoformans*
(small translucent colonies) and *Candida albicans* (large white colonies) isolated by
direct culture from pigeon droppings on creatinine-agar medium. Reproduced by
permission of E. G. V. Evans.

1964; Symmers, 1967; Evans, 1969). Evans (1969) has also shown that the fungus will persist for periods of more than two to three years in pigeon lofts and that the concentration of viable cells present in droppings, as noted by Emmons (1962), is considerable (Fig. 18). Emmons (1962), from a sidewalk sample, found 50 million viable cells, and Evans (1969), from a pigeon-loft sample, found 172,000 viable cells, each per gram dry weight of droppings. It is not difficult to envisage, therefore, that many persons will be exposed to high concentrations of *Cr. neoformans*, and this is particularly true for those who have a close association with birds or their excreta.

While the role of birds, or their excreta, in the conservation and establishment of reservoirs of *Cr. neoformans* is thus well established, they themselves are not apparently prone to infection. Emmons (1955) was unable to find the fungus in the viscera or digestive tract of any of 20 young pigeons which he examined. However, Staib (1962a) has shown that *Cr. neoformans* can persist in the intestinal tract for up to eight days, and Littman *et al.* (1965) found that intracerebral inoculation of pigeons produced a widespread fatal systemic infection involving such organs as the brain, liver, spleen, lungs and intestine. The body temperature of birds, which is higher than that of man, will clearly resist development of infection. Kuhn (1949) reports that *Cr. neoformans* will not grow at 39·4°. Moreover, Kuhn (1949) using mice, and Kligman *et al.* (1951) using chick embryos, have shown that a high ambient temperature modifies or prevents infection following inoculation with *Cr. neoformans*.

Cryptococcus neoformans is undoubtedly present in soils and other habitats without avian associations. Evenson and Lamb (1964) recovered it from the slime flux of mesquite trees, McDonough *et al.* (1961) from wood, and Emmons (1951), Ajello (1958) and others from soil. However, it is the favourable conditions for growth afforded by bird droppings, thus leading to the development of high concentrations, which clearly constitute the main reservoirs of infection for man and animals.

E. MYCOLOGICAL ASPECTS

Cryptococcus neoformans forms neither ascospores nor ballistospores and is classified in the family Cryptococcaceae. Within the family, the genera *Cryptococcus* and *Torulopsis* may be differentiated from the other members by the shape of their cells, which are round to oval, and their failure to form pigments, mycelium or pseudomycelium. These two genera differ in that, in *Torulopsis*, fermentative ability is usually present and capsule formation is exceptional while, in *Cryptococcus*, fermentative ability is absent and capsule formation is the rule. The cryptococci also form an extracellular starch-like substance which stains blue with iodine (Aschner *et al.*, 1945).

In common with most fungi pathogenic for man and animals, *Cr. neoformans* grows easily on media containing an organic source of nitrogen. Thiamine, which was noted by Reid (1949) to stimulate growth, and by Littman (1958) to be an essential growth factor, is supplied in sufficient quantity in the peptone or malt extract usually used for isolation or conservation media. It has been suggested that the thiamine requirement may explain the affinity which *Cr. neoformans* has for the central nervous system since brain and spinal fluid contain large amounts of this vitamin (Littman, 1958).

Most strains of *Cr. neoformans* form a glistening mucoid, yellow-brown colony on malt-extract agar on which capsule formation, which determines the gross morphology, is usually good. Some isolates may not have a very pronounced capsule and, in others, abnormally large capsules may be produced. On certain media, e.g. glucose-peptone agar, capsule formation, even of well capsulated strains, is reduced and the colony may be pasty and cream coloured.

Evans (1969) has found that, as previously reported by Kao and Schwarz (1957) and Emmons (1962), the capsule size, in general, is small in isolates from pigeon droppings compared to those from disease processes in man and animals. He has also found (Evans, 1969) that small-capsule strains maintained on malt extract-, creatinine- (Staib, 1963) or capsule- (Littman, 1958) agar at 28° produce sectors of large-capsule varieties after a relatively small number of transfers during a period of approximately seven to ten weeks. Apparently the temperature is an important factor, since he noted that sectoring at 37° was extremely rare but, if strains maintained at this temperature over the same period of time were subcultured to 28°, then the large-capsule varieties were produced. Production of large-capsule varieties by sectoring or spontaneously during culture transfers has also been reported by other workers (Drouhet and Couteau, 1951; Emmons, 1952; Littman and Zimmerman, 1956) but details of the conditions influencing this were not given. Littman (1958) formulated his capsule-agar on the basis of the requirements of *Cr. neoformans* for vitamins, amino acids, and carbohydrates, and found that all strains tested on his medium at 37° produced a higher proportion of large-capsule cells than when grown on other media.

Cells of *Cr. neoformans* are round or almost so, and usually measure 4–6 μ in diameter (excluding the capsule) although, in some isolates, the majority may be above or below these limits. Daughter cells appear as small projections on the side of the parent. The connection between the parent and daughter cell always remains narrow and, although the bud secretes a capsule, a single capsule frequently surrounds both cells for a period of time. Microscopic examination of cultures is best carried out in Indian ink preparations (Fig. 19) which clearly reveal the capsule. Other mountants are much less satisfactory for this purpose, although there are excellent staining methods (e.g. periodic acid-Schiff, muci-carmine) for investigation of pathological material suspected to contain *Cr. neoformans*.

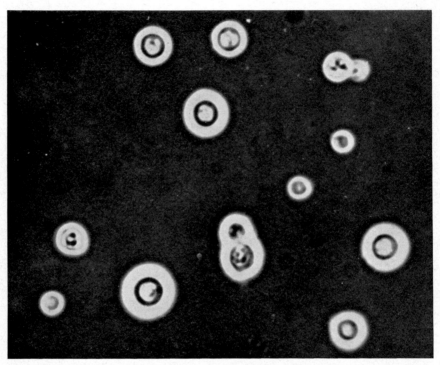

FIG. 19. *Cryptococcus neoformans* in cerebrospinal fluid. The preparation was mounted in Indian ink. Magnification, × 1,000.

Mycelium formation does not usually occur, but occasionally in ani-mal tissue (Freeman, 1931; Misch, 1955) or under extremely adverse conditions for growth *in vitro*, such as at very high or low pH values or at high temperatures, short "germ-tube-like" projections are formed Evans, 1969). Pseudomycelium formation is reported during *in vitro* growth of a variant by Drouhet and Couteau (1951). One very unusual strain in which extensive mycelium formation took place was recently described by Shadomy and Utz (1966). Occasional reports of sexual reproduction by *Cr. neoformans* (Todd and Herrmann, 1936; Redaelli *et al.*, 1937; Benham, 1955) have not been confirmed by other workers.

TABLE XI. *Assimilation Reactions of the Genus* Cryptococcus *According to Kreger-van Rij* (1963)

Yeast	Nitrate	Galactose	Sucrose	Maltose	Lactose
		Ability to Assimilate			
Cryptococcus laurentii	−	+	+	+	+
Cryptococcus neoformans	−	+	+	+	−
Cryptococcus luteolus	−	+	+	+	−
Cryptococcus gastricus	−	+	−	+	±
Cryptococcus skinneri	−	+	−	−	−
Cryptococcus albidus	+	+	+	+	+
Cryptococcus terricolus	+	+	+	+	+
Cryptococcus diffluens	+	+	+	+	−
Cryptococcus terreus	+	+	−	+	+

The assimilation pattern for *Cr. neoformans* and other species of the genus *Cryptococcus* is given in Table XI. Various other identification techniques have been devised not only to supplement the results of assimilation, when as is often the case these are not sufficiently clear-cut, but also to reduce the time required for identification of this important pathogen. In common with many other fungal pathogens, the optimum growth temperature of *Cr. neoformans* is lower than 37° (Kuhn, 1939, reports it as 29°). However, the yeast has the ability to grow at 37° and this is a property not shared by other members of the genus. It is probably the simplest test of all to perform but, since occasional strains of other species, e.g. *Cr. laurentii*, *Cr. luteolus*, may make limited growth at 37°, care, even with this simple test, must be exercised by the inexperienced worker.

One of the most useful criteria for identification of *Cr. neoformans* is probably that recently evolved by Staib (1962b, c; 1963) who, because of the established relationship between this fungus and bird droppings, carefully investigated the nutritional properties of droppings. He found that pigeon droppings, when incorporated in a medium, adequately supplied the nitrogen requirements and, following a study of the various chemical constituents, discovered that creatinine, a constituent of bird urine, could be used by *Cr. neoformans* as a sole source of nitrogen and that it was unavailable to most other yeasts including other species of the genus *Cryptococcus* (Fig. 20).

Subsequently it has been found (Kreger-van Rij and Staib, 1963; van Uden *et al.*, 1963) that a few strains of *Cr. laurentii* and certain

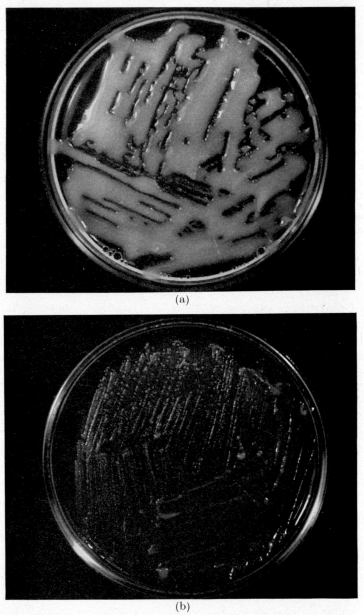

(a)

(b)

FIG. 20. Growth of *Cryptococcus neoformans* (a) and *Cryptococcus diffluens* (b) on solid medium containing creatinine as the sole nitrogen source. Plates were incubated for 5 days at 28°. Reproduced by courtesy of E. G. V. Evans.

species of *Debaryomyces*, *Torulopsis* and *Candida* can utilize creatinine as a nitrogen source. However, for some of these species, it is available only when they are grown in a liquid medium and the method remains largely selective. In addition to providing a most useful diagnostic tool, this work of Staib provides an explanation for the close association of *Cr. neoformans* and bird droppings.

Another simple procedure which is of considerable value for identification of members of the genus *Cryptococcus* is that described by Seeliger (1956) for detecting urease production. All species of *Cryptococcus* produce urease, and streak inoculations on Christensen's (1946) urea-containing medium will lead to the development of a deep red colour which gradually spreads throughout the medium within one or two days at room temperature. The test is thus most useful for separation of cryptococci from other non-fermenting yeasts as a preliminary procedure to specific identification, and its value has been confirmed by others (Lacaz *et al.*, 1958; Ajello *et al.*, 1963). Finally, although the virulence may vary from strain to strain, all isolates of *Cr. neoformans* are pathogenic and, when necessary, a test for pathogenicity should be

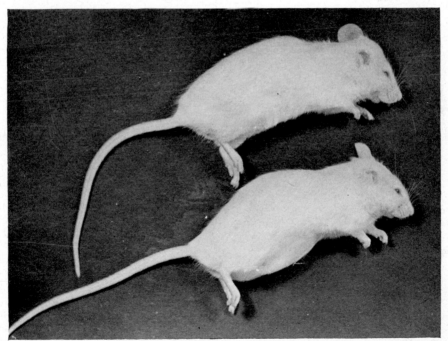

FIG. 21. Experimental infection with *Cryptococcus neoformans* in mice, showing in the upper photograph the domed skull in an infected mouse, and in the lower photograph an uninfected control mouse. Reproduced by courtesy of E. G. V. Evans.

6*

made. This should be done by inoculation of at least three 6–10 week-old mice intracerebrally with a saline suspension of the yeast. The inoculated animals may die within a few days, but it is not unusual for them to survive for a number of weeks. Development of a domed skull (Fig. 21) may or may not take place. The top of the skull of animals which die, or which at the end of 2–3 weeks have survived, should be removed. The material of the brain will contain the typical capsulated cells of the yeast if it is *Cr. neoformans*.

IV. The Genus Torulopsis

Although the pathogenicity of members of this genus which have been isolated from both human and animal sources has been much disputed, the frequency with which some of them, notably *T. glabrata* and *T. famata*, have been isolated from these sources inclines one to attribute to them, at least tentatively or until disproved, a pathogenic role.

A. TORULOPSIS GLABRATA

1. *Incidence and Distribution*

Among species of *Torulopsis* isolated in culture from skin, epidermal scales, dandruff, and nails by Batista *et al.* (1961c) in Recife, Brazil, this species comprised 4·2% and, from genital secretions and fluids, 40·8%. Petru (1965), in a survey of 6,258 women in Czechoslovakia, estimated its incidence in vaginal secretions at 21·77% of the total yeast flora of the vagina. Artargaveytia-Allende and Silveira (1961) found the organism in Uruguay in 61 patients; 20 of the isolates were from the urethra and vagina, 11 from urine, 24 from faeces, five from the appendix and one from bile. It formed 14% of 968 cultures of yeasts obtained by Stenderup and Pedersen (1962) from human sources at Aarhus in Denmark; and Mackenzie (1961) found in Northern Ireland that, amongst 247 yeast isolates from human sources representing 23 species, it formed 12·4%, and that it was the species of yeast most frequently isolated from urine specimens. Some cases of systemic infection due to *T. glabrata* are also reported in the literature. Zech *et al.* (1965) found it in seven successive blood cultures made from a patient suffering from acute renal insufficiency. Negroni *et al.* (1965) also cultured it from the blood of a girl aged 17, suffering from myasthenia gravis. She had developed a mediastinal empyema and pleural effusion after an operation for removal of the thymus and had been treated with antibiotics. Other cases of septicaemia in which this species was cultured from the blood are reported by Louria *et al.* (1960), Ahearn *et al.* (1966), Grimley *et al.* (1965), and by Minkowitz *et al.* (1963).

Although records are few in number, *T. glabrata* has also been reported from animals. Van Uden and do Carmo-Sousa (1962b) estimated that it formed 42·3% of yeasts in the caecal contents of free-living baboons (*Papio cynocephalus*), animals which live on a diet rich in starch and simple carbohydrate. It was noted by Austwick *et al.* (1966) among yeasts obtained from the digestive tract of piglets suffering from "tympany", and in milk from cows suffering from mastitis by Scholer (1963b) and by Mehnert (1963).

Attempts to inoculate healthy animals with *T. glabrata* have been, generally speaking, unsuccessful, but Hasenclever and Mitchell (1962) succeeded in infecting experimental mice after they had been physiologically altered by such chemicals as cortisone acetate, and alloxan monohydrate or by X-rays. Results showed proliferation of this yeast in the liver, spleen and kidneys of the mice so treated. While this proliferation appeared to be dependent on the continued administration of the drugs or X-rays and gradually decreased when it was discontinued, the results are not without significance since similar abnormalities can be, and not infrequently are, induced by certain therapeutic regimens.

There is little doubt that *T. glabrata* is a species of importance and fully merits consideration among the pathogenic yeasts.

2. Cultural Characteristics

Colonies of *T. glabrata* are grey to brown in colour, smooth and glistening, and with a pasty consistency. Cells are sub-globose to ovoid, and measure $2·5–3·5\ \mu \times 4–5\ \mu$. Pseudomycelium is not formed, and only glucose is fermented and assimilated.

B. TORULOPSIS FAMATA

1. Incidence and Distribution

Torulopsis famata has also been isolated frequently and often as a high proportion of yeasts obtained from human and animal sources. In man, it has been obtained from a variety of skin lesions by Vieira and Batista (1962); from genital secretions, skin samples and faeces, by Batista *et al.* (1961c); from the genital skin of males by Koch *et al.* (1959); from the urinary tract by Perez and Gil (1960); in bile by Batista *et al.* (1958); in appendices removed from patients with appendicitis by Della Torre *et al.* (1959); in tonsils removed on account of tonsillitis by Heymer (1958); and from decayed teeth by Passos (1961).

In animals, it is widespread as a commensal in captive birds, especially of the parrot tribe (Kaben and Preuss, 1967a, b), and in the faeces of mules (Batista *et al.*, 1961d). It has also been recovered from the mouths of vitamin A-deficient pigs (Hajsig *et al.*, 1962), and from the udders of cows suffering from mastitis (Funke, 1960). As regards other sources,

Lodder and Kreger-van Rij (1952) mention strains obtained from chilled beef and moist tobacco.

2. *Cultural Characteristics*

Colonies of *T. famata* are cream coloured, smooth and shiny. Cells are globose or oval, and measure 2·5–5·5 μ × 3·5–7·0 μ. Pseudomycelium is not formed, or rarely is very primitive in nature. Glucose may be fermented but only very weakly. Glucose, sucrose, maltose and galactose are assimilated.

V. The Genus Pityrosporum

A. INCIDENCE AND DISTRIBUTION OF PITYROSPORA

The genus *Pityrosporum*, in the family Cryptococcaceae, was created by Sabouraud (1904) for the organism which Malassez (1874) observed in cases of pityriasis simplex, and which he described as budding "spores" of various shapes. It includes yeasts found only in close association with the skin of man and animals. Lodder and Kreger-van Rij (1952) describe two species, *Pityrosporum ovale*, the type species, and *Pit. pachydermatis* which was not available at the time of their study and has apparently not been isolated since 1952.

The original isolation of this latter species was made from the inflamed skin of an Indian rhinoceros by Weidman (1925) who showed clearly that it differs from *Pit. ovale*. His observations were confirmed by other workers including Lodder (1934). Other strains of *Pityrosporum* have been reported by Manktelow (1959) from the external auditory canals of more than 100 dogs, and these, like *Pit. pachydermatis*, grew readily on media without a fat supplement. In 1951, Gordon described *Pit. orbiculare* which he had isolated from the skin of normal individuals and those suffering from the disease pityriasis versicolor.

As a result of the work of Martin-Scott (1952) and Rocha *et al.* (1952) it seems clear that *Pit. ovale* has no pathogenic properties. However, although still not exactly understood, there is evidently a relationship between *Pit. orbiculare* and the disease pityriasis versicolor, and this, together with the very restricted distribution of these fungi and their exacting and unusual growth requirements, render them of particular interest and worthy of consideration.

B. PITYROSPORUM OVALE

Pityrosporum ovale was first described by Rivolta (1873) under the name *Cryptococcus psoriasis* and further, more detailed, descriptions were given by, among others, Malassez (1874), Klaman (1884) and Sabouraud (1902, 1904). Many attempts were naturally made to obtain the organism in culture but with little success, and Sabouraud (1904) stated categorically that those who had reported isolation of the organism were mistaken. According to Benham (1947) who gives a detailed chronological account of the various reports on *Pit. ovale*, the work of Ota and Huang (1933) produced the essential information concerning a growth requirement for lipid and, as a result of this, they were able to confirm reports of its isolation by Castellani (1925) and Acta and Panja (1927) as authentic. Benham (1939) was able to isolate eight strains of *Pit. ovale* and to maintain five of them in subculture on a medium supplemented with lipids and lipid-like substances. In 1947 Benham, by culturing the organism on a synthetic medium to which lipids were added, conclusively showed the essential requirement for a lipid substance.

1. *Isolation and Cultural Characteristics*

Lipids are best added to the medium in the form of ether extracts of lanolin, butter, or oleic acid pipetted over the surface of the agar. The ether evaporates, leaving a thin film of the compounds on the surface of the agar (Benham, 1947). The constitution of the basal medium is apparently not of particular importance. Benham used wort agar which she states is very suitable for isolation but, as stated above, a synthetic medium may be satisfactory for subcultures, and the essential factor is the overlay of a lipid substance. A medium which has begun to dry out is not satisfactory for supporting growth of *Pit. ovale*. On suitable medium, the colonies appear after four to five days at 37°. Cultures are cream to orange coloured, and soon become wrinkled and dry in appearance. In liquid medium *Pit. ovale* grows both in the form of granules which settle to the bottom of the culture vessel and as a film or pellicle on the surface (Benham, 1947).

In addition to its requirement for a lipid or a fatty acid, *Pit. ovale* has been shown by Benham (1947) to have other specific and interesting growth requirements. She found oleic acid to be the most suitable fatty acid to supply the lipid substance requirement and used it in various synthetic media to investigate other growth requirements. In this way she showed a requirement for either asparagine or ethyl oxaloacetate. When asparagine was the nitrogen source, she found that growth was enhanced by addition of thiamine, but when ethyl oxaloacetate was supplied, ammonium chloride was suitable as a nitrogen source, and not only were asparagine and thiamine not necessary, but they did not appreciably increase growth. *Pityrosporum ovale* is not particularly sensitive to pH value and grows well over a wide range of hydrogen-ion concentration. The optimum growth temperature is 35–37°.

The cells of *Pit. ovale* vary in size and shape but are most frequently elongate, measuring 4–5 × 2–3 μ (Fig. 22). A proportion of them are two-celled with a large cell joined to a smaller by a neck-line projection to give the "bottle" form.

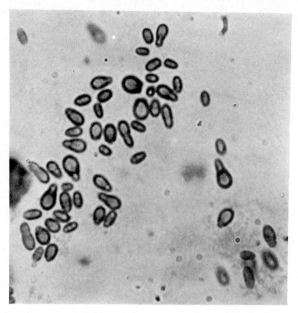

FIG. 22. *Pityrosporum ovale* isolated from a scraping of skin from a patient with seborrhoeic dermatitis. The preparation was stained with periodic acid-Schiff reagent. Magnification, × 2,000.

C. PITYROSPORUM ORBICULARE

1. *Relationship with pityriasis versicolor*

Pityrosporum orbiculare was isolated from the skin of 15 individuals by Gordon (1951) and, since he satisfied himself that no previous report of the culture of the organism existed, he described it as a new species of *Pityrosporum*. Like *Pit. ovale*, it has a growth requirement for lipid, but it differs from this species morphologically and in several physiological characteristics. Gordon points out that, although no previous report of the culture of *Pit. orbiculare* exists, it was almost certainly observed and described by a number of workers in its natural habitat, the skin of man. In particular, he believes that it is the same organism as that described by Bizzozero (1884) under the name *Saccharomyces sphaericus* since he was able to observe cells of the size and shape of those described by Bizzozero on many occasions in the skin scales of normal individuals. However, 13 of Gordon's 15 isolates were obtained from individuals suffering from pityriasis versicolor, a disease which is manifest by areas of fine-scaling lesions most frequently of the neck and trunk. In dark skins, the lesions appear to be paler than the normal skin and, in white races, have a brownish appearance (Marples, 1965). They are discrete at first but eventually coalesce to form uniform patches (Fig. 23). In normal skin, *Pit. orbiculare* is present as spherical cells, but the organism present in skin scales of pityriasis versicolor consists

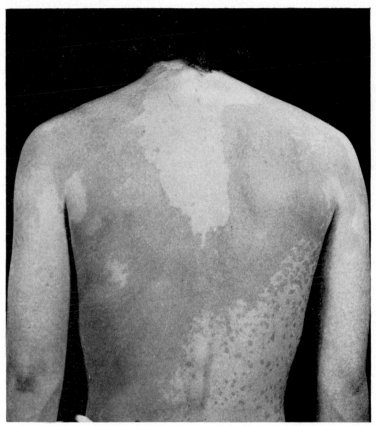

FIG. 23. Pityriasis versicolor due to *Malassezia furfur* showing discrete and confluent pigmentation. Reproduced by courtesy of the General Infirmary at Leeds, England.

of clusters of spherical spore-like structures, associated with short lengths of stout hyphae, and is known as *Malassezia furfur*. It has, as far as is known, never been obtained in culture in this form. However, there is now a considerable weight of evidence in support of the view that *Malassezia furfur* and *Pit. orbiculare* are one and the same organism. Although Gordon (1951) was unsuccessful in his attempts to produce the disease in human volunteers and laboratory animals by inoculation of pure cultures of *Pit. orbiculare* to scarified skin, and did not see the characteristic hyphal elements of *Malassezia furfur*, cells resembling *Pit. orbiculare* were seen in skin scales of the inoculated regions after three to four weeks. Some ten years later, moreover, Burke (1961), by injecting culture suspensions of *Pit. orbiculare* intradermally, was able to produce experimental infections, confirmed clinically and by microscopy of skin scrapings. She also noted that one of her isolates of *Pit. orbiculare* produced hyphae spontaneously *in vitro*, although attempts to induce this change by altering growth conditions were unsuccessful. Burke's (1961) successful inoculations were obtained in patients who were abnormal because of some underlying disease, such as tuberculosis and/or because of treatment with corticosteroids for their illnesses. The one exception later became ill. She postulated, on the basis of her results, that a basic biochemical or physiological change in the skin or its secretions renders individuals susceptible to pityriasis versicolor. Another approach to the problem is that of Keddie and Shadomy (1963) who showed that the morphology of *Pit. orbiculare* in culture and *Malassezia furfur* in skin scales to be very similar with respect to the method of spore production, both forming phialide-like conidiophores as part of their growth cycles. This method of spore production by *Malassezia furfur* was also noted by Vanbreuseghem (1954). Keddie and Shadomy (1963) also showed, by means of fluorescent antibody procedures, the presence of common antigenic material in the various growth phases of *Malassezia furfur* and *Pit. orbiculare*. There is thus considerable evidence in favour of the view that *Pit. orbiculare* in persons who, for one reason or another are abnormal, is capable of forming an invasive mycelial phase and causing pityriasis versicolor. The final decision awaits further investigation.

2. *Isolation and Cultural Characteristics*

Isolation of *Pit. orbiculare* is best made at 30–37°; no growth occurs at 25° in two weeks. It is recommended that the isolation medium be supplemented with penicillin (20 units per ml) and streptomycin (40 units per ml) to diminish bacterial contamination. Slopes of glucose-peptone agar (5 ml) overlaid with 2 ml of sterile olive oil provide a suitable medium, and it is suggested that the tubes be sloped in the incubator to keep the oil over the agar surface (Gordon, 1951). Oleic acid, which is suitable as a lipid substrate for *Pit. ovale*, will not support growth of *Pit. orbiculare* for which olive oil, linseed oil, myristic acid and stearic acid are listed by Gordon as the best sources for meeting this growth requirement. At 37°, growth is visible in four to five days and subcultures should be made at three-week intervals, ensuring that the oil covers the slant at all times. Gordon (1951) reports survival of a strain over 23 transfers during 14 months at 37°. As with *Pit. ovale*, old or dried-out media are unsatisfactory. Colonies appear first as fine, white granules and eventually spread until the entire slope is covered with a white to cream coloured layer.

The cells range from 2·1–4·8 μ in diameter, mostly 2·3–3·8 μ. They are regularly spherical with a "double-wall" appearance and with little or nothing visible in the interior of the cell. Buds, which are usually produced singly, are spherical to slightly elongate, and are attached by a narrow isthmus, a further distinction from *Pit. ovale* in which buds are attached by a broad base.

3. *Malassezia furfur* (*Robin*) *Baillon*

This fungus is known at present to occur only in skin scales of individuals suffering from pityriasis versicolor, an infection characterized by superficial white, brown or fawn-coloured lesions which are non-inflammatory, sharply marginated and appear as thin bran-like or furfuraceous scaly areas. It was first described by Robin (1853) under the

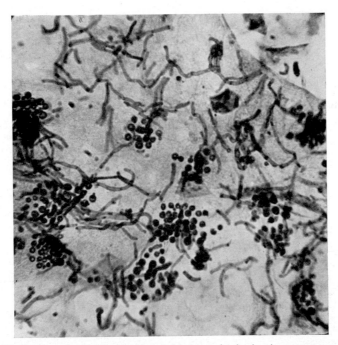

FIG. 24. Hyphae and spherical cells of *Malassezia furfur* in a preparation from the pigmented area of skin from a patient with pityriasis versicolor. The preparation was stained with periodic acid-Schiff reagent. Magnification, × 500.

name *Microsporon furfur*. The name was changed by Baillon (1889) to *Malassezia furfur*. Numerous attempts have been made by a number of workers over a long period of time to isolate the fungus. As reported above, *Pit. orbiculare* has been isolated frequently but, with the exception of Burke (1961), no one has been able to satisfy Koch's postulates and, although the evidence in favour of assuming that *Pit. orbiculare* and *Malassezia furfur* are the saprophytic and parasitic stages respectively of the same fungus is strong, it is still inconclusive and further investigations are required. *Malassezia furfur* is present in skin scales as short, stout curved hyphae, 2·5–4·0 μ in diameter, associated with clusters of spherical cells with thick walls up to 8 μ in diameter. These structures may be seen in scales in which the keratin has been cleared by mounting in potassium hydroxide solution (20%), but diagnosis is best made by staining with methylene blue or using the periodic acid-Schiff method (Fig. 24).

VI. The Genus Trichosporon

Classified in the family *Cryptococcaceae*, the genus *Trichosporon* is distinctive in producing, in addition to budding cells and pseudomycelium, true mycelium which separates to form arthrospores. The arthrospores are frequently produced in profusion and, with the mycelium, they are the predominant structures seen in slide mounts. The members of this genus are thus usually distinguished easily from other yeasts. One species produces an infection of hair known as white piedra, and certain other members of the genus are not infrequently found closely associated with man and his environment. Although of uncertain pathogenic significance, they merit consideration.

A. TRICHOSPORON CUTANEUM

1. *Incidence and Distribution*

White piedra, a condition in which soft white nodules are formed at various intervals on hair, has been generally attributed to *Trich. beigelii*. It is now generally accepted, however, that this species is identical with *Trich. cutaneum*, which name takes precedence. According to Kaplan (1959), the original description of *Trich. beigelii* was so incomplete that it could apply to any *Trichosporon* species, and it was for this reason that Lodder and Kreger-van Rij (1952) considered *Trich. cutaneum* as the valid name for the causative agent of white piedra.

In white piedra, the fungus is present on the hair shaft as a nodule composed of a mass of hyphae which are fragmented to a considerable degree into arthrospores and held together by a cement-like material. Kaplan (1959) states that careful study may disclose blastospores. The fungus also invades the cortex of the hair, causing it to fracture and break off at the point of attack. White piedra is a relatively rare disease, occurring usually in temperate zones in contrast to the tropical distribution of the more common black piedra caused by an ascomycete, *Piedraia hortai*. In man, *Trich. cutaneum* usually invades the hairs of the beard and moustache, and is occasionally found on terminal hairs situated elsewhere. It has been reported in horses, mainly on the thick hairs of mane and tail, and in the spider monkey (Kaplan, 1959).

The source of infections is uncertain, and it has been suggested (Ajello, 1962) that it is a soil organism. In support of this view are reports of isolations of *Trich. cutaneum* from soil, sewage and wood-pulp (Kaplan, 1959). However, Lodder and Kreger-van Rij (1952) note that, of the strains they studied, 18 of known origin were from human sources, and, of ten strains of unknown origin, nine were most probably from human sources. In our experience, while this fungus is encountered rather rarely, isolates, when they are obtained, originate mainly from sources such as sludge from drains of swimming pools or communal bathing places and skin scrapings from feet, suggesting a natural habitat associated closely with keratinous debris.

2. *Isolation and Cultural Characteristics*

Trichosporon cutaneum develops easily on a medium containing malt extract or glucose-peptone and, after a few days, the colony which is rapid-growing appears moist and cream-coloured. With age, the colonies become wrinkled and tough and dull greyish in appearance. Certain strains also develop a hairy surface growth composed of vertically-growing twisted hyphae.

Microscopical examination shows good development of mycelium and profuse arthrospore production. These spores are mainly rectangular in shape and measure $2-4$ μ $\times$ $3 \cdot 5-9 \cdot 0$ μ. Blastospores, which are within the same size range, are formed much less frequently, singly or in small chains or clusters.

Strains of this species have no fermentative ability but assimilate glucose, galactose, sucrose, maltose and lactose. They are unable to assimilate nitrate. Growth with ethanol as sole source of carbon is very weak or absent, and ability to split arbutin is also weakly positive or absent.

B. TRICHOSPORON CAPITATUM

1. *Incidence and Distribution*

Although only one of the five strains of this species studied by Lodder and Kreger-van Rij (1952) was of human origin, it is our experience that this species is not uncommonly isolated from sputa of bronchitics or individuals with other chest complaints. The strain of human origin studied by Lodder and Kreger-van Rij was isolated from a patient suffering from asthma; the others were from wood pulp and compressed yeast. While the presence of *Trich. capitatum* in sputa is of uncertain or even doubtful significance, its inclusion here is considered advisable since it bears some resemblance to *Geotrichum candidum*, the causal agent of a pulmonary or bronchial disease described as a chronic bronchitis (Emmons *et al.*, 1963) and may be confused with it (Gilbert and Fetter, 1962). Furthermore, Gilbert and Fetter (1962), following the misdiagnosis of their isolate which came from a patient with pulmonary tuberculosis, showed that their isolate was pathogenic by producing infection in animals experimentally. Of 15 rabbits inoculated intravenously with 1 ml of a 4% suspension of their strain of *Trich. capitatum*, nine died spontaneously and the fungus was seen in the tissues of the remainder which were killed at intervals from two to 16 days. The major sites of infection were in the kidney, brain, and heart.

2. *Isolation and Cultural Characteristics*

On malt extract-agar, *Trich. capitatum* is relatively fast-growing, forming a yellowish or pale-white colony depending on the strain. The surface of the colony may be shiny or dull, and the texture of variable toughness. True mycelium, arthrospores and blastospores are formed in profusion. It has no fermentative ability, and assimilates only glucose and galactose (weakly). Assimilation of potassium nitrate is absent, and the yeast is unable to split arbutin. Growth on ethanol as sole carbon source is absent or very weak.

VII. The Genus Rhodotorula

The genus *Rhodotorula*, in the family Cryptococcaceae, includes all yeasts which form a carotinoid pigment (Lodder and Kreger-van Rij, 1952). They develop mycelium only occasionally, and this is primitive in nature. Seven species of *Rhodotorula* are recognized, viz. *Rh. mucilaginosa* (Demme) Lodder, *Rh. rubra* (Jorg) Harrison, *Rh. glutinis* (Fres.) Harrison, *Rh. aurantiaca* (Saite) Lodder, *Rh. flava* (Saito) Lodder, *Rh. minuta* (Saito) Harrison and *Rh. pallida* Lodder. All have been obtained from human and animal sources at one time or another, and *Rh. mucilaginosa* merits particular attention since it has been frequently obtained from these sources.

Members of the genus, unfortunately not identified as species, have been reported from systemic infections of man. Louria *et al.* (1960) isolated a *Rhodotorula* species from the blood of a woman with symptoms of septicaemia, and Shelburne and Carey (1962) reported the culture of an unidentified species from the blood of a patient suffering from diabetes complicated by staphylococcal endocarditis. A member of the genus was found in a resected, healed tuberculous lung cavity by Fors and Sääf (1960).

A. RHODOTORULA MUCILAGINOSA

1. *Incidence and Distribution*

Mackenzie (1961) found this species as 6·7% of yeasts obtained from human sources, and Stenderup and Pedersen (1962) obtained 22 isolates among 968 yeasts from human sources. Sonck and Soemesalo (1963) found it together with other yeasts in 123 samples taken from the anogenital area of girls suffering from diabetes. It was repeatedly isolated by Cramer and Koch (1963) from various lesions of the skin and nails of the same patient, and Staib (1958) found it with a number of other species of *Rhodotorula*, but usually in predominant numbers, in bile associated with liver and gall-bladder disease in 169 humans.

Little is known of its association with animals, but Batista *et al.* (1961b) obtained it from poultry excreta in Recife, Brazil, and Kaben and Preuss (1967a, b) from the excreta of a variety of birds in captivity at Rostock in the Deutsche Demokratische Republik.

2. *Cultural Characteristics*

Colonies of *Rh. mucilaginosa* are flat and mostly smooth and mucilaginous. They are distinctly red in colour. Cells are short-oval or oval, and measure 2·5–5·0 μ × 4–7 μ. Pseudomycelium formation rarely occurs and, if formed, is primitive in nature. The yeast has no fermentative ability and assimilates glucose, sucrose, maltose and galactose (weakly).

VIII. Serological Aspects

The many complex aspects of serology are outside the scope of this chapter, and for more detailed information the interested reader is referred to any of the standard textbooks on bacteriology and immunology (e.g. Topley and Wilson, 1965; Cruickshank, 1965). However, the subject is of considerable importance in medical mycology, and therefore cannot be omitted entirely from any text dealing with fungi pathogenic for man and animals. It is hoped that the following brief account will be sufficient to introduce the subject and, at the same time, serve as an indication of the significant progess which has already been made in its applications to medical and veterinary mycology.

If a human or animal body becomes infected by a pathogenic micro-organism, substances such as soluble proteins and certain polysaccharides, called antigens, evoke an antagonistic response in the body. This manifests itself in the generation of substances called antibodies. In some cases, sufficient antibodies are produced to guard against a second attack. This is the basis of immunity to disease.

Immunity can also be induced artificially by injecting into the body killed or attenuated micro-organisms or their products. These act as antigens in the same way and, by stimulating the production of antibodies, establish a state of immunity. Since each pure antigen evokes the production of its corresponding specific antibody, it is possible to identify the micro-organism which is causing or has caused disease, by demonstrating the presence of certain antibodies in the blood of the animal concerned. A number of procedures have been developed for the demonstration of these antibodies *in vitro*.

Charrin and Roger (1889) discovered that, when *Bacillus pyocyaneus* was grown in the blood serum from an animal which had been previously inoculated with this bacterium, the organism did not grow diffusely but became aggregated in small clumps which sank to the bottom of the fluid. From this and similar observations by Metchnikoff (1891), the procedure termed agglutination was developed. The specificity of the procedure was demonstrated by Kraus (1897) and it is commonly used for diagnosis of bacterial infections such as those caused by *Salmonella*, *Shigella* and *Brucella*, and for the identification of these organisms. Further developments in the study of reactions between micro-organisms and sera were the complement-fixation test by Bordet and Gengou (1901) and the double-diffusion test originally devised by Bechold (1905) for his studies on the physical structure of gelatin, and later elaborated by Ouchterlony (1949) for the study of antigen-antibody reactions in agar gels. Bordet and Gengou (1901) found that, if the serum of a person who had recovered from bubonic plague was mixed with plague bacilli, the free complement (a non-specific substance normally present in all mammalian blood) was used up or fixed in destroying the bacilli. This discovery led to the development of the complement-fixation procedure which is extensively used to diagnose disease by the identification of certain antigens causing disease or their corresponding antibodies, and also for other purposes such as identification of blood stains. The double-diffusion test is based on the fact that different constituents of both antigens and antibodies are able to diffuse in agar gels and do so at different rates so that, when they meet and react, precipitin bands separated from one another appear in the medium.

A more recent development is the immunofluorescence test (Coons and Kaplan, 1950) which is a localized staining procedure depending on the combination of antibody globulin fractions in antiserum with antigen from the test organism, the globulin fraction having been previously treated (conjugated) with a fluorescent dye. Preparations viewed with a fluorescence microscope in ultraviolet light show fluorescence indicating where union between antibody globulin and the antigen has taken place.

The introduction of an antigen into the body, in addition to stimulating the production of antibodies, also alters the body tissues so that a second injection of the same antigen (given about two weeks after the first) produces an immunological response more rapidly and more vigorously than the first. In some cases, an allergic or hypersensitive state, revealed by the second exposure to the antigen, is induced. This can be of a very serious nature (anaphylactic shock), but more often is a localized reaction at the site of injection which may be used for diagnostic purposes. When a small quantity of the antigen to which the patient is sensitized is scratched or injected into the skin, a localized area of inflammation develops. One of the best known skin tests is the Mantoux test for tuberculosis (Mantoux, 1908) which involves injecting tuberculin (a suspension of dead cells of *Mycobacterium tuberculosis*) intradermally and recording the result 24 h later. A positive result is manifest by an induration of the skin measuring about 5 cm in diameter at the site of injection.

Before considering serology in relation to fungi and fungal diseases, it is necessary to indicate certain factors which may affect the results of tests or investigations. First, because of their complex structure, microorganisms are not single antigens and therefore will give rise to a number of different antibodies. Certain antigens and their corresponding antibodies may be common to more than one species of fungus, and this may result in serum containing antibodies (antiserum) evoked by one fungus reacting with antigens of another. Such cross-reactions are not unusual. In serum-reaction procedures (e.g. agglutination), the common antibodies in a heterologous serum can be eliminated by adsorbing the serum with antigens from the cross-reacting fungus, thus leaving the remaining antibodies free to react with the antigens peculiar to the fungus which produced them and are specific to them. This procedure is used both to determine antigenic similarities between organisms and to increase the specificity of antigen-antibody reactions in the diagnosis of infections. In skin testing, cross-reactions are exceedingly common and can only be obviated by purification of the antigen. This cannot always be achieved to a degree sufficient to avoid false positive results. Furthermore, their complex nature makes standardization of fungal antigens exceedingly difficult, and this is an aspect which, although by no means neglected, still requires extensive investigation.

Despite bacteriological experience, many of the inherent difficulties in serological procedures still remain to be solved and, except in a few instances, their application to fungal infections is of comparatively recent origin. In the case of histoplasmosis, coccidioidomycosis and North American blastomycosis, agglutination and complement-fixation tests, together with skin testing, have been practised routinely for the past 20 years (Campbell, 1967). More recently, too, these procedures have been used to good effect with paracoccidioidomycosis.

Notwithstanding the widespread use of skin tests for these fungal infections in their endemic areas, care is required in interpreting results which may sometimes be misleading owing to the occurrence of false positives arising from cross-reactions. This happens if some of the individuals being tested harbour, or have harboured, one of the other aetiological fungi. Thus, in testing for coccidoidomycosis, false positives may be obtained if histoplasmosis is also present in the area (Smith *et al.*, 1949).

There is much dissension regarding the usefulness of skin tests in the diagnosis of infections due to pathogenic yeasts, the consensus of opinion being unfavourable owing chiefly to yeast species occurring in a high proportion of apparently normal people. Here too there is considerable difficulty in obtaining consistent results because of the lack of standardization of the antigen preparations used in these tests (Maibach and Kligman, 1962; Lewis *et al.*, 1937; Graciansky and Puissant, 1960). Holti (1966) is, however, convinced of the usefulness of skin tests in the case of patients suffering from certain abnormal conditions involving allergic responses to the presence of *Candida* in the alimentary tract, but not attributable to invasive spread of *Candida*. Thus, in 52 of 65 patients who were suffering from mucous colitis, he obtained a positive skin test to *Candida*, and treatment with nystatin resulted in the cure of a significant number (17) of these patients, as well as relief in 31 cases.

Other serological investigations of infections due to yeasts are relatively recent in origin, and progress has been perhaps rather disappointing. This again is because of cross-reactions and because of the high proportion of normal people whose serum yields a positive reaction to *Candida albicans* and other yeasts.

Some progress has been made recently as regards infections due to *Cryptococcus neoformans*. This fungus exhibits only a very low degree of antigenicity and, consequently, antibody reaction is difficult and often impossible to demonstrate by the usual *in vitro* serological methods. Bloomfield *et al.* (1963) described a modification of the agglutination procedure, the latex-slide agglutination method, which is based on the use of latex particles coated with antibody globulin obtained from rabbit serum immunized with a small-capsule strain of *Cr. neoformans*. The cerebrospinal fluid or serum of a patient suffering from cryptococcosis contains sufficient antigen, as a rule, to react with the antibody on the latex particles, and agglutination takes place. The efficacy of this method has been confirmed by Gordon and Vedder (1966) and others. However, its further application in routine practice is required before it can be accepted as a reliable diagnostic procedure.

Agglutination and double-diffusion tests have been widely used, despite their limitations, for the diagnosis of septicaemic or systemic candidiasis (Taschdjian et al., 1964, 1967; Stallybrass, 1964; Murray and Buckley, 1966). In such infections, an increased concentration of γ-globulin antibodies occurs in the patients' blood, a condition known as hyperglobulinaemia. This concentration gradually decreases as the patients recover. With Candida yeast-phase cells as antigen, the γ-globulin concentration in the blood of patients can be assessed, and its gradual decrease during recovery monitored.

The application of in vitro serological tests to the differentiation of yeast species has been in progress for some time, and has met with more success than for diagnosis of infections (Yukawa et al., 1928, 1929; Benham, 1931; Almon and Stovall, 1934; Rawson and Norris, 1947). In 1954, Tsuchija et al., using agglutination techniques, commenced their intensive and protracted series (1954–65) of serological studies on the antigenic structure of some 140 species of yeasts, including members of the genera Candida, Torulopsis, Pichia, Hansenula, Debaryomyces and Rhodotorula. From a taxonomic aspect, such knowledge is particularly valuable since difficulty is experienced in identifying and classifying many species of yeast because of variation in morphological and biochemical characteristics between strains of the same species, as well as occasional loss in reactive ability after several successive subcultures. Similar investigations were made by Seeliger (1958) who used the double-diffusion method in addition to agglutination for his studies of the relationships between the genera Candida, Torulopsis and Cryptococcus.

By serological methods, Hasenclever and Mitchell (1960, 1961) made the important observation that there are two different serotypes of C. albicans. They designated them C. albicans A and C. albicans B, Type A being the more common of the two. The basis of distinction was the antigenic relationship which Type A has with C. tropicalis and Type B has with C. stellatoidea.

The immunofluorescence technique of Coons and Kaplan (1950) has also been applied to the study of antigenic relationships between pathogenic and non-pathogenic yeast species (Gordon, 1958, 1962; Gordon et al., 1967). It has been shown to be of use as a practical method for identifying yeasts, not only in smears of prepared suspensions but also in smears of body secretions and in fixed-tissue sections of material obtained from lesions caused by pathogenic yeasts (Gordon, 1962).

While there are still many difficulties to overcome in the application of serological procedures both to the diagnosis of infections caused by pathogenic yeasts and to the study of relationships within this group of fungi, it cannot be denied that this approach has led to many interesting advances, revealed many new and hitherto unsuspected facts, and promises much for the future.

IX. Therapeutic Aspects

Since the discovery of penicillin, a wide range of highly effective antibiotics has become available for the treatment of bacterial diseases of man and animals. Unfortunately, progress has not been as rapid or as successful with regard to the therapy of the mycoses. At the present time, only one really effective drug, the antibiotic amphotericin B, is available for the treatment of serious systemic infections. Among the few other antifungal drugs which have been discovered, some, e.g. hamycin (Padhye and Thirumalachar, 1963) and candicidin (Lechevalier, 1953) have still to be fully evaluated; others have not been suitable for therapeutic purposes, e.g. actidione (Whiffen, 1948) or have limitations, such as nystatin (Hazen and Brown, 1951) which is suitable only for topical treatment, and griseofulvin (Oxford et al., 1939) which is active only against the ringworm fungi (Gentles, 1958).

For superficial infections, such as those of the skin and mucous membranes, the therapeutic range of antifungal drugs is greater, but the topical application of therapeutic substances is, generally speaking, less successful than might be desired, and successful treatment of some apparently mild infections is often difficult to achieve. It is because of such difficulties, the major one of which is the lack of effective penetration through the keratin barrier, that orally-administered griseofulvin, which acts on ringworm infections from the inside of the body outwards, had been so widely acclaimed.

Since yeasts in general thrive best under moist conditions, many superficial infections due to pathogenic species of this group improve and even clear completely if the affected site is kept dry. To assist or initiate the clearance of superficial infections, antifungal agents commonly used are gentian violet and the antibiotic, nystatin. Nystatin, a polyene antibiotic derived from *Streptomyces noursei* (Brown and Hazen, 1949) is also used widely and with good therapeutic effect in the treatment of oral and gastro-intestinal infections when it is administered orally, and for vaginal candidiasis in the form of locally administered pessaries. There are no significant side effects associated with the administration of this drug, and it is widely used concomitantly with broad-spectrum antibiotics and corticosteroids to prevent the onset of candidiasis. Nystatin is effective against many mycelial fungi as well as yeasts, but it does not possess any antibacterial activity. It has been shown (Littman et al., 1958) that increased resistance to this drug may be induced in species of *Candida in vitro*, but the occurrence of naturally resistant strains of *C. albicans* has not been reported with any frequency. Because nystatin is not absorbed from the gastro-intestinal tract, and because it is not suitable for intravenous or intramuscular administration, it is of no value in the treatment of systemic mycoses.

Amphotericin B, another polyene antibiotic, was developed following its isolation from cultures of *Streptomyces nodosus* by Gold et al. (1956). It may also be used topically and given orally for oral and gastro-intestinal infections (Kozinn et al., 1960) but, although limited absorption has been reported (Campbell and Hill, 1960), it is clear that the blood concentrations obtained by this route are too low for appreciable therapeutic effect. For deep-seated infections, against which it is highly effective, amphotericin B is given intravenously in a carefully prepared formulation. The dose must be strictly controlled, and a careful watch maintained for the onset of adverse side effects which most patients who receive this drug experience. These include a rise in temperature, chills,

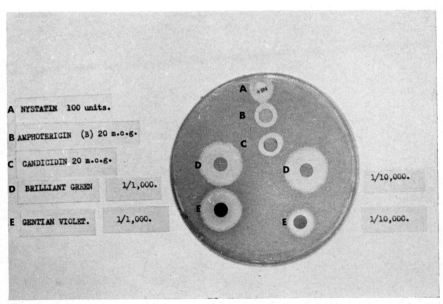

A NYSTATIN 100 units.

B AMPHOTERICIN (B) 20 m.c.g.

C CANDICIDIN 20 m.c.g.

D BRILLIANT GREEN 1/1,000.

E GENTIAN VIOLET. 1/1,000.

D 1/10,000.

E 1/10,000.

FIG. 25. Sensitivity of *Candida albicans* to various antifungal agents used in the therapy of *Candida* infections.

nausea, headache, and general malaise often accompanied by an elevation of the blood urea-nitrogen. Renal effects are most serious, and may necessitate reduction of the dose or temporary withdrawal of the drug and, in some cases, complete cessation of administration (Utz, 1966). Despite these difficulties, amphotericin B is of considerable therapeutic value, and is the only available drug at present with proven efficacy in the treatment of serious infections due to *Candida* species and *Cryptococcus neoformans*, and other fungal pathogens capable of causing systemic disease.

The *in vitro* sensitivity of *C. albicans* to a number of antifungal drugs, including those already mentioned, is shown in Fig. 25. Treatment of the mycoses is dealt with in considerable detail by Hildick-Smith *et al.* (1964), and there are also sections in certain textbooks of medical mycology and dermatology, such as Emmons *et al.* (1963), Wilson and Plunkett (1965) and Rook *et al.* (1968).

References

Acta, H. W. and Panja, G. (1927). *Indian med. Gaz.* **62**, 603–609.

Ahearn, D. G., Jannach, J. R. and Roth, F. J. Jr. (1966). *Sabouraudia* **5**, 110–119.

Ainley, R. and Smith, B. (1965). *Br. J. Ophthal.* **49**, 505–515.

Ainsworth, G. C. and Austwick, P. K. C. (1959). "Fungal Diseases of Animals". Commonwealth Agricultural Bureaux, Farnham Royal, England.

Ajello, L. (1958). *Am. J. Hyg.* **67**, 72–77.

Ajello, L. (1962). *In* "Fungi and Fungous Diseases" (G. Dalldorf, ed.), pp. 69–83. Charles C. Thomas, Springfield, Illinois.

Ajello, L. (1967). *Bact. Rev.* **31**, 6–24.

Ajello, L., Georg, L. K., Kaplan W. and Kaufman, L. (1963). "Laboratory Manual for Medical Mycology". U.S. Public Health Service Publication No. 994, Washington.

Almon, L. and Stovall, W. D. (1934). *J. infect. Dis.* **55**, 12.

Anderson, T. (1958). *In* "Fungous Diseases and their Treatment" (R. W. Riddell and G. T. Stewart, eds.), p. 84–87. Butterworths, London.

Andriole, V. T., Kravetz, H. M., Roberts, W. C. and Utz, J. P. (1962). *Am. J. Med.* **32**, 251–285.

Artargaveytia-Allende, R. C. and Silveira, J. S. (1961). *Ciencia' Méx.* **21**, 59–60.

Aschner, M., Mager, J. and Leibowitz, J. (1945). *Nature, Lond.* **156**, 295.

Austwick, P. K. C., Pepin, G. A., Thompson, J. C. and Yarrow, D. (1966). *In* "Symposium on Candida Infections" (H. I. Winner and R. Hurley, eds.), pp. 89–99. E. & S. Livingstone, Edinburgh.

Baillon, H. (1889). "Traité de Botanique Médicale Cryptogamique", p. 234. Octave Doin, Paris.

Baker, R. D. and Haugen, R. K. (1955). *Am. J. clin. Path.* **25**, 14.

Barbesier, J. (1960). *Archs Inst. Pasteur Algér.* **38**, 231–235.

Batista, A. C., Silveira, G. and Oliveira, D. (1958). *Revta Ass. méd. bras.* **4**, 360–361.

Batista, A. C., de Vasconcelos, C. T., Fischman, O. and Staib, F. (1961a). *Publções Inst. Micol. Recife.* No. 325.

Batista, A. C., Fischman, O., de Vasconcelos, C. T. and da Rocha, I. G. (1961b). *Publções Inst. Micol. Recife.* Nos. 326, 327, 329.

Batista, A. C., de Vasconcelos, C. T. and Fischman, O. (1961c). *Publções Inst. Micol. Recife,* No. 323.

Batista, A. C., de Vasconcelos, J. A., Fischman, O. and da Silva, J. O. (1961d). *Publções Inst. Micol. Recife.* No. 326.

Batista, A. C., Fischman, O., de Vasconcelos, C. T. and da Rocha, I. G. (1961e). *Publções Inst. Micol. Recife.* No. 327.

Batista, A. C., de Vasconcelos, C. T., de Lima, J. A. and Shome, S. K. (1961f). *Publções Inst. Micol. Recife.* Bull. No. 329.

Bechold, H. (1905). *Z. phys. Chem.* **52**, 185–189.

Behounkova, L., Cerna, I., Lepkova, V., Marsalek, E. and Zizka, Z. (1965). *Čslká Epidem. Mikrobiol. Imunol.* **13**, 159–164.

Beemer, A. M., Price, D. M. and Riddell, R. W. (1954). *J. Path. Bact.* **68**, 259–366.

Bendel, W. L., Jr. and Race, G. J., (1961). *Archs. intern. Med.* **108**, 916–924.

Benham, R. W. (1931). *J. infect. Dis.* **49**, 183–215.

Benham, R. W. (1935). *J. infect. Dis.* **57**, 255–274.

Benham, R. W. (1939). *J. invest. Derm.* **2**, 187–193.

Benham, R. W. (1947) *In* "Biology of Pathogenic Fungi", (W. J. Nickerson, ed.), pp. 63–70. Chronica Botanica Co., Waltham, Mass.

Benham, R. W. (1955). *Proc. Soc. exp. Biol. Med.* **89**, 243–245.

Benham, R. W. and Hopkins, A. M. (1933). *Archs. Derm. Syph.* **28**, 532.

Berge, T. and Kaplan, W. (1967). *Sabouraudia* **5**, 310–314.

Bergman, F. (1961). *Sabouraudia* **1**, 34–40.

Bergman, F. (1963). *Acta Med. scand.* **174**, 651–655.

Bergman, F. (1965). *Sabouraudia* **4**, 23–31.

Beuthe, D. (1955). *Zentbl. allg. Path. path. Anat.* **93**, 241–245.

Bisping, W., Refai, M. and Trautwein, G. (1964). *Berl. Münch. tierärztl. Wschr.* **77**, 260–262.

Bizzozero, J. (1884). *Virchows Arch. path. Anat. Physiol.* **98**, 441–447.

Blaxland, J. D. and Fincham, I. H. (1950). *Br. vet. J.* **106**, 221–231.

Bloomfield, N., Gordon, M. A. and Elmendorf, F. (1963). *Proc. Soc. exp. Biol. Med.* **114**, 64–67.

Blythe, W. (1958). *Mycopath. Mycol. appl.* **10**, 91–112.

Blythe, W. (1959). *Mycopath. Mycol. appl.* **10**, 269–282.

Böhme, H. (1965). *Derm. Wschr.* **151**, 107–117.

Bollinger, A. and Finckh, E. S. (1962). *Med. J. Aust.* **1**, 545–547.

Bordet, J. and Gengou, O. (1901). *Annls Inst. Pasteur* **15**, 289–303.

Borowski, J., Dziedziusko, A., Mierzejewski, W., Dubrzynska, T. and Iwanowski, K. (1963). *Proc. Internat. Symp. Med. Mycol.* Warsaw. pp. 133–135.

Brabander, J. O., Blank, F. and Butas, C. A. (1957). *Can. med. Ass. J.* **77**, 478–483.

Bret, J. and Coupe, C. I. (1958). *Presse méd.* **66**, 937–938.

Brown, R. and Hazen, E. L. (1949). *Annual Report, Div. Laboratories and Research, N.Y. State Dept. Hlth.* p. 19.

Buckley, H. R. and van Uden, N. (1963). *Sabouraudia* **2**, 205–208.

Burke, R. C. (1961). *J. invest. Derm.* **36**, 389–401.

Buschke, A. (1895). *Dt. med. Wschr.* **21**, 14.

Busse. O. (1894). *Zentbl. Bakt. ParasitKde* **16**, 175–180.

Busse, O. (1895). *Virchows Arch. path. Anat. Physiol.* **140**, 23–46.

Busse, O. (1896). *Virchows Arch. path. Anat. Physiol.* **144**, 360–372.

Campbell, C. C. (1967). *Sabouraudia* **5**, 240–259.

Campbell, C. C. and Hill, G. B. (1960). *Antibiotics Annual 1960*, Medical Encyclopedia, Inc. New York.

Carter, H. S. and Young, J. L. (1950). *J. Path. Bact.* **62**, 271–273.

Castellani, A. (1925). *J. trop. Med. Hyg.* **28**, 217–222.

Cawson, R. A. (1963). *Br. dent. J.* **115**, 441–448.

Cawson, R. A. (1966) *In* "Symposium on Candida Infections", (H. I. Winner and R. Hurley, eds.), pp. 138–152. E. & S. Livingstone, Edinburgh.

Charrin, A. and Roger, H. (1889). *C. r. Séanc. Soc. Biol.* **1**, 667–669.

Christensen, W. B. (1946). *J. Bact.* **52**, 461–466.

Clayton, Y. M. and Noble, W. C. (1966). *J. clin. Path.* **19**, 76–78.

Conant, N. F. (1940). *Mycopathologia* **2**, 253.

Conn, N. K., Crean, G. P., McCabe, A. F. and MacLean, N. (1959). *Br. med. J.* **1**, 944–947.

Converse, J. L., Reed, R. E., Kuller, H. W., Trautman, R. J., Snyder, E. M. and Ray, J. G. (1967). *In* "Coccidioidomycosis", (L. Ajello, ed.), pp. 397–402. University of Arizona Press.

Coons, A. H. and Kaplan, M. H. (1950). *J. exp. Med.* **91**, 1–12.

Cormane, R. H. and Goslings, W. R. O. (1963). *Sabouraudia* **3**, 52–63.

Cramer, H. J. and Koch, H. A. (1963). *Derm. Wschr.* **147**, 563–568.

Cruickshank, R. (1934). *J. Path. Bact.* **39**, 213–219.

Cruicksank, R. (1965). *In* "Medical Microbiology", (R. Cruickshank, J. P. Duguid and R. H. H. Swain, eds.), 11th. edition. E. & S. Livingstone Ltd., Edinburgh.

Debré, R., Mozziconacci, P., Drouhet, E., Drouhet, V. and Hoppeler, A. (1955) *Annls. paediat.* **184**, 129–164.

Della Torre, B., Arisi, C. and Savino, L. (1959). *Boll. Soc. med. chir. Pavia.* **3–4**, 405–413.

Dhom, G., Staib, F. and Stroder, J. (1964). *Arch. Kinderheilk.* **170**, 2–12 and 221–233.

Drouhet, E. (1960). *Bull. Soc. fr. Derm. Syph.* **67**, 646–659.

Drouhet, E. and Couteau, M. (1951). *Annls Inst. Pasteur, Paris* **80**, 456–457.

Drouhet, E., Segretain, G. and Aubert, J. P. (1950). *Annls Inst. Pasteur, Paris* **79**, 891–900.

Drouhet, E., Segretain, G. and Destombes, P. (1961). *Presse méd.* **69**, 1983–1986.

Durant, J. R., Epifano, L. D. and Eyer, S. W. (1960). *Annls int. Méd. phys. Physio-Biol.* **53**, 534–537.

Durcan, D. J., Goodchild, R. T. and Wengraft, C. (1968). *J. Lar. Otol.* **82**, 379–388.

Dutt-Choudhuri, R. and Dutt, R. (1961). *Indian J. Microbiol.* **1**, 61–64.

Emmons, C. W. (1951). *J. Bact.* **62**, 685–690.

Emmons, C. W. (1952). *Mycopath. Mycol. appl.* **6**, 231–234.

Emmons, C. W. (1954). *Trans. N.Y. Acad. Sci.* 16, 157–166.

Emmons, C. W. (1955). *Am. J. Hyg.* **62**, 227–232.

Emmons, C. W. (1960). *Publ. Hlth. Rep., (Wash.)* **75**, 362–365.

Emmons, C. W. (1962). *Lab. Invest.* **11**, 1026–1032.

Emmons, C. W., Binford, C. G. and Utz, J. P. (1963). "Medical Mycology". Henry Kimpton, London.

Evans, E. G. V. (1969). Ph.D. Thesis: University of Glasgow.

Evenson, A. E. and Lamb, J. W. (1964). *J. Bact.* **88**, 542.

Fendel, K. and Pietsch, P. (1965). *Z. Oto-Rhino- u. Lar., Tokyo* **44**, 145–154.

Fors, B. and Sääf, J. (1960). *Acta chir. scand.* **119**, 212–229.

Freeman, W. (1931). *J. Psychol Neurol Lpz* **43**, 236–345.

Frothingham, L. (1902). *J. med. Res.* **3**, 31–43.

Funke, H. (1960). *Nord. Vet/Med.* **12**, 54–62.

Garin, J. P., Monier, P., Humbert, G., Despeigne, J. and Michel, D. (1965). *J. Méd. Lyon* **46**, 1951–1965.

Gausewitz, P. L., Jones, F. S. and Worley, G. Jr. (1951). *Am. J. clin. Path.* **21**, 41–49.

Gentles, J. C., (1958). *Nature, Lond.* **182**, 476–477.

Gentles, J. C. and Holmes, J. G. (1957). *Br. J. indust. Med.* **14**, 22–29.

Gitter, M. and Austwick, P. K. C. (1959). *Vet. Rec.* **71**, 6–11.

Gilbert, W. R. and Fetter, B. F. (1962). *J. Bact.* **84**, 961–966.

Gold, W., Stout, H. A., Pagano, J. S. and Donovick, R. (1956). *Antibiotics Annual.* 1955–1956. Medical Encyclopedia Inc., New York.

Gonzáles-Ochoa, A. (1967). *In* "Coccidioidomycosis" (L. Ajello, ed.), pp. 293–299. University of Arizona Press.

Gordon, M. A. (1951). *J. invest. Derm.* **17**, 267–272.

Gordon, M. A. (1958). *Proc. Soc. exp. Biol. Med.* **97**, 694–698.

Gordon, M. A. (1962). *In* "Fungi and Fungous Diseases" (G. Dalldorf, ed.), pp. 207–217. Charles C. Thomas, Springfield, Illinois.

Gordon, M. A. and Vedder, D. K. (1966). *J. Am. med. Ass.* **197**, 961–967.

Gordon, M. A., Elliott, J. C. and Hawkins, T. W. (1967). *Sabouraudia* **5**, 323–328.

Graciansky, P. de and Puissant, S. G. (1960). *Bull. Soc. fr. Derm. Syph.* **67**, 659–673.

Graf, K. (1963). *Klin. Mbl. Augenheilk.* **143**, 356–362.

Gregson, A. E. W. and La Touche, C. J. (1961). *J. Lar. Otol.* **75**, 45–69.

Gresham, G. A. and Whittle, C. H. (1961). *Sabouraudia* **1**, 30–35.

Grimley, P. M., Wright, L. D. and Jennings, A. E. (1965). *Am. J. clin. Path.* **43**, 216–223.

Grose, E., Marinkelle, C. J. and Striegel, C. (1968). *Sabouraudia* **6**, 127 132.

Gruby, M. (1842). *C. r. hebd. Séanc. Acad. Sci., Paris* **14**, 634–636.

Guze, L. B. and Haley, L. D. (1958). *Yale J. Biol. Med.* **30**, 292–305.

Hajsig, M., Riznar, S. and Marzan, B. (1962). *Vet. Arh. Zagreb.* **32**, 276–282.

Hajsig, M., Kopljar, M. and Steficic, M. (1964). *Vet. Arh. Zagreb.* **34**, 133–137 (English Summary).

Haley, L. D. (1965). *Sabouraudia* **4**, 98–105.

Hansen, P. (1963). *In* "Hefepilze als Krankheitserreger bei Mensch und Tier" (C. Schirren and H. Rieth, eds.), p. 30. Springer-Verlag, Berlin.

Harris, L. J., Pritzken, H. G., Laski, B., Eisen, A., Steiner, J. W. and Schack, L. (1958). *Can. med. Ass. J.* **79**, 891–896.

Haugen, R. K. and Baker, R. D. (1954). *Am. J. clin. Path.* **24**, 1381–1390.

Hasenclever, H. F. and Mitchell, W. O. (1960). *J. Bact.* **79**, 677–681.

Hasenclever, H. F. and Mitchell, W. O. (1961). *J. Bact.* **82**, 578–581.

Hasenclever, H. F. and Mitchell, W. O. (1962). *Sabouraudia* **2**, 87–95.

Hasenclever, H. F. and Emmons, C. W. (1963). *Am. J. Hyg.* **78**, 227–231.

Hazen, E. L. and Brown, R. (1951). *Proc. Soc. exp. Biol Med.* **76**, 93–97.

Herbeuval, R., Herbeuval, H., Cuny, G., Debry, G. and Manciaux, M. (1955). *J. fr. Méd. Chir. thorac.* **9**, 160–172.

Heymer, T. (1958). *Arch. klin. exp. Derm.* **208**, 74–80.

Hildick-Smith, G., Blank, H. and Sarkany, I. (1964). "Fungus Diseases and Their Treatment". Little, Brown and Company, Boston.

Hoffmann, D. H. (1963). *In* "Hefepilze als Krankheitserreger bei Mensch und Tier" (C. Schirren and H. Rieth, eds.). Springer-Verlag, Berlin.

Holti, G. (1966). *In* "Symposium on Candida Infections" (H. I. Winner and R. Hurley, eds.), pp. 73–81. E. & S. Livingstone, Edinburgh.

Hurley, R. (1964). *Post-grad. med. J.* **40**, 644–653.

Hurley, R. (1966). *In* "Symposium on Candida Infections" (H. I. Winner and R. Hurley, eds.), p. 17. E. & S. Livingstone, Edinburgh.

Hurley, R. (1966). *J. Path. Bact.* **92**, 57–67.

Hurley, R. and Morris E. D. (1964). *J. Obstet. Gynaec. Br. Commonw.* **71**, 692–695.

Hurley, R. and Winner, H. I. (1964). *Mycopath. Mycol. appl.* **24**, 337–346.

Innes, J. R. M., Seibold, H. R. and Arentzen, W. P. (1952). *Am. J. vet. Res.* **13**, 469–475.

Jamshidi, A., Pope, R. H. and Friedman, N. H. (1963). *Archs intern. Med.* **112**, 370–376.

Jansson, E. (1962). *Annls Med. intern. Fenn.* **51**, 249–253.

Jennison, R. F. (1966). *In* "Symposium on Candida Infections", (H. I. Winner and R. Hurley, eds.), pp. 102–118. E. & S. Livingstone, Edinburgh.

Johnson, C. G. and Mayne, R. (1948). *Am. J. Obstet. Gynec.* **55**, 852–858.

Jones, C. P. and Martin, D. S. (1938). *Am. J. Obstet. Gynec.* **35**, 98–106.

Kaben, U. and Preuss, B. (1967a). *Zentbl. Bakt. ParasitKde.* **204**, 274–282.

Kaben, U. and Preuss, B. (1967b). *9th Internat. Symp. on Diseases of Animals.* Prague, pp. 207–212.

Kao, C. J. and Schwarz, J. (1957). *Am. J. clin. Path.* **27**, 652–663.

Kaplan (1959). *J. Am. vet. med. Ass.* **134**, 113–117.

Kaul, K. K., Shah, P. M. and Pohowalla, J. N. (1960). *Indian J. Paediat.* **27**, 115–124.

Kawakita, S. and van Uden, N. (1965). *J. gen. Microbiol.* **39**, 125–129.

Kearns, P. R. and Gray, J. E. (1963). *J. Obstet. Gynaec. Br. Commonw.* **70**, 621–625.

Keddie, F. and Shadomy, S. (1963). *Sabouraudia* **3**, 21–25.

Keymer, I. F. and Austwick, P. K. C. (1961). *Sabouraudia* **1**, 22–29.

Klaman, J. (1884). *Allg. med. ZentZtg.* No. 23.

Klein, E. (1901). *J. Hyg., Camb.* **1**, 78–95.

Kligman, A. M., Crane, A. P. and Norris, R. F. (1951). *Am. J. med. Sci.* **221**, 273–278.

Koch, H., Rieth, H. and Ruther, E. (1959). *Hautarzt* **10**, 393–397.

Kozinn, P. J., Burchall, J. J., Katz, A. and Taschdjian, C. L. (1960). *Antibiotic Med.* **7**, 749–751.

Kraus, R. (1897). *Wien. klin. Wschr.* **10**, 736–739.

Kreger-van Rij, N. J. W. (1963). *Proc. Int. Colloq. Med. Myc. Antwerp*, pp. 13–19.

Kreger-van Rij, N. J. W. and Staib, F. (1963). *Arch. Mikrobiol.* **45**, 115–118.

Kruspl, W. (1963). *In* "Hefepilze als Krankheitserreger bei Mensch und Tier", (C. Schirren and H. Rieth, eds.), p. 40. Springer-Verlag, Berlin.

Kuhn, L. R. (1939). *Proc. Soc. exp. Biol Med.* **41**, 573–574.

Kuhn, L. R. (1949). *Proc. Soc. exp. Biol Med.* **71**, 341–343.

Lacaz, D. Da S., Pereira, O. A., Fernandes, J. de C. and Ulson, C. M. (1958). *Medna Cirurg. Farm.* **262–263**, 76–78.

Langenbeck, B. (1839). *Neue Mot. Geb. Natur-u-Heilk (Froriep).* **12**, 145–147.

La Touche, C. J. (1966). *In* "Symposium on Candida Infections" (H. I. Winner and R. Hurley, eds.), pp. 154–160. E. & S. Livingstone, Edinburgh.

Lechevalier, H. (1953). *Presse méd.* **61**, 1327–1328.

Lehner, T. (1964). *Lancet* **1**, 1414–1416.

Lewis, G. M., Hopper, M. E. and Montgomery, R. M. (1937). *N.Y. St. J. Med.* **37**, 878–881.

Linden, I. H. and Steffen, C. G. (1954). *Am. Rev. Tuberc. pulm. Dis.* **69**, 116–120.

Littman, M. L. (1958). *Trans. N.Y. Acad. Sci.* **20**, 623–648.

Littman, M. L. and Zimmerman, L. E. (1956). "Cryptococcosis (Torulosis or European Blastomycosis)". Grune and Stratton, New York and London.

Littman, M. L. and Tsubura, E. (1959). *Proc. Soc. exp. Biol Med.* **101**, 773–777.

Littman, M. L., Pisano, M. A. and Lancaster, R. M. (1958). *Antibiotics A.* 1957–1958. Medical Encyclopedia Inc., New York.

Littman, M. L. and Schneierson, S. S. (1959). *Am. J. Hyg.* **69**, 49–59.

Littman, M. L., Borok, R. and Dalton, T. J. (1965). *Am. J. Epidem.* **82**, 197–207.

Lodder, J. (1934). "Die anaskosporogenen Hefen, I Hälfte". *Verk. k. Acad. Wet. Afd. Natuurhunde, Sect. II* **32**, 1–256.

Lodder, J. and Kreger-van Rij, N. J. W. (1952). "The Yeasts, a Taxonomic Study". North-Holland Pub. Co., Amsterdam.

Louria, D. B., Greenberg, S. M. and Molander, D. W. (1960). *New Engl. J. Med.* **263**, 1281–1284.

Louria, D. B., Brayton, R. G. and Finkel, G. (1963). *Sabouraudia* **2**, 271–283.

Ludlam, G. B. and Henderson, J. L. (1942). *Lancet* **1**, 64–70.

Mackenzie, D. W. R. (1961). *Sabouraudia* **1**, 8–15.

Mahnke, P. F., Zschoch, H. and Sichert, H. (1961). *Patholgia Microbiol.* **24**, 327–340.

Maibach, H. I. and Kligman, A. M. (1962). *Archs Derm. Syph.* **85**, 233–254.

Malassez, L. (1874). *Arch. Physiol. Ser. II* **1**, 451.

Manchester, P. T. and Georg, L. K. (1959). *J. Am. med. Ass.* **171**, 1339–1341.

Manktelow, W. (1959). Ph.D. Thesis: University of Otago, New Zealand.

Mantoux, C. (1908). *C. r. hebd. Séanc. Acad. Sci., Paris* **149**, 355–357.

Marples, M. J. (1965). "The Ecology of the Human Skin". Thomas, Springfield, Illinois.

Marples, M. P. and di Menna, M. E. (1952). *J. Path. Bact.* **64**, 497–502.

Marples, M. J. and Somerville, D. A. (1966). Cited by Marples (1966). *N.Z. ecol. Soc.* **13**, 29–34.

Martin-Scott, I. (1952). *Br. J. Derm.* **64**, 257–273.

McDonough, E. S., Ajello, F., Ausherman, R. J., Balows, A., McClellan, J. T. and Brinkman, S. (1961). *Am. J. Hyg.* **73**, 75–83.

Medical Research Council (1967). *Nomenclature of Fungi Pathogenic for Man and Animals.* Memorandum No. 23 (3rd. ed.) H.M.S.O., London.

Mehnert, B. (1963). *In* "Hefepilze als Krankheitserreger bei Mensch und Tier" (C. Schirren and H. Rieth, eds.), p. 120. Springer-Verlag, Berlin.

Mehnert, B. and Koch, U. (1963). *Zentbl. Bakt. ParasitKde (Abt. Orig.)* **188**, 103–119.

Mendelblatt, D. L. (1953). *Am. J. Ophth.* **36**, 379–385.

Metchnikoff, E. (1891). *Annls Inst. Pasteur, Paris* **5**, 465–478.

Minkowitz, S., Koffler, D. and Zak, F. G. (1963). *Am. J. Med.* **34**, 252–255.

Misch, K. A. (1955). *J. clin. Path.* **8**, 207–210.

Mitsui, Y. and Hanabusa, J. (1955). *Br. J. Ophthal.* **39**, 244–250.

Mizuno, S. (1961). *In* "Studies on Candidiasis in Japan" (Research Committee on Candidiasis, eds.). Education Ministry of Japan.

Murray, I. G. and Buckley, H. (1966). *In* "Symposium on Candida Infections" (H. I. Winner and R. Hurley, eds.), pp. 44–49. E. & S. Livingstone, Edinburgh.

Negroni, P. (1934). *Rev. Inst. bact. Buenos Aires* **6**, 164–169.

Negroni, R., de Obrutsky, C. W. and Gonzales, R. O. (1965). *Revta. Asoc. méd. argent.* **79**, 483–442.

7

Neill, J. M., Abrahams, I. and Kapros, C. E. (1950). *J. Bact.* **59**, 263–275.

Norton, A. H. (1927). *Am. J. Ophthal.* **10**, 357.

Osborne, A. D., McCrea, M. R. and Manners, M. J. (1960). *Vet. Rec.* **72**, 237.

Ota, M. and Huang, P. T. (1933). *Annls Parasit. hum. comp.* **11**, 49–69.

Oxford, A. E., Raistrick, H. and Simonart, P. (1939). *Biochem. J.* **33**, 240–248.

Ouchterlony, O. (1949). *Acra. path. microbiol. scand.* **26**, 505–515.

Padhye, A. A. and Thirumalachar, M. J. (1963). *Hindustan. Antibiot. Bull.* **6**, 41–43.

Pan, N. C. and Pan. I. H. (1964). *J. Formosan med. Ass.* **63**, 396–399.

Passos, G. daM. (1961). Ph.D. Thesis: University of Recife, Brazil.

Perez, J. S. and Gil, R. A. (1960). *Microbiologia esp.* **13**, 323–325.

Petru, M. (1965). *Čas. Lék. česk.* **104**, 749–753 (English Summary).

Plummer, N. S. (1966). In "Symposium on Candida Infections" (H. I. Winner and R. Hurley, eds.), pp. 214–219. E. & S. Livingstone, Edinburgh.

Powell, D. E. B., English, M. P. and Duncan, E. H. L. (1962). *J. Lar. Otol.* **76**, 12–21.

Prophet, K. (1963). *Inaug. Diss. Tierarztliche Hochschule*, Hanover.

Rahim, G. F. (1964). *J. méd. libari.* **17**, 97–114.

Rawson, A. J. and Norris, R. G. (1947). *Amer. J. clin. Path.* **17**, 807–812.

Redaelli, P., Ciferri, R. and Giordano, A. (1937). *Boll. Sez. ital. Soc. int. Microbiol.* **9**, 24–28.

Reid, J. D. (1949). *J. Bact.* **58**, 777–782.

Rivolta, S. (1873). *Dei parassiti vegetali*, Torino.

Robin, C. (1853). *Histoire Naturelle des Végétaux parasites*, pp. 436–439. Baillière, Paris.

Roberts, S. S. and Rabson, A. S. (1962). *Annls int. Méd. phys. Physio-Biol.* **56**, 610–618.

Rocha, G. L., Silva, C., Lima, A. O. and Goto, M. (1952). *J. invest. Derm.* **19**, 289–296.

Roessmann, U. and Friede, R. L. (1967). *Archs Path.* **84**, 495–498.

Rogers, A. L. and Beneke, E. S. (1964). *Mycopath. Mycol. appl.* **22**, 15–20.

Rogers, K. B. (1966). In "Symposium on Candida Infections" (H. I. Winner and R. Hurley, eds.), pp. 179–194. E. & S. Livingstone, Edinburgh.

Rook A., Wilkinson, D. S. and Ebling, F. J. G. (1968). "Textbook of Dermatology" Blackwell Scientific Publications, Oxford.

Sabouraud, R. (1902). *Séborrhée, Acnées, Calvitie*, Masson et Cie, Paris.

Sabouraud, R. (1904). *Maladies du cuir chevelu*, t. 2, Masson et Cie, Paris.

Sandhu, R. S. and Randhawa, H. S. (1962). *Mycopathologia* **18**, 179–183.

Sanfelice, F. (1894). *Annali Ig. sper.* **4**, 463–495.

Schaulow, I., Spassowa, P. and Boschikowa, A. (1967). *Arch. klin. exp. Derm.* **227**, 985–992.

Schirren, C. (1963). In "Hefepilze als Krankheitserreger bei Mensch und Tier" (C. Schirren and H. Rieth, eds.), pp. 28–29. Springer-Verlag, Berlin.

Scholer, H. (1963a). In "Hefepilze als Krankheitserreger bei Mensch und Tier" (C. Schirren and H. Rieth, eds.), pp. 115–118. Springer-Verlag, Berlin.

Scholer, H. (1963b). In "Hefepilze als Krankheitserreger bei Mensch und Tier" (C. Schirren and H. Rieth, eds.), pp. 122–124. Springer-Verlag, Berlin.

Scholer, H., Gloor, F. and Dettli, L. (1963). In "Hefepilze als Krankheitserreger bei Mensch und Tier", (C. Schirren and H. Rieth, eds.), pp. 78–85. Springer-Verlag, Berlin.

Schönborn, C. (1967). *Mykosen*, **10**, 523–536.

Schulte, F. and Scholz, H. D. (1962). *Dt.-öst. tierärztl. Wschr.* **69**, 677–680.

Schwartz, L., Tulipan, L. and Birmingham, D. J. (1951). "Occupational Diseases of the Skin", 3rd. ed. Lea & Febiger, Philadelphia.

Seeliger, H. P. R. (1956). *J. Bact.* **72**, 127–131.

Seeliger, H. (1958). *In* "Beitrage zur Hygiene und Epidemiologie" (H. Habs and J. Kathe, eds.), pp. 68–98. J. A. Barth, Leipzig.

Seeliger, H. P. R. (1964). *Annls Soc. belge Méd. trop.* **44**, 657–672.

Shadomy, H. J. and Utz, J. P. (1966). *Mycologia* **58**, 383–390.

Shelburne, P. F. and Carey, R. J. (1962). *J. Am. med. Ass.* **180**, 38–42.

Silveira, J. S. and Correia, J. U. (1960). *Hospital, Rio de J.* **58**, 333–339.

Singh, B. and Sharma, M. D. (1963). *Indian J. Med. Sci.* **17**, 143–147.

Skobel, P., Jorke, D. and Schabinski, G. (1955). *Munsche Med. Wschr.* **97**, 194–197.

Smith, A. G., Taubert, H. D. and Martin, C. W. (1963). *Am. J. Obstet. Gynec.* **87**, 455–462.

Smith, C. E. (1943). *Med. Clins N. Am.* **27**, 790–807.

Smith, C. E., Saito, M. T., Beard, R. R., Rosenberg, H. G. and Whiting, E. G. (1949). *Am. J. publ. Hlth* **39**, 722–736.

Smith, J. M. B. (1967). *Sabouraudia* **5**, 220–225.

Smits, B. J., Prior, A. P. and Arblaster, P. G. (1966). *Br. med. J.* **1**, 208–210.

Smyth, G. D. L. (1962). *J. Lar. Otol.* **76**, 797–821.

Somerville, D. A. (1964). *N.Z. med. J.* **63**, 592–596.

Sonck, C. E. and Soemesalo, O. (1963). *Archs Derm. Syph.* **88**, 846–852.

Staib, F. (1958). *Zentbl. Bakt. ParasitKde* (*Abt. I Orig.*) **172**, 142–146.

Staib, F. (1961). *Zentbl. Bakt. ParasitKde* (*Abt. I Orig.*) **182**, 562–567.

Staib, F. (1962a). *Zentbl. Bakt. ParasitKde* (*Abt. I Orig.*) **185**, 129–134.

Staib, F. (1962b). *Zentbl. Bakt. ParasitKde* (*Abt. I Orig.*) **186**, 233–247.

Staib, F. (1962c). *Zentbl. Bakt. ParasitKde* (*Abt. I Orig.*) **186**, 274–275.

Staib, F. (1963). *Zentbl. Bakt. ParasitKde* (*Abt. I Orig.*) **190**, 115–131.

Stallybrass, F. C. (1964). *J. Path. Bact.* **87**, 85–97.

Stanley, V. C. and Hurley, R. (1967). *J. path. Bact.* **94**, 301–315.

Stenderup, A. and Pedersen, G. T. (1962). *Acta path. microbiol. scand.* **54**, 462–472.

Sturde, H. C. (1956). *Arch. klin. exp. Derm.* **203**, 266–269.

Suie, T. and Havener, W. H. (1963). *Am. J. Ophthal.* **56**, 63–77.

Sutmöller, P. and Poelma, F. G. (1957). *W. Indian med. J.* **6**, 225–228.

Swatek, F. E., Becker, S. W., Wilson, J. W., Omieczynsyi, D. T. and Kazan, B. H. (1964). *Proc. 7th Internat. Congr. trop. Med. and Malaria.* **3**, 118–138.

Sykes, E. M. (1946). *Tex. St. J. Med.* **42**, 330–332.

Symmers, W. St. C. (1964). *Medica Internat. Ser.* No. 85, p. 111.

Symmers, W. St. C. (1966). *In* "Symposium on Candida Infections" (H. I. Winner and R. Hurley, eds.), pp. 196–212. E. & S. Livingstone, Edinburgh.

Symmers, W. St. C. (1967). *Lancet i*, 159.

Symmers, W. St. C. (1968). *In* "Systemic Mycoses" (G. E. W. Wolstenholme and R. Porter, eds.), pp. 26–48, J. & A. Churchill, London.

Taschdjian, C. L. and Kozinn, P. J. (1957). *J. Paediat.* **50**, 426–433.

Taschdjian, C. L. and Kozinn, P. J. (1961). *Sabouraudia* **1**, 73–82.

Taschdjian, C. L., Kozinn, P. J. and Caroline, L. (1964). *Sabouraudia* **3**, 312–320.

Taschdjian, C. L., Kozinn, P. J., Okasa, A., Caroline, L. and Halle, M. A. (1967). *J. infect. Dis.* **117**, 180–187.

Terplan, K. (1948). *Am. J. Path.* **24**, 711–712.

Todd, R. L. and Herrmann, W. W. (1936). *J. Bact.* **32**, 89–103.

Topley, W. W. C., and Wilson G. S., (1965). "Principles of Bacteriology and Immunology". 5th ed. By G. S. Wilson and A. A. Miles, Edward Arnold Ltd., London.

Tsuchija, T., Iwahara, S., Miyasaki, F. and Fukazawa, Y. (1954). *Jap. J. exp. Med.* **24**, 95–103.

Tsuchija, T., Fukazawa, Y., Miyasaki, F. and Kawakita, S. (1955). *Jap. J. exp. Med.* **25**, 75–83.

Tsuchija, T., Miyasaki, F. and Fukazawa, Y. (1955). *Jap. J. exp. Med.* **25**, 15–21.

Tsuchija, T., Fukazawa, Y. and Kawakita, S. (1961). *Sabouraudia* **1**, 145–153.

Tsuchija, T., Fukazawa, Y. and Kawakita, S. (1965). *Mycopath. Mycol. appl.* **26**, 1–15.

Utz, J. P. (1966). *In* "Symposium on Candida Infections" (H. I. Winner and R. Hurley, eds.), pp. 221–243. E. & S. Livingstone, Edinburgh.

Valée, A., Drouhet, E., Guillon, J-C. and Nazinoff, X. (1964). *Bull. Acad. Vet.* **37**, 153–156.

Vanbreuseghem, R. (1954). *Annls Soc. belge. Méd. trop.* **35**, 251–254.

van Uden, N. (1960). *Trans. N. Y. Acad. Sci.* **89**, 59–68.

van Uden, N., de Matos Faia, M. and Assis-Lopes, L. (1956). *J. gen. Microbiol.* **15**, 151–153

van Uden, N. and do Carmo-Sousa, L. (1957). *Port. Acta Biol.* **4**, 7–17.

van Uden, N. and do Carmo-Sousa, L. (1962a). *Sabouraudia* **2**, 8–11.

van Uden, N. and do Carmo-Sousa, L. (1962b). *Antonie van Leeuwenhoek* **28**, 73–77.

van Uden, N., Vidal-Leiria, M. and Buckley, H. R. (1963). *Proc. Int. Colloq. med. Myc.* Antwerp, 31–49.

Versé, M. (1914). *Verb. dt. Path. Ges.* **17**, 275–278.

Vieira, J. R. and Batista, A. C. (1962). *Publções Inst. Micol. Recife.* No. 258.

Vince, S. (1959). *Med. J. Aust.* **46**, 143–149.

Viswanathan, R. and Randhawa, H. S. (1959). *Sci. Cult.* **25**, 86–87.

von Hansemann, D. (1905). *Verh. dt. pathol. Ges.* **9**, 21–24.

Walter, J. E. and Atchinson, R. W. (1966). *J. Bact.* **92**, 82–87.

Wegman, T. (1954). *Antibiotics Chemother.* **1**, 235–275.

Weidman, F. D. (1925). Quoted in: "Fox, H., Rep. Lab. Museum Comp. Pathology Sool. Soc. Philadelphia".

Whiffen, A. J. (1948). *J. Bact.* **56**, 283–291.

Whittle, C. H. and Gresham, X. (1963). *J. invest. Derm.* **40**, 267–269.

Wilson, R. (1961). *J. Am. med. Ass.* **177**, 332–334.

Wilson, J. W. and Plunkett, O. A. (1965). "The Fungous Diseases of Man". University of California Press, Berkeley and Los Angeles.

Winner, H. I. and Hurley, R. (1964). *"Candida albicans"*. J. & A. Churchill, London.

Wolfe, E. I. and Henderson, F. W. (1951). *J. Am. med. Ass.* **147**, 1344–1347.

Young, G., Resea, H. F. and Sullivan, M. T. (1951). *J. Dent. Res.* **30**, 426–430.

Yukawa, M. and Ohta, M. (1928). *Gakugei Zasshi, Kyushu imp. Univ.* **3**, 187–199 and 200–216.

Yukawa, M., Yositome, W. and Misio, S. (1929). *Gakugei Zasshi, Kyushu imp. Univ.* **4**, 267–281.

Yukawa, M. and Ohta, M. (1929). *Bull. Inst. Pasteur, Paris* **27**, 542–544.

Zech, P., Barthe, J., Robert, M., Guerrier, G. and Traeger, J. (1965). *Lyon méd.* **214**, 313–320.

Chapter 5

Yeasts Associated with Living Plants and their Environs

F. T. LAST AND D. PRICE

Glasshouse Crops Research Institute, Littlehampton, Sussex, England

I. Introduction

In the past there has been a tendency for the differing facets of microbiology to be studied in isolation. Plant pathology, the aspect of microbial nutrition primarily concerned with organisms that damage their living sources of nutrient, has been centred upon fungi and bacteria which arouse attention because they cause the development of characteristic blemishes, e.g. necrotic/chlorotic lesions whether on roots, leaves or flowers, and vascular staining. But, is it not possible that microorganisms might affect the longevity of plant structures without causing conspicuous macroscopic symptoms? Does the degradation of leaf cutin by yeasts and yeast-like organisms, many of which secrete lipolytic enzymes, affect the longevity of foliage possibly decreasing dry matter gain by increasing (a) water losses and (b) easing invasion by recognized pathogens?

Microbial nutrition has attracted the attention of many eminent scientists, notably Pasteur (1876) during the second half of the nineteenth century. He showed that yeasts were able to convert sugars to ethanol and in so doing established the need for asepsis in much biological and chemical research. His work also provided the basis for present-day studies of processes involved in the fermentation of fruit juices—oenology. Research in the past has tended to follow a predictable pattern, starting with series of general observations followed by increasingly intensive experimentation, confined to narrower and narrower facets, until it once again becomes essential to re-establish perspective taking account of all the components of the environment. Although Hansen (1881), like Pasteur, established the presence of yeasts on the surface of fruits, a concerted attempt to broaden this study to include all aerial plant structures was not made until the 1950s. But this is not to underestimate the value of work done by Derx (1930) and others in the interim. From 1950 onwards, populations of microorganisms colonizing the surfaces, phylloplanes, of living leaves have been described in a steadily increasing number of papers. These reports indicated that the phylloplane is an ecological niche inhabited by saprophytes and parasites, together forming a microbial complex such as occurs on the rhizoplane, first defined by Hiltner in 1904. This recent interest in the details of the phylloplane happens to have followed closely the publication by Lodder and Kreger–van Rij (1952) of "The Yeasts. A Taxonomic Study" and, as a result of their more lucid methods of species separation, our knowledge of yeasts colonizing leaves is already more comprehensive than that of yeasts colonizing fruits and roots.

Yeasts are defined in Chambers' Technical Dictionary as "Microorganisms producing zymase which induces the alcoholic fermentation of carbohydrates"—a description agreeing with Pasteur's finding but which is generally recognized as too restrictive. Skinner (1947) chose to base his definition on morphological characters, yeasts being fungi whose usual and dominant growth form is unicellular, but clearly this does not conform to the generally accepted image of yeasts for it would not exclude unicellular Phycomycetes. In their book, Lodder and Kreger–van Rij (1952) concede that their own morphological concept is based upon a series of arbitrary decisions, in that they accepted hyaline organisms and those producing red or yellow pigments but rejected dematiaceous types, such as *Aureobasidium pullulans* (de Bary) Arn. syn. *Pullularia pullulans* (de Bary) Berk. and yeast-like forms of *Cladosporium* (viz. *Torula nigra*). Yeasts which reproduce by fission or budding commonly occur in fungal groupings including mycelial forms. Thus, within the ascomycetous Saccharomycetoideae and Endomycetoideae there is at one end of the scale the group of *Saccharomyces* spp. with budding cells, at the other end *Endomyces* spp. with true mycelium, and between them and forming a link are species of *Endomycopsis* with both budding and mycelial phases. Similarly, in the Sporobolomycetaceae, where ballistospores are dispersed by the drop-excretion method found among the basidiomycetes (*vide* Buller, 1933), colonies of *Bullera* grow vegetatively by budding, those of *Tilletiopsis*, *Itersonilia* and *Sporidiobulus* by hyphal extension, with species of the intermediate *Sporobolomyces* having features in common with both forms.

Except that the former produces ascospores and the latter does not, many species of *Debaryomyces* and *Torulopsis* are similar. A comparable analogy to this between hyaline members of the Saccharomycetaceae and the Cryptococcaceae occurs between pigmented members of the latter group and the Sporobolomycetaceae. Some species of *Rhodotorula* can be regarded as non-spore-forming members of the Sporobolomycetaceae, and di Menna (1959) has emphasized the close parallel between *Rhodotorula graminis* di Menna and *Sporobolomyces odorus* Derx isolated from foliage of similar grass species.

In this chapter we have chosen to concentrate on members of the Endomycetaceae and Sporobolomycetaceae that reproduce vegetatively by fission or budding, on the Cryptococcaceae and, irrespective of colour, filamentous fungi with yeast-like phases occurring commonly *in vivo*, e.g. *Aureobasidium pullulans*. By inserting *in vivo* in this statement, *Mucor* species which develop bud cells when grown in highly concentrated sugar solutions are excluded as are some plant pathogenic species of *Taphrina* and of the Ustilaginales. However, this stipulation does not exclude the relatively unknown group of pleomorphic fungi within the Ascomycetes and Fungi Imperfecti which colonize tunnels made by Ambrosia beetles. Cultures of these fungi on carbohydrate-rich media are mycelial but this growth form can be changed to a dense yeast-like type by repeated scraping, a process simulating the movement of Scolytidae, Platypodidae and Lymexylonidae along fungus-lined tunnels (*vide* Batra and Batra, 1967).

As already indicated, more is known about yeasts colonizing leaves than those colonizing any other set of plant structures. But, as flowers, fruits and leaves are similarly exposed to airborne inocula, they will be grouped together in one of the two major sections of this chapter, the other dealing with roots and their soil environment.

II. Leaves, Flowers and Fruits

A. AIRBORNE INOCULA

Since the pioneer work of Schoenauer (1876) and Miquel (1877–99) numerous attempts have been made to identify and estimate numbers of airborne micro-organisms—"the air spora". Although yeasts are very widely distributed, there are relatively few records of their being trapped. Most trapping methods preclude the possibility of obtaining living cultures and, as a result, identification is restricted to morphological characteristics—a severe limitation, for the species differentiation of yeasts largely depends on their different growth habits on a range of media. However, the records obtained from suction traps, showing that ballistospores of members of the Sporobolomycetaceae sometimes occur in large atmospheric concentrations to the seemingly virtual exclusion of other yeasts, reflect the "active" spore dispersal of the former and passive means of spread of the latter. Notwithstanding, Pady and Kelly (1954) took air samples at heights of about 600–2,750 m above sea-level and from them cultured a few yeast colonies and some of *A. pullulans*, the latter being associated more with polar air masses than with those from tropical areas.

Although Derx in 1930 indicated that *Sporobolomyces* spp. commonly occurred on many leaves, the aerial dispersion of their ballistospores, discharged violently by the drop-excretion method described by Buller (1933), was not studied in detail until Gregory (1952) associated large atmospheric concentrations of hyaline basidiospores with yeasts of the Sporobolomycetaceae. Using iso-kinetic spore traps of the cascade impactor type instead of relying on natural deposition on agar plates or sticky bands (a method selectively favouring large spores), Gregory found that concentrations of *Sporobolomyces*-type hyaline basidiospores were greatest in the early hours of the morning when the relative humidity of the atmosphere was customarily high. Concentrations of this fungus can change rapidly within comparatively short periods of time. Hirst (1953) found that they decreased from $2 \cdot 5 \times 10^5$ to $1 \cdot 0 \times 10^3$ spores per m^3 during a four-hour period. Other spore-trapping results during June–October 1952 showed that *Sporobolomyces*, on average, accounted for 31% of the spores per m^3 of air (Gregory and Hirst, 1957). Last (1955b) found, when sampling air within powdery mildew-infected cereal crops, that ballistospores of *Sporobolomyces* and of the closely related mycelial *Tilletiopsis* accounted for 70% of the total number of fungus propagules trapped at 04.00 h and only 20% at 16.00 h, mean concentrations of *Sporobolomyces* spores being 3,600 and 150 per m^3 in the morning and afternoon respectively. These observations were made on dry days when relative humidity was greatest during the early hours of the morning, but spores of *Sporobolomyces* and *Tilletiopsis* spp. were similarly abundant during precipitation. By comparing catches on successive days, it was found that the percentage of *Sporobolomyces* ballistospores, in total daily catches of spores, increased from $0 \cdot 05$ and $4 \cdot 0$ on dry days to $87 \cdot 0$ and $94 \cdot 0$ on days with rain. Because *Sporobolomyces* spp. colonize leaf surfaces, it is not surprising that numbers of spores are influenced by their hosts' nutrition. Concentrations within crops given a mixture of nitrogenous, phosphatic and potassic fertilizers were $2 \cdot 5$–$11 \cdot 8$ times greater than those within crops without fertilizers. These data apply to numbers of ballistospores trapped comparatively near to sites of release, and here, the diurnal fluctuations are readily detected. Once in the main air-mass, however, this effect lessens.

Using a volumetric spore trap, Hamilton (1959) compared air samples taken in London with those from an agricultural area and found periodicity effects similar to those recorded by Gregory and Hirst; also that fewer *Sporobolomyces* spores were trapped in urban than in rural area. Gregory and Sreeramulu (1958), during a two-week sampling period from June to July 1954, found concentrations of 200,000 and 3,000 spores of *Sporobolomyces* and *Tilletiopsis* respectively per m³ of estuarine air in the early morning. Later in the day mean concentrations decreased, the decrease being sharper with *Tilletiopsis* than with *Sporobolomyces*.

From an analysis of counts obtained at different times of the day and on different days of the year, Hamilton hinted that numbers of trapped *Sporobolomyces* spores (a) were directly associated with increasing dew-point values, and (b) reached a peak at temperatures ranging between 16 and 18°. In her work, she trapped yeast-like propagules and reliably identified those of *A. pullulans*. Unlike members of the humidity-dependent Sporobolomycetaceae, peak concentrations of *A. pullulans* usually occurred during the afternoon with the largest numbers being trapped at temperatures of 24–26°, an effect reflected in the later seasonal maximum occurring during July–September.

Hyde and Adams (1960) sampled the air spora in Wales, and noted that numbers of *Sporobolomyces* were greater in wet than in dry summers. Unlike Hamilton who used a volumetric trap, they either exposed Sabouraud's agar in Petri dishes or sucked air over adhesive surfaces, so increasing the chances of identifying yeast propagules. With the former method, peak concentrations of *Candida* and *Aureobasidium* were detected in early spring and summer respectively; with the latter, the mean daily atmospheric concentration of spores in 1958 was found to be 2,200 per m³ and of these 0·5 and 13·6% were propagules of *Aureobasidium* and *Sporobolomyces* respectively, compared with 20·6% of the saprophytic mould *Cladosporium*.

Ripe (1962), who was working with allergens in Sweden, exposed Petri dishes for periods of 30 min on weekdays from February 1959 to February 1961. With two types of media, which after exposure were incubated for one week at 27°, he found that *A. pullulans* usually formed 61% of the total number of colonies cultured (Fig. 1) and that other yeasts were responsible for 15% including numerous pigmented forms, possibly of the Rhodotoruloideae and/or Sporobolomycetaceae. Numbers of *A. pullulans* followed a well-defined seasonal pattern, with maximum concentrations from July to October. While making his survey, Ripe noticed that fewer yeasts were trapped during rainfall than afterwards. Interestingly, Turner (1966), working in Hong Kong where temperatures are considerably higher, found, like Hyde and Adams (1960) in Cardiff, Hamilton (1959) in London and Ripe (1962) in Sweden, that the seasonal peak of *A. pullulans* occurred in, and towards the end of, summer. Turner (1966) exposed Petri dishes containing Czapek-Dox agar and Rose Bengal during the day and found that

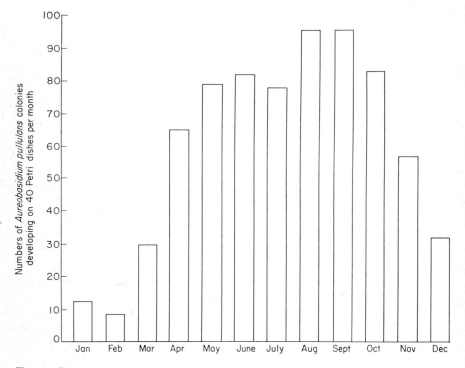

FIG. 1. Seasonal changes in numbers of colonies of *Aureobasidium pullulans* (de Bary) Arn. developing on agar media within Petri dishes exposed in Sweden to the atmosphere for 10 min periods at mid-day. Data, which are produced with permission of Ripe (1962), are monthly totals of 40 exposed Petri dishes (about 8·5 cm diam.).

A. pullulans formed 17% of the viable propagules trapped during the year. In Hong Kong, with mean daily temperatures of 16–27°, large concentrations of *A. pullulans* were sometimes associated with rainfall. Unlike the results obtained in higher latitudes, the peak values for *Sporobolomyces* spp. did not occur in late spring and early summer but instead during late summer and early winter, possibly a reflection of the inability of this fungus to sporulate at high temperatures.

Jimenez-Diaz *et al.* (1960), in Madrid, drew air through a tube lined with synthetic culture media, and identified 12·5% of the developing fungal colonies as yeasts. This proportion is less than that usually found in rural sites, possibly reflecting the arid nature of the Mediterranean environment or the selectivity of the media used.

Although not strictly applicable to plant environments, observations made by di Menna (1955b) are of relevance because she identified many yeasts. During the period March 1953 to April 1954, she took 17 air samples from within buildings in Dunedin, New Zealand, collections being made about noon. Of 2,730 fungal colonies isolated, 129 (5·7%) were identified as yeasts. In another series of samples taken with a Manning Slit Sampler, propagules were impacted on Sabouraud's agar and 96% of the developing yeasts belonged to one of four groups: 42% *Cryptococcus* spp.; 26%, perfect and imperfect *Debaryomyces* spp. including *Torulopsis candida* (Saito) Lodder and *T. famata* (Harrison) Lodder and van Rij; 18·6%, *Sporobolomyces* spp. and *Rhodotorula* spp.; 9·5%, yeast-like phases of *Cladosporium*, sometimes known as *Torula nigra* (*vide* de Vries, 1952). Connell and Skinner (1953) using a bubbler device found that 44% of viable air-borne yeast propagules developed into colonies of "black yeasts"; 33% were of *Rhodotorula* and *Sporobolomyces* species, 11·5% of colourless non-fermenting isolates of *Cryptococcus*, *Candida* and *Lipomyces*, and the remaining fermenting organisms belonged to the genera *Torulopsis* and *Saccharomyces*.

B. LEAF COLONIZERS

The term "phyllosphere" was coined more or less simultaneously by Last (1955a) and Ruinen (1956) working in the U.K. and Indonesia respectively, to refer to the milieu on surfaces of leaves. However, to be in agreement with notation used in root investigations, the word "phyllosphere" should be restricted to the zone near leaves and "phylloplane" used when referring to actual leaf surfaces (Kerling, 1958). Phylloplane organisms are sometimes so numerous that they discolour leaves. Members of the Cyanophyceae and sooty moulds give a blackish coloration, Chlorophyceae a light green or reddish tint, and filamentous fungi a greyish appearance. On a range of hosts in the tropics, Ruinen (1961) found that numbers of bacteria, usually confined at first to leaf depressions, often increased to $1·3 \times 10^7$ cells per cm^2 before significant colonization by fungi, including yeasts and filamentous types, occurred, the mixture of micro-organisms sometimes forming layers 20 μ thick.

Most of Ruinen's (1963a) work was orientated to the processes involved in nitrogen fixation and the techniques used tended to favour types that either fixed or needed only very small exogenous supplies of this element. Perhaps the limitations inherent in her selective techniques minimized the importance of yeasts as compared with that of the N_2-fixing epiphytic bacterium *Beijerinckia*. However, the yeasts that she found on foliage of *Citrus*, *Theobroma*, *Tillandsia*, *Ixora* and *Arenga* were widespread in the tropics, occurring in Indonesia, Surinam and the Ivory Coast, and they closely resemble those isolated by others in cooler climates. Of 65 isolates examined in detail, 30 belonged to the genus *Cryptococcus* followed in numerical importance by *Rhodotorula, Sporobolomyces, Candida* and *Aureobasidium*, the last appearing more frequently when her *Azotobacter* enrichment medium was slightly acidified.

When interpreting sets of data detailing populations from leaf surfaces, it is essential to bear in mind the limitations of the techniques used. This is well illustrated by some of Dickinson's (1967) data when he was identifying fungi colonizing leaves of *Pisum*. He found that numbers of yeasts obtained per unit area, from leaf washings plated on potato-dextrose agar supplemented with aureomycin, were fewer than when nail varnish replicas of the phylloplane were examined microscopically. When assessed microscopically, yeast-like cells outnumbered other fungal propagules, including those of *Cladosporium* and *Stemphyllium*. On young, green two-week old leaves, yeasts and the yeast-like *A. pullulans*, which he counted separately, accounted for 701 and nil respectively of a total fungal propagule count of 866 per cm^2. When leaves were yellowing, numbers of yeast-like cells increased to 2,740, reaching 2.9×10^4 when browning occurred. At this stage there were, for the first time, 740 propagules of *A. pullulans* and 4,000 of other fungi including spores and hyphal fragments of *Ascochyta* and *Alternaria*.

Kerling (1958), in the Netherlands, followed changes in numbers of bacteria, yeasts and filamentous fungi washed from ageing fodder beet leaves. Numbers of bacteria and yeasts usually greatly exceeded those of other fungi and, whereas large populations of yeasts occurred during May (in the spring), comparable populations of bacteria did not develop until July and August. With the onset of autumn, numbers of all microorganisms decreased (Fig. 2). The seasonal trends were erratic, possibly because the infrequent samples were too greatly affected by weather immediately beforehand. Thus, the numbers of yeasts detected on June 11th were less than those on May 21st or July 2nd and it was suggested that the decline was attributable to heavy rain washing fungus propagules from leaves immediately before sampling. During Professor Kerling's period of sampling, the numbers of filamentous fungi, yeasts and bacteria reached maxima of around 300, 2,300 and 3,000 propagules per cm^2 respectively, the yeasts including *Sp. roseus* Kluyver et van Niel, *Cryptococcus laurentii* (Kufferath) Skinner, *Cr. albidus* (Saito) Skinner, *Torula* sp. and *A. pullulans*.

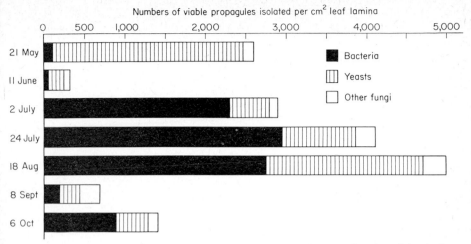

Fig. 2. Seasonal changes in numbers of yeasts, filamentous fungi and bacteria isolated from leaves of fodder beet (*Beta vulgaris* L.) first exposed in May. Reproduced with permission from Kerling (1958).

Like Ruinen, Diem (1967) agrees that bacteria are more numerous than yeasts which in turn occur in greater numbers than filamentous fungi, but the relative proportions differed greatly. He measured the concentrations of bacteria, yeasts and other fungi which developed when barley-leaf washings were plated on a range of agar-containing media. At the end of May, a total of 4,570 micro-organisms were isolated per cm^2 of leaf and of these 98·5% were bacteria. One month later, still on green leaves, the numbers of micro-organisms had increased to 71,000 per cm^2, the proportion of bacteria having decreased from 98·5% to 94·2%, the difference being mainly composed of pink yeasts, probably species of *Rhodotorula* and *Sporobolomyces*. The proportion of other fungi in the total declined from 1·5 to 0·2%, a decrease paralleled by a decrease in the proportion of *Aureobasidium* isolates which in May accounted for 90% of the fungal isolates; Diem (1967) did not include *Aureobasidium* within his yeast category. At the end of July, foliage was moribund and numbers of all types of micro-organisms increased greatly, the densities of bacteria, pink yeasts and other fungi being $5·0 \times 10^6$, $1·7 \times 10^4$ and $1·1 \times 10^4$ per cm^2 respectively.

In his survey, Diem (1967) found that the numbers of micro-organisms on foliage were affected by the position of the host within the cropping area. Fewer organisms occurred on plants sampled at the fringe than in the centre of the cropping area, possibly reflecting smaller concentrations of inocula at the fringe where spore clouds were more readily dissipated. Diem (1967) also indicated that more viable propagules were washed from shaded plants than from those exposed to the sun but, again, this effect, possibly attributable to greater exposure to ultraviolet light or related to changes in the host's metabolism, is confounded with effects on the dispersion of spore clouds.

From spot observations in different seasons of the year, it is usually impossible to characterize the factors determining the sequence of saprophytic colonization. On leaves, as suggested by Ruinen, yeast

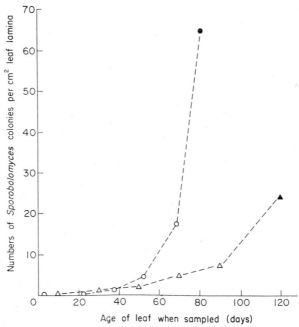

FIG. 3. Successive changes in numbers of *Sporobolomyces* colonies per cm² of leaf blade isolated from two groups of mainstem leaves taken from winter-sown wheat (after Last, 1955a). △ and ▲; leaves surviving for about 110 days and first exposed during the winter; ○ and ●, leaves surviving for about 70 days and first exposed during late spring. △ and ○, leaves alive; ▲ and ●, leaves dead.

colonization may depend on the earlier activities of bacteria which make essential nutrients available, an hypothesis not inconsistent with Diem's data. On the other hand, the progression from bacteria to yeasts may reflect changes in nutrients released unaided by the ageing host. In a study of the occurrence of micro-organisms isolated from cereal leaves, using a modification of the spore-fall technique devised by Kluyver and van Niel (1924–25), Last (1955a), in the U.K., found that numbers of *Sporobolomyces* colonies isolated per unit area depended on season and leaf age. At any one time, it was usually found, when leaves were systematically sampled from individual plants, that numbers of

Sporobolomyces colonies per cm² increased with increasing leaf age. Thus, when winter wheat was harvested on 25th June 1954, the numbers of colonies isolated from the youngest leaf (leaf 13) and from leaves 12, 11 and 10 were 0·7, 10·0, 20·2 and 41·3 respectively with 47·5 occurring per cm² on leaf 9 which was moribund. Kerling (1964) found that numbers of yeasts isolated from strawberry leaves increased from 1·8 × 10² to 1·4 × 10³ when green to 1·2 × 10³ to 3·7 × 10⁴, and 3·0 × 10⁴ to 4·6 × 10⁴ when discoloured and dead respectively. From successive samples, Last found, regardless of the time of year, that few colonies were isolated before leaves had lived half their lives, but numbers of colonies then progressively increased reaching a maximum when leaves were dead. Numbers of *Sporobolomyces* colonies isolated began to increase appeciably when leaves of winter-sown wheat first unfurled during the winter and spring, and persisting for 110 and 70 days respectively, were 55 and 35 days old (Fig. 3). More colonies were isolated in the late spring and summer than in the winter, so agreeing with observations made by di Menna (1959) in New Zealand.

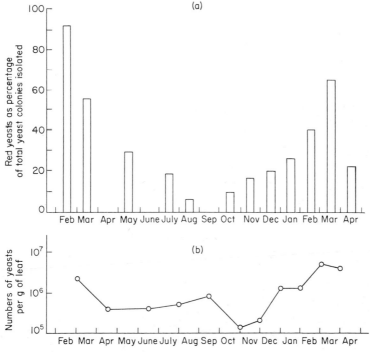

FIG. 4. Percentages of red species (a) and numbers of viable propagules per g of leaf, fresh wt (b) in yeast populations isolated from foliage of New Zealand pasture plants. Running means were plotted with permission from di Menna (1959).

Using a washing technique, di Menna found that the numbers of yeasts from surfaces of pasture grasses, including *Lolium perenne* L., increased during the year from 10^5 to $3 \cdot 0 \times 10^6$, and exceptionally 10^8, per g fresh weight of foliage, the numbers being lowest during the winter and early spring (September–November) and reaching a peak in the summer (March). During the year, the balance of species changed greatly, and the changes were associated with the change of season. The proportion of red (including species of *Rhodotorula* and *Sporobolomyces*) to total yeasts changed from 6% in winter to 92% in late autumn (Fig. 4). Although these pasture plants were growing in soils with many colonies of *Cr. albidus*, *Cr. diffluens* (Zach.) Lodder and van Rij, *Candida humicola* (Daszewska) Diddens and Lodder and *C. curvata* (Diddens and Lodder) Lodder and van Rij, these species were rarely isolated from leaves, their numbers never exceeding 8% of the total yeast flora. Instead, the numerically important phylloplane species of yeasts were the colourless *Cr. laurentii* and *T. ingeniosa* di Menna, and pink/red *Rh. marina* Phaff, Mrak and Williams, *Rh. graminis* and *Sp. roseus*. Di Menna's samples were taken from many different localities but, within wide limits, phylloplane floras seem to be independent of site. Fewer colonies of *Cr. laurentii* were isolated from young than from old leaves on which populations might contribute one per cent to the total foliar fresh weight, assuming 7×10^9 yeast cells weigh 1 g.

When studying the succession of microfungi on leaves of *Fagus sylvatica* L., Hogg and Hudson (1966) isolated numerous colonies of *Sporobolomyces*, *Bullera*, *Tilletiopsis* and *Itersonilia* from newly exposed leaves. Although ballistospores of the first two yeast-like members of the Sporobolomycetaceae occur in very high numbers in estuarine atmospheres (Gregory and Sreeramulu, 1958), colonies were not isolated from the foliage of *Halimione portulacoides* (L.) Aell., a salt-marsh plant sometimes inundated by tidal waters, but other yeasts were found (Dickinson, 1965).

Although *A. pullulans* is often associated with decaying vegetation, it is nevertheless unable to decompose cellulose (Hogg, 1966). Smit and Wieringa (1953) obtained cultures of this yeast-like organism from beech leaves, appreciable numbers of colonies being isolated from green buds and young leaves. They were unable to observe *A. pullulans* microscopically on stem sections of *Agropyron repens* Beauv. incubated for two days in humid conditions, but Hudson and Webster (1958) commonly found this micro-organism when washed segments were suspended in layers of maize extract-agar to which Rose Bengal, penicillin and streptomycin were added. Like Smit and Wieringa (1953), they isolated it from leaves just beginning to unfurl, but Pugh (1958) obtained the organism only from dead foliage of *Carex paniculata* L. When plating potato-leaf washings, made with a 0·1% solution of Tween (a surfactant), on Czapek Dox agar supplemented with yeast extract, Holloman (1967) found that viable propagules of *A. pullulans* were more abundant on the phylloplane than those of any other fungus, an observation confirmed by direct microscopy of collodion leaf imprints. As with micro-organisms isolated from cereals (Fig. 3, p. 192) and sugar beet (Fig. 5), numbers of viable propagules usually increased as potato leaves aged, from 110 to 1,100 propagules per cm², but, in contrast, the proportion of *A. pullulans* decreased, in one instance from 84 to 15% during a period of eight weeks and from 86 to 60% in another. As the percentage of *A. pullulans* decreased, that of *Cladosporium* increased. Holloman (1967) used a second technique for isolating leaf colonizers in which leaves were first surface-

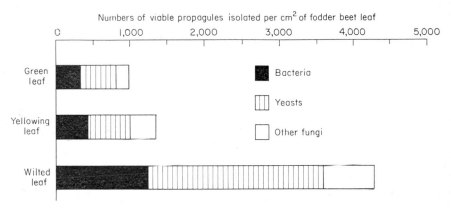

FIG. 5. Effects of leaf condition and age on numbers of bacteria, yeasts and other fungi washed from leaves of fodder beet (*Beta vulgaris* L.) sampled in the autumn. Reproduced with permission from Kerling (1958).

sterilized with hypochlorite before being incubated. But, with this method, *A. pullulans* was rarely detected, indicating that it grows superficially. Colonies of *Sporobolomyces* spp. were obtained from potato leaves when the spore-fall method was used.

To avoid over-estimating numbers of the fungi which produce conidia prolifically, Harley and Waid (1955) developed a technique in which plant fragments were washed and shaken in a series of distilled water aliquots. With this method they isolated numerous *Aureobasidium* colonies from beech-leaf petioles but rarely from root tips of the same tree, although other workers have detected this yeast-like micro-organism in surface soil.

The epiphyllic microflora on living pine needles acts as the precursor of that which subsequently plays a part in the conversion of vegetation into soil organic matter. *A. pullulans* rarely occurs on living needles during winter and spring but, during the summer, populations increase. It occurs on freshly fallen needles in the litter layer where Kendrick and Burges (1962) identified its pycnidial stage, *Sclerophoma pithyophila* (Corda) von Höhn, but it rarely occurs at greater depths. Although *A. pullulans* is found in soil, observations made by Harley and Waid (1955) indicate that foliage in different states of decay, and not the roots, provides its energy source.

Using a washing technique, Voznyakovskaya (1962) in the U.S.S.R. isolated, on cabbage-agar with added wort and maize, a range of epiphyllic micro-organisms from a diverse selection of hosts, the yeasts including colonies of *Sp. roseus*, *Rh. rubra* (Demme) Lodder, *Rh. mucilaginosa* (Jörg.) Harrison, *Rh. aurantiaca* (Saito) Lodder, *Cr. laurentii*, *Cr. diffluens*, *Cr. albidus*, *Tr. cutaneum* (de Beurm, Gougerot and Vaucher) Ota and *A. pullulans*. Her survey served to stress the ubiquity of most leaf yeasts and it also indicated the lack of host specificity among them, a feature contrasting with the restricted host range of the epiphytic *Sclerographium phyllanthicola* Deight, a member of the Moniliales (Last and Deighton, 1965). She washed *Cr. diffluens* from leaves of apple, oats, oak, horsetails and a species of fern, a general distribution also highlighted by Tubaki (1953) who isolated *Bullera alba* (Hanna) Derx, *Sp. holsaticus* Windisch, and *Sp. odorus* from many different living and dead plant substrates. *Cryptococcus laurentii* has been reported from northern and southern temperate regions and from the tropics, a distribution reflecting the ability of different isolates to grow over a wide range of temperatures. Cooke (1965) found that seven of 45 isolates of this yeast grew at 37° as did 12 of 191 isolates of *Rh. glutinis* (Fres.) Harrison. *Cryptococcus laurentii*, together with *Cr. albidus* and *Rh. minuta* (Saito) Harrison., was also found in Antarctica, colonizing the moss *Bryum antarcticum* (di Menna, 1960a).

Although its natural habitat is considered to be the digestive tract of warm-blooded animals (van Uden *et al.*, 1958), *C. albicans* (Robin) Berkh., the causative organism of moniliasis, has repeatedly been isolated from plants. In addition to its occurrence on stands of rye grass and clover (di Menna, 1958a), and on gorse and myrtle (van Uden *et al.*, 1956), appreciable populations were found on *Achillea*, *Trifolium* and *Chrysanthemum* by Keymer and Austwick (1961), records suggesting that *C. albicans* is not just a chance contaminant.

Epiphytic members of the Micropeltaceae and Chaetothyriaceae are usually found colonizing only the upper surfaces of leaves. More *Sporobolomyces* colonies were found on the lower than on the upper leaf surfaces of chrysanthemum and rowan (*Pyrus* (*Sorbus*) *aucuparia* (L.) Ehrh.), but on cereals there were no consistent differences (Last and

Deighton, 1965). Colonies of yeasts were found by di Menna (1959) usually in the depressions overlying cell walls arranged at right angles to the epidermis.

So far it has been assumed that the effects on yeast populations of factors such as changing leaf age have been exerted by undamaged hosts. But numbers of *Sporobolomyces* are increased when leaves are parasitized by rust fungi (*Puccinia menthae* Pers. on mint), plant parasitic nematodes (*Aphelencoides ritzema-bosi* (Schwartz) Steiner on chrysanthemum) and gall-forming mites (*Eriophyes macrorrhyncus*), the yeasts seeming to act as scavengers using the increased amounts of nutrient which become available when cells are ruptured by the different pathogens.

C. FLOWER AND FRUIT COLONIZERS

Yeast ecology poses many intriguing problems. With populations colonizing leaves, it is possible to trace with certainty the complete chain of events from host colonization to subsequent spore production and dispersal, at least for the violently spore-discharging species of *Sporobolomyces* and *Bullera*. The last link of this chain is, however, virtually unknown for the yeasts growing on flowers and fruits, with the outstanding exception of *Nematospora corylii* Peglion. Yeasts occurring on plants are mostly thought to be saprophytic but *N. coryli* is an important pathogen, being involved with insect stainers (*Dysdercus* spp.) in stigmatomycosis, particularly of cotton.

1. *Saprophytes*

Flowers (nectar) and fruits have for long been thought as ideal habitats for yeasts because of their high sugar content, and for many years yeasts were not believed to occur elsewhere. Hansen (1881, 1882) considered that yeasts mainly inhabited ripe sweet fruits during the summer and early autumn, having been spread by insects and birds, and that they overwintered in soil and were redistributed in the following spring by wind and rain. This picture was quickly replaced when improved techniques of identification indicated that there were innumerable yeast species, distributed among different ecological niches, some being specialized.

Lund (1954) found that a greater proportion of flowers was colonized by yeasts in July and August than in March and April. He isolated 27 different species, but only two of them produced ascospores, viz. *Saccharomyces* sp. and *Hansenula* sp.; the remainder included ten species of *Torulopsis* and nine of *Candida* with *C. reukaufii* (Grüss) Diddens and Lodder and *T. famata* (the imperfect form of *Debaryomyces kloeckeri* Guill. and Peju) being the most widespread. Martin (1954) isolated *C. reukaufii* from flowers of *Lamium album* L., *Glechoma lederaceae* L., *Trifolium pratense* L. and *Linaria vulgaris* Mill., representatives of three families of flowering plants.

Whereas Lund isolated and identified yeasts from whole flowers, Capriotti (1953, 1955a) carried this process further by dissecting flowers into their component parts before making cultures from 20 species of trees and shrubs. More yeasts were isolated from stamens and stigmas than from corollas and calyxes and, as found by others, most, including species of *Candida, Torulopsis, Kloeckera* and *Rhodotorula*, were asporogenous. Etchells *et al.* (1954) traced the cause of cucumber spoilage to damaging populations of *Rhodotorula* sp. which first colonized flowers.

Because very few yeasts are found on unripe grapes (Pasteur, 1876, 1878), it is not surprising that Capriotti was unable to isolate them from flowers of *Vitis*. However, as fruits begin to ripen, yeast populations increase. Lund detected 10^3–10^6 yeasts per ripe strawberry and gooseberry, colony types being similar to those found on flowers (*Candida, Torulopsis, Cryptococcus* and *Kloeckera* spp.). As shown by Hansen (1881), *Kloeckera apiculata* (Reess emend. Klöcker Janke syn. *Saccharomyces apiculatus* Reess) commonly occurs on ripening grapes, and *C. pulcherrima* (Lindner) Windisch was found on cherries by Windisch (1940).

The successive changes as fertilized flowers develop and ripen into mature fruits provide a slowly changing but increasingly selective range of environments to which populations of yeasts must adapt. This is well illustrated in wine production where strains of *Sacch. cerevisiae* Hansen and *Sacch. cerevisiae* var. *ellipsoideus* (Hansen) Dekker, which are rarely found on unpicked fruit, rapidly supersede the earlier more diverse range of species, including *Candida, Torulopsis, Kloeckera* and *Hansensiaspora* spp. (Mrak and McClung, 1940). As indicated by Clarke *et al.* (1954), populations fermenting fruit juices, usually containing many ascopore-formers, bear little resemblance to the aerobic types on intact fruits. Nevertheless, bearing in mind the value of the industry, it is surprising how few attempts to isolate wine yeasts from fruits have been recorded since the days of Pasteur and Hansen. Studies on the production of palm wine, an important vitamin source in areas where malnutrition is common, have shown the presence of fermenting yeasts (Ahmad *et al.*, 1954). In the preparation of Kaffir beer from sorghum, van der Walt (1956) isolated many colonies of *Sacch. cerevisiae, C. krusei* (Cast.) Berkh. and *K. apiculata* from malt, but it cannot be inferred that *Sacch. cerevisiae* was carried on grain. In California, Bioletti and Cruess (1912) isolated some colonies of *K. apiculata* and fewer of *Sacch. cerevisiae* var. *ellipsoideus* from Muscat grapes. Schanderl (1957) has suggested that some of the characteristic wine flavours associated with different European valleys are attributable to contaminating soil yeasts which are spread to trailing crops by rain splash, and which in themselves contribute little to alcohol production. In contrast, where vines in the Southern Tyrol are prevented from trailing on the ground, yeasts must be added to trigger the fermentation of pressed juice.

The changing environment, as fruits ripen, not only influences yeasts but also other types of micro-organism and doubtless causes many microbial interactions. For example, Uroma and Virtanen (1949) suggested that unsaturated fatty acids produced by yeasts might restrain the development of Gram-positive bacteria. If this were so, it might explain the differing changes in populations of yeasts and bacteria found on ripening apples by Marshall and Walkley (1951). Although it is thought by some that yeast colonization follows, and is dependent on, that by bacteria, Mossell and Ingram (1956) think otherwise, suggesting that the osmophilic yeasts, by decreasing sugar concentrations, make some fruits and juices more suitable for organisms less tolerant of high osmotic pressures. Challinor and Rose (1954), studying the dependence of bacteria on yeasts, found that the latter secreted growth-promoting substances, nicotinic acid and thiamine, essential for the growth of lactobacilli.

Just as Capriotti (1953, 1955a) showed that different floral structures carried different yeast populations, there is evidence to suggest that fruits are not uniformly contaminated, changes being caused by fruit maturity, by cultural treatments and by environmental factors operative just before sampling. The source of yeasts fermenting apple juices has intrigued many people in the past, leading some to suspect that they occurred within undamaged tissues. Marshall and Walkley (1951) examined samples of Bramley Seedling apples and found about 14,000 yeast cells per g of tissue. Further analyses indicated that most of these were associated with the epidermis (about 13,000 per g), some with the core (approx. 1,000 per g), and very few (within limits of experimental error) with previously unexposed flesh. As apples aged, initially high populations of acetic acid- and lactic-acid bacteria on Grenadier fruit, detected by washing and swabbing techniques, progressively decreased concomitantly with progressively increasing numbers of yeasts which reached a peak at the end of September and subsequently decreased (Fig. 6).

Williams et al. (1956), in Canada, isolated yeasts from apple epidermis, and found that absolute numbers tended to decrease as the crop matured; this was in contrast to the observations made by Marshall and Walkley (1951). However, between the end of September and mid-October, this trend was reversed and the mean numbers at two sites

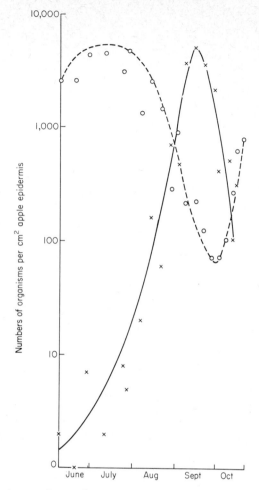

FIG. 6. Changes in numbers of yeasts ($\times$ — $\times$), and combined totals of acetic acid- and lactic acid-bacteria ($\bigcirc$—$\bigcirc$) isolated per cm^2 of epidermis sampled from ageing apples. Reproduced with permission from Marshall and Walkley (1951). Copyright © 1951 by Institute of Food Technologists.

increased from 3,250 to 10,300 per g. of apple tissue. From mid-August to mid-October, *C. malicola* Williams, Wallace and Clarke was isolated on more occasions than any other yeast, followed in order of frequency by *Rh. glutinis*, *Cr. laurentii*, *T. famata* and *T. candida*.

In the literature, there are many apparent inconsistencies with phylloplane microfloras differing before and after rain. Beech (1959) found that 35 and 50% of cultures obtained from the skin of unwashed Kingston Black cider apples were isolates of *C. pulcherrima* and *D.*

kloeckeri respectively, but, on a similar but previously washed batch of apples, these species were unrepresented. Instead, *C. catenulata* Diddens and Lodder, and *Hansenula* spp. were numerous. Although *C. pulcherrima* in one instance occurred in appreciable numbers in the eyes of unwashed fruit, populations found in this habitat usually differed from those occurring on the skin. The hyaline species found on apple epidermis by Beech (1959) differed from those recorded by Clarke *et al.* (1954) and Williams *et al.* (1956) who recorded appreciable numbers of pigmented forms, possibly because apple varieties and isolation techniques differed. Clarke *et al.* (1954) suggested that the continued presence of members of the Cryptococcoideae after rain might be attributed to their starch capsules, causing clumping of propagules and increasing their adhesion to the fruit. Unlike many workers who have studied the distribution of yeasts on plant material, Bowen and Beech (1964) used a selective medium favouring the development and recognition of a single species, the pigment-producing *C. pulcherrima*. The medium, an apple juice-yeast extract-agar, contained antibacterial agents and ferric ammonium citrate. Of the $12-95 \times 10^4$ yeasts per g obtained when macerates of bitter-sweet, sharp and bitter-sharp varieties were inoculated on this medium, at least 80% of the viable propagules belonged to *C. pulcherrima*, but on the sweet variety, Sweet Alford, this species accounted for only 8% of yeasts, it being superseded by *Rh. glutinis*. The first survey done by Bowen and Beech (1964) was made on different cider-apple varieties within the same orchard but, when other orchards were examined, the numerical superiority of *C. pulcherrima* was usually confirmed. Within an orchard, two species were usually dominant, accounting for 80% of the yeast isolates; important yeasts in addition to *C. pulcherrima* included *C. krusei*, *Rhodotorula* spp., *Sp. roseus*, *Sacch. delbruckei* Lindner, and *D. kloeckeri*, the perfect and imperfect forms of the latter both occurring.

In 1967 Bowen and Beech recorded the effects of handling practices on apple yeast floras. They found that *C. pulcherrima*, the dominant yeast on unpicked fruits, was rapidly superseded when apples were detached and either stored or allowed to rest on the ground. On the ground, it was usually displaced by other species of *Candida*, and more interestingly by *Torulopsis* spp. and *Rhodotorula* spp., none of which is readily isolated from soil. In contrast, when Dabinett apples were stored in sacks, *C. pulcherrima* was displaced by *K. apiculata* which imparts the characteristic taste to cider. After two weeks' storage, the proportion of yeasts identified as *C. pulcherrima* decreased from 60% to 5%, whereas those of *K. apiculata* increased from 10% to 39%. These examples, using a contrasting selection of apples (i.e. Bramleys Seedling, Grenadier, Kingston Black, Sweet Alford and Dabinett), suggest, contrary to the evidence obtained from flowers, that the yeast floras of different apple varieties are distinct but may be modified considerably after picking. Recca and Mrak (1952) found a greater range of yeasts, including species of *Hanseniaspora*, *Candida* and *Pichia*, on oranges than on lemons.

There are close similarities between the yeasts found on fruits, especially apples, in Europe and North America. However, when Sasaki and Yoshida (1961) sampled a range of fruits in Japan, *C. pulcherrima* was absent and instead the dominant species were *A. pullulans*, *T. candida* and *Rh. glutinis*.

As fruits ripen, sugar concentrations approaching 60% have been recorded. In these and similar environments, such as honey and maple syrup, species of *Zygosaccharomyces*, *Hanseniaspora* and *Candida* are found to occur, the species of *Zygosaccharomyces* possibly being the haploid state of *Sacch. rouxii* Bontreux. Of these, *Zygosacch. barkeri* Sacc. and Sydow, *Zygosacch. japonicus* Saito var. *soya* (Saito) Dekker and *Zygosacch. nadsonii* Guill. grew appreciably in solutions containing 60% date syrup, whereas the growth of *C. krusei*, frequently found on apples, and *C. tropicalis* (Cast.) Berkh. was inhibited by concentrations of 50% (Mrak *et al.*, 1941). Workers who have followed the pattern of micro-organisms colonizing slime fluxes exuding from trees have been impressed by the paucity of yeast species seemingly capable of utilizing this specialized substrate. Not surprisingly their species composition remains stable for many years despite the opportunities for other colonizers to invade the seasonally added new flux. Miller *et al.* (1962) recorded the presence of *Pichia silvestris* Phaff and Knapp and *Pi. pastori* (Guill.) Phaff in fluxes of *Abies magnifica* Murr., *Populus trichocarpa* Torr. and Gray, and *Quercus kelloggii* Newb.; Lund (1954) found *C. pulcherrima* and *C. krusei* in those of *Aesculus*, *Betula* and *Ulmus* spp. Others occurring in flux, viz. *Sacch. mellis* (Fabian and Quinet) Lodder and van Rij and *T. magnoliae* Lodder and van Rij, have also been isolated from dried prunes, being capable of fermenting media containing 70% sugar (Tanaka and Miller, 1963).

Increasing selectivity has probably played a part in the evolution of different associations between insects and yeasts, ranging from the casual association of common plant yeasts used for food to those forming a close symbiosis, e.g. that between figs, bacteria, yeasts and fig wasps (Phaff *et al.*, 1966). Although Sergent and Rougebief (1926), like Hansen, considered that fruit flies (*Drosophila* spp.) carried wine yeasts as surface contaminants, El Tabey Shehata *et al.* (1955) were surprised to isolate yeast populations with species of *Saccharomyces* and *Kloeckera* as numerical dominants from fruit flies feeding on plant substrates colonized by species of *Candida* and *Pichia*. However, the interrelations between insects and yeasts are beyond the scope of this chapter. Not unexpectedly, Lund (1954) found that nectar-feeding wasps (*Vespa* spp.) and bees (species of *Apis* and *Bombus*) yielded cultures of *C. reukaufii*, *C. pulcherrima* and *K. apiculata*, yeasts typical of flowers. That an obligate relationship sometimes exists is suggested by the death of some wood-destroying insects reared in the absence of their yeast symbionts (Koch, 1960, 1963).

Although a considerable range of yeasts occur on flowers and fruits, there are few records of the presence of ballistospore-producers. Lund (1956), however, when examining ripe barley grains before and immediately after threshing, detected 7.0×10^5 viable yeast propagules per g, with species of *Sporobolomyces* and *Rhodotorula* being the main colonizers; colonies of *Hansenula, Torulopsis* and *Candida* were also cultured. Teunisson (1954) found that numbers of yeasts on rice increased considerably, from 1.0×10^3 before storage to 8.5×10^6 per g when stored in containers. The dominant yeasts were *Endomycopsis chodati* (Nechitch) Wickerham and Burton, *Hansenula anomala* (Hansen) H. and P. Sydow and *Pi. farinosa* (Lindner) Hansen which were able to survive a temperature of 39° and low oxygen tensions.

Mention has been made of the fruits of higher plants, but yeasts also colonize the fruiting bodies of fungi. *Sporobolomyces coralliformis* Tubaki and *Sp. odorus* were cultured from species of *Exidia* and *Clavaria* (Tubaki, 1958, 1953), and *C. parapsilosis* (Ashf.) Lang. and Talice, *C. mesenterica* (Geiger) Diddens and Lodder and *T. colliculosa* (Hartmann) Sacc. from sporophores of *Amanita muscaria* (L.) Fr., *A. pantherina* Secr., *Lactarius subdulcis* Fr. and *Russula nigricans* Fr., numbers of yeasts per g usually increasing with age and sometimes reaching 10^4 (Lund, 1954). Anderson and Skinner (1947) attempted to isolate yeasts from the decomposing fruit bodies of agarics and ascomycetes, but few were cultured from fresh structures until decomposition changed the substrate from alkaline to acid. Most fruit bodies contain appreciable amounts of trehalose which yeasts cannot assimilate, but decay-producing bacteria convert this sugar to glucose so favouring invasion by *Sacch. cerevisiae, Sacch. steineri* Lodder (syn. *Sacch. chodati* Steiner) and *Sacch. marxianus* Hansen (syn. *Sacch. muciparis* Beijerinck).

2. *Parasites*

Although yeasts may influence plant pathogenesis by eroding leaf cutin and thus facilitating the entry of plant pathogens and/or increasing transpiration losses (Ruinen, 1966), few are regarded as plant primary pathogens. Some species of the Sporobolomycetaceae were first described from diseased plants, but the yeast-like types of *Bullera* and *Sporobolomyces* are not primary causes of plant loss; instead they colonize lesions made by pathogens. Numbers of *Sporobolomyces* colonies were increased five times when mint leaves were infected with *Puccinia menthae* Pers. (Last and Deighton, 1965). Species of the mycelial genus *Itersonilia*, within the Sporobolomycetaceae, are, however, primary pathogens, *I. pastinacea* Channon and *I. perplexans* Derx attacking parsnips and a range of flower crops respectively (Channon, 1963; Gandy, 1966).

In their taxonomic study, Lodder and Kreger–van Rij (1952) regard *Nematospora lycopersici* (Schneider), *N. phaseoli* (Wingard) and *N. nagpuri* Dastur (Dastur and Singh) as synonyms of *N. coryli* Peglion, *Nematospora* Peglion being a monospecific genus within the tribe Nadsonieae of the subfamily Saccharomycetoideae. *N. coryli* produces large asci with eight ascopores arranged in two bundles of four, and it was on the basis of spore morphology and pathogenicity, without regard to methods of vegetative growth, that Gaumann (1949) unacceptably grouped species of *Spermaphthora*, *Eremothecium*, *Ashbya* and *Nematospora* into the family Spermaphthoraceae. *Nematospora gossypii*, described as a new species by Ashby and Nowell in 1926, was switched to the non-yeast genus of *Ashbya* by Guilliermond (1928) because, although spores of *N. gossypii* and *N. coryli* were alike, *N. gossypii* usually develops plurinucleate mycelium and not yeast cells.

The four genera included in the Spermaphthoraceae by Gaumann (1949) are all associated with stigmatomycoses, invading punctures (stigmas) made by insects, and occur in tropical and Mediterranean regions (Fig. 7). Nowell (1917a, b, 1918), working with cotton, noted that the staining of lint and shedding of bolls were associated with probing damage done by cotton stainers, *Dysdercus* spp., and green bugs, *Nezara viridula* L., and subsequent colonization by *N. coryli* and *Ashbya gossypii* (Ashby and Nowell) Guill. When attacked at an early stage, bolls tended to be prematurely shed, whereas older bolls remained attached but with stained lint. These fungi are not confined to cotton, having been isolated from tomatoes (Schneider, 1916), sweet pepper pods and citrus (Weber, 1933) and lima beans (Wingard, 1922). Wingard (1922) noted that dark sunken areas, from which *N. coryli* (syn. *N. phaseoli*) could readily be isolated, developed on seeds within apparently healthy bean pods. As with cotton, early infection of lima beans stopped seed development, lesions developing most if seeds were half-grown at inoculation.

Further work on the insect transmission of stigmatomycotic fungi was summarized by Leach (1940) who showed that a range of *Dysdercus* spp., including *D. superstitiosus* F. and *D. cingulatus* A., were vectors of *N. coryli* and *Ash. gossypii*. There is still, however, some doubt as to whether nymphs as well as adults are able to transmit. Frazer (1944)

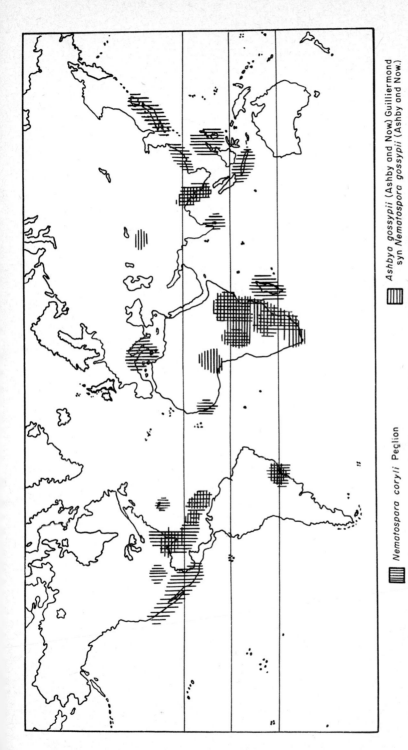

FIG. 7. World distribution of *Nematospora coryli* Peglion and *Ashbya gossypii* (Ashby and Nowell) Guill. which, with insects, stain lint and damage developing cotton bolls. Data derived from Distribution Maps of Plant Diseases Nos. 153 and 163, and published with the permission of the Director, Commonwealth Mycological Institute, U.K.

Nematospora coryli Peglion

Ashbya gossypii (Ashby and Now.) Guilliermond syn *Nematospora gossypii* (Ashby and Now.)

suggested that the fungus is transmitted on, and not in, *Dysdercus* stylets although her experiments indicated that yeasts occur both within and on the external surfaces of nymphs, the different instars being recontaminated from the exuviae.

Pearson (1947) found that *Dysdercus* spp., without contaminating *Nematospora* spp., sometimes disorganized the maturation of cotton bolls when the latter were punctured at a very early stage of development; but, if they were five weeks old or more when punctured, the damage done was usually inconsequential. Bolls injected with a suspension of *Ash. gossypii* when 3–4 weeks old continued to mature, but the lint was heavily stained and the carpels were contorted. Later inoculations were less damaging.

The relation between needle damage of *Pinus resinosus* Ait. and *A. pullulans* has not been fully investigated, but this yeast seemed to be a harmful secondary parasite colonizing tissues previously damaged by gall midges (Haddow, 1941).

Candida tropicalis, *K. apiculata*, *K. javanica* (Klöcker) Janke, *Rh. mucilaginosa* and *Lipomyces lipoferis* (den Dooren de Jong) Lodder have been associated with fruit flies, *Drosophila* spp., in the post-harvest decay of fruits, e.g. *Litchi chinensis* Sonn (Roth, 1963). *Kloeckera apiculata* has sometimes occurred as a secondary invader of strawberries damaged in the first instance by *Botrytis* spp. (Lowings, 1956).

III. Soil and Root Yeasts

Compared with those found on foliage, little is known of soil yeasts. Apart from a continuing series of observations made by di Menna during the 1950s and early 1960s and a review by Casas-Campillo (1967), most papers detail what are in effect series of isolated observations of doubtful quantitative value because of inadequate replication. This was emphsized by Lund (1954) working in Denmark where he found that populations of viable yeast propagules per gram of dry soil ranged from 150 to 5,000 when samples were taken simultaneously at a depth of 5 cm from sites less than 40 cm apart.

Hansen (1881, 1903) considered soil to be a reservoir for yeasts which grew actively on fruits. He frequently isolated *K. apiculata* from orchard soils, but rarely obtained yeast cultures from other soil types, an observation confirmed later by Lund (1954). Hansen, who concentrated his attention on fermenting yeasts, found that members of the Endomycetaceae occurred in 67% of orchard soils and, like *K. apiculata*, were less abundant elsewhere. Similarly, Lund found members of the Endomycetaceae occurring in about 60% of orchard soil samples, but this frequency was not conspicuously different from the 40–60% he recorded for meadows and cultivated fields; only bog soils and those sampled near sandy seashores were virtually without yeasts. From a very wide range of sites, Lund (1954) isolated 27 different yeasts including species of the sporogenous genera *Saccharomyces* (1), *Pichia* (2), *Hansenula* (5), *Hanseniaspora* (1) and *Lipomyces* (1), and of the asporogenous *Cryptococcus* (1), *Torulopsis* (5), *Candida* (7), *Kloeckera* (1) and *Rhodotorula* (3). Of them *H. suaveolens* (Klöcker) Dekker was the most widely distributed member of the Endomycetaceae, followed by *H. angusta* Wickerham and *H. californica* (Lodder) Wickerham with *Cr. albidus*, *C. parapsilosis* and *T. magnoliae* being the asporogenous counterparts.

The isolation of ascospore-forming and fermenting cultures of *Hansenula*, *Saccharomyces* and *Pichia* from European soils by Müller-Thurgau (1889, 1905), Capriotti (1955b) and Lund (1954) contrasts with populations recorded from New Zealand, and may reflect the influence of trees in Europe which were absent from the tussock grassland soils analysed by di Menna (1958b). In New Zealand she sometimes isolated *H. californica*, but it was usually less abundant than *Cr. albidus*, *Cr. terreus* di Menna, *Cr. diffluens*, *C. humicola* and *C. curvata* which she considers to be typical soil inhabitants, the latter five species, which accounted for 90% of the colonies isolated from tussock soils, being colourless, capsulated, non-fermenting, starch-synthesizing and nitrate-utilizing (di Menna, 1955a, b).

Some of di Menna's soil and leaf yeasts are physiologically very similar, e.g. the budding *T. ingeniosa* di Menna and pseudomycelial *C. muscorum* di Menna (1958c), but nevertheless she was able to show the contribution made by phylloplane yeasts to soil populations. In samples of the top 2–4 cm of soil taken during the winter near Marton, colonies of the leaf-inhabiting *Sp. roseus*, *Cr. laurentii*, *T. ingeniosa*, *Rh. glutinis* and *Rh. graminis* formed 54% of all isolates, but at depths of 4·5–6·8 cm the proportion decreased to 6% being superseded by *Cr. albidus* (35%), *Cr. terreus* (25%) and *C. curvata* (27%) (di Menna, 1960c). Other than this, the only significant effect of vertical distribution that she detected was the restriction of *C. humicola* to topsoil (di Menna, 1958b). When layers of litter under New Zealand native broadleaf-podocarp forest were examined, populations of *C. humicola* were found to increase appreciably as freshly fallen leaves and twigs (litter layer) aged to form the fermentation and humus layers of decomposing material and humus (di Menna, 1960c). Complementing di Menna's observations, Lund (1954) showed that numbers of viable yeast propagules per gram of dried soil decreased with increasing depth from $2·1 \times 10^5$ and $2·5 \times 10^3$ in topsoils to $3·0 \times 10^2$ and 67 at a depth of 30 cm in orchard and uncultivated sites respectively (Fig. 8).

Although Lund (1954) isolated a greater variety of species in topsoils than in soils at greater depths, individual species were not zoned in a consistent pattern. Nissen (1930), in attempting to obtain a quantitative assessment of numbers of yeasts, found that a greater proportion of soil samples from cultivated gardens and arable fields yielded yeasts than (a) moorland sites and (b) deciduous and coniferous soils. Lund (1954) isolated maxima of about 4×10^3 and 2×10^5 yeasts per g of topsoil from uncultivated and orchard sites respectively, and while

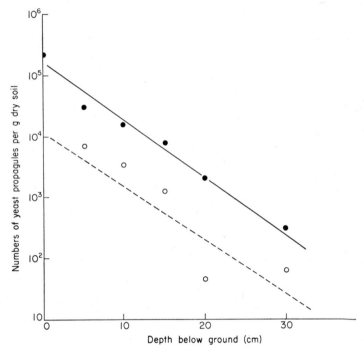

FIG. 8. Numbers of viable yeast propagules isolated per g of dry soil taken at different depths in an orchard, ●—●, and an uncultivated grassy site, ○- - -○. Data, which are means of observations made in March and August, are reproduced with permission of Lund (1954).

agreeing that cultivation usually increased yeast populations di Menna (1960c) reported one instance where the reverse occurred. Cultivating Gollans silt loam at Paraparaumu decreased yeast numbers from 1.5×10^4 in soil still under native bush to 6.4×10^3, and also eliminated appreciable numbers of the fermenting *H. californica* and *H. mrakii* Wickerham. Simultaneously the proportion of the nitrate-assimilating *Cr. albidus* increased from 2 to 11%, as did those of *C. curvata* and *Tr. cutaneum*. However Bab'eva and Savel'eva (1963) found that cultivation decreased the sizes of populations of *Candida* spp., and di Menna believes that this is the more common occurrence. Cultivating Kiwitea silt loam during the last 50–80 years has increased yeast populations from $2 \cdot 1 \times 10^3$, in soil still under native bush, to $1 \cdot 9 \times 10^4$ propagules per g. Additionally, cultivation has changed the proportions of *C. curvata* and *Tr. cutaneum* colonies from 72 and 11% to 6 and 4% respectively, while increasing the frequencies of the nitrate-assimilating *Cr. albidus* and *Cr. terreus* from about 2% to 19 and 59%. *Cryptococcus terreus* was separated from *Cr. albidus* because of its inability to assimilate sucrose (di Menna, 1964).

As with leaf yeasts, numbers of cultures obtained from soil can be affected by types of synthetic isolating media. Adding lactic acid and Rose Bengal or oxgall to potato-dextrose agar increased numbers of yeasts isolated, but even so not all soils yielded these micro-organisms. Miller and Webb (1954) detected, on average, $3 \cdot 3 \times 10^3$ propagules per g of topsoil, with numbers ranging from $2 \cdot 0 \times 10^3$ under birch to $5 \cdot 5 \times 10^4$ in an apple orchard. Their data, although subject to the dilution errors discussed by Meiklejohn (1957), suggest that for every yeast there were 10^2, 10^3 and 10^4 propagules of other fungi, actinomycetes and bacteria respectively. Di Menna (1957) isolated more colonies of *Cr. terreus*, *Cr. albidus* and *C. curvata* from a yellow-brown earth forest soil when suspensions were incubated on an acidified glucose peptone, than on soil extract-agar. Emmons (1951), when studying the development of cryptococcosis, detected the presence of *Cr. neoformans* (Sanf.) Vuill. by injecting mice with soil suspensions made in physiological saline.

In addition to being affected by the composition of isolating media, the range of species obtained has sometimes been influenced by the temperature during subsequent incubation. From a particular soil sample, *Rh. mucilaginosa* was isolated at 37° more frequently than any other yeast, followed by *C. parapsilosis*; at 18° *Cr. albidus*, *Cr. terreus* and *Tr. pullulans* (Lindner) Diddens and Lodder predominated (di Menna, 1955a). Incubating soil from the Antarctic at 4° usually favoured the isolation of *C. scottii* Diddens and Lodder, a typically colourless and mucoid yeast, smaller numbers being isolated at higher temperatures (di Menna, 1960a). At temperatures ranging from 0° to 4°, Sinclair and Stokes (1965) isolated species of *Torulopsis* and *Candida* from Antarctic soil.

Appreciable numbers of a wider range of yeasts were found in rhizosphere soils than in soil distant from roots. Thus, whereas *Cr. albidus*, *T. aeria* (Saito) Lodder, *Rh. glutinis* and *Candida* spp. were detected in both plot and rhizosphere soils, *Cr. diffluens*, *Cr. neoformans*, *T. famata*, *Rh. aurantica*, *Rh. glutinis* and *Rh. mucilaginosa* were isolated from rhizosphere soils only (Bab'eva and Belyanin, 1966). Numbers (in thousands) of yeasts isolated per g of rhizosphere soil from different hosts ranged from 5 with wheat to 30, 259, 471 and 5,200 with fodder beet, corn, and cabbage respectively (Bab'eva and Savel'eva, 1963), these fungi forming 4·3, 28·5, 25·0 and 90·9% of their respective fungal floras. Although, like many other observations, these data are again subject to dilution errors, the latter, if corrected, are unlikely to eliminate host effects. Bab'eva and Belyanin (1966) found that the multiplication of yeasts within the rhizosphere was dependent not only on host species but also on seasonal growth factors. Thus, when corn was starting to flower in late July, there were 55 times as many yeasts in rhizosphere as in field soil, but only eight times as many in September when the crop was maturing. In close agreement with Miller and Webb (1954), the two Russian authors detected $3\text{--}11 \times 10^3$ propagules per g of field soil. Additionally they found peak concentrations of $5\cdot2 \times 10^6$ propagules in rhizosphere soils, a concentration in excess of the maximum recorded for orchard soils by Lund (1954).

Because numbers of yeasts in rhizosphere soil fluctuate with the seasonally dependent development of crop plants, one would expect to find appreciable changes in numbers isolated from field soils. The published data, however, are not unanimous, but usually more yeasts were isolated in summer than during winter and spring (*vide* Lund, 1954; di Menna, 1960b; Nissen, 1930). At the end of the summer, Bab'eva and Belyanin found that the proportion of the soil-yeast population attributable to *Lipomyces* spp. increased.

Many of the results published by di Menna and by Bab'eva and her colleagues tend to cast doubts on the observations made by Starkey and Henrici (1927). The latter detected yeasts in only 39 of 87 soils sampled from the Minneapolis area in the U.S., but their media, while suppressing bacterial growth, did not retard the growth of filamentous fungi which may have obscured the presence of yeasts. Although one suspects that they underestimated populations, the reference made by Starkey and Henrici (1927) to single-species dominance at differing sites is in accord with observations made by di Menna.

In surveying soil-borne populations of micro-organisms, techniques usually depend on the ability of isolated propagules to develop on synthetic media. To attempt to separate "actively" growing soil inhabitants from "passive" contaminants, Chesters (1945) devised soil immersion tubes with which it was possible to retrieve, for further study, agar columns the exposed ends of which were colonized by "active" micro-organisms. Among the many fungi isolated was the yeast-like *A. pullulans*, also detected by Bouthilet (1953), together with the spore-forming *H. saturnus* (Klöcker) H. and P. Sydow and *H. californica* (syn. *Zygohansenula californica* Lodder) and asporogenous *C. krusei* (syn. *C.*

monosa (Kluyver) Diddens and Lodder), *Rh. mucilaginosa* and *T. candida*.

Although *A. pullulans* secretes extracellular lipids and lipolytic enzymes, properties common among epiphytic micro-organisms, it nevertheless seems to be an active inhabitant of soil, possibly associated with the early stages of organic matter degradation, populations sometimes exceeding 10^5 propagules per g of soil (Holding *et al.*, 1965). *Aureobasidium pullulans* may actively help in the conversion of plant debris to soil organic matter. Data from Metcalfe and Chayen (1954) indicate that other soil yeasts can fix nitrogen and may influence the soil nitrogen status. Using a perfusion medium without nitrogen, they isolated a range of micro-organisms from the A1 horizon of a health profile. Two of these cultures subsequently grew in an ammonia-free atmosphere, one resembling *Saccharomyces* and the other a species of *Rhodotorula*, the latter being able to fix *in vitro* 1·3 mg atmospheric N_2 per g of glucose (Roberts and Wilson, 1954).

IV. Discussion

In bringing together references from a wide variety of sources, and in attempting to determine not only qualitative but also quantitative relationships between different micro-organisms, it is essential to consider the limitations imposed by the different techniques adopted. A detailed analysis of such effects has been deferred until now because it will primarily affect the approach to future investigations, existing data being sufficient to establish, in quasi-quantitative terms, the occurrence of specialized yeast populations on different higher plant structures. It seems that continuing attempts have not been made to enumerate yeasts on less advanced hosts, e.g. mosses, liverworts and ferns. The limitations of some of the procedures such as methods of isolation (spore fall, leaf washing and direct leaf microscopy), temperatures of incubation and types of media have already been mentioned in passing, but their importance cannot be over-emphasized. By exploiting the inability of *Sacch. cerevisiae* to use lysine, Brady (1958) found, when this amino acid was supplied as the sole source of nitrogen, that supposedly pure cultures contained cells of *Pi. membranaefaciens* Hansen, *T. famata* and species of *Hansenula*, *Candida* and *Rhodotorula*. Sodium propionate has been widely used in media for isolating yeasts primarily because it restricts the growth of bacteria and filamentous fungi. But this chemical also minimizes the growth of *Sporobolomyces* spp. (Lund, 1956). Instead of using 3,000 ppm of sodium propionate, Hertz and Levine (1942) favoured 100 ppm of diphenyl, but this restricted *D. hansenii* (Zopf.) Lodder and van Rij (syn. *D. membranaefaciens* Naganishi) without affecting species of the filamentous *Rhizopus*.

8

Notwithstanding the limitations of the techniques used in past work, it is possible to categorize some yeasts as follows according to their differing plant habitats, but overlapping does occur:

(i) Leaves. *Sporobolomyces roseus, Rhodotorula glutinis, Rh. mucilaginosa, Cryptococcus laurentii, Torulopsis ingeniosa* and *Aureobasidium pullulans.*

(ii) Flowers. *Candida reukaufi* and *C. pulcherrima.*

(iii) Fruits. *Kloeckera apiculata* and *Candida pulcherrima.*

(iv) Fluxes. *Candida pulcherrima, C. krusei, Saccharomyces mellis* and *Torulopsis magnoliae.*

(v) Soil. *Cryptococcus albidus, Cr. terreus, Candida curvata, C. humicola, Hansenula saturnus* and *Trichosporon cutaneum.*

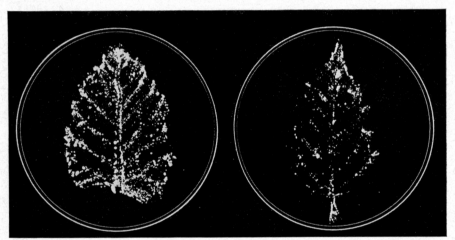

FIG. 9. Colonies of *Sporobolomyces* isolated by the spore-fall method from the lower surfaces of hazel leaves, *Corylus avellana* L.–the leaf outline and venation being mirrored.

The group of leaf yeasts differs from all others by containing a high proportion of pigmented forms (Fig. 9). Together with soil yeasts, those on leaves differ from others by their inability to ferment; fruit, flower and flux yeasts can all ferment at least one sugar weakly. Di Menna (1958b) emphasized the ability of soil yeasts to assimilate nitrate, using this criterion as a useful differentiating character. Very few flower, fruit and flux yeasts can utilize this nitrogen source (*T. magnoliae* is exceptional) but some leaf yeasts can, including the numerically important *Sp. roseus.* On the other hand, the most important leaf species of *Cryptococcus,* viz. *Cr. laurentii,* cannot, whereas the soil-inhabiting *Cr. albidus* is able. In a series of interesting *in vitro* tests, di Menna (1962) found that the growth of leaf yeasts was restricted more by some soil bacteria and moulds than by comparable micro-organisms isolated from leaves. But, because of the smallness of the differential effect, she concluded that physical and chemical effects of the host's micro-environment were more important in determining the species composition of yeast populations than were biological factors (e.g. direct microbial competition).

Many foliage yeasts when cultured in liquid media first secrete lipids, forming a slimy capsule, and then extracellular lipases, a sequence of events paralleling the activities of the epiphyllic bacterium *Beijerinckia* (Ruinen, 1966). The secretion of lipases was confirmed in tests where plant cuticle was seriously eroded within five days, the cell-wall relief becoming progressively less distinct, when *Cr. laurentii* and *Rh. glutinis* were cultured on (a) the stripped epidermis of *Aloe* sp. and (b) cuticle fragments of *Sanseviera* spp. Yeasts colonizing leaves above the cutin-re-inforced anticlinal cells walls may, as a result of their enzymic activity, detrimentally increase transpiration losses which are, to some extent, regulated by the intact cuticle (Ruinen, 1963b, 1966). The lipids produced by *Rh. graminis*, when nitrogen is no longer available, contain oleic and palmitic acids (Deinema and Landheer, 1960).

A number of papers give quantitative data based on dilution-plate procedures but, as indicated by Meiklejohn (1957) and Reddy (1962), data obtained in this way are liable to appreciable errors. When numbers of viable propagules in soil suspensions were analysed at two dilutions, estimates of numbers per g of soil proved to be greater at the higher dilution. For example, 1 g of soil plated at dilutions of 1:300 and 1:1,000 averaged 8 and 6 fungal colonies per Petri dish, numbers equivalent to 2,400 and 6,000 propagules per g. Instead of correcting colony numbers by factors of 10, 10^2 and 10^3, Reddy found, for the particular soil with which he was experimenting, that they should be multiplied by 4, 4^2 and 4^3. At the present stage of yeast ecology it seems desirable to take notice of this appreciable factor, particularly when making comparisons between different types of micro-organisms taken from different substrates. Further, although comparisons between propagule numbers of different yeasts are unlikely to be seriously affected, extension to include comparisons between budding and filamentous organisms will be influenced by the interpretation of the term "propagule". For yeasts, this is more likely to be synonymous with single cells than with Hyphomycetes, and as a result propagules of the latter will underestimate their volumes in relation to those of yeasts.

In spite of these reservations, it has been possible to show that populations of leaf and fruit yeasts are in dynamic equilibrium with their hosts. Numbers usually increase as hosts age although Holloman

(1967) found that *A. pullulans* becomes relatively less important on ageing potato leaves. The nutrition of many yeasts has been studied in considerable detail, if only as an aid to identification. The possible sources of nutrients on leaf surfaces were reviewed by Last and Deighton (1965) and Singh (1968). There is ample evidence to indicate that nutrients, absorbed by roots and subsequently metabolized, reach leaf surfaces where a variety of substances have been detected, including organic acids, sugars and amino acids (Tukey and Tukey, 1962). Possibly the siting of many yeasts above the anticlinal epidermal cell walls may coincide with one transport pathway through strands within the epidermis and cuticle to ectodesmata (Franke, 1961), other pathways being through hydathodes and secretory glands.

Materials washed from leaves may stimulate the activity of soil yeasts. Will (1959), investigating the cycling of nutrients in stands of New Zealand conifers, found that, of the elements returned to soil, potassium, sodium and phosphorus were predominantly carried in rain water. By sampling rain collected (a) beneath tree canopies and (b) in the open, he found that only small quantities of sodium were washed from foliage which was the main source of potassium and phosphorus. Beneath *Pinus radiata* Don and *Pseudotsuga taxifolia* Britt., rain water deposited between 9 and 11 kg of potassium per acre per annum.

Many epiphytic orchids (Ruinen, 1953) and the fungus *Sclerographium phyllanthicola* are host-specific, but most non-parasitic epiphytes are not. What is their function? At present an assessment of function can only be based on circumstantial evidence which suggests they act as scavengers. They possibly deprive plant pathogens of a nutrient supply which may significantly influence the incidence of disease caused by those pathogens needing an external source of energy during germination and the early stages of host penetration. Yeasts such as *Rhodotorula* spp. (Roberts and Wilson, 1954) may influence the nitrogen nutrition of their hosts. Others may protect their hosts by being antagonistic to pathogens as indicated by Azare-Nyako (1967), Waksman (1941) and Yamasakai *et al.* (1951), but it is unwise to extrapolate in the absence of *in vivo* proof. However, there seems to be no doubting the ability of many leaf yeasts to degrade cutin, an effect that may shorten leaf-life. Indirectly this may be done by changing the hydrophobic nature of the cuticle, so influencing the water relations between host and parasite during infection processes.

Having determined the part played by yeasts during fermentation, workers in the late nineteenth century were concerned with the distribution of these microbes. Where were fruit-inhabiting yeasts to be found during the winter? In the last twenty years, great strides have been made in techniques to estimate air sporas. These techniques have shown that very many ballistospores of the Sporobolomycetaceae are sometimes airborne and, at the same time, they tend to confirm that members of the Endomycetaceae and Cryptococcaceae are not adapted for aerial spread. Hansen (1881, 1903) considered that soil acted as a reservoir for fruit yeasts, and di Menna (1962) has clearly shown that many leaf yeasts can sometimes be detected in topsoils, e.g. *Sporobolomyces* spp. But the puzzle still remains. How do non-ballistospore-forming yeasts spread to foliage? Although seeds are often contaminated before planting, developing seedlings may be colonized by foliage organisms harboured in topsoil. In experiments paralleling those done by Leben (1961) with bacteria, Voznyakovskaya (1962) found, when wheat grains each carrying 10^5 propagules of *Rh. aurantiaca* were germinated in sterilized soil, that developing leaves were heavily colonized by this micro-organism. But would this transference occur in unsterilized soil? Interestingly, this Russian worker isolated yeasts over-wintering in buds in much the same way as the pathogenic *Podosphaera leucotricha* (Ell. and Everh.) Salm. does in apple buds (Woodward, 1927). These methods of spread can be added to others of rain splash and insect transfer which have been propounded without being unequivocally proved, excepting the relation between *N. coryli*, its insect vectors and the incidence of stigmatomycosis of cotton.

References

Ahmad, M., Chaudhury, A. R. and Ahmad, K. U. (1954). *Mycologia* **46**, 708–720.
Anderson, K. W. and Skinner, C. E. (1947). *Mycologia* **39**, 165–170.
Ashby, S. F. and Nowell, W. (1926). *Ann. Bot.* **40**, 69–83.
Azare-Nyako, A. (1967). *Diss. Abstr.* **27**, 4206–4207.
Bab'eva, I. P. and Savel'eva, N. D. (1963). *Mikrobiologiya*, **32**, 86–93.
Bab'eva, I. P. and Belyanin, A. I. (1966). *Mikrobiologiya*, **35**, 712–720.
Batra, S. W. T. and Batra, L. R. (1967). *Scient. Am.* **217**, 112–120.
Beech, F. W. (1959). *J. appl. Bact.* **21**, 257–266.
Bioletti, F. T. and Cruess, W. V. (1912). *Univ. Calif. Publs. agric. Sci.* **230**.
Bouthilet, J. (1953). *Mycopath. Mycol. appl.* **6**, 79–85.
Bowen, J. F. and Beech, F. W. (1964). *J. appl. Bact.* **27**, 333–341.
Bowen, J. F. and Beech, F. W. (1967). *J. appl. Bact.* **30**, 475–483.
Brady, B. L. (1958). *J. Inst. Brew.* **64**, 304–307.
Buller, A. H. R. (1933). "Researches on Fungi", Vol. **5**.
Capriotti, A. (1953). *Riv. Biol.* **45**, 367–374.
Capriotti, A. (1955a). *Riv. Biol.* **47**, 209–266.
Capriotti, A. (1955b). *Antonie van Leeuwenhoek*, **21**, 145–156.
Casas-Campillo, C. (1967). *Rev. lat.-Amer. Microbiol. Parasitol.* **9**, 91–97.
Challinor, S. W. and Rose, A. H. (1954). *Nature, Lond.* **174**, 877.
Channon, A. G. (1963). *Ann. appl. Biol.* **51**, 1–15.
Chesters, C. G. C. (1945). *Trans. Br. mycol. Soc.* **30**, 100–117.
Clarke, D. S., Wallace, R. H. and David, J. J. (1954). *Can. J. Microbiol.* **1**, 145–149.
Connell, G. H. and Skinner, C. E. (1953). *J. Bact.* **66**, 627–633.
Cooke, W. B. (1965). *Mycopath. Mycol. appl.* **25**, 195–200.

Deinema, M. H. and Landheer, C. A. (1960). *Biochim. biophys. Acta* **37**, 178–179.

Derx, H. G. (1930). *Ann. Mycol. Berl.* **28**, 1–23.

Dickinson, C. H. (1965). *Trans. Br. mycol. Soc.* **48**, 603–610.

Dickinson, C. H. (1967). *Can. J. Bot.* **45**, 915–927.

Diem, H. G. (1967). *Bull. de l'Ecole Nationale Supérieure Agronomique de Nancy* **9**, 102–108.

di Menna, M. E. (1966). *N.Z. Jl agric. Res.* **9**, 576–589.

El Tabey Shehata, A. M., Mrak, E. M. and Phaff, H. J. (1955). *Mycologia* **47**, 799–811.

Emmons, C. W. (1951). *J. Bact.* **62**, 685–690.

Etchells, J. L., Costilow, R. N., Bell, T. A. and Demain, A. L. (1954). *Appl. Microbiol.* **2**, 296–302.

Franke, W. (1961). *Nátúrwissenschaften* **48**, 227–228.

Frazer, H. L. (1944). *Ann. appl. Biol.* **31**, 271–290.

Gandy, D. G. G. (1966). *Trans. Br. mycol. Soc.* **49**, 499–507.

Gaumann, E. (1949). "Die Pilze, Grundzuge ihrer Enturicklungsgeschichte und Morphologie". Verlag Birkhäuser A. G. Basel.

Gregory, P. H. (1952). *Nature, Lond.* **170**, 475.

Gregory, P. H. and Hirst, J. M. (1957). *J. gen. Microbiol.* **17**, 135–152.

Gregory, P. H. and Sreeramulu, T. (1958). *Trans. Br. mycol. Soc.* **41**, 145–156.

Guilliermond, A. (1928). *Revue. gén. Bot.* **40**, 328–342.

Haddow, W. R. (1941). *Trans. R. Can. Inst.* **23**, 161–189.

Hamilton, E. D. (1959), *Acta allerg.* **13**, 143–175.

Hansen, E. Chr. (1881). *Medd. Carlsberg Lab.* **1**, 293.

Hansen, E. Chr. (1882). *Medd. Carlsberg Lab.* **1**, 381.

Hansen, E. Chr. (1903). *Zentbl. Bakt. ParasitKde (Abt. II)*, **10**, 1.

Harley, J. L. and Waid, J. S. (1955). *Trans. Br. mycol. Soc.* **38**, 104–118.

Hertz, M. A. and Levine, M. (1942). *Fd Res.* **7**, 430–441.

Hiltner, L. (1904). *Arb. dtsch. Landw'Ges.* **98**, 59.

Hirst, J. M. (1953). *Trans. Br. mycol. Soc.* **36**, 375–393.

Hogg, B. M. (1966). *Trans. Br. mycol. Soc.* **49**, 193–204.

Hogg, B. M. and Hudson, H. J. (1966). *Trans. Br. mycol. Soc.* **49**, 185–192.

Holding, A. J., Franklin, D. A. and Watling, R. (1965). *J. Soil Sci.* **161**, 44–47.

Holloman, D. W. (1967). *Eur. potato J.* **10**, 53–61.

Hudson, H. J. and Webster, J. (1958). *Trans. Br. mycol. Soc.* **41**, 165–177.

Hyde, H. A. and Adams, K. F. (1960). *Acta allerg. Suppl.* **7**, 159–169.

Jimenez-Diaz, C., Ales, J. M., Ortiz, F., Lahoz, F., Garcia-Puente, L. M. and Canto, G. (1960). *Acta allerg. Suppl.* **7**, 139–149.

Kendrick, W. B. and Burges, A. (1962). *Nova Hedwigia* **4**, 313–342.

Kerling, L. C. P. (1958). *Tijdschr. PlZiekt.* **64**, 402–410.

Kerling, L. C. P. (1964). *Meded. LandbHoogesch. OpzoekStns Gent* **29**, 885–895.

Keymer, I. F. and Austwick, P. K. C. (1961). *Sabouraudia* **1**, 22–29.

Kluyver, A. J. and van Niel, C. B. (1924–25). *Zentbl. Bakt. ParasitKde (Abt. II)* **63**, 1.

Koch, A. (1960). *A. Rev. Microbiol.* **14**, 121–140.

Koch, A. (1963). *In* "Recent Progress in Microbiology", Vol. III, pp. 151–161. University of Toronto Press, Toronto.

Kvasnikov, E. I., Schelokova, I. F., Masumyan, V. Ya., Kotlyar, A. N., Aristova, M. V. and Kuzimenko, L. T. (1967). *Mikrobiologiya* **36**, 1077–1082.

Last, F. T. (1955a). *Trans. Br. mycol. Soc.* **38**, 221–239.

Last, F. T. (1955b). *Trans. Br. mycol. Soc.* **38**, 453–464.

Last, F. T. and Deighton, F. C. (1965). *Trans. Br. mycol. Soc.* **48**, 83–99.

Leach, J. G. (1940). "Insect Transmission of Plant Disease". McGraw-Hill, N.Y.

Leben, C. (1961). *Phytopathology* **51**, 553–557.
Lodder, J. and Kreger-van-Rij, N. J. W. (1952). "The Yeasts, a Taxonomic Study". North-Holland Publishing Company, Amsterdam.
Lowings, P. H. (1956). *Appl. Microbiol.* **4**, 84–88.
Lund, A. (1954). "Studies on the Ecology of Yeasts". Munksgaard, Copenhagen.
Lund, A. (1956). *Friesia* **5**, 297–302.
Marshall, C. R. and Walkley, V. T. (1951). *Fd. Res.* **16**, 448–456.
Martin, H. H. (1954). *Arch. Mikrobiol.* **20**, 141–162.
Meiklejohn, J. (1957). *J. soil Sci.* **8**, 240–247.
di Menna, M. E. (1954). *J. gen. Microbiol.* **11**, 195–197.
di Menna, M. E. (1955a). *J. gen. Microbiol.* **12**, 54–62.
di Menna, M. E. (1955b). *Trans. Br. mycol. Soc.* **38**, 119–129.
di Menna, M. E. (1957). *J. gen. Microbiol.* **17**, 678–688.
di Menna, M. E. (1958a). *Nature, Lond.* **181**, 1287–1288.
di Menna, M. E. (1958b). *N.Z. Jl. agric. Res.* **1**, 939–942.
di Menna, M. E. (1958c). *J. gen. Microbiol.* **19**, 581–583.
di Menna, M. E. (1959). *N.Z. Jl. agr. Res.* **2**, 394–405.
di Menna, M. E. (1960a). *J. gen. Microbiol.* **23**, 295–300.
di Menna, M. E. (1960b). *N.Z. Jl. agric. Res.* **3**, 207–213.
di Menna, M. E. (1960c). *N.Z. Jl. agric. Res.* **3**, 623–632.
di Menna, M. E. (1962). *J. gen. Microbiol.* **27**, 249–257.
Metcalfe, G. and Chayen, S. (1954). *Nature, Lond.* **174**, 841.
Miller, M. W., Phaff, H. J. and Snyder, H. E. (1962). *Mycopath. Mycol. appl.* **16**, 1–18.
Miller, J. J. and Webb, N. S. (1954). *Soil Sci.* **77**, 197–204.
Miquel, P. (1877–99). *Annu. Obs. Montsauris.* Reports.
Mossell, D. A. H. and Ingram, M. (1956). *J. appl. Bact.* **18**, 232–268.
Mrak, E. M. and McClung, L. S. (1940). *J. Bact.* **40**, 395–407.
Mrak, E. M., Phaff, H. J. and Vaughn, R. H. (1941). *J. Bact.* **43**, 689–694.
Müller, T. H. (1889). *Ber. Verh. XI. dt. Weinbaukongr. in Trier*, p. 80–100.
Müller, T. H. (1905). *Zentbl. Bakt. ParasitKde. (Abt. I)* **14**, 296–297.
Nissen, W. (1930). *Milchw. Forsch.* **10**, 30–67.
Nowell, W. (1917a). *W. Indian Bull.* **16**, 203–235.
Nowell, W. (1917b). *W. Indian Bull.* **16**, 152–159.
Nowell, W. (1918). *W. Indian Bull.* **17**, 1–28.
Pady, S. M. and Kelly, C. D. (1954). *Can. J. Bot.* **32**, 202–212.
Pasteur, L. (1876). "Études sur la bière". Paris.
Pasteur, L. (1878). *C. r. hebd. Séanc. Acad. Sci., Paris* **87**, 813–819.
Pearson, E. O. (1947). *Ann. appl. Biol.* **34**, 527–544.
Phaff, H. J., Miller, M. W. and Mrak, E. M. (1966). "The Life of Yeasts". Harvard University Press, Cambridge, Massachusetts.
Pugh, G. J. F. (1958). *Trans. Br. mycol. Soc.* **41**, 185–195.
Reddy, M. A. R. (1962). Ph.D. Thesis: University of London.
Recca, J. and Mrak, E. M. (1952). *Fd. Technol. Champaign* **6**, 450–454.
Ripe, E. (1962). *Acta allerg.* **17**, 130–159.
Roberts, E. R. and Wilson, T. G. G. (1954). *Nature, Lond.* **174**, 842.
Roth, C. (1963). *S. Africa Dept. Agric. Tech. Serv. Tech. Comm.* vol. 11.
Ruinen, J. (1953). *Ann. Bogor* **1**, 101–158.
Ruinen, J. (1956). *Nature, Lond.* **177**, 220–221.
Ruinen, J. (1961). *Pl. Soil* **15**, 81–109.
Ruinen, J. (1963a). *Antonie van Leeuwenhoek* **29**, 425–435.
Ruinen, J. (1963b). *J. gen. Microbiol.* **32**, iv.

Ruinen, J. (1966). *Annls Inst. Pasteur, Paris* **3**, 342–346.

Sasaki, Y. and Yoshida, F. (1961). *J. Fac. Agric. Hokkaido Univ.* **51**, 194–210.

Schanderl, H. (1957). *In* "The Yeasts", (W. Roman ed.), D. and W. Junk, The Hague.

Schneider, A. (1916). *Phytopathology* **6**, 395–399.

Schoenauer, M. (1876). *Annu. Obs. Montsauris Reports.*

Sergent, E. and Rougebief, H. (1926). *C. r. hebd. Séanc Acad. Sci., Paris* **182**, 1238–1239.

Sinclair, N. A. and Stokes, J. L. (1965). *Can. J. Microbiol.* **11**, 259–269.

Singh, P. (1968). *Acta phytopathol. hung.* **3**, 13–22.

Skinner, C. E. (1947). *Bact. Rev.* **11**, 227–274.

Smit, J. and Wieringa, K. T. (1953). *Nature, Lond.* **171**, 794–795.

Starkey, R. L. and Henrici, A. T. (1927). *Soil Sci.* **23**, 33–35.

Takahashi, J., Kawabata, Y. and Yamada, K. (1965). *Agr. Biol. Chem.* **29**, 292–299.

Tanaka, H. and Miller, M. W. (1963). *Hilgardia* **34**, 167–181.

Teunisson, D. J. (1954). *Appl. Microbiol.* **2**, 215–220.

Tubaki, K. (1953). *Nagaoa* **3**, 12–21.

Tubaki, K. (1958). *Bot. Mag.*, Tokyo **71**, 133–137.

Tukey, H. B. Jr. and Tukey, H. B. Sr. (1962). "Radioisotopes in Soil-Plant Nutrition Studies". International Atomic Energy Agency, Vienna.

Turner, P. D. (1966). *Trans. Br. mycol. Soc.* **49**, 255–267.

Uden, N. van, Carmo Sousa, L. D. and Farinha, M. (1958). *J. gen. Microbiol.* **19**, 435–445.

Uden, N. van, Matos Faia, M. de and Assis-Lopes, L. (1956). *J. gen. Microbiol.* **15**, 151–153.

Uroma, E. and Virtanen, O. E. (1949). *Chem. Abstr.* **43**, 2277.

van der Walt, J. P. (1956). *J. Sci. Fd. Agric.* **7**, 105–113.

Voznyakovskaya, Yu. M. (1962). *Mikrobiologiya* **31**, 616–622.

de Vries, G. A. (1952). Ph.D. Thesis: University of Utrecht.

Waksman, S. A. (1941). *Bact. Rev.* **5**, 231–293.

Weber, G. F. (1933). *Phytopathology* **23**, 384.

Will, G. M. (1959). *N.Z. Jl. agric. Res.* **2**, 719–734.

Williams, A. J., Wallace, R. H. and Clarke, D. S. (1956). *Can. J. Microbiol.* **2**, 645–648.

Windisch, S. (1940). *Arch. Mikrobiol.* **11**, 368–390.

Wingard, S. A. (1922). *Phytopathology* **12**, 525–532.

Woodward, R. C. (1927). *Trans. Br. mycol. Soc.* **12**, 173–204.

Yamasakai, I., Satomura, Y. and Yamamoto, T. (1951). *J. agric. chem. Soc. Japan* **24**, 399.

Protein production by yeasts has stimulated interest in microbes near oil deposits. Casas-Campillo (1967) enumerated soil isolates of *Candida*, *Rhodotorula* and *Cryptococcus* able to utilize aromatic hydrocarbons and n-paraffins, the use of the latter with 15–19 carbons by *C. guillermondii* (Cast.) Langeron and Guerra and *C. tropicalis* being defined by Kvasnikov *et al.* (1968) and Takahashi *et al.* (1965). Other nutritional studies showed that soil additions of lactose-containing dairy wastes favoured the development of *C. humicola* and *Trich. cutaneum* at the expense of *C. curvata* and *Cr. terreus* (di Menna, 1966).

Chapter 6

Yeast Cytology

Ph. Matile, H. Moor and C. F. Robinow

Institut für allgemeine Botanik, Eidg. Technische Hochschule, Zurich, Switzerland, and Department of Bacteriology and Immunology, University of Western Ontario, London, Ontario, Canada

8*

I. Introduction

In the early decades of the last century, before the cell principle had been clearly developed and widely accepted, some biological microscopists became convinced that the yeasts in beer and wine were living organisms.

Of the four men who recorded this discovery one, Desmazières (1827), regarded yeast as a species of animalcule endowed with active motility. Another, Kützing (1837), believed that the yeast of every fresh fermentation arose *de novo* from the organic constituents of the substrate. Though he realized that, once created, yeast had at the same time acquired the ability to propagate itself without the intervention of further acts of creation. The two remaining men, Cagniard Latour (1838) and Schwann (1837) established simultaneously and independently that yeasts multiply by budding. The former inferred this from the microscopic study of sample populations removed at hourly intervals from a vat of fermenting wort; the latter followed the growth of buds on individual yeasts in a sample of grape juice. At this distance of time there may be merit in quoting *verbatim* the passages in which the two *micrographes* (to use Cagniard Latour's expression) recorded their findings:

"Dans la note dont je viens de parler, je fais remarquer qu'ayant examiné avec attention des échantillons de porter d'heure en heure, au fur et à mesure de leur extraction de la cuve, j'ai reconnu qu'au bout de la première heure, après la mise en levain, le moût contenait déjà des globules doubles, c'est-à-dire sur chacun desquels on apercevait un globule secondaire plus petit; qu'un peu plus tard ce dernier paraissait avoir pris de l'accroissement, puisque chez plusieurs couples les deux globules avaient à peu près la même grosseur; qu'enfin le quatrième échantillon n'offrait guère que des globules doubles. J'ajouterai que pour m'assurer que ces couples avaient leurs globules soudés et non simplement rapprochés, j'ai appliqué, à l'aide d'un petit poinçon, des chocs sur le verre recouvrant les globules placés sous le microscope, et que ces chocs, quoiqu'ils produisissent de grands ébranlemens parmi les globules, n'en détruisaient point les soudures; mais il paraîtrait que ces corps en devenant plus agés se désunissent naturellement."

Schwann (1837), having described the arrangement of the "granules" composing beer yeasts and having suggested that because of their lack of chlorophyll they probably represent some kind of fungus, continues: "In frisch ausgepresstem Traubensaft ist nichts der Art vorhanden.

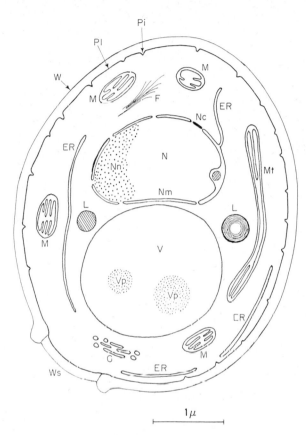

FIG. 1. Diagram of a resting cell of baker's yeast (*Saccharomyces cerevisiae*). ER indicates endoplasmic reticulum; pER, proliferating endoplasmic reticulum; ERv, vesicles derived from endoplasmic reticulum; F, filament; G, Golgi apparatus; Gl, glycogen; L, lipid granule (sphaerosome); M, mitochondrion; Mt, thread-like mitochondrion; N, nucleus; Nc, centriolar plaque; Nm, nuclear membrane; Nn, nucleolus; Ns, spindle apparatus; Pi, invagination; Pl, plasmalemma; Pp, plasmalemma particle; R, ribosome; V, vacuole; Vp, polymetaphosphate granule; W, cell wall (1, outer; 2, middle; 3, innermost layer); Wf, cell wall fibrils; Ws, bud scar.

Setzt man denselben aber einer Temperatur von ungefaehr 20°R (25°C) aus, so finden sich schon nach 36 Stunden einige solcher Pflanzen darin, die aber erst aus wenigen solcher Koerner (*cells*) bestehen. Diese wach-

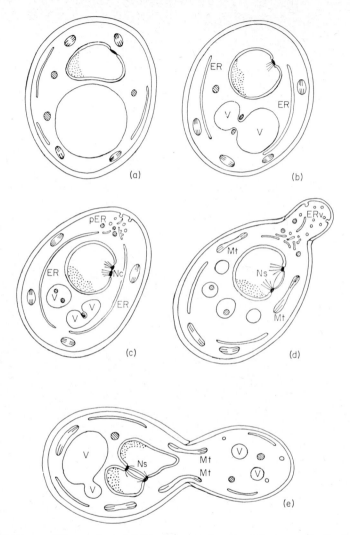

FIG. 2. Diagram of the main cytological events in a budding yeast cell (*Saccharomyces*). (a) Resting cell. (b) Fusion of elements of endoplasmic reticulum, fission of the vacuole. (c) Continued fission of the vacuoles, proliferating endoplasmic reticulum, first endoplasmic reticulum-derived vesicles secreted through the plasmalemma, division of the centriolar plaque. (d) Formation of a bud-initial, reduction of the endoplasmic reticulum, appearance of thread-like mitochondria, construction of the spindle apparatus. (e) Entry of vacuoles and

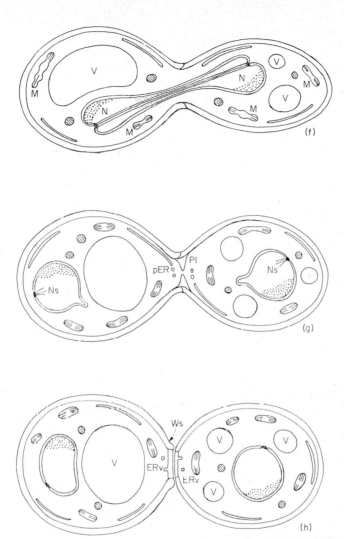

thread-like mitochondria into the bud, fusion of vacuoles in the mother cell, nucleus preparing to enter the bud. (f) Nuclear division, fragmenting mitochondria. (g) Plasmalemma closing the gap between mother cell and bud, proliferating endoplasmic reticulum, decay of the spindle. (h) Cross-wall formation, secretion of endoplasmic reticulum-derived vesicles. The mother cell contains one large, the daughter cell several small, vacuoles. See caption to Fig. 1 for explanation of symbols.

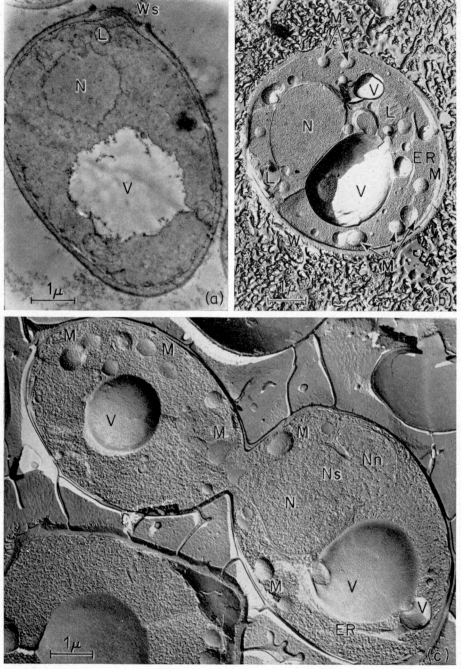

Fig. 3. (a) Electron micrograph of thin section of a permanganate-fixed cell of *Saccharomyces cerevisiae*. (b) Electron micrograph of a freeze-etched cell of *Saccharomyces cerevisiae* showing improved details of the fine structure. (*c*) Electron micrograph of a budding yeast cell (*Saccharomyces cerevisiae*); see Fig 2e. The specimen was freeze-etched. See caption to Fig. 1 for explanation of symbols.

sen sichtbar unter dem Mikroskop, so dass man schon nach $\frac{1}{2}$ bis 1 Stunde die Zunahme des Volumens eines sehr kleinen Koernchens, welches auf einem groesseren aufsitzt, beobachten kann."

It seems that at this time Schwann was still thinking of yeast in the manner of Kützing (1837), as being initiated by a heat-labile "substance" contained in atmospheric air. Schwann was obviously not aware that the yeasts in fermenting grape juice are the descendants of small numbers of preformed contaminants. The first sentence of the passage quoted above, with its categorical "nichts der Art vorhanden", is reinforced later on by: "Wird Ferment, welches schon gebildete Pflanzen enthaelt, in eine Zuckerloesung gebracht, so treten die Erscheinungen der Gaerung sehr bald ein, viel schneller, *als wenn sich die Pflanzen erst bilden muessen*" (our italics).

Cagniard Latour also anticipated by more than a hundred years the recent demonstration (Barton, 1950; Streiblová and Beran, 1963) that yeasts bear scars. Having described the budding process and the separation (désunion) of daughters from their mother cells, Latour (1838) continues:

". . . il s'est présenté des cas ou l'on distinguait chez certains globules (*cells*, Transl.) plusieurs granules, et quelquefois une tache ronde ou ovale, tantôt centrale et tantôt latérale, que, d'après la désunion dont il vient d'être question, ont peut présumer être une cicatricule ou marque ombilicale."

According to Miller (1967), who has surveyed the early history of yeast biology, the first acceptable description of the ascospores of yeast was published by de Seynes (1868). The first sighting of the yeast nucleus is usually ascribed to Schmitz (1879). With the aid of haematoxylin he succeeded in demonstrating a spherical nucleus "etwa in der Mitte der Zelle, neben den grossen Vakuolen". However, the chromatin of the yeast nucleus has no marked affinity for haematoxylin, and it seems likely that what Schmitz regarded as the nucleus was the readily stainable *nucleolus*. Modern times have seen numerous instances of the reverse of this mistake: unwise reliance on hydrolysed preparations which allow the chromatin to be stained but greatly reduce the affinity for stains of the nucleolar moiety. Realistic drawings of the budding habit and proportions of *Saccharomyces*, truer to life than in many a text of our time, are in de Bary's celebrated treatise on fungi of 1884. The life cycles of yeasts were described in detail in numerous publications by Hansen and by Guilliermond at the turn of the century. The genetical consequences of spore formation and of the fusion of pairs of germlings were explained by Winge (1935). His work, and that of the Lindegrens in the years following World War II, provided the basis on which yeast genetics has been built.

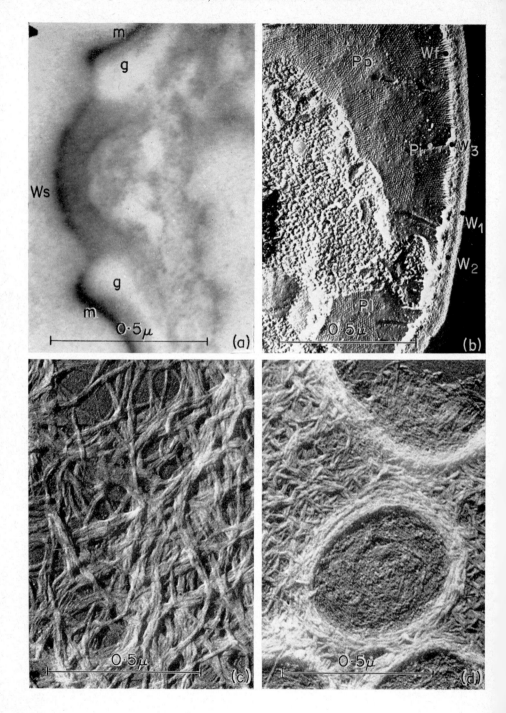

II. The Cell Wall

A. MORPHOLOGY

1. *The Strata*

A multitude of information has been compiled about the stratification of the yeast cell-wall (Windisch and Bautz, 1960; Villanueva, 1966a). In thicker walls, three different layers or strata can be distinguished while, in very thin walls and in slimy degenerated walls, these structures may be obscured. Thin sections of permanganate-fixed cells show a dark stained outer and inner plasmalemma-apposed stratum, and a more electron-transparent intermediate region (Marquardt, 1962; Vitols *et al.*, 1961). This image approaches that of freeze-etched specimens where also three strata can be observed (Fig. 4(b)). These layers most probably are not completely independent but differ only in their chemical composition. The outer stratum is claimed to contain mainly mannan-protein (Northcote and Horne, 1952) which would explain its stainability by permanganate. Following the cytochemical investigations of Mundkur (1960a), this layer also appears to contain a certain amount of chitin, while the stratum in the middle of the wall is comprised of the non-stainable glucan (Fig. 4(a)). Acid-treated cell walls show a framework of microfibrils of artificially produced hydroglucan (Figs. 4(c), (d): Houwink and Kreger, 1953). Nevertheless, this structure must resemble the native arrangement of the glucan molecules because, in freeze-etchings (Fig. 4(b)) and during the regeneration of protoplasts (Fig. 5(d)), fibrillar elements can be observed also (Moor and Mühlethaler, 1963; Eddy and Williamson, 1959; Nečas, 1965; see Uruburu *et al.*, 1968). Following the ideas of Falcone and Nickerson (1956) and Nickerson (1963), these glucan microfibrils are thought to form part of a heterogeneous polymer which would be responsible for the physical strength of the cell wall. The innermost stainable part of the cell wall again would be composed of a more proteinaceous

FIG. 4. (a) Electron micrograph of a thin section through a bud scar of *Saccharomyces cerevisiae* showing the scar plug and the strata of the cell wall, with a mannan-containing outer layer (m) and glucan-containing middle layer (g) (Mundkur, 1960a, b). (b) Electron micrograph of a freeze-etched cell of *Saccharomycodes ludwigii* showing the strata of the cell wall, the plasmalemma sculptured by invaginations, the hexagonally arranged plasmalemma particles and the glucan fibrils. Reproduced by courtesy of E. Streiblová. (c) Electron micrograph of an acid-treated cell wall (*Sacch. cerevisiae*) showing aggregations of semi-artificial hydro-glucan fibrils (Houwink and Kreger, 1953). (d) Bud scar region of a similar cell wall as shown in Fig. (4c) (Houwink and Kreger, 1953). See caption to Fig. 1 for explanation of symbols.

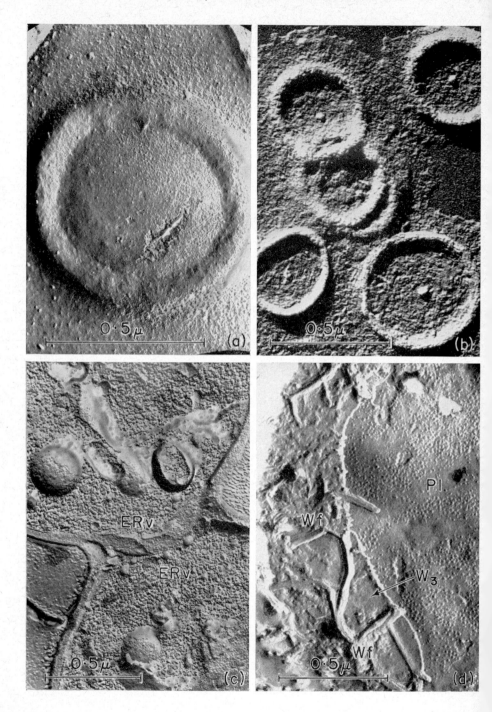

material and, as suggested by Streiblová (1968), cannot be removed by the production of "cell wall-free, naked protoplasts" (Fig. 5(d)). This innermost stratum which covers the plasmalemma as a thin film and fills up the invaginations (Vitols *et al.*, 1961) can occasionally be demonstrated in freeze-etched intact cell walls.

2. *Budding*

The first step in budding must consist in a local weakening of the existing cell wall which could be done by the secretion of a suitable enzyme (protein-disulphide reductase: Nickerson, 1963; Moor, 1967). The staining properties of the bud cell wall differ distinctly from those of the mother cell (Nečas and Svoboda, 1967; Marchant and Smith, 1967), and investigations by the use of fluorescent antibodies (Chung *et al.*, 1966) indicate that the mother-cell wall is not expanded and does not grow by intussusception to form the bud cell wall, but is "opened" or perforated with the bud wall containing only new material. This opening results in a rearrangement of the glucan fibrils around the later bud scar in the mother-cell wall (Fig. 4(d): Houwink and Kreger, 1953). The opening and the formation of the new cell wall must occur at the same time, otherwise the protoplast would slip out as it does during artificial protoplast production. Freeze-etchings confirm that, in every stage of budding, the cell wall is continuous and never shows a break.

3. *Cross-Wall Formation*

After nuclear division, the cross-wall is formed in a similar manner in budding or dividing yeasts as well as in other fungi (Nečas and Svoboda, 1967; Marchant and Smith, 1967). The plasmalemma infolds to form a ring-shaped aperture. In budding yeasts, this aperture closes completely, and two cell walls are constructed between the newly formed membranes, separating mother and daughter cell (Fig. 5(c)). The new wall material formed by the mother cell is essentially different from the pre-existing one in terms of staining properties (Mundkur, 1960a; Fig. 4(a)). This area contains much more chitin (Fig. 5(b):

FIG. 5. (a) Electron micrograph of a bud scar in surface view of a freeze-etched cell of *Saccharomyces cerevisiae*. (b) Electron micrograph showing chitin residues of the bud-scar region of *Sacch. cerevisiae* after solubilization of the mannan and glucan constituents (Bacon *et al.*, 1966). (c) Electron micrograph of a freeze-etched *Sacch. cerevisiae* showing vesicles in the region of the cross-wall formation derived from endoplasmic reticulum. (d) Electron micrograph of a cell-wall regenerating freeze-etched protoplast of *Sacch. cerevisiae* showing the plasmalemma, the innermost wall layer and the network of glucan fibrils. Reproduced by courtesy of F. Schwegler. See caption to Fig. 1 for explanation of symbols.

Bacon *et al.*, 1966) and also mannan-protein (Mundkur, 1960a). The difference between daughter cell-wall and cross-wall is much less pro-

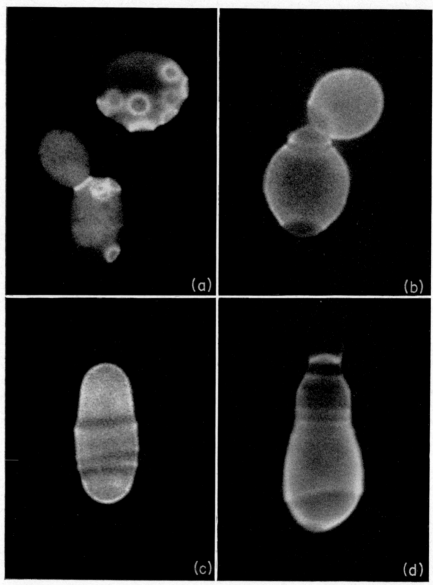

FIG. 6. Types of cell division in yeasts as revealed by primulin staining of the scar region. (a) *Saccharomyces cerevisiae*: bud scars. (b) *Saccharomycodes ludwigii*: multiple scars. (c) *Schizosaccharomyces pombe*: division scars. (d) *Endomyces magnusii*: division scars. Reproduced by courtesy of E. Streiblová.

nounced. The two cross-walls contain a framework of glucan fibrils (Marchant and Smith, 1967). The differences described above in the structures in the area of the cross-walls lead to the formation of a prominent bud scar in the mother-cell wall (Fig. 5(a)) and a morphologically nearly invisible birth scar on the daughter-cell wall.

4. *Types of Cell Division*

The mode of formation of new cell walls in dividing cells of different genera of yeasts proceeds in distinct ways. Budding yeasts and fission

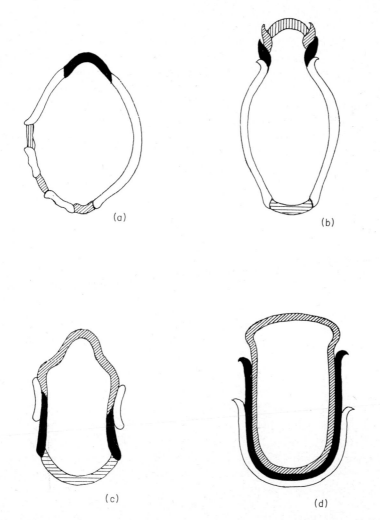

Fig. 7. Diagrams of the types of division in yeast cells shown in Fig. 6. Reproduced by courtesy of E. Streiblová.

yeasts can be distinguished (Streiblová and Beran, 1965, 1966). In budding yeasts, birth scars and bud scars arise as a result of cell division. By means of fluorescence microscopy, these structures can be observed in walls stained with primulin. In *Saccharomyces cerevisiae* multipolar budding results in the formation of brightly fluorescing rings which in old cells occupy a large proportion of the cell surface (Fig. 6(a)). The number of bud scars corresponds to the number of daughter cells produced. As many as 24 scars have been counted on an individual cell. The birth scar exhibits only weak fluorescence because the circular strongly fluorescent structure formed upon protrusion of a bud remains on the wall of the mother cell. The birth scar represents one half of the "plug" (septum) formed between mother and daughter cell.

In contrast, apiculate yeasts (e.g. *Saccharomycodes* spp.) are characterized by bipolar budding, that is, buds are produced at two opposite poles of the cell (Fig. 6(b)). As a result, multiple bud scars are formed. In contrast to *Saccharomyces* spp., apiculate yeasts do not form prominent circular structures; upon separation of the cells, equivalent moieties of the septum are present in the mother- and daughter-cell walls. In one type of fission yeast (*Schizosaccharomyces pombe*) the whole cell wall possesses rings of different age (Fig. 6 (c)) and in a second type (*Endomyces magnusii*) multilamellar cell walls are formed upon repeated division of an individual cell (Fig. 6 (d)). A diagrammatic representation of these types of cell division (Figs. 7(a)–(d)) shows the significance of the resulting structures of the cell walls.

5. Cell Wall Regeneration

The data available on the *de novo* organization of the yeast cell wall at an artificially produced "naked protoplast" are not very productive. In a summary, Villanueva (1966a) concludes that "practically all naked protoplasts are capable of regeneration into normal cells. Their capacity for regeneration is apparently determined primarily by the physical conditions in the medium". Some authors claim that gelatin is necessary for support of the first weak initial of the new wall; others regenerate cell walls in liquid media. According to Streiblová (1968) the innermost stratum is not stripped off completely during sphaeroplast formation and therefore the first components of the cell wall to be formed (i.e. a fibrillar framework; see Fig. 5(d)) are assembled outside this layer (Uruburu *et al.*, 1968). The encapsuled cells often show aberrant forms and cell-wall structures (Nečas, 1956, 1961, 1965) but later on produce normal cells by budding.

Sphaeroplasts specifically secrete a variety of cell-wall constituents including extracellular mannan and glucan (Lampen, 1968). In the case of invertase, evidence has been accumulated that secretion of the enzyme protein (synthesized in the cytoplasm) involves its conversion into a mannan-protein (Gascón *et al.*, 1968).

B. CHEMISTRY

1. *Isolation*

The isolation of clean cell walls (a prerequisite for a reliable chemical analysis) requires an appropriate disintegration of the cells by means of vigorous shaking in the presence of glass beads (e.g. Kessler and Nickerson, 1959) or by ultrasonic vibration (Miller and Phaff, 1958). The separation of free cell-walls from the cytoplasmic material is achieved by differential centrifugation at low speed followed by multiple washings in sucrose and buffer solutions (e.g. Mendoza and Villanueva, 1963). A possible source of contamination is the plasmalemma which is readily extractable only from stationary-phase cells. The isolated walls of baker's yeast account for as much as 30% of the dry cell mass (Falcone and Nickerson, 1956).

2. *Components*

The native cell-wall represents a very complex heterogeneous polymer. Upon its disintegration by the action of chemicals or enzymes, various fragments may be formed. These fragments represent complex macromolecules when the walls are subjected to mild treatments with dilute alkali or certain digestive enzymes. Severe extractions may result in the cleavage of complex cell-wall constituents into more or less homogeneous macromolecules. These ultimate building stones are glucan (yeast cellulose), mannan (yeast gum), chitin, protein, lipid and phosphate (see summaries by Phaff, 1963; Nickerson, 1963; Northcote, 1963).

Glucan represents the only polysaccharide which has been found in all yeasts examined. It is partially soluble in hot dilute alkali; subsequent partial hydrolysis by boiling in dilute hydrochloric acid makes about one half of the alkali-insoluble glucan soluble, the residual "hydroglucan" now being alkali-soluble (Houwink and Kreger, 1953). Hydroglucan is characterized by its microfibrillar structure (Fig. 4(c), p. 226). It represents a linear polymer of β (1→3)-linked glucose residues. The structure of the native glucan is more complicated; it contains about 10–20% of β (1→6)-linked glucose residues which form blocks of at least three β (1→6) linkages flanked by longer blocks of β(1→3)-linked glucose residues (Peat *et al.*, 1958; Tanaka and Phaff, 1965). However, a structure involving a β (1→6)-linked backbone with long β (1→3)-linked side chains has also been suggested (Manners and Patterson, 1966). The total amount of glucan as well as its structure vary greatly from one species or genus to another (Kreger, 1966). In the baker's yeast *Sacch. cerevisiae*, glucan accounts for about 30% of the dry cell-wall (Northcote, 1963).

Mannan, an alkali-soluble polysaccharide, is a highly branched polymer of mannose with an α $(1{\rightarrow}6)$-linked backbone and α $(1{\rightarrow}2)$- and some α $(1{\rightarrow}3)$-linked side chains (Haworth *et al.*, 1941; Phaff, 1963). In *Saccharomyces* cell walls , the amount of mannan equals that of glucan (Northcote, 1963). Mannan is absent however from cell walls of species of the genera *Schizosaccharomyces, Nadsonia, Rhodotorula* and of all hyphal fungi (Garzuly-Janke, 1940; Kreger, 1954).

Chitin, a polymer of β $(1{\rightarrow}4)$-linked N-acetyl-D-glucosamine, forms an insoluble residue after exhaustive extraction of mannan, glucan, protein and lipid. This minor constituent of yeast cell-walls is characterized by its X-ray pattern (Houwink and Kreger, 1953). Baker's yeast cell-walls contain a few percent of glucosamine, but less than 10% of the total glucosamine occurs in the form of chitin (Korn and Northcote, 1960). Thus, a large proportion of glucosamine is contained in non-chitin constituents of the wall. Chitin is absent from the cell walls of *Schizosaccharomyces octosporus* (Roelofsen and Hoette, 1951).

Protein is a conspicuous constituent of all yeast cell-walls. In baker's yeast, it accounts for about 7% of the dry cell-wall (Falcone and Nickerson, 1956) but the amount depends largely on growth conditions, age of culture and species. The wall protein contains a full complement of amino acids with a large proportion of glutamic acid and aspartic acid (Kessler and Nickerson, 1959). It has been termed "pseudokeratin" because it is characterized by its high sulphur-to-protein ratio.

Fractions of the external protein of yeast cells show enzyme activities including invertase (Frijs and Ottolenghi, 1959), acid phosphatase (McLellan and Lampen, 1963), aminopeptidase (Matile, 1968) and other hydrolases which are localized in the cell wall. The above enzymes are glycoproteins. Purified invertase contains about 50% mannan and a few percent of glucosamine (Neumann and Lampen, 1967) and hence, the external enzymes resemble the complex polymers mentioned below.

Polysaccharide-protein complexes represent the prominent fragments released upon mild treatment of isolated walls with alkaline agents (Falcone and Nickerson, 1956; Kessler and Nickerson, 1959; Korn and Northcote, 1960). Glucan-protein and glucomannan-protein complexes have been distinguished on the basis of solubility properties. It is an essential feature of these complexes that they are all heterogeneous regarding the composition of the polysaccharide moiety; small amounts of mannose are present in the glucan-protein, and the two mannan-proteins described are characterized by glucan-to-mannan ratios of about 1:1 and 1:2 respectively. Furthermore, the complexes contain small amounts of glucosamine and of phosphorus. Hence, the native cell-wall appears to be composed of copolymers of mannan and glucan linked together with structural protein. Indeed, glycoproteins are re-

leased from isolated cell walls by the action of trypsin and papain
(Eddy, 1958). Glucosamine is thought to serve as a connecting link between polysaccharides and protein.

The amount of lipid extractable from cell walls of a variety of yeasts
ranges from less than one percent to as much as 10% of the dry matter
(Nickerson, 1963). These lipids have not yet been identified but it is
interesting to note that certain species of the genera *Rhodotorula*,
Torulopsis, Candida and others secrete large amounts of unusual lipids
such as polyol esters and glycosides of hydroxy fatty acids (Stodola *et
al.*, 1967).

3. *Structural Organization*

A partial depolymerization of cell-wall components results in the
release of "naked protoplasts". It is commonly achieved by the use of
snail digestive juice (Eddy and Williamson, 1957) or preparations of
microbial origin (e.g. Garcia Mendoza and Villanueva, 1962) which
contain a variety of lytic enzymes including glucanases, mannanases
and proteases (see the review by Villanueva, 1966b).

In isolated walls, macromolecular constituents seem to be accessible
to a variety of enzymes; purified glucanase and mannanase (Millbank
and Macrae, 1964; Tanaka and Phaff, 1966) and proteolytic enzymes
(Eddy, 1958; Nickerson, 1963) dissolve isolated walls completely or
partially. However, these enzymes are unable to attack the intact cell-
wall in *Sacch. cerevisiae*. The respective macromolecules are inaccessible
unless phosphomannan complexes which cover the cell surface (Eddy,
1958) are degraded. In fact, a phosphomannanase is required for the
formation of sphaeroplasts (McLellan and Lampen, 1968). This enzyme,
isolated from a bacterial source, cleaves mannosidic bonds adjacent
to phosphodiester links between individual mannan complexes. The
release of phosphomannans is accompanied by the dissolution of invertase which therefore seems to be localized in a zone close to the wall
surface. It is likely that the mannan moieties of the cell-wall enzymes
represent a means of retaining the enzyme protein within the meshwork
of the cell-wall polysaccharides (Lampen, 1968). In the inner strata of
the cell wall, mannan-proteins and glucan-proteins are localized,
possibly linked together with disulphide bridges (Nickerson, 1963).
These disulphide bridges seem to determine largely the structure of the
wall, since their cleavage by reducing agents facilitates greatly the subsequent formation of sphaeroplasts by the use of snail enzyme (Duell *et
al.*, 1964). The above description of the structural organization of the
yeast cell-wall is summarized in Lampen's (1968) schematic representation (Fig. 8); it is valid however only regarding the wall of *Saccharomyces*. The intact wall of *Candida utilis*, for example, is degraded by

purified β (1→3)-glucanase at its equatorial zone, and sphaeroplasts are subsequently released (Villanueva and Elorza, 1968). Thus, in contrast to *Saccharomyces*, *Candida* seems to possess a cell wall with glucan localized at its periphery.

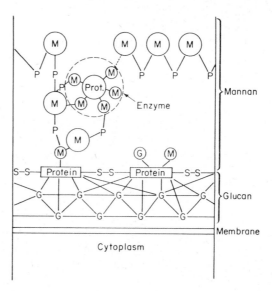

FIG. 8. Schematic structure of the yeast cell-wall. M indicates mannan; G, glucan; P, phosphate; S, sulphur. From Lampen (1968).

C. FUNCTION

The outline of isolated cell walls corresponds exactly with that of the intact cells. Obviously the rigid cell wall is responsible for the specific shape of yeast cells (Nickerson, 1963). The stability of shape is due to the existence of a large number of bonds by which the individual wall constituents are cemented into a single heterogeneous polymer. Disulphide bridges have been recognized to play a prominent role in determining the cellular shape (Nickerson and Falcone, 1959).

Another feature of yeast cell-walls is the elasticity which appears when cells shrink in the presence of high concentrations of osmotically active substances. Under normal conditions, the remarkably high turgor pressure keeps the wall under tension. Therefore, the preparation of cell wall-free sphaeroplasts has to be carried out in a solution of appropriate osmotic pressure.

As a consequence of the dense random meshwork, yeast cell-walls possess comparatively narrow pores. Dextran molecules with a molecular weight higher than 4,500 are excluded from the walls of *Saccharomyces cerevisiae* (Gerhardt and Judge, 1964). Indeed, yeasts are unable to utilize protein molecules present in the culture medium (Matile, 1968). Thus, the cell wall represents a filter which allows only micromolecules to diffuse to the sites of absorption.

Extracellular enzymes, such as invertase, melibiase and glucamylase, present in yeast cell-walls (Lampen, 1968) function in utilizing the respective carbon sources. A similar role may be attributed to the aminopeptidase. Phosphate esters may be utilized by the action of the extracellular acid phosphatase (Günther and Kattner, 1968).

III. The Plasmalemma

A. MORPHOLOGY

1. *Problems of Fixation*

In general, chemical fixation and embedding techniques encounter many difficulties in preserving the membraneous structures in yeast cells. Direct osmium tetroxide-fixation results in poor preservation (Agar and Douglas, 1957) except in the case of protoplasts (Fig. 20(a), p. 272). A previous fixation in glutaraldehyde followed by an enzymatic digestion of the cell wall lightens the osmium tetroxide-fixation of nuclear components while the membrane structures stay obscured (Robinow and Marak, 1966). Up to now, only potassium permanganate is known to preserve the membraneous architecture and show its unit-membrane structure (Figs. 3(a), p. 224 and 19(a), (b), p. 268).

2. *Invaginations*

According to Vitols *et al.* (1961) the plasmalemma (cytoplasmic membrane) of a yeast cell consists of a triple-layered unit about 8·0 mμ wide and differing in its substructure to a certain extent from all other membranes. In thin sections, this membrane and the tonoplast show a curved, invaginated outline, most probably caused by shrinkage (Fig. 3(a), p. 224). In living cells, stabilized by a suitable freezing technique (Moor, 1964) and freeze-etched (Moor and Mühlethaler, 1963), both membrane systems are fully expanded (Figs. 3(b), (c)); but nevertheless the plasmalemma is sculptured by invaginations (Figs. 9(a), (b), (d)), a characteristic structure which is also present on protoplasts. These invaginations, being about 30 mμ wide and extending not more than about 50 mμ into the cytoplasm, exhibit a variable length depending on the species, age and growth phase of the cell. As a rule, the plasmalemma of small buds is unsculptured. During the bud maturation more and more invaginations may appear and become a prominent feature of

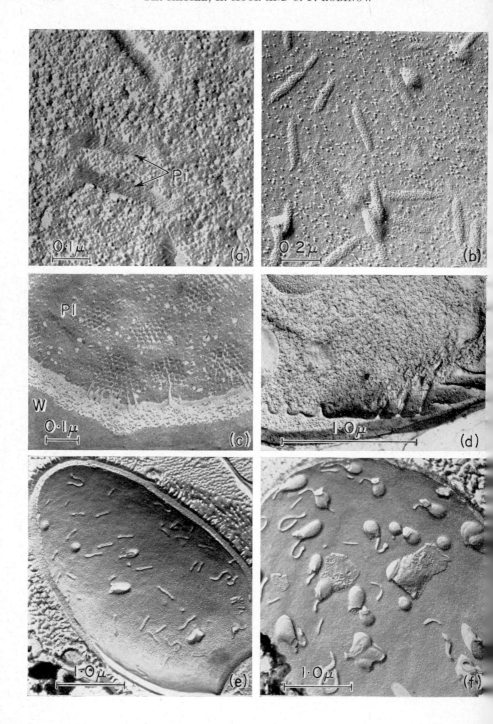

stationary-phase cells (Moor and Mühlethaler, 1963). Acridine orange-treated cells show aberrant invaginations with a curved, sack-like shape (Figs. 9(e), (f)). Similar structures have been found in the plasmalemma of *Neurospora* during secretion of proteolytic enzymes (Matile *et al.*, 1965). Therefore, the acridine orange-induced aberrant shape of the invaginations might indicate a structural process for the elimination of cytoplasmic substances which occurs during this treatment (Bogen and Keser, 1954).

3. Fine Structure

In freeze-etched specimens, which exhibit surface views of all membrane systems (Moor, 1966), the plasmalemma of yeast cells (*Saccharomyces, Schizosaccharomyces, Candida* spp.) shows a particulate structure on the outer surface. The particles have a diameter of about 15 mμ and may be concentrated in hexagonal arrays showing a lattice period of 18 mμ (Fig. 9(c); Moor and Mühlethaler, 1963). These units are embedded in the membrane in a variable depth suggesting that they are moving through the plasmalemma. The inner surface of the cytoplasmic membrane is covered by fewer particles of a similar size except in the region of the invaginations, where they appear to be densely concentrated (Fig. 9(b)). An outer surface view of the same region exhibits a particle-free membrane being structured by 4 mμ-subunits (Fig. 9(a)). Isolation and biochemical investigation has shown the 15 mμ-particles to be composed of mannan and protein in a ratio of 5:2. No enzymic activity could be detected in this fraction (Matile *et al.*, 1967). These findings indicate that the particles represent building stones of the glycoprotein meshwork of the cell wall (Nickerson, 1963), and to be produced by, or secreted through, the plasmalemma. In fact, a close association of the particles and the glucan fibrils has been demonstrated (Fig. 9(c); Moor and Mühlethaler, 1963). At first, this image leads to the assumption that the particles could be the sites of synthesis of the glucan fibrils, but subsequent investigations have not revealed appropriate enzymic properties.

Fig. 9. Electron micrographs of plasmalemma structures in freeze-etched cells of *Saccharomyces cerevisiae*. (a) Outer membrane surface, showing the structure of the invaginations. (b) Particulate structure of the inner membrane surface. (c) Hexagonally arranged plasmalemma particles (outer surface), and glucan fibrils extending from these areas to the adjacent cross-fractured cell wall (Moor and Mühlethaler, 1963). (d) Invaginations, on the left cross-fractured, on the right in surface view (Moor and Mühlethaler, 1963). (e) Aberrant invaginations of the outer surface of the plasmalemma of acridine orange-treated cells. (f) An inner surface view of the same invaginations as shown in Fig. 9(e).

4. *Ontogeny*

During budding or fission, the new plasmalemma is produced by growth in area of the pre-existing membrane. An uptake of endoplasmic reticulum- or Golgi-derived membrane material might be suggested (see Section IV, p. 242). As already mentioned, the protoplasts of the mother cell and daughter cell are separated by the infolding plasma-lemma which produces a closing aperture.

As known so far, the plasmalemma of ascospores is produced in a completely different manner, as suggested in the case of *Saccharomyces* (Marquardt, 1963) and more precisely described for *Saccobolus* (Carroll, 1967). Each nucleus produced by the meiotic divisions is surrounded by elements of endoplasmic reticulum enclosing only a part of the cyto-plasm available, while a considerable amount of it stays outside in the ascus. Consequently, the ascospores are encapsuled by two membranes derived from endoplasmic reticulum both of which are thought to be involved in the production of the spore cell-wall and the inner one alone in the formation of the new plasmalemma of the ascospore.

B. CHEMISTRY

1. *Isolation*

As shown above, the outer surface of the plasmalemma is sculptured with particles, a morphological feature which has been used for the identification of membranes obtained upon fractionating cell-free ex-tracts. Fragments of plasmalemma bearing this marker could be isolated only from stationary-phase cells of *Saccharomyces cerevisiae* which were grown anaerobically in a medium lacking the growth factors necessary for anaerobic growth (ergosterol and a Tween). The isolation of such "complete" plasmalemmae requires the subfractionation of a micro-somal preparation using isopycnic gradients of Urografin (the methyl-glucamine salt of N,N'-diacetyl-3,5-diamino-2,4,6-triiodobenzoic acid). Preparation of plasmalemma particles involves lysis of the membranes using detergents and ultrasonic vibration, and subsequent isolation of the particles by density-gradient centrifugation (Matile *et al.*, 1967).

2. *Components*

The analysis of isolated fragments of plasmalemma has shown the presence of lipids (including phospholipids), protein and polysaccharide. Hydrolysis of the polysaccharide yields mannose only. This mannan, together with protein, is localized in the plasmalemma particles. Iso-lated plasmalemma is furthermore characterized by the presence of an ATPase which is active only in the presence of Mg^{2+}. This enzyme is distinct from the mitochondrial ATPase in being oligomycin-insensitive (Matile *et al.*, 1967).

C. FUNCTION

The osmotic properties of yeast cells and sphaeroplasts are due to the impermeability of the plasmalemma with respect to solutes. In intact cells, the high internal osmotic pressure (0·6 molar and even higher depending on growth conditions and species) is compensated by the rigid and elastic cell wall, whereas sphaeroplasts must be stabilized by providing adequate external osmotic pressures in the suspending medium. Shrinkage of yeast cells and sphaeroplasts in the presence of high concentrations of solute is reversible; adaptation, that is an increase in the internal osmotic pressure, is caused by the use of intracellular macromolecular material rather than by the uptake of solute from the medium (Lillehoj and Ottolenghi, 1966).

In a study on the effect of anions and cations on the lysis of osmotically shocked yeast sphaeroplasts, Indge (1968a) infers the presence on the plasmalemma of a cation-binding site which is involved in the maintenance of its structural integrity. Under normal conditions, this site is occupied by inorganic cations such as Mg^{2+} and Ca^{2+}. However, under certain conditions (low ionic strength in the medium, appropriate pH values), positively charged proteins, like bovine serum albumin or ribonuclease, are bound to the plasmalemma of whole yeast cells. As a consequence, the cells eventually undergo lysis, possible because the binding of protein results in a structurally altered membrane and in the loss of its permeability properties (Alper *et al.*, 1967; Ottolenghi, 1967).

Repeated freezing and thawing of yeast cells causes the leakage of solutes into the suspending medium; this phenomenon is due to intracellular formation of ice crystals resulting in a damaged plasma membrane (Mazur, 1965). The same treatment has often been used in order to render the cytoplasm and its organelles accessible for substrates which are not allowed to move across the intact permeability barrier. Thus, the plasmalemma undoubtedly plays a prominent role in controlling the movement of compounds from the medium into the cell and *vice versa*. On the one hand it is involved in the uptake of various nutrients present in the medium, such as ions, sugars, amino acids and vitamins; on the other hand it controls the release of ethanol and a variety of other compounds which upon fermentation are excreted into the growth medium (Suomalainen, 1968). Furthermore, it must be assumed that cell-wall synthesis requires the passage of precursors and even of macromolecules such as exoenzymes across the plasmalemma.

Some of the mechanisms involved in plasmalemma-mediated transport have been studied in great detail. It has been recognized that the surface of the cytoplasm is occupied by binding sites specific for certain substrates or groups of structurally related compounds. These binding sites belong to transporting systems which, by means of movable carriers, transport the molecules from one surface of the plasmalemma to the other. The activity of these transportases, permeases or carriers has features characteristic of enzymes. It shows saturation kinetics, substrate specificity and stereospecificity; the velocity of transport depends on temperature; substrates using common binding-sites compete with each other according to their affinities. Both active (energy-requiring) and passive (downhill) transport of molecules have been reported (see reviews by Cirillo, 1961; Rothstein, 1961).

It is likely that the ATPase present in isolated plasmalemma is involved in the energy-requiring movements of compounds against concentration gradients. In *Saccharomyces*, the uptake of a variety of monosaccharides does not require energy provided by metabolism (Cirillo, 1962). The respective carriers perform a rapid establishment of equilibrium between internal and external sugar concentration (facilitated diffusion). One sugar carrier with a high affinity for glucose and a broad specificity transports a variety of metabolizable and non-metabolizable sugars (Kotyk, 1967). Isolated plasmalemma binds glucose and other sugars with the same affinity constants as found in intact cells (Kotyk and Matile, 1968). This finding conclusively establishes the localization of transporting systems in the plasmalemma.

IV. The Endoplasmic Reticulum and Related Organelles

A. MORPHOLOGY

1. *Characterization*

In the cytoplasm of yeast (as in other ascomycetes) a double-membrane system is present which resembles the endoplasmic reticulum of higher plant and animal cells (Vitols *et al.*, 1961; Marquardt, 1962; Moor and Mühlethaler, 1963). Occasionally these membranes are connected

---

FIG. 10. Electron micrographs of the endoplasmic reticulum in freeze-etched *Saccharomyces cerevisiae*. (a) A cross-fractured cell exhibiting nucleus-connected, intermediate and plasmalemma-associated strands of endoplasmic reticulum (Moor, 1967b). (b) A surface view of an endoplasmic reticulum membrane covered by groups of ribosomes. (c) A micrograph showing the nuclear membrane which persists during division; on the right, cross-fractured; on the left, in surface view showing the nuclear pores.

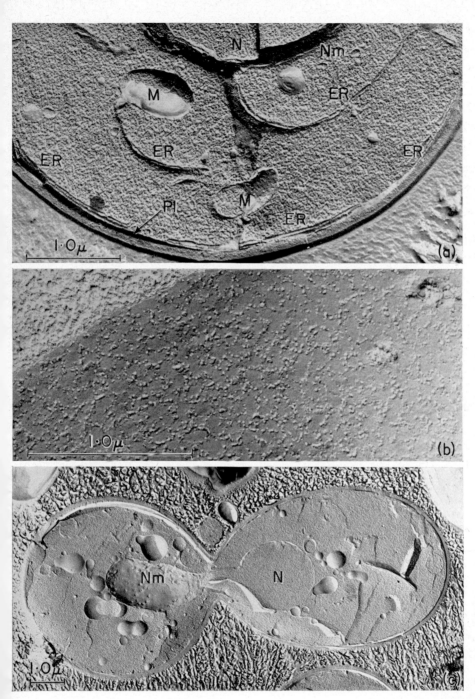

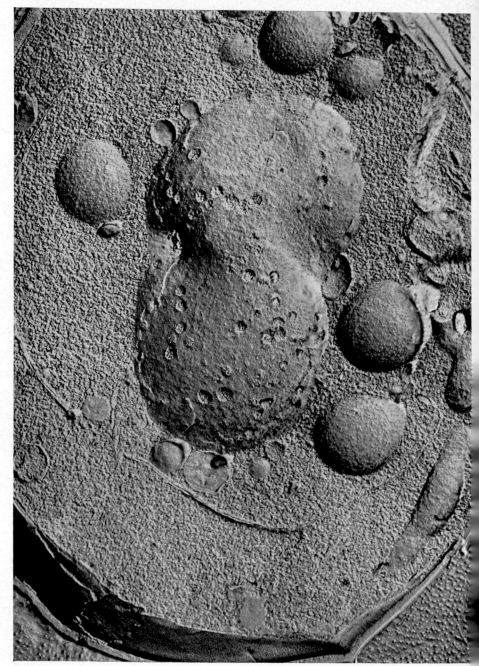

FIG. 11. A nucleus in freeze-etched *Saccharomyces cerevisiae*, showing pores in the nuclear envelope. Magnification × 21,500. From Moor (1966).

with the outer nuclear membrane or may be closely associated to the plasmalemma (Fig. 10(a)). As a rule the lumen between the two membranes is about 20 mμ wide and contains a watery sap (the so-called "enchylema") except when dense globular bodies with a diameter of up to 100 mμ are enclosed. The cytoplasmic surface of this membrane system shows a particulate structure (Fig. 10(b)) consisting of single or grouped granules having a size of 10–15 mμ (Moor and Mühlethaler, 1963; Moor, 1967a, b). The largest grouped units of this particle population could be polysomes. In protoplasts fixed with osmium tetroxide dark-stained granules of ribosomal size can be seen in the ground-plasm and associated with the double membrane system (Fig. 21(a), p. 272). As a result, the major characteristics show this system to be morphologically identical with the endoplasmic reticulum of higher plant cells.

2. *Nuclear Envelope*

As indicated by the common breakdown of the nuclear envelope into pieces which cannot be distinguished from elements of endoplasmic reticulum (Porter and Machado, 1960), the nuclear membranes can be interpreted as a part of the endoplasmic reticulum system having a special function. In *Saccharomyces*, its fine structure resembles closely that described for the endoplasmic reticulum yet concerning the occasionally enclosed globular particles. The nuclear pores, exhibiting the same size of 80–100 mμ as those occurring in endoplasmic reticulum elements, may be concentrated in certain areas as well as uniformly spread over the whole surface (Fig. 11). This phenomenon, combined with the observation of circular areas of pore-size on the nuclear envelope, leads to the conclusion that a pore is an ephemeral structure which may disappear (close up) and be formed at another place (Moor and Mühlethaler, 1963).

In *Saccharomyces* and other yeasts, the nuclear envelope does not break down during division. Part of it enters the newly formed bud in an ameoboid manner to enclose the nuclear material of the daughter cell. Before the final separation, a long narrow neck is formed between the two nuclei (Figs. 2(f), p. 223; 3(c), p. 224; 10(c), p. 243).

3. *Activities*

A prominent activity of the endoplasmic reticulum can be shown in budding yeast cells (Moor, 1967a,b). The endoplasmic reticulum elements of a cell, preparing to form a bud, fuse and form a nearly closed envelope containing the nucleus and vacuoles. In the area of the opening in this envelope, a proliferation process starts producing a mass of vesicular strands which segregate spherical vesicles with a

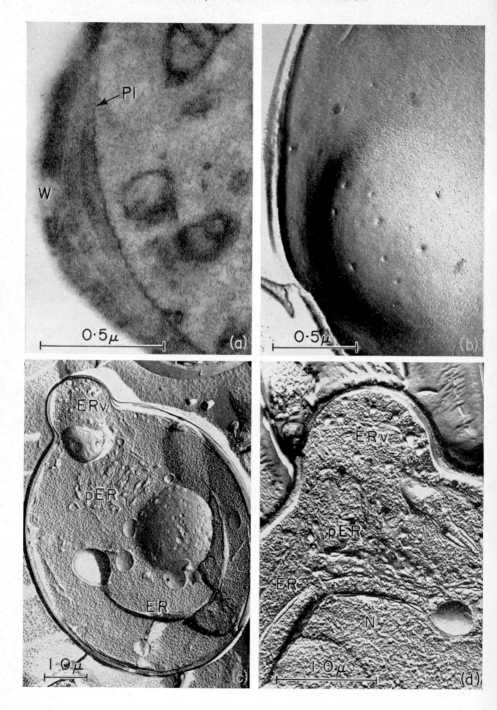

diameter of about 100 mμ (Figs. 12(c), (d)). These vesicles penetrate the plasmalemma in a limited region (Fig. 12(b)) and probably represent the local accumulation of secreted products observed by Marchant and Smith (1967; Fig. 12(a)). This is the region where the bud starts to grow explosively by a local evagination of the cell wall. The endoplasmic reticulum-derived vesicles are thought to contain protein-disulphide reductase, an enzyme capable of decreasing the cell-wall rigidity (Nickerson and Falcone, 1956). Beside this, the secreted product present in the vesicles most probably contains also additional enzymes and proteinaceous building stones necessary for the production of the new bud-wall.

Morphologically identical vesicles can be observed during cross-wall formation (Fig. 5(c), p. 228). These vesicles line up along the newly formed plasmalemmae of the mother cell and daughter cell and, after the cell walls have been built up, they cannot be observed any more (H. Moor, unpublished results). It can be concluded that the vesicles penetrate the plasmalemma also and release their contents into the gap between the two cells, a phenomenon which is difficult to observe in this very limited area.

An additional phenomenon suggests the involvement of endoplasmic reticulum in the formation of the cell-wall also. In regenerating protoplasts, endoplasmic reticulum elements that are lined up along the plasmalemma are a very common feature; but products derived from these elements passing through the cytoplasmic membrane have not yet been observed.

Besides these activities, parts of the endoplasmic reticulum may differentiate to form distinct organelles. The formation of the ascospore plasmalemma has already been mentioned; the production of sphaerosomes and Golgi bodies will be the subject of the following sections (see Fig. 14).

Fig. 12. Electron micrographs showing events during the start of budding in yeasts. (a) A thin section of permanganate-fixed *Rhodotorula glutinis*. The region of bud formation shows the development of the new wall layer (a secretion product of the endoplasmic reticulum) between the parent cell-wall and the plasmalemma (Marchant and Smith, 1967). (b) Freeze-etched plasmalemma of *Saccharomyces cerevisiae* showing an early stage of bud formation. The spherical invaginations indicate the secretion of the endoplasmic reticulum-derived vesicles (Moor, 1967a, b). (c) Freeze-etched cell of *Saccharomyces cerevisiae* showing a bud initial, proliferating endoplasmic reticulum and derived vesicles (Moor, 1967a, b). (d) A bud initial at higher magnification (Moor, 1967a, b). See caption to Fig. 1 for explanation of symbols.

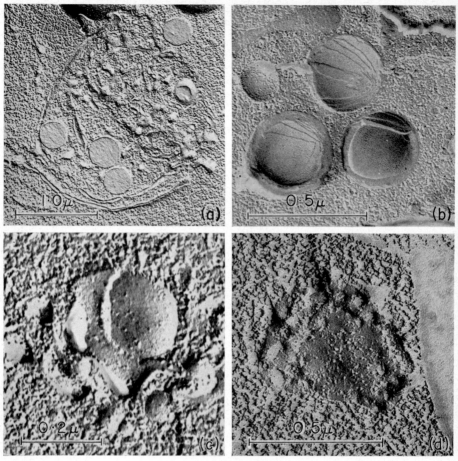

FIG. 13. Elements derived from endoplasmic reticulum in freeze-etched *Saccharomyces cerevisiae*. (a) Proliferating endoplasmic-reticulum elements producing lipid granules (sphaerosomes). (b) Lamellated lipid granules partly split up by the freeze-etching procedure (Moor and Mühlethaler, 1963). (c) A Golgi body showing a central pile of flattened sacs surrounded by small vesicles (Moor and Mühlethaler, 1963). (d) A Golgi body of the porous type. See caption to Fig. 1 for explanation of symbols.

4. *Lipid Granules (Sphaerosomes)*

Lipid granules have been described by light-microscopists as a common structure in yeast cells (Guilliermond, 1923). Electron microscope investigations with thin sections have shown them to be "empty" vesicles (Vitols *et al.*, 1961; Marquardt, 1962b) and in freeze-etched specimens to be membrane-bound spherical granules enclosing a struc-

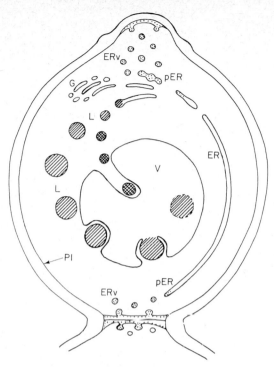

Fɪɢ. 14. A diagram of the relationship between the endoplasmic reticulum and endoplasmic reticulum-derived elements with the vacuole and with the plasma-lemma and cell-wall synthesis. See caption to Fig. 1 for explanation of symbols.

tured content (Moor and Mühlethaler, 1963). These granules, which occur predominantly in stationary-phase cells, exhibit a concentrically lamellated or amorphous content depending on whether the stored substances are polar and/or apolar lipids (Figs. 13(a), (b)).

During budding, these lipid granules appear in the region of the pro-liferating endoplasmic reticulum and successively grow from about 100 mμ to 700 mμ in diameter (Fig. 13(a)). The image of this process resembles closely the differentiation of endoplasmic reticulum elements described in higher plant cells which results in the formation of the sphaerosomes (Frey-Wyssling et al., 1963; Grieshaber, 1964). In principle, the structure and ontogeny of these two particles are morphologically identical, a fact which is responsible for the use of both terms (lipid granules, sphaerosomes) for the same organelle (Marquardt, 1962b).

For the purpose of mobilization of the stored material, the lipid granules will be incorporated into the vacuole. (See Section V, p. 255 and Fig. 14.)

5. *Microbodies*

The occurrence of a membrane-bound vesicle, having about the same size and shape as the sphaerosomes but containing stainable granular material, was first described by Marquardt (1962a, b). This organelle resembles structures in animal cells (Afzelius, 1965) which have been termed microbodies. Indeed, a microbody-like structure containing catalase, which is a key enzyme of animal microbodies, has been isolated from yeast cells (Avers and Federman, 1968). It may be suggested here that this structure could also be a differentiation product of the endoplasmic reticulum.

6. *Golgi Apparatus*

The existence of a Golgi apparatus in yeast cells is controversial. A Golgi-like structure was first described by Moor and Mühlethaler (1963) to occur occasionally in *Saccharomyces* sp. Havelkova and Mensik (1966) have found Golgi structures in regenerating protoplasts of *Schizosaccharomyces* sp. and suggested a relation between the Golgi apparatus and cell-wall formation in agreement with the findings in higher plant cells (Whaley and Mollenhauer, 1963; Frey-Wyssling *et al.*, 1964).

In freeze-etched *Saccharomyces cerevisiae* cells, the Golgi-like structures (Fig. 13(c)) may consist of a central pile of about three flattened sacs surrounded by many very small vesicles (Moor and Mühlethaler, 1963) or only of one disc-like element perforated by a few pores (Fig. 13(d)) resembling Golgi elements found in higher plant cells (Branton and Moor, 1964). Between these different structures, all types of transitional stage can be found. Considering the fact that in yeast also the Golgi apparatus should be involved in cell-wall construction, the mass of vesicular strands produced by the proliferating endoplasmic reticulum during budding could be interpreted as Golgi structure. In such a case, the Golgi apparatus would release these vesicles which would penetrate the plasmalemma and contribute to cell-wall synthesis. The Golgi apparatus most probably is totally "consumed" by this process and therefore cannot be found in most of the cells. A disc-like perforated element that can often be found in the region of endoplasmic reticulum proliferation and bud formation confirms these ideas.

Thus, the above hypothesis about the ontogeny and disappearance of the Golgi apparatus in yeast cells would characterize this organelle as an ephemeral structure which is produced by the endoplasmic reticulum and is "consumed" during cell-wall synthesis.

7. *Ontogeny*

A structural and possibly precursor relation between elements of endoplasmic reticulum and the plasmalemma has been searched for intensively in freeze-etched specimens but never found. Therefore, the endoplasmic reticulum can be described as an original organelle of the cytoplasm. It grows in area, and fusion and fission of its elements occur. During cell division these elements are considered to be distributed equally to mother cell and daughter cell.

B. CHEMISTRY AND FUNCTION

Structures described in this section are present in the postmitochondrial pellet obtained upon differential centrifugation of cell-free extracts. This microsomal fraction is heterogeneous and little work has been done on its subfractionation. It contains fragments of the plasmalemma which can be isolated by centrifugation in suitable density gradients (Matile et al., 1967). Using density gradients of sucrose, Schatz and Klima (1964) have been able to characterize another microsomal membrane fraction by the presence of a specific oxidoreductase, $NADPH_2$-cytochrome c-reductase, which does not participate in a respiratory chain. Similar enzymes are associated with isolated fragments of endoplasmic reticulum from animal cells. It is not known, however, whether endoplasmic reticulum membranes of yeast resemble those of other organisms regarding the existence of a common superstructure of membranes and ribosomes. Evidence presented by Klein et al. (1967a) suggests that this is not the case. Therefore, a prominent function attributed commonly to the endoplasmic reticulum, the synthesis of protein, may be assumed only by analogy.

Another particulate constituent present in microsomal fractions incorporates acetate into a variety of lipids. Fatty acid synthase activity is associated with a large protein particle (Klein et al., 1967b); other enzymes involved in the synthesis of lipids seem to be associated with membranes, presumably fragments of the endoplasmic reticulum (Klein et al., 1967c).

V. The Vacuole

A. MORPHOLOGY

1. Characterization

The vacuoles form the most conspicuous organelles of the yeast cell visible in the light microscope. Nevertheless, many of the reported data about ontogeny, chemistry and function of yeast vacuoles are controversial, partly caused by an occasional confusion of vacuole and nucleus (lit. cit. in Guilliermond et al., 1933; Bautz, 1960). Only the essential light-microscope results will be cited here, greater emphasis being given to electron microscope findings.

In thin sections, the vacuole appears mostly as an electron-trans-
9*

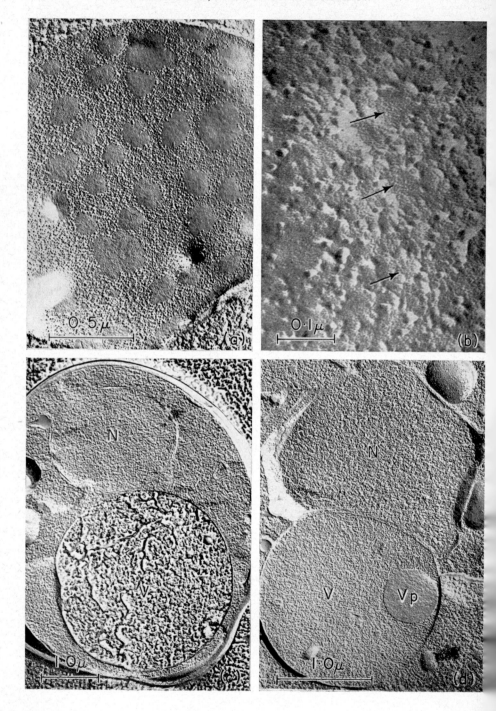

parent area limited by a single membrane (Vitols *et al.*, 1961). In some cells, the shape of the vacuole may be quite irregular, with narrow "diverticula" extending far into the cytoplasm. In freeze-etched specimens, any shrinkage artefacts are excluded and no similar forms can be detected. The vacuoles are fully expanded, spherical structures having a diameter of about 0·3 to 3 μ (Moor and Mühlethaler, 1963). In yeast cells, as a rule, one large vacuole or several smaller units of similar size can be observed.

The inner and outer surfaces of the vacuolar membrane are covered by a variety of particles 8–12 mμ wide. These particles may be uniformly spread over the whole surface, but very often circular uncovered spots appear on the membranes with a diameter of about 100–200 mμ (Fig. 15(a)). The presence of RNA in isolated membrane fractions (A. Wiemken, unpublished results) seems to confirm the suggestion that ribosomes could be located on the vacuolar membrane (Moor and Mühlethaler, 1963), but the membrane-bound particles have not been isolated and analysed up to the present time. At high magnifications and optimum resolution, the replicas of vacuolar membranes exhibit a granular fine structure of about 4 mμ (Fig. 15 (b)) which might represent the structural protein found to constitute a large fraction of the total mass of the membranes (Criddle and Willemot, 1967).

Observations under the light microscope have revealed many structural components in the vacuolar sap (Guilliermond *et al.*, 1933). However, these structures can easily be confused with other "granulations" in yeast cells due to the limitations of resolution in the light microscope (Bautz, 1960). Conventional fixation and embedding techniques used for electron microscope investigations have not contributed much to knowledge of the vacuolar content because most of the water- and lipid-soluble substances are lost during preparation. Only by freeze-etching can the whole multitude of substances and structures be preserved. In exponentially growing cells, the vacuolar sap is very watery and contains no structural elements (Fig. 15(c)). According to Brachet (1957) an addition of inorganic phosphate to starved cultures leads to the formation of relatively large globules in the vacuoles (Fig. 15(d)).

FIG. 15. Electron micrographs of vacuolar structures in freeze-etched cells of *Saccharomyces cerevisiae*. (a) Inner surface of a vacuolar membrane exhibiting a particulate structure that leaves circular areas uncovered. (b) Fine granular structure of the vacuolar membrane which is supposed to represent the structural protein (arrows). (c) Vacuole filled with a liquid containing relatively large ice crystals. (d) Vacuole containing a polymetaphosphate globule. See caption to Fig. 1 for explanation of symbols.

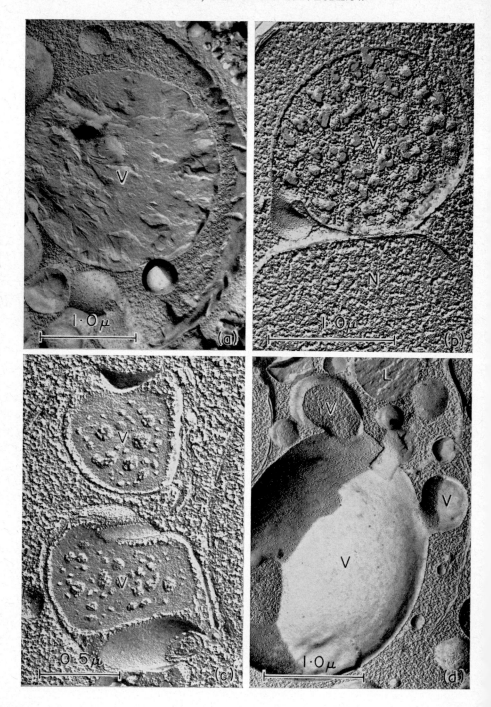

These globules contain mainly polymetaphosphate (Wiame, 1947). Because of their staining properties Guilliermond *et al.* (1933) called these globules *corpuscules métachromatiques*. In freeze-etched specimens, they show a fine-granular substructure (Moor and Mühlethaler, 1963). In cultures kept under anaerobic conditions, a morphologically different kind of inclusion can be found (Fig. 16(b)). As cultures enter the stationary phase of growth the cells show vacuoles containing an increasing amount of granular material which may be concentrated to a porous mass (Fig. 16(c)). In isolated fractions, an increased storage of intermediates of lipid synthesis has been observed. Under very unusual culture conditions (liquid media containing 20% glycerol and sparged with oxygen), a lipophanerosis takes place resulting in an accumulation of sphaerosomes and a formation of vacuoles completely filled with lipid (Fig. 16(a)).

2. *Activities*

In budding yeasts, a newly formed cell normally contains many small vacuoles (see Fig. 2, p. 223). During the budding of such a cell, some of the vacuoles enter the daughter cell and the rest gather at the distal end of the mother cell where they start to fuse (Fig. 16(d)). The products of this process consist in two differently structured cells, one containing a large vacuole the other comprising many smaller ones (Fig. 2(h), p. 223). The fusion process, followed by swelling of the vacuole, evokes the suggestion that the rapidly growing organelle has to compensate for the loss of cytoplasmic material being transferred into the bud. In addition, this morphological feature of expanding vacuole could be a force that pushes the dividing nucleus against the neck between mother cell and daughter cell to cause its partial entry into the bud.

The opposite process can be observed in cells which prepare for budding. Light-microscope observations indicate that the large vacuole may disappear completely and reappear during bud formation (Narayana, 1956). Freeze-etched specimens show that this "disappearance" is caused by a fragmentation or fission of the large vacuole into several much smaller units which are invisible under the light microscope. In

---

FIG. 16. Electron micrographs of vacuolar structures in freeze-etched cells of *Saccharomyces cerevisiae*. (a) Large vacuole completely filled with lipid. (b) Granular material of lipoidic nature in the vacuole of an anaerobically grown cell. (c) The same material as shown in Fig. 17(b) concentrated to form a porous mass. (d) Two small vacuoles fusing with a large one during cell division. See caption to Fig. 1 for explanation of symbols.

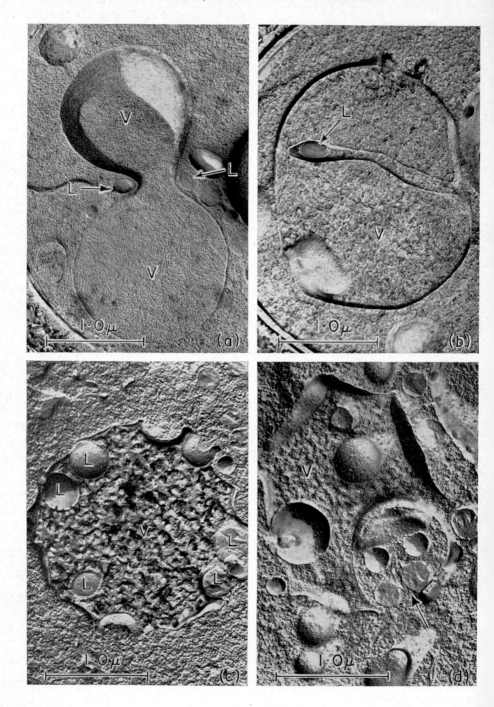

the course of the fragmentation, the indentations in the profile of a constricted vacuole regularly contain lipid granules (Fig. 17(a)). Sometimes these granules can be observed to form a beaded chain around the neck of such a vacuole. In other specimens, the lipid granules may be found at the ends of long tunnels extending from the vacuolar surface into the contents (Fig. 17(b)). These findings suggest an active role for the sphaerosomes during vacuolar fragmentation. The surface area of the fission products is considered to be much larger than that of the original large vacuole and, therefore, the role of the sphaerosomes may lie in the consumption of stored lipid for membrane production. The fragmentation process described above has also been observed under the light microscope when a vital stain accumulated in the vacuole can be found in the corresponding organelles of the descendants of the stained cell (Svihla and Schlenk, 1960).

A drastic decrease in the number of sphaerosomes can be observed in stationary-phase cells of *Saccharomyces cerevisiae* transferred to growth medium. These organelles attach themselves to the vacuolar membrane, which forms indentations and surrounds the sphaerosomes (Fig. 17(c)). During this uptake, the membranes of the sphaerosomes and the vacuole most probably fuse, as has been suggested by Smith and Marchant (1968). Together with the sphaerosomes, other vesicular constituents of the cytoplasm and ground substance may be incorporated in the vacuole (Fig. 17(d)). A morphologically identical process takes place in the root-tip cells of corn (Matile and Moor, 1968). The circular spots on the vacuolar membrane, which do not show any particulate structure, are considered to represent areas where such an uptake occurred. Usually, the incorporated structured elements are dissolved and disappear very rapidly, except during lipophanerosis when the large vacuole may be filled completely by sphaerosomal material. No information about the enzymic properties of this material is available up to now, and therefore the origin of the lytic enzymes present in the vacuole (Matile and Wiemken, 1967) is still unknown. Studies on root-tip cells (Matile and Moor, 1968) suggest that these enzymes could also be produced by the endoplasmic reticulum and transferred to the

---

FIG. 17. Electron micrographs of vacuolar structures in freeze-etched cells of *Saccharomyces cerevisiae*. (a) Fragmenting vacuole exhibiting sphaerosomes around the notch. (b) Sphaerosome contained in the end of a long tunnel formed by the vacuolar membrane. (c) Uptake of lipid granules into the vacuole. (d) Uptake of different cytoplasmic constituents into the vacuole (arrow). See caption to Fig. 1 for explanation of symbols.

vacuoles by means of sphaerosomes derived from the endoplasmic reticulum. In this case, the sphaerosomes would represent primary, and the vacuoles secondary, lysosomes, following the interpretations of de Duve and Wattiaux (1966).

3. *Ontogeny*

During the division cycle of a yeast cell, the vacuole multiplies by fission, and the products of this fragmentation process are distributed between the mother cell and daughter cell. At least part of the vacuolar membrane could originate from the endoplasmic reticulum and would be transferred via sphaerosomes.

B. CHEMISTRY

1. *Isolation*

Yeast vacuoles are liberated upon lysis of sphaeroplasts under suitable conditions. Stable vacuoles are released from osmotically shocked sphaeroplasts of *Sacch. cerevisiae* if the tonicity of the medium is lowered from 0·7 M to 0·233 M mannitol (Matile and Wiemken, 1967). Although they are osmotically active, the vacuoles seem to be less sensitive to rapid changes of the external concentrations than are sphaeroplasts. Taking advantage of the metabolic lysis of sphaeroplasts in the presence of a metabolizable sugar, Indge (1968b, c) has been able to liberate perfectly preserved vacuoles of *Sacch. carlsbergensis*. The medium in which sphaeroplasts were suspended contained glucose, a chelating agent and mannitol as an osmoticum. In the presence of Ficoll (about 8%), free vacuoles assume a density lower than that of the medium and can be separated from the other cell constituents by flotation upon low speed centrifugation (Matile and Wiemken, 1967).

2. *Components*

Vacuoles of yeast cells accumulate cationic dyes such as neutral red (Guilliermond, 1930; Indge, 1968c). It has been concluded that this phenomenon is due to the presence in the vacuoles of polyphosphate, formerly also termed volutin. Preparations rich in free vacuoles do indeed contain large amounts of acid-soluble phosphorus which is present largely as high molecular-weight polyphosphate (Indge, 1968d). On cytochemical grounds, Pfeiffer (1963) has suggested the occurrence of lipids in yeast vacuoles. Still another phenomenon of accumulation in vacuoles concerns a pink pigment of certain adenine-dependent strains of *Sacch. cerevisiae* (Brown and McClary, 1963).

3. *Enzymes*

Isolated vacuoles contain remarkably high activities of several hydrolytic enzymes, including proteases, ribonuclease and esterase. These enzymes are soluble in extracts from yeast cells subjected to mechanical disintegration which results in the disruption of vacuoles. Evidence has been presented that the hydrolases are localized exclusively in the vacuoles (Matile and Wiemken, 1967). An aminopeptidase, with a slightly alkaline pH-optimum, is localized mainly outside the plasmalemma (Matile, 1968), but some activity is also present in the vacuole. In isolated vacuoles, acid phosphatase, α-glycerophosphatase and α-glucosidase are absent (A. Wiemken, unpublished observations).

C. FUNCTION

Turnover of macromolecules, like proteins and nucleic acids, involves the simultaneous synthesis and breakdown in an individual cell. This phenomenon has been demonstrated repeatedly to occur in yeast, particularly in stationary-phase cells (see the reviews by Mandelstam, 1960; Fukuhara, 1967a; Sarkar and Poddar, 1965). Turnover requires a cell compartment (termed a lysosome) in which the breakdown reactions are separated from the synthetically active structures of the cytoplasm. The localization of acid hydrolases suggests that the vacuole represents the lysosome of the yeast cell (Matile and Wiemken, 1967). In fact, the activities of lysosomal enzymes are remarkably higher in rapidly turning-over stationary-phase cells as compared with exponentially growing cells; in addition, growth phases in which turnover reactions result in the adaptation of metabolism are characterized by an increased lytic activity (A. Wiemken, unpublished data).

Another prominent function of the yeast vacuole concerns the accumulation of various purines which are used as a nitrogen source by *Candida utilis*. Crystals of uric acid, isoguanine and other compounds may be formed in vacuoles if the concentration exceeds the saturation of the vacuolar sap (Roush, 1961). Using ultaviolet microscopy, Svihla *et al.* (1963) demonstrated the accumulation of *S*-adenosylmethionine, lysine, purines and uric acid in vacuoles of *Sacch. cerevisiae* and *C. utilis*. It appears from these findings that active transport of certain compounds is catalysed by permeases present in the vacuolar membrane. Many of the substances accumulated in the vacuole are transported back into the cytoplasm following the formation of inducible enzymes which are involved in the utilization of these compounds.

In addition to polyphosphates, Indge (1968d) found considerable proportions of the total pools of potassium ion and amino acids localized in the vacuoles. Hence, it appears that this organelle functions as a repository for reserves which temporarily are not metabolized.

VI. The Mitochondria

A. MORPHOLOGY

1. *Introduction*

Since Guilliermond (1913) described the "chondriosomes" of yeast cells, the identification of mitochondria as these objects has been controversial because the "specific" staining techniques may fail or stain different kinds of "cytoplasmic granula" (Bautz, 1955, 1960). The first electron-microscope evidence for mitochondrial structures in *Sacch. cerevisiae* was presented by Agar and Douglas (1957) and Hagedorn (1957). Later on, supported by subsequent improvement in the preparation technique, many electron microscopists have shown mitochondria to be present in all yeast species investigated. Most of the structures described by these authors can also be found after freeze-etching (Moor and Mühlethaler, 1963). In addition, this new technique (Moor, 1964) has shown the mitochondria to be a persistent organelle (at least in *Sacch. cerevisiae*) which cannot be suppressed by any environmental influences. In anaerobically-grown cells, in *petites* and in glucose-repressed cultures of *Sacch. cerevisiae*, mitochondrial structures have been observed (H. Moor, unpublished results), a finding which is in contradiction to many investigations based on ordinary fixation and embedding techniques. As a consequence, we have to infer that the structure of inactivated mitochondria is much less resistant to chemical treatment than that of the respiring organelle.

2. *Characterization*

The mitochondria in yeast cells have been described as spherical, rod-shaped or thread-like and branched bodies having a diameter of about 0.3–1 μ and a length of up to 3 μ (Vitols and Linnane, 1961; Hirano and Lindegren, 1961; Kawakami, 1961; Marquardt, 1962a, b; Yotsuyanagi, 1962a). All of these authors agree that the structure of these bodies corresponds in principle to that of mitochondria of higher plant and animal cells. The organelles are surrounded by two membranes, the inner membrane forming the few cristae which extend into the mitochondrial stroma.

FIG. 18. Electron micrographs of mitochondria in freeze-etched cells of *Saccharomyces cerevisiae*. (a) Spherical mitochondria showing the wart-like architecture of the outer membrane surface. (b) A spherical mitochondrion developing a thread. (c) A thread-like mitochondrion. (d) Thread-like mitochondria passing through the neck between the mother cell and the bud. (e) A thread-like branched mitochondrion. (f) Mitochondrial threads fragmenting to form rod-shaped or spherical units after cell division. See caption to Fig. 1 for explanation of symbols.

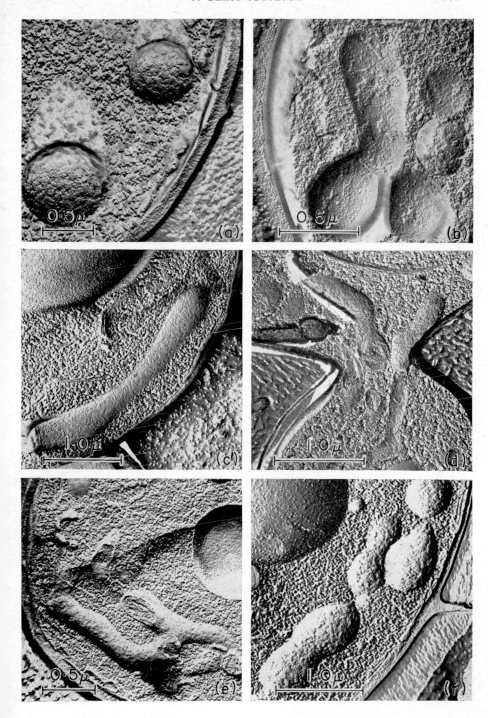

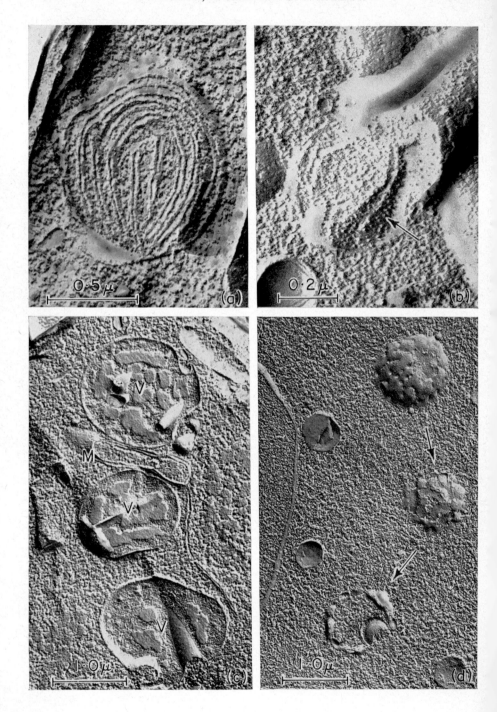

In freeze-etched cells from synchronized cultures, transitional stages between all of the differently shaped mitochondria can be observed. Stationary-phase cells regularly contain relatively small spherical units (Fig. 18(a)) which grow out to form threads when budding starts following the supply of a suitable cultivation medium (Figs. 18(b), (c)). A similar process has been demonstrated by Kawakami (1961) to occur in germinating ascospores. During cell division, many of these thread-like mitochondria pass though the neck between the mother cell and the daughter cell (Fig. 18(d)) and split up to form rod-shaped units in the bud (Fig. 18(f)). Certain of these processes have also been observed by Thyagarajan et al. (1961) and Yotsuyanagi (1962a). Cells entering the stationary phase exhibit branched mitochondrial threads (Fig. 18(e)) which split up to form small spherical units.

In contradiction to many observations based on conventional preparation techniques (Yotsuyanagi, 1962b; Polakis et al., 1964; McClary and Bowers, 1967), freeze-etched cells of the first exponential growth phase (aerobic fermentation; glucose-repressed respiration) show large "swollen" mitochondria which seem to be completely filled with a few large cristae (Fig. 19(a)). They present an image which might be compared with structures found in respiration-deficient yeasts (Yotsuyanagi, 1962b; Federman and Avers, 1967). Cells of the second exponential growth phase (respiration) exhibit more, but much less developed, cristae mitochondriales.

On the inner surface of the cristae, a particulate structure has been demonstrated by the use of negative staining (Shinagawa et al., 1966) and by freeze-etching (Fig. 19(b)) which resembles the elementary particles found in the mitochondria of animal cells (Fernández-Morán, 1962; Stoeckenius, 1963; Moor, 1964). The outer mitochondrial membrane shows a very rough surface sculptured by warts and pore- and slit-like depressions (Fig. 18(a); Moor and Mühlethaler, 1963).

3. *Degeneration*

A degeneration, decay or even a complete disappearance of mitochondria has been reported to take place in glucose-repressed cultures

---

FIG. 19. Electron micrographs of mitochondrial structures in freeze-etched *Saccharomyces cerevisiae*. (a) A "swollen" mitochondrion in a glucose-repressed cell exhibiting large cristae. (b) Surface view of mitochondrial cristae showing the elementary particles (arrow). (c) Aberrant forms of vacuoles and mitochondria in anaerobically cultivated cells. (d) Stages of mitochondrial decay. See caption to Fig. 1 for explanation of symbols.

(Yotsuyanagi, 1962a; Polakis *et al.*, 1964; McClary and Bowers, 1967), in respiration-deficient mutants (Yotsuyanagi, 1962b; Federman and Avers, 1967), and in anaerobically-grown cells of yeast (Linnane *et al.*, 1962; Hirano and Lindegren, 1963; Wallace and Linnane, 1964; Wallace *et al.*, 1968).

In *Sacch. cerevisiae*, we have never been able to find a decay or disappearance of mitochondrial structures either in glucose-repressed or in acridine orange-treated respiration-deficient cells prepared by the freeze-etching method. Only certain structural modifications can be observed, as has already been described. Thin sections of glucose-repressed cells, used as contols, have also not shown conspicuous mitochondrial profiles (see Fig. 3(a), p. 224)!

Under anaerobic conditions, cultures grow *ad infinitum* if the medium contains the appropriate growth factors, ergosterol and unsaturated fatty acids. The cells contain the usual organelles; however, the vacuoles seem to be enlarged and exhibit a very watery content while the size of the mitochondria is decreased (Fig. 15(c)). If the medium lacks ergosterol, the inoculated cells are able to pass only a few division cycles and then stop growing. During entry into the stationary phase, a fragmentation of the vacuoles can be observed leading to smaller units of an aberrant shape containing a dense granular material of lipoidic nature (Fig. 19(c)). This process is accompanied by a degeneration of the mitochondria which leads to tiny spherical or rod-shaped bodies exhibiting a fine granular content, without cristae and with an altered surface structure. Nevertheless, these degenerated mitochondria persist and cannot be confused with the vacuoles. During such an anaerobic cultivation, an increasing difficulty in the fixation of the membranous structures has been established by Wallace *et al.* (1968), a phenomenon which might be related to the accompanying change in the lipid composition of the membranes (Jollow *et al.*, 1968). Therefore, we have to conclude that the "disapperance" of yeast mitochondria is a fixation artifact. The myelin-like internal structures or "vesicles", evident in anaerobic cells (Yotsuyanagi, 1962b; Linnane *et al.*, 1962; Hirano and Lindegren, 1963; Federman and Avers, 1967), could be a degeneration product in mitochondria, or artificially evoked in the lipoidic content of the vacuoles.

4. *Regeneration*

In cells which have been brought back to aerobic conditions, the anaerobically-degenerated mitochondria regenerate and gain a normal appearance (Ph. Matile and H. Moor, unpublished data). Therefore, the "degenerated" mitochondria could also be termed "promitochondria" (an expression already used by Marquardt, 1962b), because they exhibit a primitive structure and possess the ability to differentiate into normally structured organelles.

The theory about a *de novo* genesis of mitochondria in cells changing from anaerobiosis to aerobiosis (Linnane *et al.*, 1962; Wallace and Linnane, 1964) is based on a confusion of vacuolar and mitochondrial structures. The "electron transparent vesicles" and "dense particles", described as precursors of mitochondria, are identical in structure, size and position with the aberrant vacuoles already referred to (Figs. 16(b), (c), p. 254; 19(c), p. 262). The promitochondria, which are invisible in the fixed specimens, most probably have been destroyed by the chemical treatment.

Light-microscope observations relating to a *de novo* genesis of yeast mitochondria (Müller, 1956a) have gone astray, most probably because promitochondria or tiny mitochondria are invisible or inactivated and do not show a Nadi- or Janus green-positive reaction.

5. *Decay*

As in animal cells, a turnover of mitochondrial proteins has to be taken into account, i.e. an uptake of mitochondria or mitochondrial constituents into the lysosomal apparatus (de Duve and Wattiaux, 1966). Up to now, neither thin sectioned nor freeze-etched specimens have shown an inclusion of mitochondria in the vacuoles. But two distinct observations indicate at least a possible mechanism of mitochondrial turnover in yeast cells. Firstly, mitochondria may decay *in situ* to much smaller vesicular elements (Fig. 19(d)); and secondly not only sphaerosomes but also further vesicular elements may be incorporated in the vacuole (Fig. 16(d), p. 256) leading to the suggestion that the debris of partly or totally decaying mitochondria might be transferred to the lysosomal apparatus (Matile and Wiemken, 1967).

B. CHEMISTRY

1. *Isolation*

The extraction of mitochondria can be achieved by vigorous shaking of yeast cells in the presence of small glass beads and a buffered sucrose-containing medium (e.g. Linnane and Still, 1955; Nossal *et al.*, 1956). In order to avoid the disintegration of mitochondria, the duration of shaking should be as short as possible (Schatz *et al.*, 1963). Mackler *et al.* (1962) and Mahler *et al.* (1964), taking advantage of the disintegration of mitochondria after several minutes of shaking, have prepared submitochondrial particles instead of intact mitochondria. An elegant, although somewhat laborious, technique for isolating perfectly preserved mitochondria has been based on the gentle lysis of yeast sphaeroplasts (Duell *et al.*, 1964; Ohnishi *et al.*, 1966). Since the preparation of sphaeroplasts requires the incubation of cells for at least 30 minutes, these investigators are dealing with mitochondria not necessarily identical with those in freshly grown cells.

The purification of mitochondria has been achieved by differential centrifugation (e.g. Vitols and Linnane, 1961). Since yeast cells possibly contain no structures with the size and density of mitochondria, differential centrifugation yields comparatively pure preparations. Density-gradient centrifugation using isopycnic gradients (Schatz et al.,1963) may provide information on integrity, density and homogeneity of mitochondrial populations, and in addition may yield highly purified preparations.

2. Components

Little information on the overall composition of yeast mitochondria is available because more attention has been paid to certain specific constituents, such as respiratory proteins and nucleic acids.

Yeast mitochondria are characterized by a high lipid content; in baker's yeast, lipids account for 25·4%, and phospholipids for 12·8% of the mitochondrial dry mass. Phosphatidylcholine, -ethanolamine, -inositol and -serine have been identified (Letters, 1968). Ergosterol is another lipid of *Saccharomyces* mitochondria (Ph. Matile and A.Wiemken, unpublished observations). Undoubtedly these lipids are constituents of the mitochondrial membranes.

Yeast mitochondria contain a small amount of DNA which amounts to about 1–4 μg per mg of mitochondrial protein (Schatz et al., 1964). Mitochondrial DNA is distinct from nuclear DNA in having a lower buoyant density (Tewari et al., 1965; Moustacchi and Williamson, 1966) and a lower molecular weight (Mahler et al., 1968). DNA seems to be associated, not only with whole mitochondria, but also with submitochondrial particles (Mahler et al., 1968) and cytochrome b_2 (Montagne and Morton, 1960). In both mitochondrial and submitochondrial preparations, RNA is present in considerably higher amounts than DNA (Mahler et al., 1968). Presumably a large proportion of this RNA is ribosomal.

C. FUNCTION

Yeast mitochondria are equipped with enzymes capable of oxidizing various substrates, with electron-transporting systems, and with enzymes that convert free energy of oxidative reactions into ATP. Submitochondrial particles, presumably representing fragments of the inner mitochondrial membrane, contain flavin, coenzyme Q, nonhaem iron, copper and the cytochromes b, c_1, c and a. Thus, the prominent function of yeast mitochondria is the aerobic energy conversion.

Another metabolic activity is the synthesis of RNA and protein (Mahler *et al.*, 1968), and this is discussed in the following section dealing with the ontogeny of mitochondria. The localization in yeast mitochondria of functions which have not been investigated so far (e.g. accumulation of ions) may be deduced from the close resemblance between mitochondria of yeasts and other plant and animal cells.

D. ONTOGENY

The respiratory system of many yeasts does not represent a constitutive entity of the cells. Several external and internal (genetic) factors induce or repress on the one hand, or control its synthesis on the other hand. These factors in turn are responsible for the formation of normal cristate or modified mitochondria.

Many yeasts are facultative anaerobes. They contrast with obligate aerobes whose mitochondria have an essentially constant structure and composition. If these yeasts obtain their energy mainly from fermentation, then mitochondria with a low respiratory activity and with an altered organization are formed.

Among the facultative anaerobes *Sacch. cerevisiae* represents a special type whose respiration is repressed, not only under anaerobic conditions, but also in the presence of glucose (the so-called Crabtree effect). Glucose specifically inhibits the synthesis of respiratory enzymes (Polakis *et al.*, 1964, 1965; Tustanoff and Bartley, 1964) and causes the formation of only "rudimentary" mitochondria (Figs. 20(a), (b); McClary and Bowers, 1967). As a consequence, *Sacch. cerevisiae* inoculated into a glucose-containing medium shows a diauxic growth pattern. A first exponential-growth phase is characterized by aerobic fermentation of glucose. After the exhaustion of glucose, respiration is induced and ethanol is respired during a second exponential-growth phase. During aerobic fermentation, only a small number of mitochondria with few cristae are present, though normal cristate mitochondria appear when the cells begin to respire (Yotsuyanagi, 1962a). When cells of *Sacch. cerevisiae* having a fully developed respiratory system are placed in a glucose-containing medium, the respiratory activity declines very rapidly. Glucose not only inhibits the synthesis of new mitochondrial enzymes, but causes the degeneration of existing mitochondria. Many of the activities of primary dehydrogenases and the electron-transport capacity disappear from the mitochondria before the cells even start to divide (Utter *et al.*, 1968). In growing cells which adapt themselves to fermentative metabolism in the presence of high concentrations of glucose, the loss of activity of several respiratory enzymes is much faster than would be expected assuming a total repression of synthesis of new enzyme molecules. Thus, not only are mitochondria degenerat-

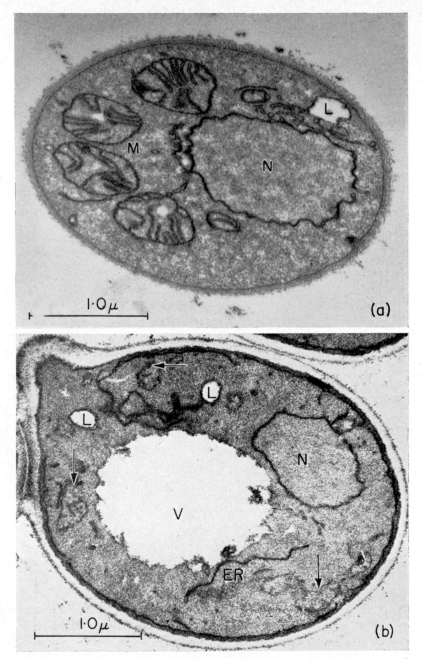

FIG. 20. Electron micrographs of mitochondrial structures in permanganate-fixed cells. (a) Mitochondria in an obligate aerobe (*Rhodotorula gracilis*). (b) "Disappearance" of mitochondrial profiles (arrows) in a glucose-repressed facultative anaerobe (*Saccharomyces cerevisiae*; McClary and Bowers, 1967). See caption to Fig. 1 for explanation of symbols.

ing under these conditions but, in addition, mitochondrial constituents are inactivated (Chapman and Bartley, 1968). The exact mechanism of this type of metabolic control is not known. It is likely however that it involves turnover of protein, that is, the digestion in the lysosomal compartment of enzyme molecules which were detached from mitochondrial membranes. The recovery period which follows the exhaustion of glucose in the medium is characterized by the rapid synthesis of respiratory enzymes. The rates at which primary dehydrogenases and electron-transporting enzymes are formed and integrated into the mitochondrial membranes seem to be considerably different (Utter et al., 1968). This observation suggests that yeast mitochondria are not synthesized as a single entity.

In the absence of oxygen, the degeneration of mitochondria may proceed to such an extent that no morphological equivalents can be observed in thin sections (Polakis et al., 1964; see Sections VI.A.3 and 4, pp. 263 and 264). Biochemical equivalents of these structural changes involve the virtual absence of many respiratory enzymes in anaerobically-grown cells, and the induction of the biosynthesis of these mitochondrial constituents in the presence of molecular oxygen (Slonimski, 1953; Somlo and Fukuhara, 1965).

The apparent absence of morphologically and biochemically normal mitochondria in anaerobically-grown yeasts raises the question of the biogenesis of mitochondria. It has been suggested that these organelles may be formed de novo from non-mitochondrial sources. On the other hand, the differentiation of non-mitochondrial structures or of mitochondrial remnants into cristate mitochondria has been assumed (Wallace et al., 1968). The fine structural appearance of anaerobically grown yeast cells seems to depend largely on the presence of the essential growth factors ergosterol and unsaturated fatty acids. Indeed, sub-cellular particles which carry certain mitochondrial enzymes, whose synthesis is not completely repressed, are present in extracts from anaerobically-grown yeast cells (Schatz, 1963, 1965). Although these possible mitochondrial precursors have not yet been characterized sufficiently, they may be termed "promitochondria" which are thought to differentiate eventually into mitochondria.

The problem of ontogeny is intimately related to the genetic determinants of mitochondrial constituents (proteins), that is to DNA, especially to mitochondrial DNA (m-DNA). Indeed, m-DMA is present in anaerobically-grown yeast, although in much smaller quantities than in aerobically-grown cells (Swift et al., 1967). However, RNA extracted from these cells hybridizes poorly with m-DNA (Fukuhara, 1967b) Thus, under conditions resulting in repressed respiration, the m-DNA-dependent synthesis of RNA seems to be inhibited. It takes place, however, in mitochondria isolated from aerobically-grown cells (Tuppy and Wintersberger, 1966; Mahler *et al.*, 1968). In fact, metabolically stable RNA from aerobically-grown yeast cells hybridizes with m-DNA (Fukuhara, 1967b). In turn, isolated mitochondria are capable of synthesizing protein which is thought to be coded by m-DNA (Mahler *et al.*, 1968; Tuppy and Wintersberger, 1966).

These properties of yeast mitochondria point to the regulation of respiratory activity and of the formation of mitochondrial structures at the level of m-DNA. According to this view, the multiplication of mitochondria in growing respiring cells would involve the replication of m-DNA and a continuous synthesis of mitochondrial proteins, followed by the structural assembly of these and other constituents into mitochondrial membranes, and finally the numerical growth of the mitochondrial population by division of mitochondria. Repression of respiration in turn appears as the result of partial or complete inhibition of m-DNA-dependent protein synthesis. A small amount of m-DNA present under anaerobic conditions shows that the replication of m-DNA proceeds at a comparatively slow rate. Therefore, the mitochondrial genome, and not its apparent product, the respiratory system, seems to represent a constitutive entity of the yeast cell.

This image of an autonomous self-replicating organelle must however be modified. A special type of promitochondrion is formed in mutants of *Sacch. cerevisiae* which are unable to utilize non-fermentable substrates. The mitochondria formed in these respiratory-deficient mutants lack cristae (Yotsuyanagi, 1962b). They resemble the organelles formed in the presence of high concentrations of glucose. The mutational change to respiratory deficiency occurs at a comparatively high natural frequency (about 1%); it is greatly enhanced to almost complete mutation by acridines and by a large number of other agents. The mutants have been termed *petites* because they divide more slowly than normal cells and hence form smaller colonies (Ephrussi, 1953). Respiratory-deficient mutants fall into two categories. If mutants in the first category are crossed with normal cells, the mutant character fails to appear among the progeny; this non-Mendelian pattern of inheritance concerns mutation or loss of genes localized in the cytoplasm. In fact, cytoplasmic petites possess an altered m-DNA (Mounolou *et al.*, 1966). As a consequence of this alteration, they lack the cytochromes a_3, a and b and certain dehydrogenases (Sherman and Slonimsky, 1964). In contrast, respiratory-deficient mutants of the second category show Mendelian inheritance of respiratory deficiency. In this case, the mutant character is a consequence of lesions in chromosomal genes. The existence of such chromosomal mutants indicates that the genetical autonomy of yeast mitochondria is not an absolute one. Cytochrome *c*- deficient segregating petites demonstrate that the synthesis of the respiratory system is at least partially controlled by nuclear genes. In contrast to other cytochromes, cytochrome c seems to be synthesized in the cytoplasm and is subsequently incorporated into the inner mitochondrial membrane; respiratory-inactive mitochondria isolated from a cytochrome c-deficient strain readily incorporate purified cytochrome c and assume the properties of normal mitochondria (Mattoon and Sherman, 1966). Further evidence in favour of a semi-autonomous biogenesis of yeast mitochondria has been obtained from studies with chloramphenicol and other antibiotics which specifically inhibit mitochondrial protein synthesis. The respiratory system synthesized in the presence of these drugs has similar properties to those of cytoplasmic petites. The inhibitory effect concerns cytochromes a and b and other structural constituents of the inner mitochondrial membrane. Synthesis of cytochrome c, fumarate hydratase and malate dehydrogenase, which does not proceed in the mitochondria, is only little affected (Clark-Walker and Linnane, 1967; Mahler et $al.$, 1968). Cytochemical tests suggest that cytochrome c peroxidase, another mitochondrial constituent, is synthesized in the endoplasmic reticulum (Avers, 1967). Hence, mitochondrial growth and division involves the protein-synthesizing capacity of the organelle itself and that of a non-mitochondrial system. According to current concepts, mitochondrial protein synthesis concerns only certain constituents of the inner membrane, whereas the outer membrane and many components of the cristae are formed in the cytoplasm (Roodyn and Wilkie, 1968). The ontogeny of mitochondria therefore requires the transfer through the outer membrane of those components of the cristae which are synthesized outside the organelle.

VII. The Cytoplasmic Ground Substance

A. RIBOSOMES

Large quantities of ribosomes are present in crude microsomal preparations (Klein, 1965). These particles are characterized by a sedimentation constant somewhat higher than 80 S, and a ratio of RNA to protein of $1 \cdot 04 : 1 \cdot 12$ (Horstmann, 1968). Since they are composed of two subunits (40 S and 60 S), they closely resemble the ribosomes of other organisms. Upon cautious extraction of cells, a large proportion of ribosomes can be isolated as polysomal aggregates (Marcus et $al.$, 1967). Since the preparation of polysomes does not require the dissolution of membranes, it seems that a large quantity of these centres of protein synthesis are localized in the ground substance (Fig. 21(a)).

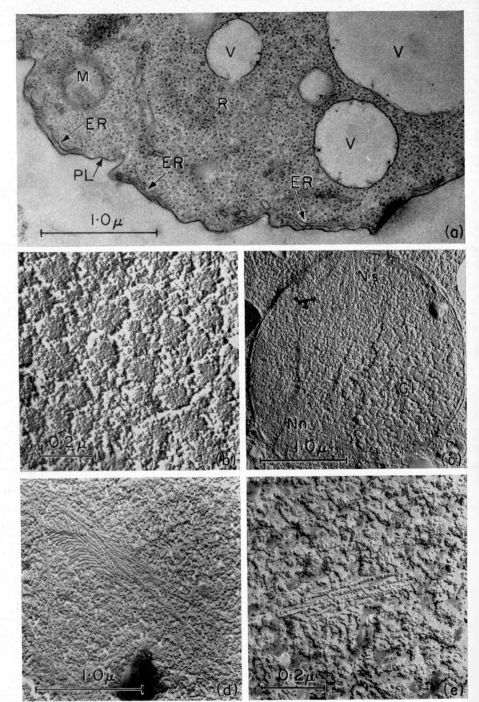

B. ENZYMES

According to common assumptions, enzymes and components recoverable in a soluble cell-fraction are localized in the ground substance. However, structurally-bound cell constituents may well become soluble if the respective organelle is destroyed during extraction of cells. This kind of artifact has already been discussed in connection with the vacuole and its lysosomal enzymes. Extracellular constituents associated with the cell wall may also contribute to the soluble fraction. Soluble enzymes, which are neither associated with fragile structures nor with the cell wall, comprise ethanol dehydrogenase, enzymes of the glycolytic pathway and of the oxidative pentose phosphate cycle (glucose 6-phosphate dehydrogenase). Yet, it has been claimed that some of the phosphorylative activity coupled with glycolysis is particulate in yeast extracted under certain conditions (Green et al., 1965). It should be emphasized that extractions of cells using rapid changes of pressure result in a chaos of artificially produced vesicles which may contain compounds originally located in the cytoplasmic ground substance.

Another enzyme of the ground substance is trehalase, and it is interesting to note that its substrate, trehalose, is completely absent from the soluble fraction (Souza and Panek, 1968). This disaccharide seems to be enclosed in a membrane-bounded cell compartment. This kind of differential localization of enzymes and respective substrates, together with appropriate transporting systems, appears to represent a means by which the cell is able to regulate the synthesis and breakdown of certain components.

C. GLYCOGEN

The main carbohydrate in the cytoplasmic ground substance and in the nucleoplasm is the so-called "yeast glycogen". Its molecular weight ranges from 2 million to about 10 million, and its molecular architecture cannot be distinguished from that of "animal glycogen" (Silbereisen, 1960). This reserve carbohydrate accumulates in stationary-

---

FIG. 21. Electron micrographs of structures contained in the cytoplasmic ground substance of *Saccharomyces cerevisiae*. (a) Thin section of a protoplast fixed with osmium tetroxide. Note the ribosomes in the ground plasm and the plasmalemma-associated strands of endoplasmic reticulum. Reproduced by courtesy of F. Schwegler. (b) Aggregates ("lumps") of glycogen molecules. (c) Aggregates of glycogen in the nucleoplasm. (d) Feather-like arranged cytoplasmic filaments. (e) Substructure of the cytoplasmic filaments. See caption to Fig. 1 for explanation of symbols.

phase cells and may predominate over the remaining constituents of the ground substance (Figs. 21(b), (c)). It consists of lumps of spherical bodies having a diameter of about 40 mμ. The fine granular structure of such a body, representing one glycogen molecule, indicates the presence of subunits about 6 mμ in diameter (Moor and Mühlethaler, 1963). These lumps, spherical bodies and their subunits resemble closely the α-, β-, and γ-particles observed in liver glycogen (Drochmans, 1962).

D. FILAMENTS

In freeze-etched specimens of *Sacch. cerevisiae*, filamentous structures have occasionally been observed (H. Moor, unpublished observations). The filaments appear to be arranged in bundles which may split up to give feather-like forms (Fig. 21(d)). The diameter of the filaments is about 10 mμ; their length cannot be determined exactly. The bundles may be up to 1 μ in length. In some cases, a particulate substructure can be distinguished exhibiting a periodicity of about 13 mμ (Fig. 21(e)). The relationship between these structures and certain organelles and certain activities of the cells has not been reported.

VIII. The Nucleus

A. YEAST CARYOLOGY IN RETROSPECT

Most of our knowledge of the behaviour, fine structure and composition of the cell wall, mitochondria and vacuoles of yeasts is of very recent date. By contrast, the nucleus of the yeast cell has long been the subject of a voluminous and controversial literature. Detailed surveys of the history of yeast caryology have been published by Lindegren and Rafalko (1950), Bautz (1960) and McClary (1964). In recent years, mutually supporting results of the phase-contrast microscopy of living cells, selective staining with acid fuchsin, thin sectioning, and freeze-etching have yielded a clear-cut picture of the anatomy of the yeast nucleus which is not without its oddities and blank spots, but which is nevertheless amenable to description in the accepted terms of ordinary cytology. It increases confidence in the soundness of this picture that many of its features had been correctly seen, if misinterpreted, by earlier workers.

In approaching the study of the yeast nucleus, it is helpful to remind ourselves that *Sacch. cerevisiae*, the most frequently examined species, is an ascomycete and that its nucleus is, *a priori*, likely to resemble the hyphal (vegetative) nuclei of other ascomycetes. A few orienting remarks on hyphal nuclei of ascomycetes will therefore not be out of place. Most papers on yeast caryology have appeared during a period

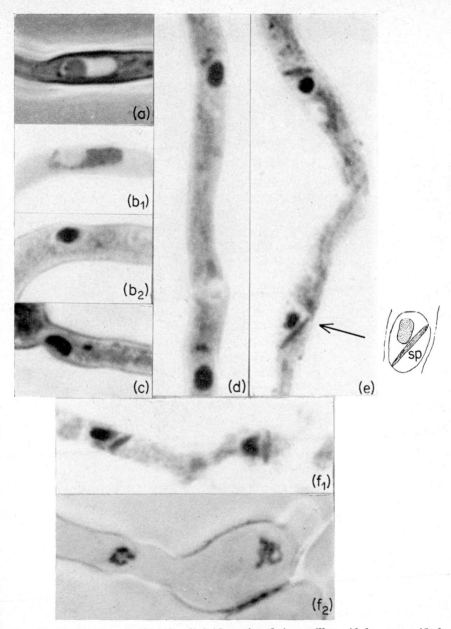

Fig. 22. Photomicrographs of a diploid strain of *Aspergillus nidulans* magnified 3,600 times. (a) Nucleus in a living hypha, with a nucleolus on the left. (b$_1$ and b$_2$) Photographs of the same Helly-fixed nucleus. In (b$_1$), the chromatin has been stained with aceto-orcein; in (b$_2$) the nucleolus and the lateral granule have been stained with acid fuchsin. (c) Another nucleus in which the nucleolus and the lateral granule have been selectively stained with acid fuchsin. (d, e) Transformation of the lateral granule into the spindle of mitosis (sp); f$_1$ was stained for nucleoli and spindles; f$_2$ shows the same specimen after it had been stained for chromosomes (f$_1$ and f$_2$ are duplicate photographs of the same two nuclei).

when little attention was paid to the structure and behaviour of hyphal nuclei. However, important observations have been published by Harper (1905) and Colson (1938) which may serve as a starting point for our investigation. Harper's and Colson's principal concern was with meiosis. Both authors noted in the margin of the relatively large resting nucleus in ascogenous hyphae and young asci of *Phyllactinia corylea* (the fungus causing hazelnut mildew) a readily stainable short bar or granule which divided before meiosis and the postmeiotic mitoses to provide the centrioles or centrosomes of the spindle apparatus. Harper (1905) refers to this element as the "central body", even between divisions when it is attached to the membrane of the resting nucleus. It was seen in that position also in *Ceratostomella* by Andrus and Harter (1937). Colson (1938) calls it the "lateral granule". Harper (1905) states, and Colson's (1938) illustrations bear out, that not only ascus nuclei but "the nuclei throughout the entire plant" conform to this pattern. Both workers depict resting nuclei of vegetative hyphae as possessing, in addition to the lateral granule, evenly distributed finely granular chromatin and a spherical or lens-shaped nucleolus which usually occupies an eccentric position close to the nuclear membrane and opposite the lateral granule. Mitosis of hyphal nuclei was not discussed by the older workers, but recent work on mitosis in *Ceratocystis* (Wilson and Aist, 1967) and *Aspergillus nidulans* (Robinow and Caten, 1969) has revealed that, in resting hyphal nuclei of these ascomycetes, too, a lateral granule is demonstrable and that, as in ascus nuclei, it is transformed during mitosis into an intranuclear spindle apparatus. Similar and remarkably clear observations have been made on the imperfect fungus *Fusarium oxysporum* (Aist, 1969). The intranuclear spindle is visible during life under the phase-contrast microscope and, in *Aspergillus*, both lateral granule and spindle are clearly visible in fixed preparations lightly stained with low concentrations of acid fuchsin in acid solution. In both species, the lateral granule and the fibrillar spindle arising from it have also been seen in electron micrographs of sections (Robinow and Caten, 1969; J. R. Aist, personal communication, 1968). Fig. 22 illustrates the lateral granule of a resting nucleus of *Aspergillus nidulans*, the spindle to which it gives rise during mitosis, and the relationship of the spindle to the chromosomes.

When the methods which have given these results are applied to *Sacch. cerevisiae* and other yeasts, then the structures commonly seen in hyphal nuclei of ascomycetes, namely lateral granule, eccentric nucleolus and intranuclear spindle of division are readily demonstrated. In the following pages, nuclear behaviour in a few representative yeasts will be illustrated and briefly discussed. As will be seen, numerous problems of yeast caryology remain to be solved.

B. LIGHT MICROSCOPY OF THE YEAST NUCLEUS

1. *The Nucleus in the Living Yeast Cell*

The nucleus is hard to see in living yeast cells examined in aqueous media. Phase-contrast microscopy gives a clear image of the nucleus in yeasts growing in the presence of air in a nutrient medium containing 18–21% gelatin. In budding *Sacch. cerevisiae*, the nucleus is usually found between the vacuole and the bud, except immediately after division when it moves to the opposite pole of the mother cell. The nucleus is composed of two distinct portions, an optically dense cup- or lens-shaped lower half, the nucleolus, and a less dense, more translucent, dome-shaped upper portion, the nucleoplasm. Within this latter

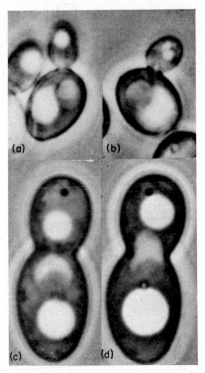

FIG. 23. Phase-contrast micrographs of budding yeasts. (a) and (b) *Saccharomyces cerevisiae*. In both cells, the nucleus is seen between the vacuole and the bud. The lower half of each nucleus is occupied by a cap of dense nucleolar material. In (b), the lateral granule is seen as a dark patch in the bright portion of the nucleus. (c) and (d) Successive views, taken minutes apart, of *Wickerhamia fluorescens*. A third view of the same cell, taken some 15 minutes later, is seen in Fig. 26 (b), p. 281. The nucleus is above the vacuole. The dense eccentric nucleolus contrasts with the clear nucleoplasm. In (d), the lateral granule is faintly visible at the upper pole of the nucleus. All figures are magnified 2,700 times.

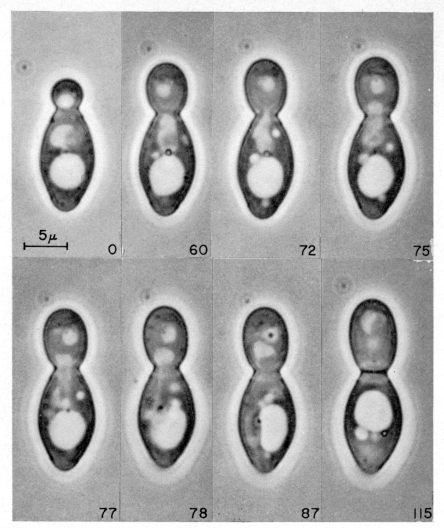

5μ

| 0 | 60 | 72 | 75 |

| 77 | 78 | 87 | 115 |

FIG. 24. Time-lapse phase-contrast micrographs of a dividing cell of *Wickerhamia fluorescens*. Numerals give minutes elapsed since the first picture was taken. The nucleus, with a distinct eccentric nucleolus, is above the vacuole at zero time. Further explanation is in the text.

region and close to its margin, a small dense granule is often discernible in resting nuclei (Fig. 23(d)). A thin grey line traversing the translucent portion of dividing nuclei in *Sacch. cerevisiae* has been recorded by Robinow and Marak (1966). As will be seen presently, the granule and the grey line represent respectively the "lateral granule" and intranuclear spindle of the yeast nucleus.

The same features, on a larger scale, are possessed by the nucleus of *Wickerhamia fluorescens* Soneda, an apiculate sporogenous yeast (Soneda, 1960). It will be helpful to describe first the organization of the nucleus of this relatively large yeast before we continue with our description of the nucleus of *Sacch. cerevisiae*.

2. *Phase-Contrast Microscopy of the Dividing Yeast Nucleus*

Phase-contrast microscopy of living yeasts, and the electron microscopy of chemically-fixed as well as freeze-etched yeast cells, has confirmed the view, held by many since Guilliermond's work of 1903, that the yeast nucleus divides by elongation and constriction within the intact nuclear envelope (Moor and Mühlethaler, 1963; see the review by Hawker, 1965). A time-lapse sequence of nuclear and cell division in baker's yeast is given in the paper by Robinow and Marak (1966). The essential features of the process are shown with particular clarity in the division of the large nucleus of *Wickerhamia fluorescens* (Fig. 24). Note the stretching of the nucleus after 77 minutes, the constriction of the nucleolus within, the flattening of one side of the vacuole by the lower daughter nucleus on its way to the lower pole of the mother cell after 87 minutes, and the return of the vacuole to a rounded shape at the end of the sequence. A lens-shaped eccentrically placed nucleolus is clearly visible in the nucleus at zero time, and in the upper daughter nucleus after 115 minutes. A stage intermediate between those seen after 77 and 78 minutes is seen (in another cell) in Fig. 26(b), p. 281.

3. *The Organization of the Nucleus as Seen in Stained Preparations*

a. The Nucleus in Wickerhamia fluorescens. In a Helly-fixed preparation of *Wickerhamia fluorescens* briefly stained with acid fuchsin (1:40,000 in 1% acetic acid), the nucleus has the same proportions as it has in the living cell. The amounts of stain taken up by the two portions of the nucleus, much of it by the nucleolus and little by the nucleoplasm, reflect the different optical densities of these components as seen in live preparations. At the upper pole of the nucleus, a small bead or tiny dumbbell is regularly seen (Figs. 25(a), (b)). The reader will recognize it as the equivalent of the "lateral granule" of the nucleus in *Aspergillus nidulans* in Fig. 22(c) (p. 275). In that mould, and also in *Wickerhamia fluorescens*, the lateral granule gives rise to a gradually elongating intranuclear *fibre* during the division of the nucleus (Figs. 25(d), (e)). Under favourable conditions, as pointed out above, the granule and the fibre to which it gives rise can be seen in the nuclei of living yeast cells (Robinow and Marak, 1966; see also Fig. 2 of Robinow and Bakerspigel, 1965, for views of the intranuclear fibre in the dividing nucleus of a fission yeast). The image of the nucleus is the same in acid-fuchsin preparations and in preparations stained with the formerly much used iron haematoxylin of Heidenhain.

It remains to demonstrate the chromatin of the nucleus in *Wicker-hamia fluorescens*. It has been stained in Fig. 25(c). The cell illustrated

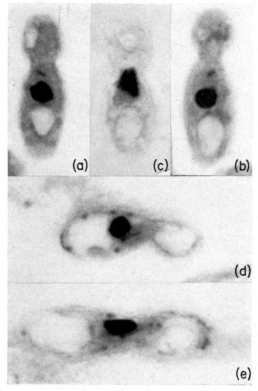

FIG. 25. Helly-fixed, stained cells of *Wickerhamia fluorescens*. (a), (b), (d) and (e) were stained with acid fuchsin: (c) with Giemsa stain after hydrolysis. The deeply stained body in (a) and (b) is the nucleolus. Above it in the ill-defined nucleoplasm lies a small rodlet, the "lateral granule". (c) shows a cell comparable to (a) and (b) but stained for chromatin. The nucleolus is now only faintly stained, and the main body of the nucleus is filled with chromatin. The slight nick at the tip of the nucleus is the site of the lateral granule. Compare (a), (b) and (c) with views of the living nucleus in Figs. 23 (c) and (d). In (d) and (e), the lateral granule has expanded into an intranuclear fibre ("spindle") which in both cells is seen just beneath the nucleolus. All micrographs are magnified 2,700 times.

is comparable to the acid fuchsin-stained ones flanking it. Comparison reveals that chromatin completely fills that portion of the nucleus which appears translucent under the phase-contrast microscope, and has little affinity for acid fuchsin under the conditions used in staining

the preparations of Figs. 25(a) and (b). The notch in the tip of the
chromatinic portion corresponds to the site of the lateral granule.
The behaviour of the chromatin during division is illustrated in Fig.
26(a), lower left. The chromatin of the dividing nucleus forms a thin
shell around the elongated and constricted nucleolus. The latter, containing no cytochemically detectable DNA, is invisible in the aceto-

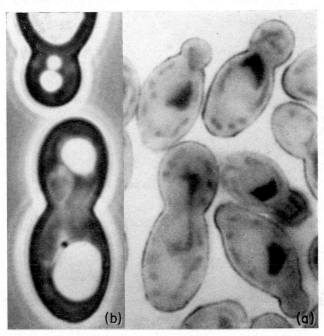

FIG. 26. Nuclear division in *Wickerhamia fluorescens*. (a) Helly fixation, hydrochloric acid and aceto-orcein. The dividing nucleus is at the lower left. Note the
dense chromatin areas with concave bases in the remaining interphase nuclei.
The concavity is filled by the (unstained) nucleolus. Compare with Figs. 23 (c)
and (d), (p. 277). (b) Phase-contrast micrograph of a living dividing cell. The
elongated and constricted nucleus with a bulky core of nucleolar material is at
the same stage of division as the nucleus at the lower left in (a). Both micrographs
are magnified 2,700 times.

orcein preparation (Fig. 26(a)) but clearly visible in the dividing
nucleus of a comparable live cell shown in Fig. 26(b) (the same as in
Figs. 23(c) and (d), p. 277). The distribution of chromatin and the
geometry of the dividing nuclei is the same in *Sacch. cerevisiae* as in
the larger model yeast here illustrated. The paleness of the dividing
mass of chromatin in the central nucleus of Fig. 26(a) is in agreement
with the finding of Williamson (1966) in synchronized yeast cultures,

that DNA is duplicated in the nuclei of cells with small buds, and that
there is no further increase of DNA before the next round of budding
has started.

b. *The Nucleus of* Saccharomyces cerevisiae. The nucleus of baker's
yeast (Fig. 27) is constructed in the same way as that in *Wickerhamia
fluorescens,* and differs from the latter only in being smaller and in

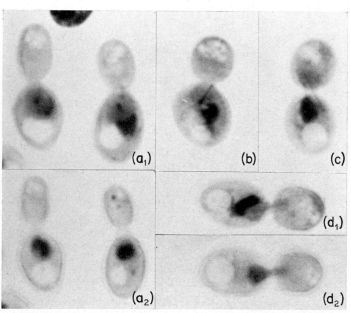

Fig. 27. Photomicrographs of *Saccharomyces cerevisiae* Fleischmann strain.
Micrographs (a_1)–(d_1) were stained with acid fuchsin. Micrographs (a_2) and (d_2)
are duplicate photographs of the cells shown in (a_1) and (d_1) as they appear after
hydrolysis followed by Giemsa staining. Note the lateral granule at the pole opposite to the nucleolar cap in (a_1) and its expansion into the intranuclear fibre in
the dividing nuclei of (b), (c) and (d_1). The duplicate pictures show that the lateral
granule and the intranuclear fibre are in the chromatin portion of the nucleus.
Micrographs are magnified 2,700 times.

having its nucleolar material cast in the shape of a cap or irregular
hump rather than a sphere or lens. Opposite the nuclear cap, which
appears as a crescent in optical section, is again found a typical lateral
granule which turns into a straight, thin, long fibre during nuclear
division (Figs. 27(b), (c)). A stage in the development of the fibre intermediate between Figs. 27(a) and (b) is seen in the inset to Fig. 30 (p.
286). Figs. 27(a_1) and (a_2) are duplicate photographs of the same pair
of yeasts. The first picture shows the nucleolus and the lateral granule

or intranuclear fibre within the indistinctly stained chromatin portion of the nucleus. In the second picture of each set, the chromatin is seen stained with Giemsa solution after hydrolysis, and the nucleolar material is no longer visible.

C. ELECTRON MICROSCOPY OF THE YEAST NUCLEUS

There is much information on the fine structure of the acidophil intranuclear fibre which will henceforth, for the sake of convenience, be referred to as "the spindle", although we do not know whether it plays the same role in mitosis as the spindle in dividing nuclei of higher organisms, and have reasons for doubting it. Electron microscopy of chemically-fixed as well as freeze-etched yeast cells reveals that the nucleus is traversed by a narrow bundle of microtubules that runs in a straight line between enlarged pores in the nuclear envelope (Robinow and Marak, 1966; Moor, 1966, 1967a) (Figs. 28, 30 and 31). The pores are filled with what appears to be a many-layered disc, clearly seen in unpublished micrographs of sections of the spindle of *Sacch. carlsbergensis* by Dr. B. Afzelius which are reproduced in Fig. 28 with the kind permission of the author. Robinow and Marak (1966), who refer to the discs as "centriolar plaques", stress in a passage evidently overlooked by de Harven (1968) *the lack* of similarity between these plaques and the cylindrical centrioles found in animal cells and in fungi which are capable of forming zoospores. Recent observations by Girbardt (1968) on dividing nuclei of the basidiomycete *Polystictus* suggest that the plaques are more reasonably interpreted as the equivalents of *kinetochores*. This view brings electron microscopy in line with Harper's belief of 1905 that the chromosomes are permanently attached to the lateral granule. An enlarged pore which may represent a centriolar plaque (or kinetochore) is seen among ordinary pores of the nuclear envelope in Fig. 29.

The centriolar plaques of *Sacch. cerevisiae* have, so far, been more distinct in sections of chemically-fixed than in replicas of freeze-etched cells (Fig. 30) but, in the yeast *Wickerhamia fluorescens* (Fig. 31), a cone-shaped plug is regularly seen at the ends of the spindle, indicating that some formed element does exist at this site. Its fine structure remains to be revealed. The mode of duplication of centriolar plaques has also not yet been ascertained.

The discovery of a spindle-like organelle has not explained the mode of division of the yeast nucleus. Robinow and Marak (1966) and Moor (1966) point out that the variable orientation of the spindle relative to the axis of the elongating dividing nucleus makes it unlikely that the spindle of *Sacch. cerevisiae* plays an active mechanical part in the process of division. On the other hand, there is good evidence that the elongating

10*

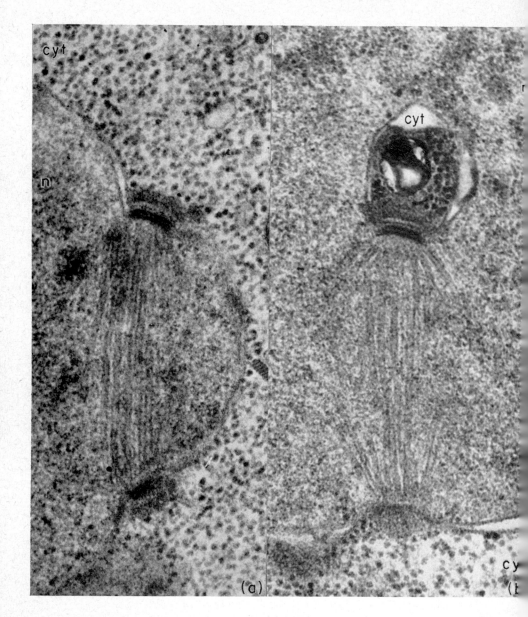

FIG. 28. Intranuclear spindles—the electron microscopial aspect of the "lateral granules" and "intranuclear fibre" of light microscopy—between centriolar plaques in nuclei of *Saccharomyces carlsbergensis*. The membrane-lined cavity above the plaque in (b) represents a finger of cytoplasm indenting the nucleus. Magnified about 100,000 times. n indicates the nucleus; cyt the cytoplasm. Micrographs provided by Dr. Bjoern Afzelius.

FIG. 29. Surface view of a nucleus of *Saccharomyces cervisiae*. The recessed disc in the large pore presumably represents a centriolar plaque. A freeze-etched specimen prepared by Dr. H. Moor. Magnified about 40,000 times; shadowed from the upper right.

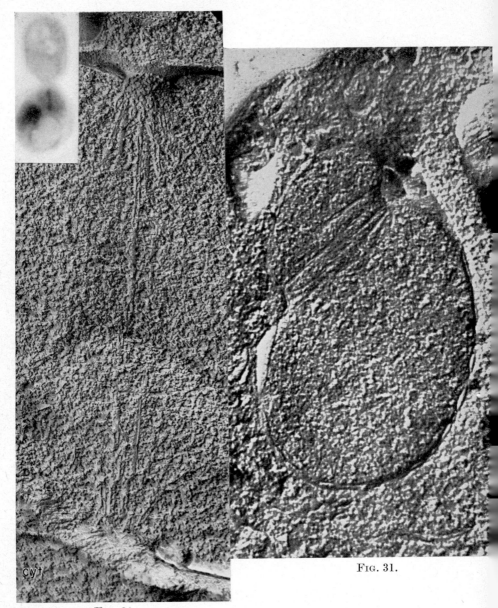

Fig. 30.

Fig. 31.

spindle is responsible for the rhomboid deformation and eventual constriction of the dividing nucleus of *fission* yeasts (Robinow and Bakerspigel, 1965, and unpublished observations).

Remarkable is also the discrepancy between the numbers of micro-tubules in the spindle and the numbers of chromosomes (linkage groups) in the nucleus. Robinow and Marak (1966) worked with a tetraploid strain which would have a minimum of 56 chromosomes, and they estimated the number of spindle tubules as close to 15. Moor (1967a), working with a commercial baker's yeast likely to have been a triploid (personal communication from Dr. H. Gutz), arrived independently at the same figure. These figures contrast markedly with the proportion of tubules to chromosomes in mitotically dividing nuclei of higher organisms, where the former usually far outnumber the latter.

In addition to the spindle apparatus, which is an integral part of the yeast nucleus, another structure composed of microtubules may, under certain conditions, arise in the nucleus. In yeast cells which have ceased to grow owing to lack of oxygen, admittance of air initiates a well synchronized round of growth and nuclear division. In the nuclei of the first generation of recuperating cells, Moor (1966, 1967a) has discovered a fibrous protein crystal of considerable size which elongates in step with the elongation of the nucleus. This object, which is composed of tightly packed microtubules with diameters of $21 \pm 0\cdot 8$ mμ, differs from the spindle in not being connected with centriolar plaques. It elongates in step with the elongation of the nucleus and at its longest draws the nuclear envelope out into pointed pockets in contrast to the spindle whose points of origin are invariably at the bottom of more or less deep depressions or invaginations of the nuclear envelope. It has not been

---

FIG. 30. Electron micrograph showing an intranuclear spindle in *Saccharomyces cerevisiae*. A freeze-etched specimen prepared by Dr. H. Moor. Magnified 34,000 times. The direction of shadowing was from the lower right. The inset is a light micrograph of a yeast cell in which the spindle has been stained with acid fuchsin. Note the dumbbell-shaped contours. It is thought that they reflect the divergence of the spindle tubules from their points of origin. Magnified 3,600 times.

FIG. 31. Electron micrograph showing a short spindle in a fractured nucleus of a freeze-etched cell of *Wickerhamia fluorescens*. Note the solid plug at the upper end of the spindle and that the spindle origins lie in depressions of the nuclear en-velope. Compare with Fig. 28 (a) of *Saccharomyces carlsbergensis*. It is thought that spindles of this length give rise to images of rodlet-like "lateral granules" in the light microscope as illustrated in Figs. 25 (a) and (b). Magnified about 50,000 times. The micrograph was made by Dr. Heinz Bauer.

reported whether fibrous crystals with these dimensions continue to reappear in the nuclei of the descendants of the first aerated generation. It seems likely that their emergence depends on the special conditions of these synchronizing experiments. Crystals with these properties have not been encountered in the nuclei of randomly multiplying aerobically-grown cells of *Sacch. cerevisiae*, *Sacch. carlsbergensis*, *Lipomyces lipofer* or *Wickerhamia fluorescens*.

D. CHROMOSOMES OF THE YEAST NUCLEUS

The numerous chromosomes known from genetical work to exist in the nucleus of *Sacch. cerevisiae* are not visible during life in budding yeast cells and do not become visible as separate, reliably countable

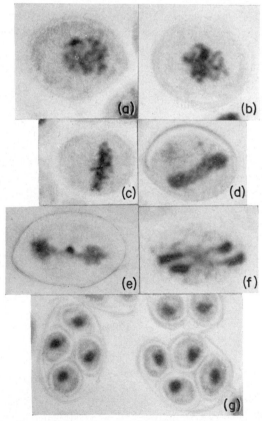

FIG. 32. Photomicrographs showing meiosis in *Saccharomyces cerevisiae*. (a) and (b) show chromosomes of the first division: (d)–(g) show steps in the formation of ascus nuclei. Helly, hydrochloric acid and Giemsa stain. Micrographs (a) and (b) are magnified 3,600 times, the rest 2,700 times.

elements in chemically-fixed preparations in either the light microscope or the electron microscope. They have not been identified in cells prepared for electron microscopy by the method of freeze etching. The most valuable observations are probably still those of Levan (1946, 1947) on benzene- and camphor-treated yeasts. Interesting electron micrographs, possibly but not certainly showing profiles of chromosomes in the nuclei of permanganate-fixed synchronized yeast cells, have been published by Williamson (1966).

1. *The Chromosomes at Meiosis of* Saccharomyces

Elements having the appearance of chromosomes become visible in the nucleus of sporulating cells before the completion of the first meiotic division. Our illustrations (Fig. 32) do not go beyond what has been seen by others at this stage of the life cycle (McClary *et al.*, 1957; Pontefract and Miller, 1962). The "chromosomes" have never been found in the numbers to be expected from yeast genetics. More morphological work on yeast meiosis is evidently needed. Of great interest is the discovery of typical syneptinemal complexes in yeast by Engels and Cross (1968).

In the opinion of one of us (C. R.), it is doubtful whether the micrographs published by Tamaki (1965) represent chromosomes. The large granules shown are scattered evenly over the cells to their margin, and leave unstained a more or less central rounded area which, in our experience, contains the nucleus. Confidence in the author's interpretations is undermined by the identification of two closely similar patterns of randomly distributed granules once as a "diffuse nucleus" and next as "anaphase I". Meiotic cells are known to be crammed with a variety of granules. It seems possible that, in this instance, some of them have been mistaken for chromosomes.

2. *Yeasts in Whose Nuclei Chromosomes Become Visible During Mitosis*

It is not true of all yeasts that their chromosomes become visible only at meiosis. In *Lipomyces lipofer*, the chromosomes of mitosis are distinct in aceto-orecin-stained preparations of Helly-fixed hydrolysed cells (Figs. 33(a)–(d)). Seeing the chromosomes, however, is not the same as understanding their behaviour, and the course of mitosis in *L. lipofer* (Robinow, 1961) needs re-investigating.

In the fission yeast, *Schizosacch. pombe*, the chromosomes of mitosis and meiosis have been most recently described by Schopfer *et al.* (1963). Our illustrations of the yeast (Figs. 34(a)–(d)) show once again the type of nucleus common in ascomycetes. There is a large nucleolus and, at mitosis, an intranuclear spindle. The latter presumably accounts for the presence of tubules in electron micrographs of sections of the nucleus in *Schizosacch. pombe* described by Schmitter and Barker (1967). For further cytological information on *Schizosacch. pombe*, see Mitchison (1969).

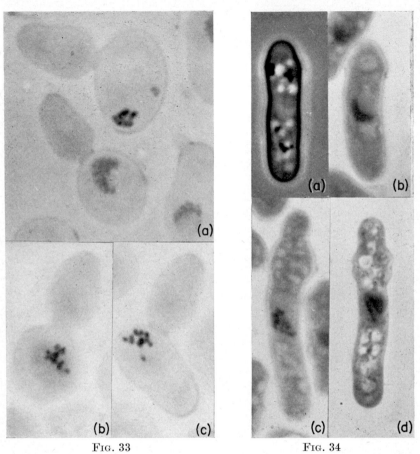

FIG. 33   FIG. 34

FIG. 33(a), (b), (c). Photomicrographs showing chromosomes at mitosis in *Lipomyces lipofer*. Helly, hydrochloric acid and aceto-orcein. Two of the five nuclei shown are in interphase. The concavity of the chromatin portion of these nuclei is due to the nucleoli. The remaining nuclei are in mitosis. Magnification 3,600 times.

FIG. 34(a)–(d). Photomicrographs of *Schizosaccharomyces pombe*. (a) living cell, phase-contrast. The upper half of the nucleus is occupied by the nucleolus. (b) Helly, hydrochloric acid and aceto-orcein. The interphase nucleus shows dense diffuse chromatin. The cell is orientated as in (a) with the nucleolus (unstained) towards the upper right. (c) nucleus in mitosis. (d) shows a comparable nucleus stained with acid fuchsin to show the intranuclear spindle and, at the upper right, the nucleolus. This preparation was obtained by Dr. F. Treichler. All micrographs are magnified 2,700 times.

E. HISTORY AND CRITIQUE OF LINDEGREN'S SCHEMES OF THE YEAST
NUCLEUS

In retrospect, the shifting and nebulous character of the yeast caryology of the past can be seen to be due to three principal causes:

1. Poor contrast between nucleus and cytoplasm in live preparations of
most species of yeasts deprived earlier generations of microscopists
of a standard by which to judge the reliability of their fixation and
staining procedures. The difficulty is not resolved by phase-contrast
microscopy unless the yeasts are examined in nutrient media rendered
suitably refractive by the addition of gelatin, albumin or other ingred-
ients (Müller, 1956; Robinow and Marak, 1966).
2. The unexpectedly low affinity for ordinary nuclear stains of the
DNA-containing chromosome-filled portion of the nucleus at all stages
of the nuclear cycle. The difficulties created by this peculiarity were
compounded by the confusion of the strongly basiphilic (as well as
acidophilic) Feulgen-negative *nucleolus* with chromatin by the school
of Lindegren.
3. The reckless manner in which C. C. Lindegren called various components of the nucleus "centriole", "centrosome" or "solid spindle"
in the absence of evidence that the parts in question have the structure
and perform the functions commonly associated with the terms he
bestowed upon them.
Points (1) and (2) have been sufficiently illustrated in the preceding
pages. Point (3) deserves to be explained in some detail.

The persuasive neatness and solidarity of Lindegren's diagrams,
reinforced by his deserved reputation as an eminent geneticist, has
ensured their survival in the pages of contemporary texts and reviews
where they continue to bewilder the novice and astound the experienced.
Their derivation and explanation is arrived at by the perusal of some
of the earliest, as well as some of the most recent, studies of yeast
nuclei. Although Lindegren's idiosyncratic schemes are now known to
have been unsound, they are still worth explaining because their geometry is frequently based on correct and reproducible anatomical
observations. Stripped of their wilful interpretation and ill-chosen term-
inology, Lindegren's schemes cease to confuse, and provide support for
modern concepts of the yeast nucleus which have removed it from the
cabinet of curiosities, where it has dwelt so long, to its proper place among
the somatic (hyphal) nuclei of other ascomycetes.

The interpretations which Lindegren has placed upon his observa-
tions on yeast nuclei have been strongly influenced by the work of
Harper (1905), already referred to above, and that of Wager and
Peniston (1910). The former's careful drawings displayed bouquets of

filamentous chromosomes suspended in the ample cavity of ascus nuclei of *Phyllactinia*. Between divisions, the chromosomes are shown attached

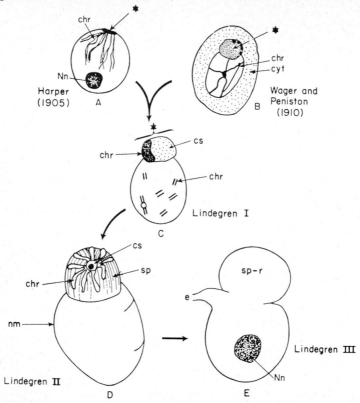

FIG. 35. Models of the yeast nucleus and their origins, based on diagrams published by Lindegren and collaborators and designated Marks I, II and III by the present authors. (A) View of an ascus nucleus of *Phyllactinia* based on a *camera lucida* drawing by Harper (1905). (B) View of a yeast cell based on a drawing by Wager and Peniston (1910). Most texts reproduce the authors' idealized *diagram*, but the comparison with Harper's work must start from comparable evidence, namely drawings made from real specimens. (C) Combined from Lindegren (1952) and Townsend and Lindegren (1953). In drawings from real specimens accompanying the latter paper the centrosome is correctly shown to contain an additional dot-like element, the "centriole". This does not appear in the author's diagram. Starred features in A, B and C are considered homologous by Lindegren. (D) Combined from Lindegren *et al.* (1956) and Yuasa and Lindegren (1958). cs indicates the centrosome; chr, the chromosomes; sp, the spindle; nm, the nuclear membrane; Nn, the nucleolus. (E) After a communication by C. C. Lindegren to McClary (1964). Some details are omitted. sp-r indicates the "spindle reservoir", which also contains "a solution of homogeneously dispersed DNA". The lower compartment contains the chromosomes and the nucleolus (n). e indicates the "extruder through which nedoplasmic reticulum enters the cytoplasm from the vacuole".

to a densely staining bar or plaque in close contact with the nuclear membrane (Fig. 35A). Some of the nuclei contained a round, relatively small nucleolus. Lindegren holds that Harper's illustrations reflect "the salient features of the fungal nucleus" (Lindegren and Rafalko, 1950). Turning to yeast, he was captivated by Wager and Peniston's (1910) notorious diagram of the yeast cell, and equated the beaded strands which there traverse the vacuole (the nucleus, in the view of the authors) with the chromosomes of Harper's drawings of ascus nuclei of *Phyllactinia*. Lindegren maintains to this day that the chromosomes of the yeast cell are free in the vacuole. It would be necessary to discuss this if evidence other than neat diagrams had ever been provided that the alleged chromosomes occur in regular numbers and go through an orderly cycle of duplication and separation. The recently accomplished isolation of intact yeast vacuoles by Matile and Wiemken (1967) and Indge (1968c) shows how the truth of these speculations may be ascertained.

In close contact with the outside of the vacuole, Wager and Peniston (1910) depict a fair sized round body possessing a peripheral cluster of tenaciously basiphil granules (Fig. 35(b)). This body they identified, most oddly, as the *nucleolus* of their vacuole-nucleus. Lindegren rejected this bizarre interpretation only to replace it by the equally arbitrary and trouble-laden notion that the relatively large and complex body in question, in which some chromatinic matter is combined with a much larger volume of non-chromatinic material, was the equivalent of the wholly chromatinic centrosome ("central body" or "lateral granule") of the ascus nuclei of *Phyllactinia*. Fig. 35, adapted from representative illustrations, summarizes the origin of what may be termed the Mark I of Lindegren's nuclear schemes.

This scheme (Mark I) suffers from several obvious weaknesses. First, the nuclei of budding yeast cells are *somatic* nuclei, and do not correspond to ascus nuclei about to undergo meiosis. They are truly comparable only to *hyphal* nuclei. Harper (1905) noted that the latter were much smaller than ascus nuclei. Speaking of their contents, he advisedly uses the term "chromatin" rather than "chromosomes", a reflection of the fact that hematoxylin-stained hyphal nuclei appear filled with indistinct chromatin, rendered in uniformly light stippling by both Harper (1905) and Colson (1938), and lack the ample non-chromatinic sap and freely floating filamentous chromosomes of ascus nuclei. Hyphal nuclei, as explained in an earlier section, possess in common with ascus nuclei distinct "lateral granules" (centrosomes, central bodies). *Nucleoli* of hyphal nuclei are relatively larger than those of ascus nuclei; they do not lie free in the nuclear sap but are jammed against the envelope in the form of a cap or lens of dense material.

Numerous illustrations in the literature conform to this description of the hyphal nuclei of ascomycetes, and one wonders why Lindegren did not choose these nuclei rather than ascus nuclei as obvious and proper models for the yeast nucleus. Fig. 36 illustrates relevant features of meiotic and post-meiotic ascus nuclei and hyphal nuclei of a generalized ascomycete. The reader is also referred back to Fig. 22 (p. 275) showing hyphal nuclei of *Aspergillus nidulans*.

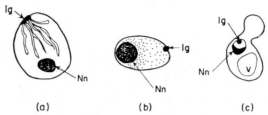

          (a)   (b)   (c)

FIG. 36. Homologous structures in an ascus nucleus (a), a hyphal nucleus of an ascomycete (b) and the nucleus of a growing yeast cell (c) as seen under the light microscope. lg indicates lateral granule (the "central body" of Harper, 1905) and Nn the nucleolus.

A further weakness of Lindegren's scheme (Mark I) derives from the lack of evidence that his "centrosome" *performs the same functions* as the "central body" of an ascus nucleus. At the beginning of karyo-kinesis in the ascus, writes Harper (1905), "the central body divides and the daughter centres migrate away from each other on the surface of the nuclear membrane". They then become the centres of broad mitotic half spindles connected to the metaphase chromosomes. How-ever, broad fibrous mitotic spindles extending across the yeast vacuole between daughter centrosomes have never been demonstrated. They should be a common sight if the huge "centrosome plus vacuole" system of the yeast cell corresponds morphologically to a mildew ascus nucleus as Lindegren would have us believe. No evidence of the alleged continuity between "centrosome" and the vacuole (Lindegren's "nu-clear sap") has been found in numerous electron micrographs of sec-tions of chemically fixed yeast cells examined in many different laboratories. On the contrary, instructive micrographs of living yeast cells made with ultraviolet radiation have clearly shown that "the nuclear apparatus (Lindegren's "centrosome") lies outside the vacuole" (McClary *et al.*, 1962). This has been confirmed at a higher level of resolu-tion by the electron microscopy of replicas of frozen-etched ("living") yeast cells (Moor and Mühlethaler, 1963). The membrane of the vacuole (more correctly "the system of vacuoles", because a cell may contain many vacuoles of different sizes) is a typical triple-layered unit membrane, whereas the nuclear envelope is composed of *two* such membranes and pierced by characteristic pores (Fig. 29, p. 285). This makes it unlikely that what is now commonly regarded as the nucleus, the "centrosome" of Lindegren's earlier models, is a compartment or reservoir arising from, and continuous with, the vacuole. Lastly, the separateness of "centrosome" and vacuole is strongly suggested by the absence of vesicular extensions from the *isolated* intact yeast vacuoles recently described from two different laboratories (Matile and Wiemken, 1967; Indge, 1968c).

Lately, aware that the chromatin of dividing yeast nuclei is, as a rule, arranged in anything but a crescent shape, Lindegren *et al.* (1956), in presenting what might be termed their Mark II model, have suggested that, at least during division, the chromatin (now depicted as a set of chromosomes) is not confined to the region formerly called "centro-chromatin" (see Fig. 35) but is spread over the surface of the whole of the former "centrosome" which has now turned into a "solid spindle". The use of this term, which persists in Lindegren's writings to the present day, confuses the ordinary reader because Lindegren speaks of "the spindle" without a trace of evidence that the thing so designated is composed of fibres. He justifies this (Lindegren and Townsend, 1954) by quoting a remark of Schrader's (1953) that mitotic spindles of *living* cells often lack visible structure, which is true unless they are looked at with a polarizing or phase-contrast microscope. But Lindegren's accounts were based on *fixed* preparations and, in these, it would have been possible to see the component fibres of a "solid spindle" had one been there. There is no evidence acceptable to ordinary cytologists that the preparations of Lindegren *et al.* (1956) and of Yuasa and Lindegren (1959) show chromosomes, and that these are attached by centromeres to putative centrioles lying within a large flat centrosome on the surface of a bulky spindle. In reality, the amounts of space occupied by chromosomes and spindle are the reverse of Lindegren's proposal. The "solid spindle" of the Mark II model is a mass of chromatin not visibly differentiated into chromosomes and traversed by a single delicate fibre not previously recognized in the light microscope except, perhaps, by Renaud (1938) until its behaviour was described from acid fuchsin-stained preparations by Robinow and Marak (1966).

An elaborate new model (Mark III) of the yeast nucleus has been contributed by Lindegren to a scholarly, unbiased, exhaustive review of yeast cytology by McClary (1964). The "solid spindle" has become a "reservoir" which now contains in addition to the spindle (which is mentioned in the legend but not drawn in the diagram) "a solution of homogeneously disposed DNA". The chromosomes themselves have been moved back into the vacuole, which also contains a nucleolus. The open passage from the reservoir (the former centrosome) to the (nuclear) vacuole has been retained, and a special organelle "the ex-truder" has been fitted to the vacuole. Wisely perhaps, diagrams of the behaviour of the whole intricate machine during mitosis are not provided.

To summarize this tortuous story: Lindegren was right in expecting the yeast nucleus to have features in common with the nuclei of mycelial fungi, especially those of ascomycetes. It has. But the resemblances are with *hyphal nuclei* not with ascus nuclei in meiosis, which Lindegren took as his models. The components which Lindegren thought he could

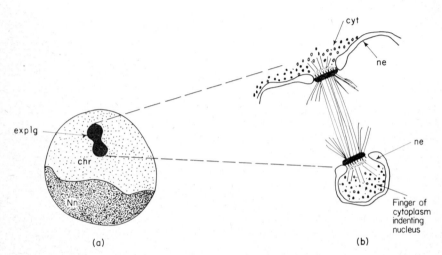

Fig. 37. The "homologies of first sight" of Fig. 36 need to be modified when electron microscopy is applied to the nucleus of the yeast cell. (a) shows a nucleus of *Saccharomyces cerevisiae* in an early phase of division as it would appear if it were stained with acid fuchsin. The "lateral granule" of Fig. 36 is in reality a complex structure. The results of electron microscopy, diagrammatically represented in (b), show that only the "centriolar plaques" at either end of the intranuclear bundle of microtubules are homologous to the lateral granule of Fig. 36(a). Based on Robinow and Marak (1966) and Fig. 28 supplied by Dr. B. Afzelius. exp lg indicates an expanding "lateral granule"; chr, chromatin; Nn, nucleolus; ne, nuclear envelope; cyt, cytoplasm.

recognize in the yeast nucleus, namely centrosome (alias "lateral granule"), nucleolus, chromosomes and spindle, are indeed found in it. *But they are all inside the small round body between vacuole and bud* which Lindegren formerly regarded as the centrosome and which he sees now as the DNA-filled "spindle reservoir" of a large vacuole-nucleus.

The diagrams of Figs. 35–37 permit a comparison of Harper's (1905) view of an ascus nucleus, several of Lindegren's nuclear models, and our own concept of the yeast nucleus which has emerged from the mutually re-inforcing results of phase-contrast microscopy, Feulgen-, aceto-orcein- and acid fuchsin-staining, the electron microscopy of sections and the life-like preservation utilized in the method of freeze-etching.

It is obvious that much remains to be discovered, but it would appear that sound foundations for future work on the yeast nucleus have now been laid.

References

Afzelius, B. A. (1965). *J. Cell Biol.* **26**, 835–843.
Agar, H. D. and Douglas, H. C. (1957). *J. Bact.* **73**, 365–375.
Aist, J. R. (1969). *J. Cell Biol.* **40**, 120–135.
Alper, R. E., Dainko, J. L. and Schlenk, F. (1967). *J. Bact.* **93**, 739–765.
Andrus, C. F. and Harter, L. L. (1937). *J. agric. Res.* **54**, 19–46.
Avers, C. J. (1967). *J. Bact.* **94**, 1225–1235.
Avers, C. J. and Federman, M. (1968). *J. Cell Biol.* **37**, 555–559.
Bacon, J. S. D., Davidson, E. D., Jones D. and Taylor, I. F. (1966). *Biochem. J.* **101**, 36C–38C.
Barton, A. A. (1950). *J. gen. Microbiol.* **4**, 84–86.
de Bary, A. (1884). "Vergleichende Morphologie und Biologie der Pilze, Mycetozoen und Bacterien", XVI, 558, Leipzig.
Bautz, E. (1955). *Ber. dt. bot. Ges.* **68**, 197–204.
Bautz, E. (1960). *In* "Die Hefen" (F. Reiff, R. Kautzmann, H. Lüers, and M. Lindemann eds.), Vol. 1, pp. 41–67. Hans Carl, Nürnberg.
Bogen, H. J. and Keser, M. (1954). *Physiologia Pl.* **7**, 446–462.
Brachet, J. (1957). "Biochemical Cytology", pp. 70–71. Academic Press. New York.
Branton, D. and Moor, H. (1964). *J. Ultrastruct. Res.* **11**, 401–411.
Brown, C. and McClary, D. O. (1963) *Bact. Proc.* 97.
Cagniard-Latour, C. (1838). *Ann. Chim. (Phys.)* **68**, (2me série), 206–223.
Carroll, G. C. (1967). *J. Cell Biol.* **33**, 218–224.
Chapman, C. and Bartley, W. (1968). *Biochem. J.* **107**, 455–465.
Chung, K. L., Hawirko, R. Z. and Isaac, P. K. (1966). *Can. J. Microbiol.* **11**, 953–957.
Cirillo, V. P. (1961). *A. Rev. Microbiol.* **15**, 197–218.
Cirillo, V. P. (1962) *J. Bact.* **84**, 485–491.
Clark-Walker, G. D. and Linnane, A. W. (1967). *J. Cell Biol.* **34**, 1–14.
Colson, B. (1938). *Ann. Bot.* **2**, 381–402.
Criddle, R. S. and Willemot, J. (1967). *Prot. Biol. Fluids.* **15**, 55–67.
Desmazières, M. J. B. (1827). *Annls Sci. nat. (Bot.)* **10**, 42–67.
Drochmans, P. (1962). *J. Ultrastruct. Res.* **6**, 141–163.
Duell, E. A., Inoue, D. and Utter, M. F. (1964). *J. Bact.* **88**, 1762–1773.
de Duve, C. and Wattiaux, R. (1966). *A. Rev. Physiol.* **28**, 435–492.
Eddy, A. A. (1958). *Proc. R. Soc. B.* **149**, 425–440.
Eddy, A. A. and Williamson, D. H. (1957). *Nature, Lond.* **179**, 1252–1253.
Eddy, A. A. and Williamson, D. H. (1959). *Nature, Lond.* **183**, 1101–1104.
Engels, F. M. and Cross A. F. (1968). *Chromosoma* **25**, 104–106.
Ephrussi, B. (1953). "Nucleo-Cytoplasmic Relations in Micro-organisms". Oxford University Press, New York.

Falcone, G. and Nickerson, W. J. (1956). *Science, N.Y.* **124**, 272–273.

Federman, M. and Avers, C. J. (1967). *J. Bact.* **94**, 1236–1243.

Fernández-Morán, H. (1962). *Circulation* **26**, 1039–1065.

Frey-Wyssling, A., Grieshaber, E. and Mühlethaler, K. (1963). *J. Ultrastruct. Res.* **8**, 506–516.

Frey-Wyssling, A., López-Sáez, J. F. and Mühlethaler, K. (1964). *J. Ultrastruct. Res.* **10**, 422–432.

Frijs, J. and Ottolenghi, P. (1959). *C. r. Trav. Lab. Carlsberg* **31**, 259–271.

Fukuhara, H. (1967a). *Biochim. biophys. Acta* **134**, 143–164.

Fukuhara, H. (1967b). *Proc. natn. Acad. Sci. U.S.A.* **58**, 1065–1072.

Garcia-Mendoza, C. and Villanueva, J. R. (1962). *Nature, Lond.* **195**, 1326–1327.

Garzuly-Janke, R. (1940). *Zentbl. Bakt. ParasitKde* (*Abt II*) **102**, 361–365.

Gascon, S., Neumann, N. P. and Lampen, J. O. (1968). *J. biol. Chem.* **243**, 1573–1577.

Gerhardt, P. and Judge, J. A. (1964). *J. Bact.* **87**, 945–951.

Girbardt, M. (1968) *Symp. Soc. exp. Biol.* **22**, 249–259.

Green, D. E., Murer, E., Hultin, H. O., Richardson, S. H., Salmon, B., Brierly, G. P. and Baum, H. (1965). *Archs Biochem. Biophys.* **112**, 635–647.

Grieshaber, E. (1964). *Viertelj. schr. naturforsch. Ges. Zürich*, **109**, 1–23.

Guilliermond, A. (1903). *Revue gén. Bot.* **15**, 49–66.

Guilliermond, A. (1913). *C. r. Séanc. Soc. Biol.* **74**, 618–620.

Guilliermond, A. (1923). *C. r. Séanc. Soc. Biol.* **88**, 517–520.

Guilliermond, A. (1930). *Protoplasma* **9**, 133–139.

Guilliermond, A., Mangenot, G. et Plantefol, L. (1933). "Traité de Cytologie Végétale". Le François, Paris.

Günther, T. and Kattner, W. (1968). *Z. Naturf.* **23b**, 77–80.

Hagedorn, H. (1957). *Náturwissenschaften* **44**, 641–642.

Hansen, E. C. (1902). *C. r. Lab. Carlsberg* **5**, 68–107.

Harper, R. A. (1905). *Carnegie Inst. Wash. Publ.* **37**, 104.

de Harven, E. (1968). *In* "The Nucleus" (A. J. Dalton and F. Haguenau, eds.), pp. 197–227. Academic Press, New York.

Havelkova, M. and Mensik, P. (1966). *Náturwissenschaften* **21**, 562.

Hawker, L. E. (1965). *Biol. Rev.* **40**, 52–92.

Haworth, W. N., Heath R. L. and Peat, S. (1941). *J. chem. Soc.* pp. 833–842.

Hirano, T. and Lindegren, C. C. (1961). *J. Ultrastruct. Res.* **5**, 321–327.

Hirano, T. and Lindegren, C. C. (1963). *J. Ultrastruct. Res.* **8**, 322–326.

Horstmann, H. J. (1968). *Z. Biol.* **116**, 1–9.

Houwink, A. L. and Kreger, D. R. (1953). *Antonie van Leeuwenhoek* **19**, 1–24.

Indge, K. J. (1968a). *J. gen. Microbiol.* **51**, 425–432.

Indge, K. J. (1968b). *J. gen. Microbiol.* **51**, 433–440.

Indge, K. J. (1968c). *J. gen. Microbiol.* **51**, 441–446.

Indge, K. J. (1968d). *J. gen. Microbiol.* **51**, 447–455.

Jollow, D., Kellerman, G. M. and Linnane, A. W. (1968). *J. Cell Biol.* **37**, 221–230.

Kawakami, N. (1961). *Expl Cell Res.* **25**, 179–181.

Kessler, G. and Nickerson, W. J. (1959). *J. biol. Chem.* **234**, 2281–2285.

Klein, H. P. (1965). *J. Bact.* **90**, 227–243.

Klein, H. P., Volkmann, C. M. and Weibel, J. (1967a). *J. Bact.* **94**, 475–481.

Klein, H. P., Volkmann, C. M. and Chao, F. C. (1967b). *J. Bact.* **93**, 1966–1971.

Klein, H. P., Volkmann, C. M. and Leaffer, M. A. (1967c). *J. Bact.* **94**, 61–65.

Korn, E. D. and Northcote, D. H. (1960). *Biochem. J.* **75**, 12–17.

Kotyk, A. (1967). *Folia microbiol., Praha* **12**, 121–131.

Kotyk, A. and Matile, P. (1969). *Proc. J. E. Purkyně University, Brno.* in press.
Kreger, D. R. (1954). *Biochim. biophys. Acta* **13**, 1–9.
Kreger, D. R. (1966). *Abh. dt. Akad. Wiss. Berl.* Klasse f. Medizin, Nr. 6, pp. 81–88.
Kützing, F. (1837). *J. prakt. Chem.* **11**, 385–409.
Lampen, J. O. (1968). *Antonie van Leeuwenhoek* **34**, 1–18.
Letters, R. (1968). *In* "Aspects of Yeast Metabolism" (A. K. Mills, ed.), pp. 303–319. Blackwell Scientific Publications, Oxford and Edinburgh.
Levan, A. (1946). *Nature, Lond.* **158**, 626.
Levan, A. (1947). *Hereditas* **33**, 457–514.
Lillehoj, E. B. and Ottolenghi, P. (1966). *Abh. dt. Akad. Wiss. Berl.* Klasse f. Medizin, Nr. 6, pp. 145–152.
Lindegren, C. C. (1952). *Symp. Soc. exp. Biol.* **6**, 277–289.
Lindegren, C. C. and Rafalko, M. M. (1950). *Expl Cell Res.* **1**, 169–187.
Lindegren, C. C. and Townsend, G. F. (1954). *Cytologia* **19**, 104–109.
Lindegren, C. C., Williams, M. A. and McClary, D. O. (1956). *Antonie van Leeuwenhoek* **22**, 1–20.
Linnane, A. W. and Still, J. L. (1955). *Archs Biochem. Biophys.* **59**, 383–392.
Linnane, A. W., Vitols, E. V. and Nowland, P. G. (1962). *J. Cell Biol.* **13**, 345–350.
Mackler, B., Collip, P. J., Duncan, H. M., Rao, N. A. and Huennekens, F. M. (1962). *J. biol. Chem.* **237**, 2968–2974.
Mahler, H. R., Perlman, P., Henson, C. and Weber, C. (1968). *Biochem. biophys. Res. Commun.* **31**, 474–480.
Mahler, H. R., Mackler, B., Grandchamp, S. and Slonimski, P. P. (1964). *Biochemistry* **3**, 668–677.
Mahler, H. R., Tewari, K. K. and Jayaraman, J. (1968). *In* "Aspects of Yeast Metabolism" (A. K. Mills, ed.), pp. 247–268. Blackwell Scientific Publications, Oxford and Edinburgh.
Mandelstam, J. (1960). *Bact. Rev.* **24**, 289–308.
Manners, D. J. and Patterson, J. C. (1966). *Biochem. J.* **98**, 19C–20C.
Marchant, R. and Smith, D. G. (1967). *Arch. Mikrobiol.* **58**, 248–256.
Marcus, L., Ris, H., Halvorson, H. O., Bretthauer, R. K. and Bock, R. M. (1967). *J. Cell Biol.* **34**, 505–512.
Marquardt, H. (1962a). *Z. Naturf.* **17b**, 42–48.
Marquardt, H. (1962b). *Z. Naturf.* **17b**, 689–695.
Marquardt, H. (1963). *Arch. Mikrobiol.* **46**, 308–320.
Matile, P. (1968). *Proc. 2nd. Symposium on Yeasts*, Bratislava. in press.
Matile, P. and Wiemken, A. (1967). *Arch. Mikrobiol.* **56**, 148–155.
Matile, P. and Moor, H. (1968). *Planta* **80**, 159–175.
Matile, P., Jost, M. and Moor, H. (1965). *Z. Zellforsch. mikrosk. Anat.* **68**, 205–216.
Matile, P., Moor, H. and Mühlethaler, K. (1967). *Arch. Mikrobiol.* **58**, 201–211.
Mattoon, J. R. and Sherman, F. (1966). *J. biol. Chem.* **241**, 4330–4338.
Mazur, P. (1965). *Trans. N.Y. Acad. Sci.* **125**, 658–676.
McClary, D. O. (1964). *Bot. Rev.* **30**, 167–225.
McClary, D. O. and Bowers, W. D. (1967). *J. Cell Biol.* **32**, 519–524.
McClary, D. O., Williams, M. A., Lindegren, C. C. and Ogur, M. (1957). *J. Bact.* **73**, 360–364.
McClary, D. O., Bowers, W. D. and Miller, G. R. (1962). *J. Bact.* **83**, 276–283.
McLellan, W. L. and Lampen, J. O. (1963). *Biochim. biophys. Acta* **67**, 324–326.
McLellan, W. L. and Lampen, J. O. (1968). *J. Bact.* **95**, 967–974.
Mendoza, C. G. and Villanueva, J. R. (1963). *Can. J. Microbiol.* **9**, 141–142.
Millbank, J. W. and Macrae, R. M. (1964). *Nature, Lond.* **201**, 1347.

Miller, J. J. (1967). *Bull. Soc. Hist. Nat., Toulouse*, **103**, 327–339.

Miller, M. W. and Phaff, H. J. (1958). *Antonie van Leeuwenhoek* **24**, 255–238.

Mitchison, J. M. (1969). *In* "Methods in Cell Physiology" (D. M. Prescott, ed.), Academic Press, New York.

Montagne, M. P. and Morton, R. K. (1960). *Nature, Lond.* **187**, 916–917.

Moor, H. (1964). *Z. Zellforsch. mikrosk. Anat.* **62**, 546–580.

Moor, H. (1966). *J. Cell Biol.* **29**, 153–154.

Moor, H. (1966). *High Vacuum Report*, nr. 9. Ed. Balzers AG. Balzers. Lichtenstein.

Moor, H. (1967a). *Protoplasma* **64**, 89–103.

Moor, H. (1967b). *Arch. Mikrobiol.* **57**, 135–146.

Moor, H. and Mühlethaler, K. (1963). *J. Cell Biol.* **17**, 609–628.

Mounolou, J. C., Jakob, H. and Slonimski, P. P. (1966). *Biochem. biophys. Res. Commun.* **24**, 218–224.

Moustacchi, E. and Williamson, D. H. (1966). *Biochem. biophys. Res. Commun.* **23**, 56–61.

Müller, R. (1956a). *Náturwissenschaften* **43**, 86.

Müller, R. (1956b). *Náturwissenschaften* **43**, 428–429.

Mundkur, B. (1960a). *Expl. Cell Res.* **20**, 28–42.

Mundkur, B. (1960b). *Expl. Cell Res.* **21**, 201–205.

Narayana, N. V. A. (1956). *Proc. Indian Acad. Sci. B* **43**, 314–324.

Nečas, O. (1956). *Nature, Lond.* **177**, 898–899.

Nečas, O. (1961). *Nature, Lond.* **192**, 580–581.

Nečas, O. (1965). *Folio microbiol., Praha* **11**, 97–102.

Nečas, O. and Svoboda, A. (1967). *Folio microbiol., Praha* **13**, 379–385.

Neumann, N. P. and Lampen, J. O. (1967). *Biochemistry* **6**, 468–475.

Nickerson, W. J. (1963). *Bact. Rev.* **27**, 305–324.

Nickerson, W. J. and Falcone, G. (1956). *Science, N.Y.* **124**, 722.

Nickerson, W. J. and Falcone, G. (1959). *In* "Sulfur in Proteins" (R. Benesch, ed.), pp. 409–424. Academic Press, New York.

Northcote, W. J. (1963). *Pure Appl. Chem.* **7**, 669–675.

Northcote, D. H. and Horne, R. W. (1952). *Biochem. J.* **51**, 232–236.

Nossal, P. M., Keech, D. B. and Morton, D. J. (1956). *Biochim. biophys. Acta* **22**, 412–420.

Ohnishi, T., Kawaguchi, K. and Hagihara, B. (1966). *J. biol. Chem.* **241**, 1797–1806.

Ottolenghi, P. (1967). *C. r. Lab. Carlsberg* **36**, 95–111.

Peat, S., Whelan, W. J. and Edwards, T. E. (1958). *J. chem. Soc.* pp. 3862–3868.

Pfeiffer, H. H. (1963). *Protoplasma* **57**, 636–642.

Phaff, H. J. (1963). *A. Rev. Microbiol.* **17**, 15–30.

Polakis, E. S., Bartley, W. and Meek, G. A. (1964). *Biochem. J.* **90**, 369–374.

Polakis, E. S., Bartley, W. and Meek, G. A. (1965). *Biochem. J.* **97**, 298–302.

Pontefract, R. D. and Miller, J. J. (1962). *Can. J. Microbiol.* **8**, 573–584.

Porter, K. R. and Machado, R. D. (1960). *J. biophys. biochem. Cytol.* **7**, 167–180.

Renaud, J. (1938). *C. r. hebd. Séanc. Acad. Sci., Paris* **206**, 1918–1920.

Robinow, C. F. (1961). *J. biophys. biochem. Cytol.* **9**, 879–892.

Robinow, C. F. and Caten, C. E. (1969). *J. Cell Sci.* in press.

Robinow, C. F. and Bakerspigel, A. (1965). *In* "The Fungi" (C. G. Ainsworth and A. S. Sussman eds.), vol. 1, pp. 119–142. Academic Press, New York.

Robinow, C. F. and Marak, J. (1966). *J. Cell Biol.* **29**, 129–151.

Roelofsen, P. A. and Hoette, I. (1951). *Antonie van Leeuwenhoek* **17**, 297–313.

Roodyn, D. B. and Wilkie, D. (1968). "The Biogenesis of Mitochondria". Methuen, London.

Rothstein, A. (1961). *In* "Membrane Transport and Metabolism" (A. Kleinzeller and A. Kotyk, eds.), pp. 270–284. Academic Press, New York.

Roush, A. H. (1961). *Nature, Lond.* **190**, 449.

Sarkar, S. K. and Poddar, R. K. (1965). *Nature, Lond.* **207**, 550–551.

Schatz, G. (1963). *Biochem. biophys. Res. Commun.* **12**, 448–451.

Schatz, G. (1965). *Biochim. biophys. Acta* **96**, 342–345.

Schatz, G. and Klima, J. (1964). *Biochim. biophys. Acta* **81**, 448–461.

Schatz, G., Haslbrunner, E. and Tuppy, H. (1964). *Biochem. biophys. Res. Commun.* **15**, 127–132.

Schatz, G., Tuppy, H. and Klima, J. (1963). *Z. Naturf.* **18b**, 145–153.

Schmitter, R. E. and Barker, D. C. (1967), *Expl Cell Res.* **46**, 215–220.

Schmitz, F. (1879). *Sitzber. der Niederrh. Ges. in Bonn* **18**, 345–377.

Schopfer, W. H., Wustenfeld, D. and Turian, G. (1963). *Arch. Mikrobiol.* **45**, 304–313.

Schrader, F. (1953). "Mitosis", 2nd ed. Columbia Univ. Press, New York.

Schwann, Th. (1837). *Ann. Phys.* **41**, 184–193.

de Seynes, J. (1868). *C. r. hebd. Séanc. Acad. Sci., Paris* **67**, 105–109.

Sherman, F. and Slonimski, P. P. (1964). *Biochim. biophys. Acta* **90**, 1–15.

Shinagawa, Y., Inouye, A., Ohnishi, T. and Hagihara, B. (1966). *Expl Cell Res.* **43**, 301–310.

Silbereisen, K. (1960). *In* "Die Hefen" (F. Reiff, R. Kauzmann, H. Lüers and M. Lindemann, eds.), Vol. 1, pp. 357–368. Hans Carl, Nürnberg.

Slonimski, P. P. (1953). "Recherches sur la Formation des Enzymes respiratoires chez la Levure". Editions Desoer, Liège.

Smith, D. G. and Marchant, R. (1968). *Arch. Mikrobiol.* **60**, 340–347.

Somlo, M. and Fukuhara, H. (1965). *Biochem. biophys. Res. Commun.* **19**, 587–591.

Soneda, M. (1960). *Nagaoa* **7**, 9–13.

Souza, N. O. and Panek, A. D. (1968). *Archs Biochem. Biophys.* **125**, 22–28.

Stoeckenius, W. (1963). *J. Cell Biol.* **17**, 443–454.

Stodola, F. H., Deinema, M. H. and Spencer, J. F. T. (1967). *Bact. Rev.* **31**, 194–213.

Streiblová, E. (1968). *J. Bact.* **95**, 700–707.

Streiblová, E. and Beran, K. (1963). *Expl Cell Res.* **30**, 603–605.

Streiblová, E. and Beran, K. (1965). *Folio microbiol., Praha* **10**, 352–356.

Streiblová, E. and Beran, K. (1966). *Abh. dt. Akad. Wiss. Berl.* Klasse f. Medizin, Nr. 6, pp. 77–79.

Suomalainen, H. (1968). *In* "Aspects of Yeast Metabolism" (A. K. Mills, ed.), pp. 1–29. Blackwell Scientific Publications, Oxford and Edinburgh.

Svihla, G. and Schlenk, F. (1960). *J. Bact.* **79**, 841–848.

Svihla, G., Dainko, J. L. and Schlenk, F. (1963). *J. Bact.* **85**, 399–409.

Swift, H., Rabinowitz, M. and Getz, G. (1967). *J. Cell Biol.* **35**, 131A–132A.

Tamaki, H. (1965). *J. gen. Microbiol.* **41**, 93–98.

Tanaka, H. and Phaff, H. J. (1965). *J. Bact.* **89**, 1570–1580.

Tanaka, H. and Phaff, H. J. (1966). *Abh. dt. Akad. Wiss. Berl.* Klasse f. Medizin, Nr. 6. pp. 113–129.

Tewari, K. K., Jayaraman, J. and Mahler, M. R. (1965). *Biochem. biophys. Res. Commun.* **21**, 141–148.

Thyagarajan, T. R., Conti, S. F. and Naylor, H. B. (1961). *Expl Cell Res.* **25**, 216–218.

Townsend, G. F. and Lindegren, C. C. (1953). *Cytologia* 18, 183–201.

Tuppy, H. and Wintersberger, E. (1966). *In* "Probleme der biologischen Reduplikation" (P. Sitte, ed.) pp. 325–335. Springer-Verlag, Berlin.

Tustanoff, E. R. and Bartley, W. (1964). *Biochem. J.* 91, 595–600.

Uruburu, F., Elorza, V. and Villanueva, J. R. (1968). *J. gen. Microbiol.* 51, 195–198.

Utter, M. F., Duell, E. A. and Bernofsky, C. (1968). *In* "Aspects of Yeast Metabolism" (A. K. Mills, ed.), pp. 197–212. Blackwell Scientific Publications, Oxford and Edinburgh.

Villanueva, J. R. (1966a). *Abh. dt. Akad. Wiss. Berl.* Klasse f. Medizin, Nr. 6, pp. 1–13.

Villanueva, J. R. (1966b). *In* "The Fungi" (G. C. Ainsworth and A. S. Sussman, eds), Vol. 2, pp. 3–62. Academic Press, New York.

Villanueva, J. R. and Elorza, M. V. (1968). *Proc. 2nd. Symposium on Yeasts*, Bratislava. in press.

Vitols, E. V. and Linnane, A. W. (1961). *J. biophys. biochem. Cytol.* 9, 701–710.

Vitols, E. V., North, R. J. and Linnane, A. W. (1961). *J. biophys. biochem. Cytol.* 9, 689–699.

Wager, H. and Peniston, A. (1910). *Ann. Bot.* 24, 45–84.

Wallace, P. G. and Linnane, A. W. (1964). *Nature, Lond.* 201, 1191–1194.

Wallace, P. G., Huang, M. and Linnane, A. W. (1968). *J. Cell Biol.* 37, 207–220.

Whaley, W. G. and Mollenhauer, H. H. (1963). *J. Cell Biol.* 17, 216–221.

Wiame, J. M. (1947). *Biochim. biophys. Acta* 1, 234–255.

Williamson, D. H. (1966). *In* "Cell Synchrony" (I. L. Cameron and G. M. Padilla, eds.), pp. 81–101. Academic Press, New York.

Wilson, C. L. and Aist, J. R. (1967). *Phytopathology* 57, 769–771.

Windisch, S. and Bautz, E. (1960). *In* "Die Hefen" (F. Reiff, R. Kauzmann, H. Lüers and M. Lindemann, eds.), Vol. 1, pp. 68–72. Hans Carl, Nürnberg.

Winge, Ö. (1935). *C. r. Lab. Carlsberg, Ser. Physiol.* 21, 77–111.

Yotsuyanagi, Y. (1962a). *J. Ultrastruct. Res.* 7, 121–140.

Yotsuyanagi, Y. (1962b). *J. Ultrastruct. Res.* 7, 141–158.

Yuasa, A. and Lindegren, C. C. (1959). *Antonie van Leeuwenhoek* 25, 73–87.

Chapter 7

Sporulation and Hybridization of Yeasts

R. R. Fowell

The Distillers Co. (Yeast) Ltd., Great Burgh, Epsom, Surrey, England

I. Introduction

Sporulation, as applied to yeasts, usually refers to the formation of ascospores inside modified cells (asci) by a process involving reduction division (meiosis). Although considerable advances have been made in recent years in our knowledge and understanding of this process, and improved methods for inducing sporulation have become available, the classical review of Phaff and Mrak (1948, 1949) can still be recommended for introductory reading.

Sporulation is of particular interest to the taxonomist and the geneticist. The ability of a yeast to sporulate is the main feature which distinguishes the ascosporogenous yeasts (Saccharomycetaceae) from other yeasts. The number of spores per ascus, their shape and appearance, and the way in which asci and spores are formed (e.g. from diploid cells, zygotes, or the larger of a pair of conjugating cells) are characteristics used in the subdivision of the Saccharomycetaceae.

The maximum number of spores per ascus in most yeasts is four, e.g. *Saccharomyces, Saccharomycodes, Endomycopsis, Pichia, Hansenula* and *Hanseniaspora*. A few yeasts, such as *Schizosaccharomyces octosporus* and members of the genera *Nematospora* and *Coccidiascus*, resemble higher Ascomycetes in regularly producing a maximum of eight spores per ascus. Certain strains of *Saccharomyces cerevisiae* (Winge and Roberts, 1950a; Lindegren and Lindegren, 1953; Pomper *et al.*, 1954; Fowell, 1956a; Santa María, 1957a, b; 1958), other species of *Saccharomyces* (Santa María, 1957b, 1959), *Lipomyces* (Lodder and Kreger–van Rij, 1952), and *Schizosaccharomyces pombe* (Leupold, 1956; Gutz, 1967) have a tendency to produce asci with a maximum of eight or more spores. Large numbers of spores occur in the asci of strains of *Kluyveromyces*, estimated to be of the order of 100 in *Kluyveromyces polysporus* (van der Walt, 1956). The asci of most yeasts contain variable numbers of spores, e.g. one to four in *Sacch. cerevisiae*, due to failure of certain nuclei, resulting from nuclear division in the ascus, to become included in spores (Nagel, 1946; Pontefract and Miller, 1962) or due to inclusion of two nuclei in certain spores (Winge and Roberts, 1950b, 1954). In some yeasts, the maximum number of spores per ascus is less than four. Certain species, such as *Sacch. bisporus, Endomycopsis javanensis, Hansenula saturnus*, and many strains of brewer's yeasts (*Sacch. cerevisiae, Sacch. carlsbergensis*), produce predominantly one- or two-spored asci, asci with three or four spores being rare or absent. In strains of *Nadsonia, Monosporella, Schwanniomyces* and *Debaryomyces*, there is seldom more than one spore in an ascus.

For the characterization of strains within a species, the degree of sporulation (percentage of cells transformed into asci) under standard conditions and the spore diagram (proportions of asci with different numbers of spores) are features worthy of consideration (Fowell, 1955, 1967; Oppenoorth, 1957). Spore viability and percentage of mating types among spores have also been suggested for strain differentiation (Koninklijke Nederlandsche Gist- en Spiritusfabriek N.V., 1962).

Shape and surface sculpture of spores (Fig. 1) are of some taxonomic importance. These spore characteristics are extremely constant in some genera or species; but there may be considerable variation between species of the same genus, and even within a single species. Spores are commonly smooth and spherical or oval as in strains of *Saccharomyces, Zygosaccharomyces* and *Saccharomycodes*. In *Schizosaccharomyces pombe*,

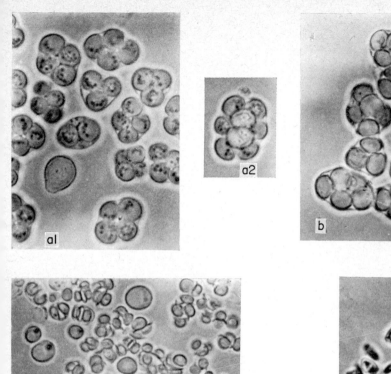
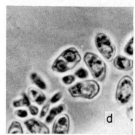
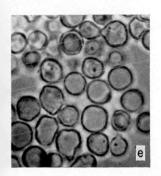
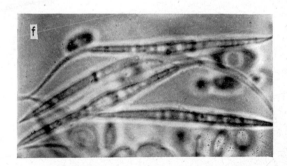

FIG. 1. Asci and ascospores of yeasts. (a) *Saccharomyces cerevisiae* showing asci with four or fewer spores (a1), and an eight-spored ascus (a2); (b) *Schizosaccharomyces octosporus* showing asci with a maximum of eight bean-shaped spores; (c) *Hansenula anomala* with hat-shaped spores; (d) *Hansenula saturnus* showing spores with equatorial brims; (e) *Debaryomyces hansenii* showing asci with solitary warty spores; (f) *Nematospora coryli* showing fusiform spores with filaments. Micrographs are magnified × 1425.

the spores are elliptical while in *Schizosacch. octosporus* they are bean-shaped; in strains of *Saccharomyces*, the association between weak fermentative properties and production of bean-shaped spores is considered by Kudriavzev (1954) to justify the creation of a new genus, *Fabospora*, to accommodate *Sacch. fragilis* and *Sacch. macedoniensis*. The helmet-shaped spores of *Pichia* spp. may be regarded as stunted forms of the hat-shaped spores (with prominent brims), most characteristic of *Hansenula* (most species) though found in a few species of other genera, e.g. *Endomycopsis fibuliger*. Saturn-shaped spores, with an equatorial brim, occur in a few species of *Hansenula*, e.g. *H. saturnus*, and also in *Endomycopsis javanensis*, the spores of which may be hat- or Saturn-shaped. Highly characteristic are the warty, spherical spores of strains of *Debaryomyces* and *Nadsonia*, and the warty, walnut-shaped spores of *Schwanniomyces* spp. Needle-like spores are produced by strains of *Coccidiascus* (slightly twisted), *Monosporella* (pointed at one or both ends) and *Nematospora* (one end drawn into a filament).

Production of spores is essential for the hybridization of yeasts because this depends on the fusion in pairs of mater (usually haploid) spores or cells produced by the germination of spores. Several difficulties, however, are likely to be encountered in an attempt to hybridize yeasts. Some yeasts do not sporulate very well or they produce only small numbers of four-spored asci which are essential for tetrad analysis in genetical investigations. It is sometimes necessary to isolate large numbers of spores because only a few may be viable and, as low spore viability is usually associated with poor sporulation, the task of obtaining sufficient numbers of spores may appear to be insuperable. Finally, only certain spores or spore-cultures may be maters, and the mating activity of some, or even all, of these may be too weak to permit successful hybridization. Lindegren (1949) has shown that, for the pursuit of academic studies in yeast genetics, it is possible to overcome these difficulties by the use of selected yeasts for the development of fertile breeding stocks. In an attempt to produce improved strains of yeasts by hybridization for use in industry, e.g. for baking, brewing and spirit production, only yeasts with certain characteristics can be considered as suitable breeding stocks, and many of these, particularly brewer's yeasts, unfortunately suffer from one or more of the above defects.

In recent years, a good deal of attention has been devoted to the task of facilitating the hybridization of yeasts, particularly in industry (for reviews, see Fowell, 1955, 1966). In the following account of sporulation and sexual reproduction, consideration is given to the controlling factors involved and the nature of the underlying mechanisms. Increased knowledge of these processes is proving useful in the formulation of more successful methods of hybridization.

Production of ascospores by yeasts in nature probably plays a part in adaptation and survival under changed environmental conditions. Fusion of spores and haploid cells could provide a variety of genotypes on which selective forces would operate to ensure the survival of the fittest. It is often assumed that yeasts are able to survive unfavourable conditions in the form of ascospores, but, as Ingram (1955) has pointed out, unlike bacterial spores, these are only slightly more resistant than vegetative cells to heat (Hansen, 1883), alcohol (Hansen, 1907) and other agents. Nevertheless, it is interesting to note that differences in heat resistance form the basis of a successful method of isolating spores from vegetative cells (Wickerham and Burton, 1954; see Section V.B, p. 368).

II. The Sporogenic Process

Our knowledge of the sporogenic process is based mainly on studies of strains of *Saccharomyces cerevisiae*. Schwann (1839) first observed that yeast cells could form small interior cells which were liberated by the bursting of the parent cell. Investigations by de Seynes (1868) and Reess (1869, 1870) revealed that these interior cells were spores, and the similarity of their formation to that in the lower Ascomycetes led Reess to term the spores "ascospores" and the parent cells "asci". The discovery that spores are formed when yeast cells are deprived of nutrients is due to de Seynes.

Little attention has been given to the cytological aspects of sporulation. The occurrence of reduction division (meiosis) has been firmly established by numerous genetical investigations, chiefly on *Saccharomyces* spp. (see Section IV.A, p. 340), but cytological confirmation has been sorely lacking until recently.

A. PRELIMINARY CHANGES

Soon after yeast cells are transferred to a sporulation medium, there is an increase in cell volume (Reess, 1870; Pontefract and Miller, 1962; Deysson and Lau, 1963; Croes, 1967a), and an increase in protoplasmic movement (Hagedorn, 1964). The original vacuole of the vegetative cell becomes fragmented into numerous small vacuoles (Reess, 1870; Mundkur, 1961b; Miller *et al.*, 1963; Svihla *et al.*, 1964; Croes, 1967a); these vacuoles increase in size and tend to coalesce, and there is a con-comitant increase in small granules (Miller *et al.*, 1963; Svihla *et al.*, 1964) which interferes with observation of the vacuoles. However, by the incorporation in the vacuoles of ultraviolet-absorbing S-adenosyl-methionine and examination by ultraviolet microscopy (Svihla *et al.*, 1964), it is possible to distinguish more clearly between granules and

11

vacuoles, and to observe the eventual disappearance of the vacuoles during actual spore formation.

The granules increase both in number and size until they fill the entire cell, and this accounts for the granular appearance of cells about to sporulate that has been noted by many workers such as Nagel (1946) and Kleyn (1954). The granules appear to be of different kinds including reserves of glycogen and fat (Pontefract and Miller, 1962; Miller *et al.*, 1963) and also mitochondria (Brandt, 1941). These granules finally condense into a compact mass in the centre of the cell, connected by protoplasmic strands to the periphery. The compact granular mass contains the nucleus and constitutes the spore apparatus, the region in which the spores are formed. In contrast to the above granules, there is initially a heavy concentration of ribosomes in the vegetative cell, but these rapidly decrease in numbers in cells on the sporulation medium and the cytoplasm becomes less basophilic (Mundkur, 1961b); vacuolar granules (polyphosphate) also disappear.

The preliminary changes also involve the nucleus which becomes enlarged and less compact soon after cells are transferred to a sporulation medium, although the change affects only a few cells initially (Nagel, 1946; Pontefract and Miller, 1962), for example after 5 h in the presence of acetate.

B. NUCLEAR DIVISION (MEIOSIS)

Little success has attended efforts to demonstrate the characteristic stages of meiosis in sporulating yeast cells. The gross changes in strains of *Saccharomyces* (e.g. Winge and Roberts, 1954; Pontefract and Miller, 1962) are familiar enough: the nucleus elongates and becomes dumb-bell shaped, then constricts into two nuclei; shortly afterwards, the two daughter nuclei divide in the same way, the original nucleus being replaced by a peripheral cluster of four smaller nuclei. Nuclear changes follow a similar pattern in *Schizosacch. octosporus*, except that an additional set of nuclear divisions results in the formation of eight nuclei (Widra and DeLamater, 1955). Electron micrography has failed to reveal chromosomes in sporulating cells of *Saccharomyces* (Hashimoto *et al.*, 1960; Mundkur, 1961b; Marquardt, 1963), but dense bodies resembling chromosomes have been demonstrated in strains of *Saccharomyces*, *Schizosaccharomyces* and other yeasts by different staining methods, including Giemsa staining which is particularly suitable for the purpose (McClary *et al.*,1957a,b; Pontefract and Miller, 1962; Schopfer *et al.*, 1963; Tamaki, 1965).

Most workers are agreed that the nucleus assumes a coarse reticular appearance in prophase I, and this is succeeded by resolution into chromosome-like structures. Later stages, however, have only been satisfactorily demonstrated in *Sacch. cerevisiae* by Tamaki (1965) whose success can be attributed to the use of snail digestive juice for dissolving the cell walls, a process which facilitated spreading of cells and chromosomes. One strain of yeast examined by this worker exhibited 18 bivalents at metaphase I; this suggests that the number of chromosomes in *Saccharomyces* is similar to that indicated by recent genetical analysis.

The development of asci with more than four spores necessitates additional nuclear divisions. In yeasts such as *Schizosaccharomyces octosporus* (Widra and DeLamater, 1955) and *Kluyveromyces polysporus* (Roberts and van der Walt, 1959), which regularly produce multispored asci, meiosis is almost certainly followed by mitotic divisions. There is some doubt, however, about the nature of the nuclear divisions which precede the formation of multispored asci by yeasts which normally produce asci with a maximum of four spores. Winge and Roberts (1950b, 1954) have provided genetical and cytological evidence for the occurrence of extra mitotic divisions following meiosis in certain asci of *Sacch. cerevisiae* × *Sacch. chevalieri* hybrids. Asci in strains of *Sacch. cerevisiae* containing 5–8 spores frequently bear small buds, and it is possible that these asci are derived from cells which, after becoming binucleate in preparation for budding, are suddenly diverted to sporulation (Fowell, 1956a). The formation of eight nuclei in an ascus could therefore result from meiosis of two nuclei. The existence of more than four genotypes among spores of the same ascus (Lindegren and Lindegren, 1953; Pomper *et al.*, 1954) is evidence of more than one meiotic process. Lindegren and Lindegren (1953) observed asci containing up to eighteen spores, and they concluded that such asci arise by fusion between cells which occur in rows on the sporulation medium. Finally, Gutz (1967) has shown that, in the conjugation of diploid cells of *Schizosacch. pombe*, the nuclei sometimes fail to fuse, and "twin meiosis" results in the production of eight haploid spores (see Section IV.B.2, p. 346).

C. SPORE FORMATION

Following nuclear division, spore boundaries become marked out round the clear areas enclosing the nuclei. Outer and inner spore walls are then laid down: the outer wall is a dense layer with a rough surface, while the inner wall, although thicker, is less dense and constitutes the new vegetative cell wall (Hashimoto *et al.*, 1958; Mundkur, 1961b; Marquardt, 1963; Hagedorn, 1964). During maturation, there is a marked increase in the volume of the spore cytoplasm.

After the spores are formed, there is a considerable amount of residual cytoplasm (epiplasm) lying between the spores and the ascus wall (Fig. 2). The epiplasm contains mitochondria, lipid and glycogen (Hashimoto

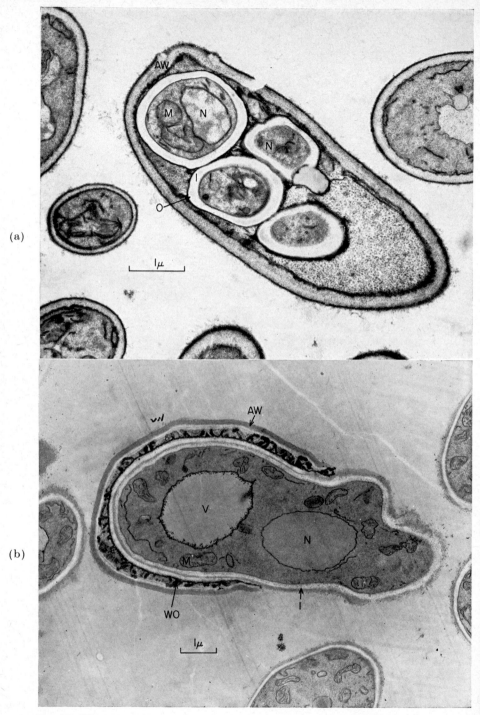

FIG. 2. Electron micrographs of thin sections of ascospores. (a) *Saccharomyces fragilis*, showing an ascus with four ascospores, and (b) *Nadsonia fulvescens*, showing germination of warty spore. AW indicates ascal wall; I, inner spore coat; M, mitochondrion; N, nucleus; O, outer spore coat; V, vacuole; WO, warty outer spore coat. Reproduced by permission of Dr. S. F. Conti.

et al., 1960; Marquardt, 1963), and sometimes one or more nuclei which have failed to become included in spores (Nagel, 1946; Pontefract and Miller, 1962). During spore maturation, the epiplasm largely disappears (Miller *et al.*, 1963). There is evidence that epiplasmic material contributes to the formation of the outer spore wall, and to the building up of food reserves in the spore (Marquardt, 1963). Typical mitochondria have not been observed in spores, but they appear rapidly when spores are transferred to a nutrient medium (Hashimoto *et al.*, 1958, 1960). When the spore germinates, the outer spore wall breaks, usually at more than one point, and the contents emerge enclosed by the inner spore wall (Fig. 2); early in germination, the cytoplasm becomes vacuolated.

III. Factors Controlling Sporulation

A. HISTORICAL RETROSPECT

Following Schwann's (1839) discovery of spore formation in yeasts, it was found that spores are only formed when yeasts are deprived of nutrients (de Seynes, 1868; Reess, 1869, 1870; Hansen, 1902) and this led to the view that starvation alone is the cause of sporulation. However, subsequent investigations showed that this interpretation is incorrect as the presence of nutrients in small amounts is either essential for sporulation or necessary for optimum sporulation (Barker, 1902; Saito, 1916; Stantial, 1935; Elder, 1937; Adams, 1949; Adams and Miller, 1954; Miller and Halpern, 1956; Miller, 1957; Fowell, 1967). The pioneer investigations of Hansen (1883, 1889, 1899, 1900, 1902) showed also that certain conditions, besides a deficiency of nutrients, are necessary for sporulation, namely the use of young, vigorous cells, presence of oxygen and moisture, and a suitable temperature both in the prior cultural (presporulation) phase and the sporulation phase.

Although a knowledge of these conditions proved useful in later investigations, it soon became evident that other factors must be involved in sporulation. Yeasts were found to differ in their ability to sporulate on different media, and no single medium proved suitable for all types of yeasts; in fact, an ideal sporulation medium of this nature has yet to be developed. The classical method of obtaining spores, by inoculation of a Plaster-of-Paris (gypsum) block moistened with water (Engel, 1872), was used by Hansen and others, and found to support good sporulation in a wide variety of yeasts (Phaff and Mrak, 1949). The efficiency of a gypsum block as a sporulation medium is probably due, in large measure, to suppression of budding by a saturated solution of calcium sulphate as budding and spore formation are mutually exclusive processes (Hansen, 1899, 1902). Some yeasts were found to sporulate better on other media than gypsum blocks. Reess (1869, 1870) discovered that spores are formed on vegetable wedges such as carrot and potato, and carrot was later found to be superior to gypsum blocks for inducing sporulation in *Hansenula* (Bedford, 1941) and other yeasts such as strains of *Schizosaccharomyces* and *Debaryomyces* (Phaff and Mrak, 1949). The realization that different media suit different yeasts led taxonomists to devise media of a very diverse nature in their efforts to induce yeasts to sporulate. The approach to this problem was largely empirical, but some workers attempted to incorporate a sufficient number of materials to induce sporulation in a wide range of yeasts, e.g. a combination of vegetable extracts (Mrak *et al.*, 1942; Wickerham *et al.*, 1946). Further details of sporulation media are given in Section V.A, p. 358; although some of the older types are of historic interest only, others may still be used to considerable advantage.

In recent years, a more scientific approach has been made to the study of sporulation, and attempts have been made to establish the precise chemical and physical conditions which govern the process. The factors at present known to influence (or suspected of influencing) sporulation are discussed below. It has become clear that sporulation is governed by events in the cultural or presporulation phase, and conditions in both the presporulation and sporulation phases must be suitably adjusted to permit of optimum sporulation. Recent work (Fowell, 1967) has shown that, by adjustment of the cell population density, it is possible to promote good sporulation on a variety of media and in a wide range of yeasts.

B. GENETICAL CONTROL

A difficulty which has always confronted the taxonomist, and is apt to be a nuisance to the industrial breeder, is the tendency for yeasts to lose their sporulating ability, especially when maintained on artificial media (Lindner, 1896); some yeasts have been known to lose this ability soon after isolation, while others can be maintained for years and still be induced to form spores. Stelling-Dekker (1931) found, on re-examination of the Dutch collection of yeasts, that representatives of nearly every genus had completely lost the power to form spores. Hansen (1889) isolated cells from a culture of *Saccharomycodes ludwigii*, and obtained three strains with different growth characteristics which, when inoculated on gypsum blocks, produced respectively numerous spores, only a few, and no spores at all. The presence of sporogenous and asporogenous cell types with different growth characteristics has been observed in other yeasts, e.g. *Schizosaccharomyces octosporus* (Beijerinck, 1897). Hansen (1889, 1900) also showed that species of *Saccharomyces*, cultured for long periods near the maximum temperature for budding, gradually lost the capacity to form spores. Some of the asporogenous types obtained in this way showed a more rapid rate of growth than the original sporogenous type, and the conclusion that decline in sporulating ability is due to mutation and may be accelerated by the more vigorous growth of sporeless mutants is supported by the observations of Olenov (1936) on *Sacch. acidificiens* (*Zygosacch. mandshuricus*) and Leupold (1950) on *Schizosacch. pombe*.

It is often recommended that yeasts should be tested for sporulating ability as soon as possible after their isolation, but this is not always feasible, especially in breeding projects. Methods for reviving sporulating ability are dealt with in Section V.A, p. 362. If yeasts maintained over a long period are used to study the conditions affecting sporulation, it is possible that changes in these conditions may operate by altering the proportion of cells with different powers of sporulation. It is strongly recommended, therefore, that single-cell isolates of yeasts, selected for high sporulating ability, should be employed in sporulation studies (Fowell and Moorse, 1960).

C. PHYSIOLOGY OF SPORULATION

1. *Presporulation Phase*

Certain types of media used for obtaining spores, e.g. V-8 agar (Wickerham *et al.*, 1946), and Gorodkowa agar (Gorodkowa, 1908), contain sugars and other nutrients in amounts sufficient to support appreciable growth by budding, and sporulation is accordingly delayed, sometimes for days, until the nutrients are nearing exhaustion. They are quite unsuitable for investigations of the conditions governing sporulation, since the presporulation and sporulation phases are entirely different metabolic processes. Distinct separation of these phases is necessary, not only for rapid induction of spore formation, but also for valid assessment of the factors that operate in the two phases.

a. Chemical composition of growth medium. As Hansen (1883) first observed, good sporulation depends on cells being in an active physiological state, and this implies that they must be cultivated on a medium which ensures that they are well nourished. Lindegren and Lindegren (1944a) devised a special presporulation agar medium, consisting of a mixture of fruit and vegetable juices, for the invigoration of strains of *Sacch. cerevisiae* prior to sporulation on gypsum slants, and media such as V-8 agar are based on the same conception. Sando (1956), however, found that preculture of baker's yeast in tomato juice or Wickerham's (1951) medium ensured the same high degree of sporulation as Lindegren's presporulation medium. The use of 16% (w/v) malt-extract wort (Fowell and Moorse, 1960) for the preculture of a baker's yeast was found to give much better sporulation on sodium acetate-agar than that obtained with tomato juice supplemented with glucose, yeast extract and peptone.

The effect of different chemical constituents in a presporulation medium on subsequent sporulation necessitates the use of a chemically-defined synthetic medium. The presence of an assimilable sugar such as glucose in the growth medium is necessary for good sporulation (Zetlin, 1914; Wagner, 1928; Adams, 1949. Wagner's (1928) observation that sporulation may or may not be affected by sugar concentration is invalidated by faulty technique, as pointed out by Phaff and Mrak (1949). We have shown (Fowell and Moorse, 1960) that optimum sporulation of a baker's yeast is obtained following preculture in Lodder and Kreger–van Rij's (1952) synthetic medium containing 4–8% (w/v) glucose, the degree of sporulation being similar to that after growth in 16% (w/v) malt-extract wort which contains 8–9% (w/v) sugar (mainly maltose; Fig. 3).

Good sporulation also depends on the presence of nitrogenous compounds in the growth medium (Zetlin, 1914; Ochmann, 1932; Kleyn, 1954; Tremaine and Miller, 1956). Ochmann (1932), unfortunately,

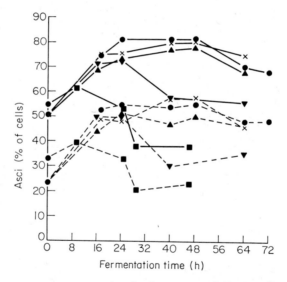

Fig. 3. Effect of sugar concentration in the presporulation medium (Lodder and Kreger-van Rij, 1952) on the sporulation of a strain of *Saccharomyces cerevisiae* (baker's yeast, strain 740). The yeast was grown in 16% (w/v) malt-extract wort at 30°. Continuous lines indicate the total number of asci; broken lines the number of four-spored asci. Glucose concentrations (%, w/v): 0·5 (squares), 4·0 (crosses), 8·0 (erect triangles), 16·0 (inverted triangles). Circles indicate data obtained using malt wort containing 8–9% (w/v) sugar. Reproduced from the *Journal of Applied Bacteriology* with permission.

used a basal medium lacking vitamins which markedly influence utilization of nitrogenous compounds, and it is therefore difficult to draw
definite conclusions from his results. Tremaine and Miller (1956) found
that, in general, nitrogenous compounds capable of supporting good
growth are the most effective for subsequent sporulation.

Stantial (1935) first demonstrated the importance of vitamins for
sporulation of *Sacch. cerevisiae*. Subsequent workers (Lindegren and
Hamilton, 1944; Lindegren and Lindegren, 1944a) found that the
presence of yeast extract or autolysate, rich in B vitamins, in the growth
medium promoted sporulation. Tremaine and Miller (1954) showed that
maximum sporulation, and the production of large numbers of four-
spored asci, depends on the presence of a full complement of B vitamins
in the presporulation medium; under these conditions no advantage
accrues from addition of vitamins to the sporulation medium, with the
exception of pantothenate which, it was claimed, gives higher yields of
spores. But Fowell (1967) found that sporulation of yeasts on sodium
acetate-agar is unaffected by inclusion of pantothenate. Addition of
nitrogenous compounds or vitamins to a sporulation medium only
partially overcomes the adverse effect of their deficiency in the presporulation medium (Tremaine and Miller, 1954, 1956).

b. *Liquid versus solid media.* Adams (1949) observed no difference in
sporulation of a baker's yeast following growth in grape juice and on

TABLE I. *Influence of Oxygen and Carbon Dioxide in the Presporulation
Phase on Sporulation of* Saccharomyces cerevisiae (*strain* 740)

Gas Mixture*			Asci (% of cells)	
N_2	CO_2	O_2	Total	Four-Spored
400	0	0	13·0	5·3
380	20	0	38·6	23·2
380	0	20	43·8	24·8
325†	0	75	44·0	28·6
0	400	0	53·4	29·0
0	380	20	52·6	27·4
0	275	125	66·8	38·0
Control (unsparged)			81·6	54·0

Gases were bubbled through medium for 1 h before inoculation, and for
an ensuing period of 40 h at 20°. The sporulation medium was sodium
acetate-agar.
  * Volumes (ml) used to sparge synthetic presporulation medium.
Rate of flow: 400 ml/100 ml medium/min.
  † This mixture is equivalent to normal air.
Reproduced with permission from Fowell and Moorse (1960).
11*

grape juice-agar, and Kleyn (1954) reported similarly for a hybrid strain of *Sacch. cerevisiae* although some agar media gave better sporulation than the corresponding liquid media. Tremaine and Miller (1954), however, found that liquid-grown cells of *Sacch. cerevisiae* sporulate much better than those grown on solid media, and this has been confirmed by Fowell and Moorse (1960). Aeration of presporulation liquid cultures was also found to be detrimental to sporulation of *Sacch. cerevisiae* (Pazonyi, 1954; Fowell, 1955) which indicates that yeasts grown under anaerobic conditions show the highest degree of sporulation.

Experiments (Table I) in which carbon dioxide, oxygen and nitrogen, singly and in various combinations, were bubbled through a synthetic growth medium inoculated with a baker's yeast revealed that complete anaerobiosis (obtained with a flow of nitrogen gas) is markedly detrimental to sporulation (Fowell and Moorse, 1960). A mixture of oxygen and nitrogen in the same proportion as in air increases sporulation considerably, but not to the same extent as pure carbon dioxide; further improvement results from admixture of oxygen which is probably necessary to ensure adequate growth, but it is evident that carbon dioxide must play an active role in conditioning cells for sporulation.

c. Age of culture. Hansen (1883) found that cells of *Saccharomyces pastorianus* grown on beer wort at 26–27° for two days sporulated poorly compared with those grown for only one day, an effect which he attributed to the accumulation of alcohol. Some yeasts, however, can be cultivated for periods up to four days without showing any decline in sporulating ability, probably due to their slow rate of development

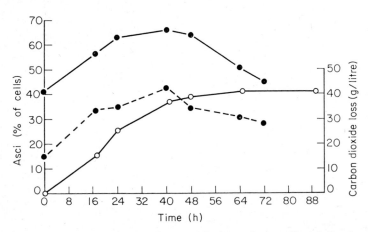

Fig. 4. Effect of culture age on the sporulation of a strain of *Saccharomyces cerevisiae* (baker's yeast, strain 568). Full circles indicate number of asci; continuous lines, total asci; broken lines, four-spored asci; open circles, carbon dioxide loss. Reproduced from the *Journal of Applied Bacteriology* with permission.

(Barker, 1902; Miller *et al.*, 1955; Fowell and Moorse, 1960). A baker's yeast propagated on 16% (w/v) malt-extract wort at 30° required an incubation period of 24–48 h for maximum sporulation on sodium acetate-agar (Fig. 4); after 48 h, sporulating ability declined steadily (Fowell and Moorse, 1960). This decline, however, was not due to an increase in the number of dead cells, as the cells remained fully viable for at least 90 h. Miller *et al.* (1955) also observed that decrease in sporulation of baker's yeast is not due to the death of cells, and it is evident that the effect is caused by physiological changes in the cells.

Croes (1967b), who used an aerated liquid presporulation medium, found that the highest numbers of asci were produced by a strain of *Sacch. cerevisiae* after only 18 h at 30°; but a few hours later sporulating ability declined very rapidly (compare Fowell and Moorse, 1960), presumably due to removal of carbon dioxide which, as indicated above, is essential for good sporulation.

d. Temperature. The effect of presporulation temperature has not, in my opinion, been properly investigated as account should be taken of its effect on growth rate and ageing of different yeasts. Growth of a baker's yeast at a high temperature (about 37°) has been reported to decrease sporulating ability (Adams and Miller, 1954) though apparently necessary for optimum sporulation of saké yeasts (Tsukahara and Yamada, 1953). These observations, however, were based on cultures incubated at different temperatures for standard periods, so that differences in sporulation could be attributed to variations in the physiological condition of the cells. Stantial's (1935) claim that, in order to obtain the highest number of spores, yeast should be grown at the same temperature as that employed for sporulation is open to similar criticism.

A baker's yeast incubated on 16% (w/v) malt-extract wort for 43 h at 30° was found to give the same degree of sporulation on sodium acetate-agar as a culture incubated for 72 h at 21° (Fowell and Moorse, 1960). Thus it would appear that the temperature of presporulation growth can be varied, at least to some extent, without any effect *per se* on sporulation.

e. Growth regulators. Auxin and gibberellic acid have been reported to stimulate sporulation of *Sacch. cerevisiae* var. *ellipsoideus* (Kamisaka *et al.*, 1967a, b). Preculture in the presence of indole acetic acid (IAA) increased sporulation. Gibberellic acid (GA) had little effect but, in combination with IAA, the stimulatory action of the latter compound was enhanced. These conclusions appear to be based on insufficient evidence (see Section III.C.2 (j), p. 333, for discussion of effects reported for addition of these substances to sporulation media). Yanagishima and Shimoda (1967) have recently found that auxin induces the formation of stable variants differing in sporulating ability.

2. Sporulation Phase

As already mentioned, Hansen (1883–1902) stressed the importance of oxygen, temperature, water and limitation of nutrients for yeast sporulation, as well as presporulation conditioning which has been discussed above. Several other factors are now known to influence sporulation, including cell population density, the importance of which has only been realized quite recently.

a. *Water.* The necessity of water for sporulation is clearly evident from the observation that desiccated cells are incapable of forming spores (Barker, 1902). Several workers, however, have claimed that essentially dry conditions are necessary for good sporulation (Nägeli and Loew, 1880; Stovall and Bulotz, 1932a, b; Todd and Hermann, 1936; Windisch, 1938; Beavens, 1940; Kleyn, 1954). Hansen (1902) found that sporulation occurred less readily on gypsum blocks placed in dry dishes covered only with filter paper. Also, a high degree of sporulation in *Sacch. cerevisiae* can be obtained in liquid media provided they are adequately aerated, either by the shaking of shallow layers (Stantial, 1928, 1935; Kirsop, 1954; Miller *et al.*, 1959), or by bubbling through air (Pazonyi, 1954), a method which has been used to show that a baker's yeast can sporulate as well in liquid acetate medium as on an agar medium of the same composition (Fowell, 1967).

Adams (1949) obtained higher yields of asci on solid media than in shallow liquid media, but this was probably due to slight anaerobiosis in the liquid media which were not shaken. A similar condition could account for decreased sporulation on excessively moist solid media (Morris, 1958). Another possibility is that dry conditions may favour sporulation indirectly by retarding growth, particularly on a medium, such as that of Kleyn (1954), containing nutrients in amounts sufficient to support some growth under moist conditions.

b. *Oxygen and carbon dioxide.* The necessity of air for sporulation, first noted by Reess (1870), accounts for the development of spores in the superficial layers of yeast colonies (Saito, 1916; Lindegren and Hamilton, 1944) and stored yeast blocks (Maneval, 1924). Hansen (1902) showed that oxygen is essential as no spores are formed in its absence: this has been confirmed by Miller *et al.* (1957b). The amount of oxygen in air appears to be near the optimum for sporulation, but high concentrations are inhibitory (Adams and Miller, 1954).

The importance of an oxidative mechanism for sporulation is evident from other observations. Bautz (1955) using the Nadi reagent found a strong increase of oxidative activity during spore formation; and this process is depressed by respiratory inhibitors, e.g. dilute solutions of dyes such as Janus Green (McClary and Nulty, 1957; Lindegren, 1958), cyanide and fluoroacetate (Miller and Halpern, 1956). It is also interesting to note that respiratory-deficient yeast mutants are unable to sporulate (Ephrussi and Hottinguer, 1951).

Carbon dioxide in high concentrations strongly inhibits spore formation (Barker, 1902; Bright *et al.*, 1949; Adams and Miller, 1954). However, the view that aeration influences sporulation, not only by exposing yeast cells to oxygen, but also by removing carbon dioxide produced by the cells (Barker, 1902), probably needs qualification. Low concentrations of carbon dioxide (about 5%, v/v) do not appear to be inhibitory (Adams and Miller, 1954), and more recently Bettelheim and Gay (1963) have shown that removal of carbon dioxide from air in contact with acetate-agar results in decreased sporulation, although this is affected to a lesser extent if glyoxylate is present. There is, therefore, a requirement for carbon dioxide both in the presporulation and the sporulation phase.

c. *Temperature.* Hansen (1883, 1902) determined the maximum, minimum and optimum temperatures for sporulation in different strains of *Saccharomyces*, the optimum temperature being that at which spores appear in the shortest time. He found that the temperature range for sporulation, 3°–37°, was narrower than that for growth, namely 0·5°–47°. The optimum temperature, which was fairly close to the maximum, was in the range 25°–30°. Stantial (1935) found that the highest degree of sporulation was obtained at 25°, and the experience of other workers such as Adams and Miller (1954) indicates that this temperature is more or less optimum for most yeasts.

d. *Carbon substrates.* As mentioned previously, limitation of sugars and other nutrients is regarded as one of the main conditions for sporulation. Restriction of budding is the essential prerequisite for spore formation, as this process can be induced by growth inhibitors such as calcium sulphate and ethanol, even in the presence of an ample supply of nutrients (Hansen, 1899, 1902).

Barker (1902) showed that the presence of nutrients was not necessary for the sporulation of yeasts such as *Saccharomyces*, and the yield of spores on gypsum blocks, immersed in water, was inversely proportional to the amount of nutrients added, as beer wort, to the water. On the other hand, in *Hansenula anomala*, *Zygosaccharomyces* sp. and *Schizosaccharomyces octosporus*, it appeared that no spores were formed in the absence of nutrients, and the yield of spores was directly proportional to the amount of added nutrients. Little attention has been devoted to the factors controlling sporulation in yeasts such as *Schizosaccharomyces*, but it should be noted that ample supply of nutrients is necessary for conjugation (Brock, 1961) which immediately precedes sporulation; nevertheless, spores are formed in abundance at a time when there is still an ample supply of nutrients (Saito, 1916).

Later investigations have shown that small amounts of utilizable carbon compounds such as sugars and salts of organic acids are essential for optimum sporulation in *Saccharomyces* (Saito, 1916; Stantial, 1928, 1935; Elder, 1937; Adams, 1949; Fowell, 1952; Kirsop, 1954; Adams and Miller, 1954; Miller and Halpern, 1956; Miller, 1957; Fowell, 1967). Saito (1916) carried out an exhaustive investigation of the effect of different carbohydrates, polyhydric alcohols, nitrogen compounds, organic acids and salts on the sporulation of yeasts. It was found that, for *Saccharomyces*, the greater the nutritive value of compounds for growth the smaller was the concentration required for abundant sporulation. In addition, some compounds were more favourable for sporulation than others; best results were obtained with glucose, fructose, mannose, sucrose, glycerol and mannitol. In *Schiz. octosporus*, quite low concentrations of sugars and polyhydric alcohols were sufficient to induce sporulation, but in agreement with Barker (1902) the degree of sporulation was found to rise with an increase in concentration of these compounds.

The stimulatory effect on sporulation of sugars in small concentrations, observed by Saito (1916), has been confirmed by other workers. In general, maximum sporulation is given by concentrations in the range 0·01–0·1% (w/v). There is some lack of agreement about the relative efficiency of different sugars, which may be attributed to the use of different yeasts or to faults in technique, as sporulation is critically dependent both on sugar concentration in the optimum range (Kirsop, 1954; Miller *et al.*, 1955; Miller, 1957) and on cell population density (Kirsop, 1954; Fowell, 1967). There is, however, considerable evidence that fructose and mannose stimulate sporulation to a greater extent than glucose (Stantial, 1935; Kirsop, 1954; Miller *et al.*, 1955; Miller, 1957). The reason for inhibition of sporulation by sugars in concentrations greater than 0·1% (w/v) is not clear. In the absence of a nitrogen source, growth would not be expected to occur; and it has been suggested that some intermediates of glucose metabolism, normally used up in growth, accumulate in amounts large enough to suppress sporulation (Miller and Halpern, 1956). However, limited cell multiplication appears to occur, even with small amounts of sugars (Saito, 1916; Kirsop, 1954; Miller and Halpern, 1956), presumably at the expense of nitrogen reserves in the cells, and a depletion of these reserves could account for decreased sporulation.

Acetate has proved to be particularly efficient for inducing a high degree of sporulation. Saito (1916) found that 0·1–1·0% (w/v) sodium acetate gave abundant sporulation with a strain of *Sacch. carlsbergensis*.

Stantial (1935) claimed that most yeasts sporulated best in the presence of acetate supplemented with glucose or mannose; later, Adams (1949) advocated the use of 0·14% (w/v) sodium acetate, 0·04% (w/v) glucose and 2% (w/v) agar. However, addition of low concentrations of glucose to acetate-containing media has since been found detrimental to sporulation (Sando, 1956; Fowell, 1967). Of various organic salts tested, acetate has generally proved to be the most stimulatory to sporulation (Adams, 1949; Kleyn, 1954) and this accords with our own observations. Miller (1957) found that pyruvate gave higher yields of asci than acetate in a baker's yeast; but of four yeasts studied by Fowell (1967), three sporulated better with acetate.

Much less growth occurs in the presence of acetate than with sugar (Kirsop, 1954; Miller and Halpern, 1956), and this no doubt accounts for the ability of acetate to support good sporulation at higher concentrations than is possible with sugars. A baker's yeast has been reported to show maximum sporulation over a wide range of sodium acetate concentration (0·04–1·5% (w/v); Fowell, 1967); however, the upper limit for most yeasts appears to be 1·0% (w/v). Acetate has proved to be more effective than other carbon substrates for promoting the production of large numbers of four-spored asci (Fowell, 1952, 1967). It is not surprising, therefore, that acetate-containing media are now used extensively in genetical studies and for industrial breeding projects.

As mentioned by Miller and Hoffmann-Ostenhof (1964), some carbon substrates cannot be used for growth and, while some of these have little or no effect on sporulation, others such as dihydroxyacetone (Miller, 1957) promote sporulation to a considerable degree. The stimulatory properties of these compounds is probably due, not only to their inability to support growth, but also to their direct intervention in the metabolism of yeast cells.

Yeasts resemble bacteria in showing early commitment to sporulation (Miller, 1959). Saito (1916) first observed that cells of *Sacch. carlsbergensis* after being exposed to conditions favourable for sporulation for a certain time, cannot be converted back into an active budding state by addition of fresh nutrients. The minimum time necessary for exposure of cells of *Saccharomyces* sp. to sodium acetate or glucose to ensure maximum sporulation is about 10 h (Kirsop, 1954; Ganesan *et al.*, 1958; Fowell, 1967). At a high pH value, however, a period of 5 h is sufficient on sodium acetate-agar but, for maximum production of four-spored asci, about 12 h exposure is necessary (Fowell, 1967). Other factors such as pH value (Sando, 1960b) also influence sporulation but only in the early stages.

The importance of cell population density and pH value in determining the degree of sporulation in the presence of different concentrations of carbon substrates is discussed below. It is sufficient to point out here that optimum sporulation in the presence of a carbon substrate, such as glucose, depends on suitable adjustment of substrate concentration, pH value and cell population density, and all these factors must be considered in assessing the value of carbon substrates as promoters of sporulation.

e. Nitrogenous compounds and vitamins. Media containing nitrogenous compounds but lacking sugar, such as nutrient agar, which support only limited growth are effective as sporulation media (Guilliermond, 1920b; Fowell, 1955, 1967). Amino acids, such as glycine and lysine, which singly cannot support growth of several yeasts (mainly *Saccharomyces* spp.) may stimulate sporulation more effectively than acetate (Sando, 1960a).

Most of the investigations carried out with nitrogenous compounds, however, have been concerned with their effect on sporulation in the presence of carbon substrates such as glucose, acetate and dihydroxyacetone. The available evidence (Saito, 1916; Kleyn, 1954; Sando, 1956, 1959; Miller, 1963a,b; Miller and Hoffmann-Ostenhof, 1964; Fowell, 1967) indicates that nitrogenous compounds, such as ammonium salts and amino acids, in combination with carbon substrates such as glucose or acetate, usually depress and delay sporulation of *Saccharomyces* and *Schizosaccharomyces* spp. On the other hand, it seems equally clear that some nitrogenous compounds may have a special role in sporulation, and in certain circumstances appear capable of stimulating sporulation in the presence of a carbon substrate. Tremaine and Miller (1956) grew baker's yeasts in a nitrogen-deficient medium, and then compared the effect on sporulation of different nitrogenous compounds added to an acetate-containing medium. No spores were formed in the absence of a nitrogenous compound. Addition of ammonium salts, amino acids, yeast extract and peptone stimulated sporulation to different extents, but in no case was the degree of sporulation as good as that obtained by transferring well-nourished cells to acetate-containing medium. Compounds capable of supporting good growth would be expected to stimulate sporulation under the above conditions by revitalizing the nitrogen-starved cells and conditioning them for sporulation. Some compounds had this effect but not all. Thus ammonium sulphate and aspartic acid, which support growth, had little effect on sporulation. Conversely, glycine which does not support growth actually stimulated sporulation. These results indicate some difference in the nutritional roles played by nitrogenous compounds during growth and sporulation. Further evidence for this view is provided by Sando (1960a) and Miller (1963a) who have shown that there is no correlation between the effects of amino acids on sporulation and their value as nitrogen sources for growth (Table II).

Both Saito (1916) and Adams (1949) claimed that small amounts of peptone are beneficial to sporulation of *Saccharomyces*, but not according to Saito (1916) of *Schizosaccharomyces* strains. Later, Sando (1956,

TABLE II. *Effect of Nitrogenous Compounds on Yeast Growth and Sporulation*

Compound*	Value† as nitrogen source for growth	Sporulation index‡		
		With acetate	With dihydroxy-acetone	With acetate + dihydroxy-acetone
Series A				
NaNO$_2$	0	0	0	
Hydroxylamine	0	0	1·5	
NaNO$_3$	6	107	97	
NH$_4$NO$_3$	109	63	115	
(NH$_4$)$_2$SO$_4$	100	23	118	
NH$_4$Cl	117	54	124	
(NH$_4$)$_2$HPO$_4$	174	26	128	
(NH$_4$)$_2$CO$_3$	94	77	136	
Series B				
L-Cysteine	0	0	0	0
DL-Methionine	3	1	3	2
Casein hydrolysate	130	2	17	2
DL-Isoleucine	38	18	21	24
L-Glutamic acid	143	13	26	35
L-Aspartic acid	154	19	35	42
Ethyl urethane	1	9	43	10
Glutathione, reduced	0	5	47	14
L-Leucine	38	21	58	24
L-Proline	141	39	59	38
Carbamoyl phosphate§	7	0·5	63	5
L-Arginine	94	17	71	52
DL-Threonine	11	32	74	37
DL-Serine	108	27	82	35
Urea	127	92	86	93
Glutathione, oxidized	20	84	91	62
L-Histidine	1	42	108	40
L-Valine	42	62	108	84
L-Tyrosine	80	27	116	35
(NH$_4$)$_2$CO$_3$	96	35	116	44
(NH$_4$)$_2$SO$_4$	100	8	118	9
Glycine	3	33	139	57
L-Phenylalanine	90	51	140	53
L-Alanine	135	45	155	57
L-Lysine	1	23	184	27

TABLE II—*continued*

Compound*	Value† as nitrogen source for growth	Sporulation index‡		
		With acetate	With dihydroxy-acetone	With acetate + dihydroxy-acetone
Series C				
β-Alanine	0	55	60	
Ureidosuccinic acid	1·5	68	79	
Adenine‖	0·2	72	83	
Uracil	76	86	94	
Guanine‖	0·3	71	98	
Uridine	0	100	100	
Thymine	0·2	89	103	
Cytosine‖	90	56	113	

* All compounds in $0·01M$ concentration, except ethyl urethane $(0·1M)$, oxidized glutathione $(0·005M)$, and casein hydrolysate $(0·1\%)$. Substances that did not dissolve completely are indicated thus ‖. The casein hydrolysate was "Vitamin Free Casein Hydrolysate" from Nutritional Biochemicals Corp.

† Amount of growth in medium containing $0·02M$-acetate as sole carbon source related to arbitrary value of 100 for control with $(NH_4)_2SO_4$ as nitrogen source.

‡ Degree of sporulation related to an arbitrary value of 100 for sporulation in the controls containing only carbon source ($0·02$ M-acetate, $0·02$ M-dihydroxy-acetone, or $0·02$ M-acetate + $0·02$ M-dihydroxyacetone). The average percentage of cells that sporulated in the controls was acetate, 57·7; dihydroxyacetone, 36·8; acetate + dihydroxyacetone, 62·9.

§ Added as the dilithium salt. Lithium chloride $(0·02M)$ did not affect growth or sporulation.

Reproduced with permission from Miller (1963a).

1959) showed that peptone or casein hydrolysate considerably improves sporulation of most saké yeasts on an acetate-containing medium, in spite of stimulating vegetative growth, although sporulation of baker's yeasts is unaffected by very low concentrations and indeed higher concentrations depress spore formation. It was subsequently found (Sando, 1960a) that certain amino acids, particularly glycine and lysine, stimulate sporulation of saké yeasts in the same striking manner; *Hansenula saturnus* has also been reported to sporulate well in a medium containing acetate and lysine (Okuda, 1961). According to Okuda (1959), ammonium sulphate and casamino acids exhibit a maximum stimulatory effect on sporulation of a saké yeast at pH 7; a similar degree of sporulation can, however, be obtained in their absence at higher pH values.

Miller (1963a) found that, for a strain of *Sacch. cerevisiae*, amino acids and most other nitrogenous compounds inhibit sporulation on media containing glucose, acetate, pyruvate, lactate or ethanol. The inability of glycine to promote sporulation in the presence of acetate is contrary to Kleyn's (1954) findings, and is confirmed by recent observations (Fowell, 1967). It is interesting to note that fewer nitrogenous compounds inhibit sporulation in the presence of dihydroxyacetone, and in fact sporulation is increased by alanine, phenylalanine, glycine and, in particular, lysine (Table II); accordingly, Miller has recommended a sporulation medium consisting of 2% (w/v) agar supplemented with 0·02 M-dihydroxyacetone and 0·01 M-lysine.

Addition of yeast extract to acetate-containing medium has been reported to improve sporulation of *Saccharomyces* spp. (McClary *et al.*, 1959). This effect could be attributed to its nitrogenous components, salts or vitamins. Tremaine and Miller (1954) have shown that, provided a yeast is grown on a presporulation medium containing adequate amounts of B vitamins, there is no need for their inclusion in a sporulation medium. They claimed, however, that addition of pantothenate to acetate-containing medium improves sporulation of *Sacch. cerevisiae*, although Sando (1956) and Fowell (1967) have been unable to confirm this. In the experiments of Kamisaka *et al.* (1967a), acetate concentration and cell population density were not properly adjusted, and their claim that vitamins are necessary for optimum sporulation of *Sacch. cerevisiae* var. *ellipsoideus* cannot be substantiated. My own experiments (Fowell, 1967) have shown that addition of yeast extract to acetate-containing media usually stimulates growth of yeasts, even at pH 8·4, and causes a decrease in sporulation. Stimulation of sporulation in a brewer's yeast on unbuffered sodium acetate-agar by yeast extract was found to be due to its potassium content.

f. Ions. Saito (1916) tested the effect of 0·1–0·25% (w/v) of various inorganic salts on the sporulation of *Sacch. acidifaciens* (*Zygosacch. mandshuricus*) and *Sacch. carlsbergensis* (*Sacch. mandshuricus*) on gypsum blocks saturated with 1% (w/v) glucose solutions. Ammonium salts completely inhibited sporulation under these conditions, although in the absence of a carbon substrate low concentrations permitted spore formation. Potassium dihydrogen phosphate was particularly effective in promoting sporulation of *Sacch. acidifaciens*, less favourable results being obtained with KCl, K_2HPO_4 and Na_2HPO_4. The effect of KH_2PO_4 may be attributed to a decrease in the high pH value of a moist gypsum block to a level more favourable to sporulation. It is interesting to note that potassium salts were found to be more stimulatory to sporulation of *Sacch. acidifaciens* than sodium salts. No apparent differences were observed, however, in the effects of sodium and potassium salts on the sporulation of *Sacch. carlsbergensis*. High concentrations of salts proved detrimental to the sporulation of both yeasts, either because of osmotic effects or due to the toxicity of the metal cations.

According to Saito (1916), ions do not appear to be necessary for the sporulation of *Schizosacch. octosporus*; sporulation in 5% (w/v) glucose solutions was unaffected by KH_2PO_4 and delayed by K_2HPO_4, while ammonium salts proved inhibitory both in the presence and absence of glucose.

Adams (1949), investigating the effect of different cations, in association with acetate, on the sporulation of a baker's yeast, found Na^+ superior to K^+, Ca^{2+} and Mg^{2+}. Sporulation was depressed, however, to varying extents by all the cations, except K^+, at the highest concentrations tested. Other workers (Stantial, 1935; Kleyn, 1954; Kirsop, 1957) observed no difference in sporulation of strains of *Sacch. cerevisiae*

TABLE III. *Effect on Sporulation of Yeasts of Addition of Potassium Ions to Sodium Acetate-Agar*

Yeast strain	Asci (% of cells)			
	Sodium acetate-agar		Sodium acetate-agar + 1% (w/v) KCl	
	Total	Four-Spored	Total	Four-Spored
Saccharomyces cerevisiae 74 (brewer's yeast)	7·3	0	28·2†	0·9
Saccharomyces carlsbergensis S8	9·2	4·2	32·9†	16·3
Saccharomyces cerevisiae 131	11·0	0·6	56·3†	19·6
Saccharomyces cerevisiae 369	19·2	6·3	44·4	21·6
Saccharomyces cerevisiae 222	22·1	3·8	58·0	29·4
Saccharomyces cerevisiae 341 (brewer's yeast)	34·4	0·2	63·4	1·6
Saccharomyces cerevisiae 746	43·5	28·3	58·4	45·6
Saccharomyces cerevisiae 757	50·2	3·5	73·0	12·9
Saccharomyces carlsbergensis 57	62·6	43·1	70·0	36·4
Saccharomyces cerevisiae 752 (brewer's yeast)	66·9	16·0	77·9	7·2
Saccharomyces diastaticus 501	69·2	37·3	85·9	54·1
Saccharomyces cerevisiae 739	71·8	38·7	82·0	52·4
Saccharomyces cerevisiae 740	80·6	52·8	84·3	53·4

All figures are based on optimal cell population density values for sporulation, each figure being the mean of three or more tests. The standard error for five yeasts each tested 10 times on acetate-containing media averaged 2·2 so that a difference of more than 5% sporulation may be regarded as significant.

† In presence of 0·23% (w/v) KCl.

Reproduced with permission from Fowell (1967).

with sodium and potassium acetates. McClary *et al.* (1959) reported that K^+ was superior to Na^+ in stimulating sporulation of several strains of *Saccharomyces*, and no advantage was obtained by including other cations such as Mg^{2+} with K^+. These results were confirmed by Fowell (1967) who showed that, in strains of *Sacch. cerevisiae*, *Sacch. carlsbergensis* and *Sacch. diastaticus*, the effect of K^+, added as 1% (w/v) KCl to 0·5% (w/v) sodium acetate + 1·5% (w/v) agar, was most marked for yeasts that sporulated poorly on this medium (Table III); some of the yeasts in this category produced more four-spored asci in the presence of K^+ than total asci on sodium acetate-agar alone. Yeasts that sporulated well on this medium benefited only slightly from the addition of K^+. In agreement with Adams (1949), it was found that K^+ showed no evidence of toxicity at the highest concentrations tested, but Na^+ and in particular Ca^{2+} depressed sporulation at much lower concentrations. Addition of K^+ as KCl to sodium pyruvate-agar or glucose-agar also improved sporulation, but not to such a marked extent as with sodium acetate-agar, and the effect was most pronounced with populations below the optimum density.

Little attention has been devoted to the effect of anions, but the available evidence indicates that Cl^- has no effect on sporulation, and PO_4^{3-} only a slight adverse effect at low pH values (Fowell, 1967). Uptake of PO_4^{3-} by yeast cells declines to practically zero at pH 8–10 (Goodman and Rothstein, 1957) and sporulation, therefore, should not be affected in this range.

g. Cell population density. The discovery that sporulation may be markedly influenced by cell population density is comparatively recent, and it is probable that many of the earlier studies of sporulation are unreliable because this effect was not appreciated. Stantial (1928) found that there was an optimum cell population density for sporulation of *Sacch. cerevisiae*, fewer spores being formed above and below this level, and this was confirmed by Elder (1937) and Adams (1949).

Kirsop (1954) studied sporulation of *Saccharomyces* spp. in glucose and acetate solutions, buffered at the optimum pH values, and aerated by shaking. The optimum cell population density was found to vary according to the concentration of the substrate, the degree of sporulation being determined by the amount of substrate per cell. When yeast was suspended in an aerated buffer solution at the optimum pH value for sporulation, it was possible by adjusting the cell population to a low level to obtain sporulation similar to that given by a dense cell suspension at the optimum concentration of carbon substrate. In short, when yeast cells were sufficiently well spaced, a carbon substrate was unnecessary for optimum sporulation. These findings, however, need qualification in the light of recent research (see page 330).

Miller *et al.* (1955) found that, in a baker's yeast, acetate concentration had little effect on the optimum cell population density in unshaken, unbuffered liquid media. A density of 4×10^6 cells/ml resulted in the maximum numbers of asci possible under these conditions with acetate concentrations in the range $0.01 - 1.0\%$ (w/v). However, more asci were produced with 0.1% (w/v) acetate than other concentrations. Sando (1960a) obtained comparable results with a saké yeast using unshaken, buffered acetate medium; but, in shaken cultures with the same medium, the optimum cell population density increased with acetate concentration in approximately a $3:1$ ratio, and the degree of sporulation also increased slightly with acetate concentration (Fig. 5).

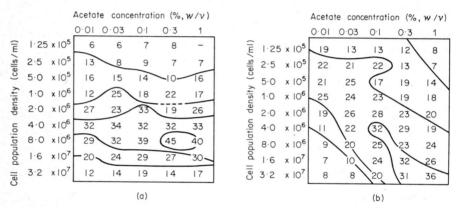

(a)   (b)

FIG. 5. Effect of cell population density and acetate concentration on sporulation of a saké yeast (*Saccharomyces cerevisiae*) in (a) static and (b) shaken liquid media composed of sodium acetate, salts and vitamins. Contour lines show that in unshaken cultures cell population density has the greater influence on sporulation, whereas in shaken cultures acetate concentration as well as cell population density determine the degree of sporulation. Reproduced with permission from Sando (1960a).

Investigations of the effect of cell population density on sporulation of various yeasts (strains of *Sacch. cerevisiae*, *Sacch. carlsbergensis* and *Sacch. diastaticus*) on agar media (Fowell, 1967) have shown that the degree of sporulation on acetate-agar is maximum over a wide range of cell densities; this applies to both unbuffered and buffered acetate (Fig. 6). Critical control of cell population density is therefore not necessary with acetate agar, but in general it should not be less than 0.5×10^6 cells/ml agar. Previous observations (Kirsop, 1954; Sando, 1960a) that sporulation in aerated liquid acetate media declines fairly rapidly above and below a narrow range of cell population densities was confirmed. In contrast to acetate-agar, sporulation on glucose-agar, pyruvate-agar and unsupplemented agar was found to be critically dependent, with

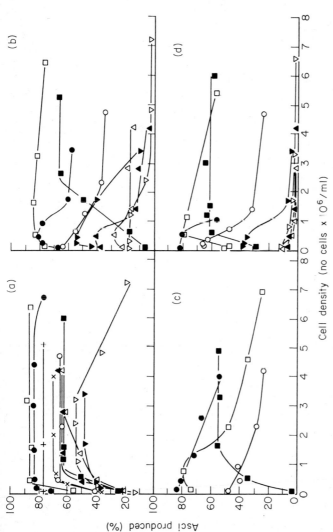

Fig. 6. Effect of cell population density on the sporulation of strains of *Saccharomyces* on agar media containing (a) acetate, (b) glucose, (c) pyruvate, (d) none of these compounds (unsupplemented agar). Media were adjusted to pH 8·4 with potassium phosphate buffer (pH 7·0 with strains S 8 and 131). Closed circles indicate behaviour of strain 740; open circles, strain 57; closed squares, strain 131; open squares, strain 501; closed erect triangles, strain 222; open erect triangles, strain 746; closed inverted triangles, strain S 8; open inverted triangles, strain 369; plus signs, strain 752; crosses, strain 757. Reproduced from the *Journal of Applied Bacteriology* with permission.

certain exceptions, on cell density (Fig. 6). To account for these observations, it is suggested that there is a threshold concentration of carbon
dioxide, slightly in excess of the amount in air, necessary for optimum
sporulation on acetate-agar. If the cells are close enough, carbon dioxide
given off by the cells will be sufficient to build up to the threshold value.
As carbon dioxide is continually diffusing away into the air above the
agar, it is unlikely that its concentration will build up to an inhibitory
level even when the cells are very close together. In liquid media,
however, aeration becomes progressively ineffective in removing carbon
dioxide as the cell density is raised, and in consequence sporulation
declines. To account for the decline in sporulation on other agar media,
it is suggested that sporulation on these is much more sensitive to a
slight build up of carbon dioxide.

Miller and Hoffmann-Ostenhof (1964) concluded that no exogenous
nutrient is required for spore formation in *Saccharomyces* spp. since this
can be observed in distilled water, in buffer solutions, on gypsum blocks
and plain agar. Substances which stimulate sporulation, according to
these authors, do not serve primarily as carbon sources, but bring about
the metabolic shift necessary for the onset of sporulation. In their view,
withdrawal of essential nutrients for growth lowers the concentration of
repressors for the synthesis of enzymes involved in spore formation
below the threshold value. The main effect of carbon substrates is to
overcome the adverse effect on sporulation caused by the cells being too
close together; their use makes it possible to obtain large numbers of
spores by heavy inoculation of sporulation media. My experiments
(Fowell, 1967) have shown that, though some strains of *Sacch. cerevisiae*
and *Sacch. carlsbergensis* can be induced to sporulate as well on unsupplemented agar as on agar supplemented with carbon substrates by
suitable adjustment of cell population density under optimum pH conditions, the sporulation of many yeasts is definitely promoted by carbon
substrates, especially acetate (Fig. 6). Thus a baker's yeast (strain 369)
was only capable of 10·8% sporulation on unsupplemented agar, compared with 54·6% on acetate-agar, both figures being based on optimum
conditions of cell population density, pH value and K$^+$ concentration.

h. Hydrogen ion concentration. Phaff and Mrak (1949) remarked on the
contradictory reports of earlier workers pertaining to the effect of pH
value on sporulation. Welten (1914) and Baltatu (1939) claimed that
sporulation only occurred in acid or neutral media, while Wagner (1928)
found that pH value had little effect, over a wide range, on sporulation
of yeasts inoculated on beechwood blocks. The general consensus of
opinion among more recent workers, however, is that an alkaline
medium is necessary for optimum sporulation (Oehlkers, 1923; Kufferath,
1929, 1930; Stantial, 1935; Hartelius and Ditlevsen, 1953; Kleyn, 1954;

Fowell, 1955; Miller and Hoffmann-Ostenhof, 1964); but Kirsop (1954) reported an optimum pH value for a strain of *Sacch. cerevisiae* of 6·4–6·8 in acetate-containing solutions, and according to Miller and Hoffmann-Ostenhof (1964) pyruvate solutions are unique in permitting optimum sporulation of *Saccharomyces* spp. at pH 4. It was shown by Sando (1960b) that sporulation in a liquid acetate-containing medium is only sensitive to pH value in the early stages (up to 26 h).

Recent evidence suggests that the optimum pH value depends on the composition of the sporulation medium (Okuda, 1959; Miller and Hoffmann-Ostenhof, 1964) and on buffer components (Kleyn, 1954); this has been confirmed by Fowell (1967) who also established that the optimum may be influenced by cell population density. For acetate-agar media, it was found that many strains of *Saccharomyces* have an optimum pH range of 8·4–10·6 which is independent of cell population density, but two strains sporulated best at pH 7·2–8·4 and 6·2–7·6 respectively (Fig. 7a, b). On 0·01% (w/v) glucose-agar, the optimum pH value for sporulation of a baker's yeast shifted from 8·4 down to 7·0 when the cell population density was raised by a factor of 20, i.e. from $0·07 \times 10^6$ to $1·3 \times 10^6$ cells/ml agar, but considerably more four-spored asci were produced at the lower cell density (Fig. 7c). Raising the glucose concentration from 0·01 to 0·1% (w/v) resulted in the optimum pH value being shifted from 7·6 to 10·6 for a cell density intermediate between the above two values (Fig. 7d). Contrary to the observations of Miller and Hoffmann-Ostenhof (1964), sporulation of a baker's yeast on sodium pyruvate-agar showed a steady rise from pH 4·8 to 8·4; a similar increase was recorded for unsupplemented agar.

Experiments have shown that Sörensen phosphate buffers should not be used for testing the effect of pH value on sporulation (Fowell, 1967). With these buffers, the pH value is raised by decreasing the proportion of KH_2PO_4 and increasing that of Na_2HPO_4, i.e. it involves a rise in Na^+ concentration and a fall in K^+ concentration. Above a pH of 7·6, Na^+ is present in amounts sufficient to depress sporulation and there is insufficient K^+ to stimulate sporulation; as a result, such buffers give a spurious optimum pH value of 7·6–7·8. In the above investigations, the effect of pH value was based on buffer mixtures composed of different proportions of 0·133 M-KH_2PO_4, 0·067 M-K_2HPO_4 and 0·133 M-KOH (containing a constant amount of K^+ and lacking Na^+) for a pH range of 5·6–12·2; phthalate buffers were used for lower pH values.

Buffering of media such as glucose- and pyruvate-agar at pH values appropriate for different yeast strains is essential for optimum sporulation, but experiments have shown that buffering of acetate-containing media is not usually necessary (Fowell, 1967). The pH value of acetate-agar increases after inoculation to a value directly related to the cell

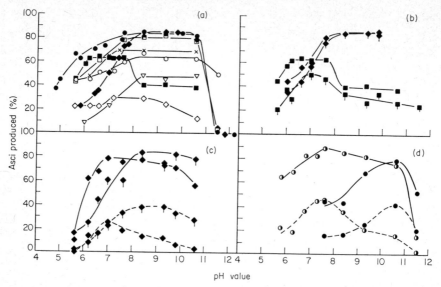

FIG. 7. Effect of pH value on sporulation of strains of *Saccharomyces*. (a) pH curves for various strains on 0·5% (w/v) sodium acetate-agar at optimum cell population densities. Closed circles indicate behaviour of strain 740; open circles, strain 57; closed squares, strain 131; open squares, strain 501; closed diamonds, strain 739; open diamonds, strain 74; crosses, strain 757; open inverted triangles, strain 369. (b) Effect of cell population density on pH curves for strains 131 and 739 on 0·5% (w/v) sodium acetate-agar. Squares indicate behaviour of strain 131, 1·6 × 10⁶ cells/ml; squares with tag, strain 131, 0·6 × 10⁶ cells/ml (suboptimum); diamonds, strain 739, 3·3 × 10⁶ cells/ml; diamonds with tag, strain 739, 0·6 × 10⁶ cells/ml. (c) Effect of cell population density on pH curves for strain 739 on 0·01% (w/v) glucose-agar. Diamonds, 1·3 × 10⁶ cells/ml; diamonds with tag, 0·07 × 10⁶ cells/ml; continuous lines, total numbers of asci; broken lines, numbers of four-spored asci. (d) Effect of glucose concentration on pH curves for strain 740 on glucose-agar. Closed circles, 0·1% (w/v); half-open circles, 0·01% (w/v); continuous lines, total numbers of asci; broken lines, numbers of four-spored asci. Buffers used were phthalate (pH 4·8–5·6) and potassium phosphate (pH 5·6–12·2). Reproduced from the *Journal of Applied Bacteriology* with permission.

population density. Provided the cell population density is above a certain value, the pH value shifts to a level more or less optimum for sporulation. These findings accord with Duggan (1964) who showed that, when yeast is shaken up in air with potassium acetate solution, K^+ is absorbed and H^+ excreted in exchange; however, the pH value of the external medium rises because acetate ions are taken up and replaced by HCO_3^- and CO_3^{2-} in concentrations greater than the amount of H^+ excreted. We have found that the pH value of 0·3% (w/v) pyruvate-agar after inoculation falls from 8·2 to 7·0 irrespective of cell population density (presumably due to greater excretion of H^+ than of

HCO_3^- and CO_3^{2-}); this fall in pH value accounts for sporulation being less on pyruvate-agar than on acetate-agar in the absence of buffer.

i. Visible light. Many years ago, Purvis and Warwick (1908) reported that red rays accelerate spore formation in *Saccharomyces* spp., while rays of shorter wavelength have a retarding effect. More recently, Oppenoorth (1956, 1957) has claimed that visible light can affect the sporulation of bottom-fermentation strains of *Sacch. cerevisiae* in different ways: some strains require more or less continuous exposure to light for maximum sporulation (long-day type), others sporulate best when daily exposure to light is restricted (short-day type), and finally some strains are unaffected by light. Experiments with light of different wavelengths indicated that only the red region is effective. Unfortunately, Oppenoorth obtained somewhat variable results for sporulation under apparently similar conditions. His method of inducing sporulation involved the transfer of 1–3 ml of yeast suspension in wort to a moistened gypsum block. Two variables were introduced by this procedure, since the cell density was not controlled and different amounts of sugars and other nutrients would be carried over with the wort on different occasions. Our experiments with one of Oppenoorth's yeasts, *Sacch. cerevisiae* S 8 (a short-day type), which was incubated in continuous darkness, have shown that its sporulation is critically dependent on cell population density, especially in the presence of small amounts of glucose. The effect of light on sporulation is therefore in urgent need of re-investigation under carefully controlled conditions.

j. Growth regulators. Reference has already been made to the claims of Japanese workers that sporulation of *Sacch. cerevisiae* var. *ellipsoideus* is affected by addition of auxins and gibberellic acid to the presporulation medium. In their main paper (Kamisaka *et al.*, 1967a), gibberellic acid was reputed to increase sporulation when added to acetate-containing medium; auxins such as indole acetic acid decreased sporulation, but this effect was reversed by gibberellic acid. Sporulation, however, was only affected when these compounds were added prior to meiotic division. In my opinion, their results are too variable and inconsistent to permit of definite conclusions. In a recent experiment, we found that sporulation of a strain of *Sacch. cerevisiae* on potassium acetate-agar was depressed by gibberellic acid even in very low concentrations.

Kamisaka *et al.* (1967b, c) also claim to have shown that pretreatment of yeast with gibberellic acid was necessary to obtain an RNA preparation active in promoting sporulation. Inhibitors of protein synthesis, such as ethionine, eliminated gibberellic acid-induced sporulation less specifically than inhibitors of RNA synthesis such as 2-thiouracil. It was concluded that the effect of gibberellic acid is primarily associated with RNA synthesis responsible for regulation of sporulation.

D. BIOCHEMISTRY OF SPORULATION

1. *Differences in Chemical Composition of Vegetative Cells and Spores*

Miller and Hoffmann-Ostenhof (1964) have summarized our knowledge of the biochemistry of sporulation in *Saccharomyces* spp. The spore wall possesses a superficial lipid layer which is absent from the wall of a vegetative cell (Langeron and Luteraan, 1947; Emeis, 1958a; Schumacher, 1926); it stains deeply with Sudan Black B (J. J. Miller and O. Eelnurme, unpublished data). Further differences in the composition of the spore wall are indicated by its greater electron density and its power to stain deeply with gallocyanin (Mundkur, 1961a,b); this author attributes depth of staining to the presence of free amino groups in the large acetylglucosamine fraction of the chitin-like mucopolysaccharides in the spore wall. The resistance of spore walls to snail enzyme extract, which readily digests asci walls (Johnston and Mortimer, 1959), is further evidence of differences in their composition.

Spores of yeast are richer in carbohydrate content than vegetative cells. They contain considerably more glucan and mannan, and more trehalose, but the glycogen content is about the same (Pazonyi and Márkus, 1955). There is also evidence that spores contain more lipid than vegetative cells (Hashimoto *et al.*, 1958, 1959; Pontefract and Miller, 1962). By contrast, both the protein and amino acid content is diminished, although the proline content is considerably greater (Ramirez and Miller, 1963 and unpublished observations). Spores also contain less RNA (Croes, 1967a) and a lower free ribonucleotide content (Abdel-Wahab *et al.*, 1961). Differences in chemical composition due to sporulation on different media are indicated by staining tests (Miller, 1963a; Bettelheim and Gay, 1963).

2. *Correlation of Cytological and Biochemical Changes during Sporulation*

After cells are transferred to an acetate-containing medium, there is an immediate rise in dry weight, followed 4 h later by a rapid increase in DNA content which precedes meiosis; after 10 h, increasing numbers of nuclei are to be found in metaphase I and mature asci, containing spores, first appear after 14 h (Croes, 1967a, b; Fig. 8). Accumulation of glycogen and lipid commences in the early stages (Pontefract and Miller, 1962; Miller *et al.*, 1963). Croes (1967a) has shown that there is a steady decline in RNA content, especially in the later stages (Fig. 9); protein content increases by about 10% in the first 2 h and this is followed by a decline which becomes more marked as spores attain maturity; the amino-acid pool becomes rapidly depleted after only 4 h, the fall in alanine content being especially pronounced. These observations agree with those of Ramirez and Miller (1963, 1964) who found that, except for proline which accumulates in the cells, amino acids are released into the medium. Lewis and Phaff (1964) observed a similar decline in the size of the amino-acid pool in cells exposed to glucose.

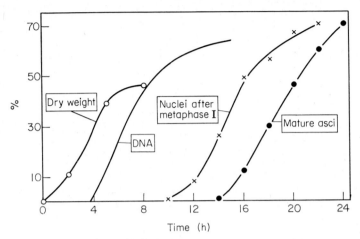

FIG. 8. Biochemical and cytological characteristics of meiosis and ascus formation in *Saccharomyces cerevisiae* (CBS 5525). Changes in dry weight and DNA content are indicated as percentages of increase above the zero-time values. Numbers of nuclei after metaphase I and of mature asci are as percentages from observations on 500 randomly selected cells. Reproduced with permission from Croes (1967a).

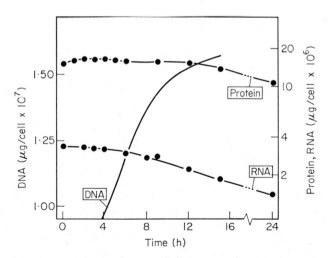

FIG. 9. Average protein, DNA and RNA contents of *Saccharomyces cerevisiae* (CBS 5525) during meiosis and ascus formation. Reproduced with permission from Croes (1967a).

The decrease in protein is considered by Croes (1967a) to be due to breakdown of cytoplasm left out of the spores.

3. Metabolic Changes

a. Changes in respiratory activity. Early experiments (Miller *et al.*, 1957a, b; 1959; Scheiber *et al.*, 1957) showed that, during the first 4 h in 0·3% (w/v) acetate solution, the respiratory quotient falls from about 2·6 to below 1·5, oxygen uptake being decreased by one-half and carbon dioxide evolution by two-thirds. The respiratory quotient is depressed by other carbon substrates such as glucose, but dihydroxyacetone has little effect. Later it was found (Pontefract and Miller, 1962) that sporulating cells do not appear to differ from vegetative cells in their ability to respire any of the carbon substrates so far tested, and mature spores respire at a rate only slightly below that of vegetative cells. Cells adapt very rapidly to acetate, and oxygen uptake reaches its maximum after only 2 h; then it declines steadily (Croes, 1967a).

Although sporulating cells do not differ from vegetative cells in ability to utilize external substrates, their endogenous respiratory capacity, when transferred from acetate-containing medium to buffer, becomes greater after 5–10 h than in vegetative cells. Glycogen appears to be the main substrate for endogenous respiration (Eaton, 1960), but nitrogenous compounds also appear to be involved (Gronlund and Campbell, 1961).

b. Changes in enzyme activity. Miller and Hoffmann-Ostenhof (1964) suggested that, prior to sporulation, some regulatory mechanism comes into action which brings about a marked change in the activity of various enzymes. Enzymes in the glycolytic pathway may be inhibited; others involved in biosynthetic processes are probably enhanced in activity. Also differences in the composition of the spore wall, compared with the vegetative cell wall, support the view that synthesis of certain enzymes may be induced by the conditions which lead to sporulation.

Glucose which is known to inhibit the formation of induced enzymes (Magasanik, 1961) also inhibits spore formation. Further evidence for induction of enzymes in the early stages of sporulation is the observation that ethionine and 2,6-diaminopurine inhibit sporulation if added within 12 h of transfer to the sporulation medium (Miller, 1959; Sando, 1960c). Oxygen, which promotes induced enzyme formation in yeast (Ingram, 1955), is evidently required for sporulation. Respiratory inhibitors, such as 2,4-dinitrophenol, which suppress induced enzyme synthesis also inhibit sporulation (Miller and Halpern, 1956). There is also evidence that polyphosphate, which disappears during sporulation (Mundkur, 1961a, b), is used up in the formation of induced enzymes.

Croes (1967b) used ethionine to test the theory that enzymes are formed during the period of high metabolic activity preceding sporulation. Methionine, which neutralizes the inhibitory effect of ethionine on sporulation, was used to arrest its action at different stages. Meiosis was blocked by ethionine at a stage before anaphase I; protein synthesis was not affected by ethionine, but DNA synthesis was completely blocked. The strongest effect of the inhibitor on DNA synthesis occurred in the first 2 h; when added later to the sporulation medium, it had no effect so that it probably interferes with enzyme formation and not with DNA synthesis. A second site of ethionine action was found after meiosis I, and maturation of spores was prevented. Both the above processes were therefore blocked by the inhibitor a considerable time in advance.

c. *Metabolic factors leading to meiosis.* Croes (1967a, b) carried out experiments which indicated that initiation of meiosis, though dependent on exposure of cells to a sporulation medium, is started by events which occur in the last stages of growth in the presporulation medium. Addition of ethionine to the growth medium 2 h before transfer to the sporulation medium resulted in almost complete suppression of sporulation, although growth was only slightly depressed. Methionine added to the sporulation medium only reversed this effect to a slight degree, thus excluding the possibility that inhibition was due to an ethionine pool accumulated during growth.

In his second paper, Croes (1967b) put forward evidence which suggests that preparation for meiosis in the late presporulation phase involves development of ability of the yeast to metabolize acetate, and the establishment and maintenance of high concentrations of protein and RNA sufficient to support a high degree of biosynthetic activity. Cells grown in a glucose-containing medium are unable to utilize acetate in the log phase (Eaton and Klein, 1954), but acquire the ability to do so in the later stages when ethanol produced earlier by fermentation is oxidized via acetate (Kornberg and Elsden, 1961). This probably accounts for the attainment of maximum sporulating ability by cells after 18 h of presporulation growth, which coincides with maximum ethanol concentration in the culture and a change from fermentation to respiration under the conditions employed by Croes, i.e. the use of an aerated presporulation phase. In support of this theory, Croes found that addition of 1% (v/v) ethanol to the growth medium, inoculated with cells previously grown for 12 h, resulted in a significant increase in sporulation. This, of course, explains why anaerobic growth is more favourable to sporulation than aerobic growth.

Under the conditions used by Croes, sporulating ability was found to decline when cells in the aerated presporulation medium were incubated beyond 18 h. After this period, the protein, DNA and RNA contents of the cells declined. There must therefore, have been a build up of these constituents in the log phase of growth in preparation for enzyme synthesis, which necessitates a large turnover of RNA and protein (Miller, 1959). Under anaerobic conditions of growth, maximum sporulating ability is maintained for a much longer period, up to 48 h or even longer (Fowell and Moorse, 1960). The rapid decline in sporulating ability observed by Croes was probably due to the effect of aeration in decreasing carbon dioxide concentration below the level necessary for maximal sporulation. Carbon dioxide fixation is an essential reaction for yeast metabolism as it supplies carbon for certain synthetic processes (Stoppani *et al.*, 1958).

Croes found that when yeast is transferred to acetate-containing medium there is an immediate increase in metabolism which results in enhanced protein synthesis. The intracellular pool of most amino acids becomes depleted and it is inferred that acetate is unable to provide the

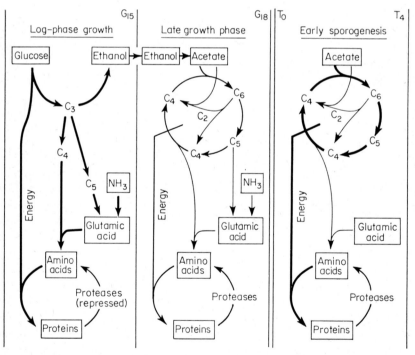

Fig. 10. Carbohydrate and protein metabolism during log-phase growth, late growth phase and early sporogenesis in *Saccharomyces cerevisiae* (CBS 5525). The metabolic changes at G_{15} (presporulation growth for 15 h) and T_0 (zero time in the sporulation medium) are considered to induce meiosis. Lines of different thickness mark the relative reaction rates of the processes shown. Reproduced by permission of Croes (1967b).

carbon moieties of these amino acids during this period. Since production of amino acids from acetate depends on the operation of the glyoxylate cycle (Kornberg, 1965), their absence indicates that this cycle is insufficient. The idea that this insufficiency is the essential element in a trigger mechanism is supported by Bettelheim and Gay (1963), who found that inhibition of growth (involving mitosis) on transfer to a sporulation medium can be overcome by the addition of a small amount of glyoxylate. The triggering action of acetate appears to be a speeding up of energy-supply processes resulting in insufficiency of the glyoxylate cycle.

The changes in metabolism which are considered, according to Croes, to induce meiosis are shown in Fig. 10. Metabolic changes occurring after 15 h presporulation growth, and at the time of transfer to the sporulation medium, are regarded as the points of induction.

IV. Sexual Reproduction

A. SEXUALITY IN YEASTS

Fusion of ascal spores in pairs was first observed by Hansen (1891) in *Saccharomycodes ludwigii* and other yeasts, but it was not considered a sexual process. By the application of staining, Hoffmeister (1900) established the existence of sexuality in yeasts by showing that fusion of cells, observed by Schiönning (1895) to precede ascus formation in *Schizosaccharomyces octosporus*, is accompanied by nuclear fusion. This was confirmed by Guilliermond (1901), and also by Barker (1901) for *Zygosaccharomyces*, and later for other yeasts by several investigators. Certain species of *Pichia* and *Zygosaccharomyces* were found to exhibit heterogamy (Guilliermond, 1911a, 1918, 1920a), the contents of a small cell (male gamete) passing into a larger cell (female gamete) which then functioned as an ascus. Other early records of heterogamy include conjugation between a mother cell and its bud characteristic of *Debaryomyces* (Guilliermond, 1911b) and *Nadsonia* (Nadson and Konokotina, 1911).

In spite of these observations, the belief prevailed for some years that sexual reproduction is unusual in yeasts, and fusion of spores was considered an abnormal process replacing cell fusion leading to ascus formation (Guilliermond, 1920b); and the view that asci are commonly formed by a parthenogenetic process was maintained until comparatively recently (Guilliermond, 1928; Stelling-Dekker, 1931).

The existence of a definite life cycle in yeasts, involving a sexual process linking a haplophase with a diplophase, was finally established by Kruis and Šatava (1918), Šatava (1918, 1934) and Winge (1935) for several species of *Saccharomyces*. In a typical strain of *Sacch. cerevisiae*,

12

the vegetative cells are diploid and, on transfer to a sporulation medium, become transformed into asci each of which contains four or fewer haploid spores. Under suitable nutrient conditions, spores may fuse in pairs forming zygotes in which, following nuclear fusion, the diploid condition is restored. Alternatively, each spore develops separately by budding off haploid cells and fusion between these may occur at a later period. Cell fusion, therefore, represents a postponement of spore fusion. Lindegren and Lindegren (1943a) were the first to show that yeast spores are of two mating types, fusion only taking place between spores or haploid cells of opposite mating type.

The apparent parthenogenetic formation of asci, without preceding fusion of spores or cells, observed by earlier workers, was found to be the result of a modified sexual process by Winge and Laustsen (1937). They found that a strain of *Sacch. cerevisiae* var. *ellipsoideus* segregated two types of spores, usually in equal numbers. One type produced small, round cells with "short-shoot" growth, their haploid nature being confirmed by their inability to sporulate. The other type of spore germinated to form elongated diploid cells ("long-shoot" growth) capable of forming spores on a gypsum block. All spores when first formed were uninucleate, but diploidization of many spores was indicated by their binucleate condition at the onset of germination, i.e. diploidy was due to nuclear fusion in certain spores.

In a series of classical papers, Winge and Laustsen (1937, 1938, 1939a,b), Winge (1944), and Winge and Roberts (1948) provided genetical proof of the occurrence of meiosis in yeasts by showing that colonial characteristics, cell shape and fermentative ability are under the control of genes which segregate at spore formation. These investigations showed that most species of *Saccharomyces* and *Saccharomycodes* are heterozygous, and the demonstration that yeasts formed by crossing spores are also heterozygous (Winge and Laustsen, 1939a) showed that fusion of spores is accompanied by nuclear fusion resulting in diploidy. The genetical evidence for the nuclear events associated with sexuality in yeasts is supported by the cytological investigations of the Winge school as well as those of later workers (Section II.B, p. 309).

Yeasts tend to lose their powers of sexual reproduction when cultivated under artificial conditions. This decline is manifested in a gradual decrease and final loss of sporulating ability (Section III.B, p. 312), spore viability and mating activity.

Spore viability is an expression which refers to the proportion of spores capable of germinating to produce definite cultures. In our experience, spore viability is a relatively fixed character in many yeasts, but as Winge and Laustsen (1939a, 1940) first showed it can be altered to a considerable degree by hybridization and inbreeding. In extreme cases, none of the spores of a hybrid or inbred yeast may be viable; on the other hand, spore viability of such yeasts is sometimes better than that of the parents. Spore viability is clearly under genetical control, and a diminution in this property can be attributed to aneuploidy, triploidy, translocations, lethal mutations and genic interaction (Mortimer and Hawthorne, 1966). Winge and Laustsen (1940) found that inbred strains formed by fusion of haploid cells budded off from a single spore showed 30% spore viability; but, when formed by nuclear fusion in the spore at the onset of germination, less than 1% of the spores were viable. The effect of this difference in mode of diploidization was accounted for in terms of cytoplasmic factors. Low spore viability is found in certain culture yeasts, especially strains of brewer's yeasts; under more natural conditions, such yeasts would probably be eliminated in competition with strains of high spore viability.

Decline in mating activity, resulting in eventual sterility, is common among haploid cultures of *Saccharomyces* maintained by subculture over long periods, and is attributed to mutation of the mating-type genes or of modifying genes controlling conjugation (Lindegren and Lindegren, 1944b). It should be noted, however, that not all haploid cultures show this behaviour; some haploids in our collection have retained their fertility unimpaired over a period of 14 years. Loss of mating activity in haploids is commonly attributed to diploidization, as evidenced by changes in the appearance of the cells which are larger and not in groups, and by their ability to sporulate. However, some sterile haploids retain a typical haploid form indefinitely (small, usually round cells in groups) and are incapable of sporulation. In my opinion, the dicaryotic condition observed in the zygotes of certain yeasts (Guilliermond, 1910; Renaud, 1937, 1938, 1946; Fowell, 1951; Lindegren and Lindegren, 1954) is an indication of weakening in mating activity. Although nuclear fusion is delayed in this condition, occurring in the cells budded off from zygotes, we have found that in some yeasts zygote formation is often not followed by nuclear fusion and the zygotes bud off haploid instead of diploid cells.

B. THE MATING SYSTEM

1. *Heterothallism*

Winge (1935) maintained that yeasts are homothallic as fusion can occur between haploid cells from a common spore, and some spores develop directly, without zygote formation, into diploid cultures capable of sporulation. Heterothallism in yeasts was first reported by Lindegren and Lindegren (1943a) for *Saccharomyces cerevisiae*. As already mentioned, they found that the spores are of two mating types: these are produced in equal numbers, indicating the operation of a single pair of alleles. The apparently contradictory findings of the above workers were resolved by Ahmad (1953) who showed that, although many strains of *Saccharomyces* are heterothallic, some are homothallic.

The nomenclature used for mating types and their alleles in yeasts has undergone several changes. They were originally designated as + and − by Lindegren and Lindegren (1943b, c, d), and these symbols have been used by Ahmad (1948), Fowell (1951) and others. Catcheside (1951) preferred the use of mt$^+$ and mt$^-$ for the mating types. Leupold (1950) used + and − to denote the mating types in *Schizosaccharomyces pombe*, and h$^+$ and h$^-$ for the corresponding alleles which, as discussed below, form part of a multiple allelic system in this yeast. As the mating system in yeasts is different in mechanism from that of other organisms, Lindegren (1945) introduced the symbols *a* and *α* (in place of + and −), and these are now widely used by yeast geneticists in spite of the confusion likely to be caused by their similarity. Winge and Roberts (1949) have used A and a to denote the mating types and their alleles; this is unfortunate as the use of a capital for one allele implies dominance whereas the two alleles are to be regarded as incompatibility factors (Ahmad, 1953; Raper, 1954).

Deviations from the expected 2:2 ascal segregation of mating-type alleles can be attributed to a variety of causes including polyploidy (Roman *et al.*, 1951; Lindegren and Lindegren, 1951; Pomper *et al.*, 1954) and additional nuclear divisions in the ascus, sometimes followed by the formation of binucleate spores (Winge and Roberts, 1950, 1954). The extra nuclear divisions, which may result in asci containing up to eight spores in certain strains of *Sacch. cerevisiae*, were considered to be mitotic by Winge and Roberts, but cytological (Fowell, 1956a), as well as genetical (Pomper *et al.*, 1954), evidence suggests that supernumerary nuclei in certain asci result from meiosis in cells which have become binucleate in preparation for budding (which, however, is arrested on the sporulation medium). According to Lindegren and Lindegren (1953), asci with more than four spores may arise through the fusion of adjacent sporulating cells.

The mating system appears to be subject to frequent mutation. As mentioned above, Lindegren and Lindegren (1944b) observed that haploid cultures, when subcultured over a long period, tend to become sterile and they ascribed this to mutation of one or other of the mating-type alleles to an inert form or mutation of modifying genes resulting in inhibition of mating activity. The concept of modifying genes arose from the observation that, while some haploid strains are completely sterile, others are excessively fertile and capable of conjugation with both *a* and *α* mating types (Lindegren and Lindegren, 1943c). The *a* types of one yeast were found incapable of mating with *α* types of the same yeast although they readily mated with α types of another yeast (Lindegren, 1945). We have observed the reverse condition: mating occurred between a and α types in a strain of *Sacch. diastaticus* and similarly in a strain of *Sacch. cerevisiae*, but opposite mating types of the two yeasts were completely incompatible, i.e. no hybrids could be obtained. Finally, it is worth noting that, where hybridization is rendered difficult because of weak mating strength among haploid cultures, this property may be considerably enhanced by inbreeding (Fowell, 1956b).

Lindegren and Lindegren (1943b, c, d) distinguished between legitimate diploids, formed by fusion of a and α haploid cells, and illegitimate diploids which apparently arose from fusion of cells of the same mating type, especially in old haploid cultures. Legitimate diploids produced viable four-spored asci whereas illegitimate diploids, which tended to be inferior in growth rate and other properties, formed mostly two-spored asci, the spores of which were seldom viable; the diminished vigour of such diploids can be attributed to homozygosity of deleterious recessive mutations (Catcheside, 1951).

Diploidization of haploid cultures is now known to be due to mutation of one mating-type allele to the other, i.e. $a\rightarrow\alpha$ or $\alpha\rightarrow a$ (Ahmad, 1952, 1953; Roman and Sands, 1953), followed by fusion of cells of opposite mating type. Analysis by Ahmad of the spore cultures from a Danish baker's yeast showed that these could be grouped into three classes: (i) maters, which give a strong mating reaction with tester strains; only a few of these are capable of sporulation; (ii) weak maters, many of which can sporulate; and (iii) non-maters, most of which sporulate. His results (Table IV) reveal an inverse relationship between mating activity and sporulating ability. Most, if not all, non-maters are diploid as they segregate a and α spores. Maters, especially weak maters, are mixtures of cells which give a mating reaction but cannot sporulate and cells which sporulate but lack mating activity, i.e. they are mixtures of haploid and diploid cells, and repeated subculture eventually results in complete replacement of haploid cells by diploid cells.

The assertion that there are marked differences in the mutation rate of the two mating-type alleles in different strains of yeast (Ahmad, 1965) should be treated with reserve, as there is evidence that production of non-mater diploid spore cultures together with active maters is due to triploidy or tetraploidy in certain yeasts (Emeis and Windisch, 1960). The Danish baker's yeast analysed by Ahmad (1953), as well as the strain discussed by Fowell (1956a), are almost certainly triploid, for the proportions of a mater, α mater and non-mater cultures recorded by these workers accord well with the segregation ratios expected on the basis of triploidy. Ahmad's (1965) contention that α mutates to a

at about twice the rate that a mutates to α is probably a misconception as a triploid yeast of the constitution, $aa\alpha$, segregates 2·6–3 times as many a maters as α maters (one-third of the former are aa, i.e. diploids, homozygous for mating type). It would appear that mutation as well as triploidy accounts for the production of diploid spore cultures in the above yeasts; triploidy accounts for about 33% of the spores being diploid non-maters at the time of their formation, while mutation is responsible for diploidization of cultures which are initially haploid. Examination of Ahmad's data (Table IV) reveals that the Danish baker's yeast segregated 25–30% diploid non-mater spores; but, as the result of diploidization in some of the mater cultures, a total of 41·5% of the spore cultures were capable of sporulation.

TABLE IV. *Inverse Correlation Between Presence of Mating Reaction and Sporulation Capacity in Single-Ascospore Cultures of Danish Baker's Yeast*

Generation	No. of single-spore cultures	Maters		Weak maters		Non-maters	
		Sporers	Non-sporers	Sporers	Non-sporers	Sporers	Non-sporers
Parent	9	0	7	0	0	2	0
F_1	61	7	23	6	3	18	4
F_2	53	6	29	4	3	10	1
F_3	36	4	16	2	0	10	4
F_4	24	0	17	0	0	7	0
Total	183	17	92	12	6	47	9
Percentage of sporers		15·6		66·6		83·9	
				41·5			

Reproduced with permission from Ahmad (1953).

The tendency for diploidization in haploid cultures is certainly more marked in some yeasts than in others. In *Saccharomyces diastaticus*, all or nearly all the spores appear to be initially haploid, but up to 50% of the spore cultures have been observed to diploidize soon after spore isolation. A marked tendency for diploidization, as in this yeast, may be attributed to delayed action of a gene D which acts as a mutator of the mating-type locus (Section IV.B2, p. 346); if this view is correct, the production of 50% diploid spore cultures by *Sacch. diastaticus* indicates that this yeast is heterozygous (Dd) for this gene. In other yeasts, e.g. some strains of *Sacch. cerevisiae*, haploid cultures show little signs of diploidization. It would appear difficult to obtain data for rates of mutation of mating-type alleles, since mutation may not be followed by diploidization because of weak mating activity.

The frequency of irregular mating-type segregation in diploid hybrids studied by Takahashi (1964) was 1·8%, which was similar to the frequency of gene conversion at various loci and much higher than would be expected on the basis of known mutation rates for nutritional markers (Magni and von Borstel, 1962). Hawthorne (1963) had earlier shown that a change in mating type could be caused by a deletion affecting part of the mating-type locus, which is considered to be a complex including *a* and *α* cistrons. The high frequency of change in mating type was therefore considered by Takahashi to be due gene conversion, involving recombination of these cistrons, and resulting in the production of non-mater spore cultures.

2. *Homothallism*

There appear to be two types of homothallism in yeasts, namely balanced homothallism characterized by a neutral mating type, and unbalanced (secondary) homothallism which is superimposed on heterothallism (Takahashi, 1961).

Schizosaccharomyces pombe is a haploid yeast represented by homothallic and heterothallic strains (Leupold, 1950). Sexual behaviour is controlled by a series of alleles at the same locus. Two of the alleles, h^+ and h^-, determine the mating types *a* and *α* (+ and −) in heterothallic strains. Two further alleles, h^{90} and h^{40}, are responsible for secondary homothallism; cells with the same alleles can mate with one another and also with h^+ and h^- cells. Clones of constitution h^{90} produce about 90% spores (based on vegetative cells + spores) and h^{40} clones about 40% spores. There is a fifth type of clone which is completely sterile. Mutations can take place from one allele to another. Finally, in crosses between strains with different alleles, e.g. $h^+ \times h^-$, $h^+ \times h^{90}$, the parental types always segregate in a 2:2 ratio, indicating monofactorial inheritance and the existence of several alleles at the same locus.

Deviations from the normal segregations for alleles controlling homothallism and heterothallism can be attributed to polyploidy (Leupold, 1956), and possibly also to the formation of extra nuclei and binucleate spores in certain asci. Recently Gutz (1967), who recognizes the presence of only three mating-type alleles (h^{90}, h^+, h^-) in *Schizosacch. pombe*, has shown that diploid cells (h^{90}/h^{90}) of this species not only sporulate directly (azygotic asci) but frequently conjugate to form zygotes, but their nuclei do not always fuse. Fusion results in a tetraploid nucleus which undergoes meiosis, the zygote becoming transformed into an ascus with four large diploid spores. When nuclear fusion fails to occur, "twin meiosis" results in the formation of asci with eight or, less frequently, seven or six haploid spores. Nuclei homozygous for the mating-type alleles h^+ or h^-, normally incapable of meiosis, can undergo this process in zygotic cytoplasm shared by a diploid nucleus of compatible mating type. Evidently, complementary gene products are formed by h^+/h^+ and h^-/h^- nuclei and these, when combined, enable the nuclei to divide by meiosis regardless of whether or not they are fused. Whether "twin meiosis" or karyogamy occurs seems to depend on the "meiotic" gene products complementing each other to a sufficient degree before the nuclei have fused.

A good example of unbalanced homothallism is *Saccharomyces chevalieri*. Winge and Roberts (1949) showed that this yeast possesses a gene D which is responsible for diploidization of its spore cultures. A hybrid formed by crossing *Sacch. chevalieri* (DD) with a heterothallic strain of *Sacch. cerevisiae* (dd) segregated out D and d alleles independently of the mating-type alleles (a, α); two spores in each ascus produced diploid cultures, and two produced haploids which were a, a; a, α; or α, α. It was later found that some homothallic strains of yeast, including *Sacch. chevalieri*, give a concealed mating reaction, not revealed by ordinary methods, which is only detectable by a minimal plate mating technique (Takahashi *et al.*, 1958; Takahashi and Ikeda, 1959). This is a special application of Pomper and Burkholder's (1949) prototroph technique and involves streaking homothallic diploid cells and heterothallic haploid cells (a or α) with complementary nutritional deficiencies on a minimal medium; the presence of vigorously growing colonies indicates that mating has taken place. It was found that *Sacch. chevalieri* is bisexual, being able to mate with a and α strains of other yeasts. Recent evidence (Oeser, 1962; Hawthorne, 1963) indicates that gene D behaves as a mutator gene that causes the mating-type gene to mutate to its allele during the first few divisions of haploid cells produced by a homothallic spore; cells of opposite mating type then fuse to form a clone of diploid cells.

Additional genes with a complementary action controlling homothallism have been reported by Takahashi (1958). In the tetrads obtained from a hybrid prepared by crossing two haploids, one derived from a cross of a homothallic and a heterothallic strain, every character studied showed a 2:2 segregation but a 1:3 ratio was obtained for homothallism:heterothallism. It is claimed that this type of homothallism is controlled by three complementary genes, HM_1, HM_2 and HM_3, with close linkage between HM_1 and the mating-type locus. None of the HM genes appears to be identical with gene D (Takahashi and Ikeda,

1959). Strains with one HM gene are haploid and give a normal mating reaction, but strains with two or more, e.g. HM_1, HM_2, hm_3, are diploid and either a or α.

Herman and Roman (1966) have described specific genes for homothallism in *Saccharomyces lactis*. They consist of two unlinked factors which exert their effect by changing the mating type in a certain percentage of cells. The two genes are allele-specific, one locus effecting a change from $a \to \alpha$, the other $\alpha \to a$. A similar, though slightly different, mating system has recently been reported for *Sacch. oviformis* (Takano and Oshima, 1967). Of the two unlinked genes, HO_α acts specifically on the α mating-type allele which it converts to a shortly after spore germination; the other gene, HM, has no effect by itself, but in combination with HO_α gene it converts the a type of culture to homothallism.

To summarize, the mating system in yeasts appears to be based on a mating-type locus which is a complex of cistrons for a and α products. Additional genes may be present in some yeasts. These include regulator genes such as D and HO_α at other loci, and a set of three suppressor HM alleles at yet another locus.

C. THE CONJUGATION PROCESS

Our knowledge of the conjugation process is based mainly on *Hansenula wingei*, although some attention has been devoted to other yeasts including *Saccharomyces cerevisiae*. It is convenient to distinguish between sexual agglutination, a preconjugation device for ensuring intimate contact between cells of opposite mating type, and actual conjugation.

1. *Sexual Agglutination*

Wickerham (1956) discovered that, in some strains of *Hansenula wingei*, intense agglutination occurs when haploid cultures of opposite mating type are mixed. Agglutinated cells cohere so firmly that they can be moulded into a rigid ball which will stand on a surface without flattening; microscopic observation shows that the cells are distorted into a polygonal shape (Brock, 1958a). After agglutination, there is extensive zygote formation and the percentage of conjugating cells may reach 80% of the total population within 2 h. The zygotes bud off diploid cells and, as the numbers of these increase, the agglutinability of the cell mass lessens and disappears. In spite of abundant zygote formation, however, the sporulating ability of the diploid cells is considerably less than that of the parent strain.

Other strains of *H. wingei* produce haploid cultures which are non-agglutinative. When first isolated, they fail to mate on mixing; but, if

12*

they are subcultured for some months and then mixed, a few cells conjugate and up to 20% of the resultant diploid cells may form spores.
This increase in mating activity, following prolonged subculture, is in
direct contrast to the behaviour of haploid cultures of *Saccharomyces*
previously discussed (Section IV.B1, p. 341), and may be ascribed to
mutation followed by selection of more active mater cells.

The agglutinative reaction shown by some strains of *H. wingei* is regarded by Wickerham as a device to compensate for weak mating
ability. It brings cells of opposite mating type into intimate contact,
thereby promoting zygote formation. In *Saccharomyces* spp., agglutination is usually weak or absent but it is very marked in *Sacch. kluyveri*
(Wickerham, 1958); this species produces unisexual diploids which
agglutinate and mate profusely with haploids or unisexual diploids of
opposite mating type, polyploids being formed more abundantly than
in any other known yeast. Polyploidy has also been demonstrated in
certain strains of *H. wingei* with strong agglutinative properties
(Herman *et al.*, 1966). Brock (1965a) observed occasional conjugation
of three or four cells in *H. wingei*, and it is probable that agglutination,
by bringing cells into intimate contact, favours multiple fusions which
could result in a polyploid condition (Crandall and Brock, 1968c).
Wickerham and Burton (1962) consider that the principal function of
sexual agglutination is to increase the ploidy level, this increase being
the most important process involved in the evolution of *Hansenula* and
other yeasts; our knowledge of phylogenetic pathways indicates that
agglutinative forms occur in yeasts which are recently evolved.

Two agglutinative haploid strains of *H. wingei* of opposite mating
type, strains 5 and 21 (Wickerham, 1956), have been extensively used
for investigations into the mechanism of sexual agglutination. Brock
(1958a) showed that agglutination occurs strongly only when the two
strains are mixed in fairly high concentrations. Washed cells do not
react unless cations or proteins are present. The thermostability of the
factors involved in agglutination was demonstrated by heating washed
cells of the two strains at 100° for five minutes; they agglutinated when
mixed even in the absence of cations and proteins, the reaction being
stronger than that between unheated cells. As discussed below, it was
later found that heat treatment removes a nonspecific inhibitor of
agglutination from cells of both mating types (Crandall and Brock,
1968a,c).

Evidence for the existence of specific mating components was first
presented by Brock (1958b, 1959a,b) who compared agglutination to an
antibody–antigen reaction. He showed that the ability of strain 21 to
agglutinate with the opposite mating type was destroyed by proteolytic
enzymes such as trypsin; but the opposite type, strain 5, was not affected.

Treatment with 80% (w/v) phenol or hot dilute hydrochloric acid inactivated strain 21 but not strain 5. Sodium periodate destroyed agglutination of strain 5 while affecting strain 21 much less. These observations were erroneously interpreted as evidence that the active mating components of strains 5 and 21 consist of protein and carbohydrate respectively. A similar specificity was found for mating activity in a hybrid strain of *Saccharomyces* (Richards, 1965), but the suggestion that a surface protein in mating type *a* reacts with a specific mannan site on α cells is based on inadequate evidence. The periodate treatment used by Brock (1959a) was carried out at 37°, a temperature at which it is not specific for carbohydrate since it causes splitting of disulphide bonds (Clamp and Hough, 1965). Mercaptoethanol, which has the same effect, also destroys the agglutinability of strain 5 cells in *H. wingei* (Taylor, 1964b). Periodate treatment at 0°, however, is specific for carbohydrate and does not then affect agglutinability of strain 5 (Brock, 1965b).

Crandall and Brock (1968c) found that it is important to distinguish between two possible effects of reagents on the agglutinability of mating types, namely inactivation of the agglutination factor *in situ* or release of the factor from the cell surface in undegraded form. The effects of reagents on agglutinability, therefore, do not provide conclusive evidence of the chemical nature of the factors involved in agglutination. As an example, it was found that destruction of the agglutinability of strain 21 cells in *H. wingei* by trypsin is due to release of the agglutination factor from the cell surface; the factor itself is remarkably resistant to digestion by trypsin.

Taylor (1964b) investigated the effect of several reagents on the agglutination of mating types in *Hansenula wingei* and *Saccharomyces kluyveri*. He found that variations in agglutination intensity, especially dependent on the age of cultures, could be overcome with various solvents, in particular 8 *M*-lithium bromide, which activated mating types of the above yeasts to produce uniformly strong agglutination on mixing, due presumably to extraction of material from cell walls which inhibits agglutination. Cultures of opposite mating type were referred to as types instead of strains by Taylor. Type (strain) 5 of *H. wingei* and type 3 of *Sacch. kluyveri* were inactivated by disulphide-cleaving agents such as mercaptoethanol; type 21 of *H. wingei* was not affected, while type 26 of *Sacch. kluyveri* was inactivated to a slight extent. Comparison of the sensitivities of these types to different treatments indicated the presence of a similar agglutination mechanism in the two yeasts. It was concluded that the mating components in opposite mating types were both proteins, but different in composition, the protein of types 5 and 3 respectively of the above two species being distinguished by the presence

of disulphide bonds. Richards (1965) reported differences in the composition of the protein in the cell walls of the two mating types in a
hybrid strain of *Saccharomyces*; treatment of walls with mercaptoethanol and a bacterial enzyme preparation resulted in release of glucose from *a* type cells only. The mannan fraction in the walls of *a* and *α*
cells also appeared to be different.

Further studies of the agglutination mechanism in *H. wingei* were
promoted by the discovery that the agglutination factor of strain 5,
termed 5-factor by Crandall and Brock (1968a,b,c), could be readily
extracted from the cell surface with snail digestive enzyme (Taylor,
1964a). The specificity of this 5-factor preparation was confirmed by
its ability to adsorb to, and agglutinate, only strain 21 cells, and its
biological activity was destroyed by mercaptoethanol. The 5-factor was
separated by specific adsorption on strain 21 cells and then elution with
0·04% (w/v) sodium carbonate.

The investigations of Taylor (1964b, 1965) showed that the agglutination factor of strain 5 was heterogeneous and capable of being
fractionated. In column chromatography on phosphocellulose, three
major peaks of agglutinating activity were observed; and, in sedimentation in guanidine hydrochloride gradients, the distribution of activity
showed a single peak and a broad leading edge of fast-sedimenting
material. Fractionation depended mainly on particle size, the more
active fractions having higher particle weight. Analyses of different
fractions indicated that the agglutinating factor is a protein-mannan
complex, the protein content of which varies from 10% to 30% (w/v).
Particles in fast-sedimenting fractions had a molecular weight of 120–
300 × 10^6 and an average of 1·7 particles/cell was sufficient for agglutination (Taylor and Tobin, 1966). This is comparable to the value of
one particle/cell for other agglutinating systems, and the above workers
stressed this similarity by terming the factor from strain 5 cells 5-
agglutinin. A purer, more homogeneous form of 5-agglutinin was later
obtained by digestion of strain 5 cells with subtilisin (Taylor and Orton,
1967), more than 85% being adsorbed by strain 21 cells; it has a much
smaller molecular weight than previous preparations derived by digestion with snail enzyme, namely 570,000, and contains only 4% (w/v)
protein.

Brock (1965b) found that 5-factor of low molecular weight, consisting
of protein and carbohydrate in equal amounts, could be isolated from
cytoplasmic extracts prepared by alumina grinding or ballistic disintegration; and material of variable molecular size was observed in
sucrose-gradient centrifugation. Crandall (1968) has observed 5-factor
in culture supernatants of growing cells. She also found that cytoplasmic
extract of 5-factor could be purified by dissolving in saturated ammonium

sulphate solution, followed by precipitation with 50% (w/v) ethanol. The purified material was further fractionated by passage through a column of Sephadex G-200; although the main fraction had a molecular weight of more than 200,000, one small peak was estimated to have a molecular weight of about 15,000. Preparations of high molecular weight consisted of 10% (w/v) protein and 90% (w/v) carbohydrate. From these observations, it may be concluded that the heterogeneity of 5-factor is well established; its particles differ in molecular size and in the proportion of protein to carbohydrate. As the activity of 5-factor is destroyed by proteolytic enzymes (Brock, 1965b) and also by mercapto-ethanol which disrupts disulphide bonds (Taylor, 1964b; Taylor et al., 1968), it is evident that a protein component with disulphide bonds is essential for agglutination. The only sugar present is mannose. Boric acid, which combines with cis-hydroxyl groups such as are found in mannose, does not prevent agglutination (Brock, 1959b). The mannan component of 5-factor, therefore, appears to serve merely a structural role.

No agglutinin can be detected in strain 21 cells (Taylor, 1964a; Brock, 1965b). It is now known, however, that cells of this strain produce a specific cell-surface 21-factor which inhibits the activity of the 5-factor; but it does not cause agglutination of strain 5 cells (Crandall and Brock, 1968a,b,c): 21-factor has been detected in cytoplasmic extracts, but it can be isolated in larger amounts from the cell surface by trypsin digestion which causes its release; a previous report (Brock, 1958b) that trypsin destroys the agglutinability of strain 21 cells by breaking down the 21-factor is now known to be incorrect. The 21-factor can be purified by adsorption on strain 5 cells, elution with 8 M-urea and then application of methods used in protein chemistry. Purified 21-factor contains protein and carbohydrate (probably mannan) in the ratio of 65:35. Thus both agglutination factors appear to be protein-mannan complexes (glycoproteins). The 21-factor is inactivated by alkali, heat, concentrated salt solutions and other protein denaturants. Protein, therefore, appears to be the active component in agglutination in both factors.

Crandall and Brock (1968a,c) have also shown that a non-specific inhibitor of agglutination (NSI) is present on the cell surface of both mating types in H. wingei. It is eluted by heating (but not by trypsin), and its removal accentuates agglutination. The available evidence indicates that it is a protein-mannan complex. Apart from the specific mating components discussed above, sterols have been reported to function as non-specific components in the agglutination of strains 5 and 21 in H. wingei (Hunt and Carpenter, 1963).

Genetical analysis (Herman et al., 1966) indicates that the agglutination factor is always segregated with the mating type; the loci for these properties, therefore, must be identical or closely linked. Herman and Griffin (1967) have shown that, by crossing agglutinative with non-agglutinative strains, agglutinative isolates can be obtained with the same mating type and agglutinative specificity of the non-agglutinative strains; this suggests that a non-agglutinative strain possesses a gene for agglutination but its expression is masked by the action of a repressor gene.

The diploid hybrid of the agglutinative mating types in *H. wingei* is non-agglutinative. This is not due to increased production of the non-specific inhibitor (NSI), but to repression of synthesis of the 5- and 21-factors (Crandall and Brock, 1968b). The mating-type locus appears to contain two genes, one controlling synthesis of a sex-specific agglutination factor and the other synthesis of a repressor which inhibits production of the factor of the opposite mating type; when the genes for both mating types are present together in the diploid stage, neither factor is synthesized. Under certain conditions, however, the repression mechanism breaks down in diploid cells which are then capable of synthesis of the 5-factor; they agglutinate with strain 21 cells only. This was found to occur in diploid cells following growth in a well-aerated medium containing glucose, phosphate and Difco yeast extract; but only certain batches of the yeast extract proved effective.

Sexual agglutination in yeasts has possible applications in medicine and industry. Brock (1959b) observed that serum globulin inhibits agglutination of the mating types in *H. wingei*. Steinberg and Giles (1959) found that some types of human sera decrease agglutination to a greater extent than other types, and as mating types of baker's or brewer's yeasts can replace those of *H. wingei*, Race (1960) has recommended inhibition of sexual agglutination in yeasts as a useful method of blood typing.

Wickerham (1960) has patented a method for rapid and effective separation of yeasts from liquid media for use in the manufacture of baker's and food yeast, and in brewing, distillation and wine production. It involves separate propagation of opposite mating types of heterothallic yeasts such as *Saccharomyces kluyveri* and *Sacch. besseyi*, mixing the propagated mating types in a common tank, and settling out of mutually agglutinated yeast cells.

2. Conjugation

The cellular changes involved in conjugation vary slightly in different yeasts. Spores, and sometimes haploid cells, swell up prior to fusion. Recent workers have emphasized that intimate contact between cells is essential for conjugation in *Schizosaccharomyces octosporus* (Conti and

Naylor, 1960), *Hansenula wingei* (Brock, 1961) and *Saccharomyces cerevisiae* (Iguti and Ônobu, 1964). In *H. wingei*, each cell forms a conjugation tube in contact with that of its partner, the main bodies of the two cells being pushed apart; shortly afterwards, the wall between the tips of the conjugation tubes dissolves away and, even before this process is complete, nuclear fusion occurs in the connecting canal (Conti and Brock, 1965). No conjugation tubes are formed in *Schizosacch. octosporus* according to Conti and Naylor (1960) who found that, following cell fusion, a short conjugation canal appears and this increases both in length and width; the nuclei of the cells then move towards one another and fuse in the conjugation canal.

Guilliermond (1901) reported that conjugation in *Schizosacch. octosporus* and *Schizosacch. pombe* usually occurred between cells lying close together but not in direct contact. Each cell developed a short conjugation tube, and the nucleus moved into its tip; fusion occurred between the tips of the conjugation tubes, followed shortly afterwards by nuclear fusion. Barker (1901, 1902) made similar observations for *Zygosaccharomyces* sp. Levi (1956) showed that direct contact of haploid cells of *Sacch. cerevisiae* is not necessary for conjugation which could take place between cells up to 50 μ apart (J. D. Levi, unpublished data). When cells were paired on a malt-agar film, only cells of mating type *a* developed conjugation tubes (in liquid media, cells of both mating types usually develop conjugation tubes). Fresh haploid cells of mating type *a*, placed on a mating site from which all the cells had been removed, were induced to form conjugation tubes; but no tubes were formed on a site on which cells of opposite mating type were allowed to grow alternately for short periods (J. D. Levi, unpublished data). The mating reaction, therefore, involves mutual interaction between cells of opposite mating type. Levi (1956) also showed that a mating reaction could be induced in cells of mating type *a* separated from α cells by a collodion membrane, which indicates that diffusible agents are involved in conjugation.

The earliest work on the physiology of conjugation is that of Nickerson and Thimann (1941, 1943). They found that conjugation in *Zygosaccharomyces* was promoted by *Aspergillus* culture filtrate, which had little effect on growth, and also by diluted yeast autolysate. The *Aspergillus* culture filtrate was separated into two active fractions, one of which was acidic while the other exhibited properties, such as heat stability in acid solutions, shared by vitamins of the B group. Experiments with various organic acids and vitamins revealed that the effect of *Aspergillus* culture filtrate could be duplicated by a mixture of riboflavin and sodium glutarate, J. D. Levi (unpublished observations), however, found that conjugation in *Sacch. cerevisiae* on a synthetic medium was strongly inhibited by addition of riboflavin and sodium glutarate in the concentrations used by the above workers.

Brock (1961, 1965a) has made an intensive study of conjugation under aerobic conditions in *Hansenula wingei*. In order to study the effect of various conditions on conjugation in this yeast, it is necessary to de-agglutinate clumped mating cells to permit microscopic observations on the percentages of cells which conjugate after different treatments. De-agglutination was originally effected (Brock, 1961) by suspending cells in 8 M-urea and autoclaving for 10–15 min at 120°. This method has been replaced by a procedure (Brock, 1965a) better suited to the study of the early stages in conjugation; it involves taking samples of agglutinated cells at intervals, washing twice in distilled water, and finally submitting to sonic treatment for 5 min to break up the remaining small clumps of cells. There was found to be a lag of about one hour before conjugation began. Later, cells were observed in different stages of conjugation. After 3–4 h, over 50% of the cells had conjugated and, in about 10% of cases, three or even four cells fused together. Thus the conjugation response appears to be a strictly localized event of a small region of the cell wall, and the cell is not prevented from conjugating in other regions.

The mutual interaction necessary for conjugation of cells, first observed by Levi (1956) for *Sacch. cerevisiae*, was confirmed by Brock (1961) for *H. wingei*. Conjugation occurred in the presence of cycloheximide (actidione) only between haploid strains which were both resistant to this antibiotic; and inactivation of one haploid strain by ultraviolet light, prior to mixing, prevented conjugation, although irradiated cells still responded by forming protuberances directed towards the irradiated cells.

The conjugation medium used by Brock (1961) consisted of 1% (w/v) glucose, 0·1% (w/v) $MgSO_4.7H_2O$ and 0·05 M-potassium phosphate buffer (pH 5·7). The magnesium salt is necessary in order to keep the cells agglutinated. Experiments showed that an energy source is essential for conjugation and, of various substrates tested, glucose gave the highest percentage of conjugating cells, namely 69% of the cell population after 5 h at 30°. Crandall and Brock (1968c) have recently shown that maximum conjugation in *H. wingei* can be obtained with a glucose concentration as low as 0·05% (w/v). Addition of nitrogen as ammonium nitrate, casein hydrolysate or yeast extract had no effect on conjugation. Brock assumed that in each cell type there is a synthesis of a wall-softening enzyme and, since an external source of nitrogen is unnecessary for conjugation, this enzyme must be formed from the amino-acid pool; depletion of this pool by starvation resulted in diminished conjugation. The wall-softening enzyme was presumed to be identical with, or similar to, the enzyme involved in normal budding and it was suggested that it is formed in larger amounts during conjugation. As both cells are involved in conjugation, he postulated that an inducing agent diffuses from one cell into the other where it promotes the synthesis of a wall-softening enzyme which acts on the cell wall at the point of contact with the other cell. Brock (1965a) made a detailed study of changes in various chemical constituents and in enzyme activity during conjugation, but none of these could be related to conjugation as they were similar to those that took place when cells of only one mating type were incubated under similar conditions.

Iguti and Ônobu (1964) have recently studied the conditions governing conjugation and nuclear fusion in *Saccharomyces cerevisiae*. Two haploid strains, with complementary nutritional deficiencies, were used for this purpose. The conjugation medium was based on Lindegren's (1949) standard broth medium. By omission of constituents one at a time it was found that glucose and inorganic salts were essential for conjugation but, in contrast to *H. wingei* (Brock, 1961), addition of yeast extract caused marked stimulation of zygote formation.

Maximum frequency of conjugation was observed at 20° but, as the conjugation medium supported some growth by budding, it was necessary to maintain the temperature at 16°–18° in order to keep the cell population more or less unaltered for 24 h, thus making it possible to obtain reliable figures for percentages of conjugating cells under different conditions.

Zygote formation was not evident until 5 h after mixing the haploid strains, reaching a maximum after 15 h. The extent of diploidization at different stages was investigated by a prototroph recovery method (Pomper and Burkholder, 1949). Samples of the mating mixture were diluted (to prevent further conjugation), and then plated on a minimal agar medium to promote selective development of diploid colonies. The conditions governing diploidization (nuclear fusion) differed in one important respect from those for conjugation; yeast extract, in contrast to its stimulatory effect on conjugation, had an inhibitory effect on nuclear fusion, the highest numbers of recoverable diploids being obtained by omission of yeast extract from the conjugation medium. Chloramphenicol, which inhibits the growth of certain yeasts (Motoc and Dimitriu, 1966; Shevchenko *et al.*, 1966) retarded conjugation, but was without effect on nuclear fusion.

The virtual lack of agglutinative ability in *Sacch. cerevisiae* may account for the low percentages of conjugating cells observed in many strains. Iguti and Ônobu (1964) attempted to increase efficiency of contact between cells of opposite mating-type by mixing *a* and *α* cells in the ratio of 3–4:1. Frequency of conjugation and diploidization was expressed as a percentage of the cells in the minority, and even on this basis the highest figures recorded for conjugation and diploidization were only 10·3% and 4·2% respectively. These low figures were probably due to the use of non-aerated liquid cultures. Jakob (1962) first demonstrated the stimulatory effect of oxygen on conjugation. She devised a hybridization technique which involves preliminary aeration of mixed haploids of opposite mating-type in liquid medium, compacting of cells by centrifugation to ensure intimate contact, and finally vigorous aeration of cells to promote zygote formation which occurs about 3 h after resuspension in fresh liquid medium. About two and a half times as many zygotes were obtained as by the classical method of mixing haploid cells in unshaken liquid media (Lindegren and Linde-

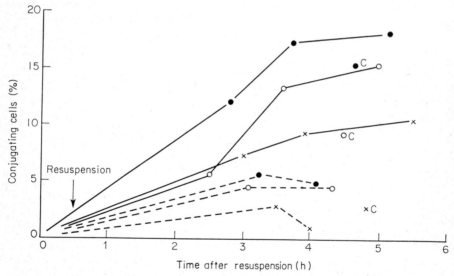

FIG. 11. Effect of glucose concentration and different methods of aeration on zygote formation in a hybrid strain of *Saccharomyces cerevisiae*. The procedure used was based on a modification of Jakob's (1962) technique. Haploids of opposite mating-type were grown on malt wort for 48 h at 30°. A portion (0·5 ml) of each haploid culture was added to 4·5 ml of 0·85% (w/v) NaCl. The haploids were mixed by adding 0·5 ml of the saline suspension of each haploid to 5 ml liquid medium. The haploid mixture was aerated on a shaker for 2-h at 22°, and then centrifuged for 2–3 min at 2,000 rev/min. The centrifugate was left standing for 30 min. Most of the supernatant was then decanted, leaving a small amount of liquid which was shaken to give a dense suspension of cells; this was used for spot inoculation of a 2% (w/v) agar film. The suspension was then diluted with 5 ml liquid medium and aerated by shaking. Brock's (1961) conjugation medium containing 1 (×), 5 (○), or 10 (●)% (w/v) glucose was used for liquid and agar media throughout each of the three separate tests. Microscopical observations were made of zygote formation at 22° at intervals up to 5–6 h. In control tests, centrifugates (C) were left standing for 4–5 h and then examined for zygotes.

gren, 1943a). Haefner (1965) showed that, when dense growths of aerobically-grown haploids were mixed on glucose-yeast extract-peptone agar, the percentage of fused cells averaged 25% but, in some haploid combinations, it reached more than 50%, which is comparable to the degree of conjugation observed, under aerobic conditions, in *H. wingei* (Brock, 1961).

In recent experiments (R. R. Fowell, unpublished observations), it has been shown that a high concentration of glucose (around 10%, w/v) is essential for maximum zygote formation in *Sacch. cerevisiae* (Fig. 11). In a study of Jakob's three-stage process (see p. 373), involving the use of Brock's (1961) conjugation medium, it was found that far more zygotes were produced if a mixture of haploid cells, after preliminary aeration and centrifugation, was transferred to an agar medium instead of being resuspended in a liquid medium and aerated (Fig. 11). This confirmed previous observations which indicated that aeration in the final stage tended to separate many of the cells to the detriment of zygote formation, probably because the yeasts employed were devoid of agglutinative properties. The importance of using haploid cells adapted to aerobic conditions was demonstrated by aerating a mixture of haploids in liquid medium and, after separation of cells by centrifugation, leaving the cells packed together for 4 h (the final aeration phase being omitted). The percentage of conjugating cells formed in the centrifugate was nearly as high as that recorded on the agar medium in the above test, 15·2% compared with 18·0% in the presence of 10% (w/v) glucose, and much higher than in a centrifugate prepared from an unshaken mixture of haploids (less than 1%).

V. Main Steps in Hybridization

The crossing of different yeasts to produce hybrids involves three main steps: induction of sporulation, isolation of spores or at least their separation from vegetative diploid cells, and finally actual hybridization which depends on the fusion of spores or of haploid cells produced, by budding, from the spores accompanied in both cases by nuclear fusion.

Attempts to produce hybrids are not always successful because of difficulties which may be encountered at one or more of the above steps. The main purpose of this section is to discuss the different methods available for hybridization, and to indicate the best means of overcoming the above difficulties.

A. INDUCTION OF SPORULATION

The factors controlling sporulation, including the conditions necessary for optimum spore formation, have been dealt with in Section III (p. 313). For purposes of hybridization, it is not essential as a rule to obtain the highest possible numbers of spores; the aim should be to obtain reasonable numbers of four-spored asci as these are not only easier to dissect (if the microneedle method is used) but they are also important for tetrad analysis in genetical studies. Some yeasts, unfortunately, sporulate poorly and produce few, if any, four-spored asci; isolation of spores from two- and three-spored asci is unavoidable in such cases.

1. *Presporulation Treatment*

Although a special presporulation medium, containing a mixture of fruit and vegetable juices, has been recommended for the invigoration of yeast cultures in preparation for sporulation (Lindegren and Lindegren, 1944a; Lindegren, 1945, 1949), it is clear from the literature survey given in Section III (p. 313) that several other media are just as effective in conditioning yeasts for sporulation.

The main requirements of a presporulation medium are a high content of glucose (Fowell and Moorse, 1960), nitrogenous compounds capable of supporting good growth (Tremaine and Miller, 1956) and a full complement of B vitamins (Tremaine and Miller, 1954) which can be provided as yeast extract. Non-aerated liquid cultures should preferably be used as they give higher yields of asci on sporulation media than solid (agar) media (Tremaine and Miller, 1954; Pazonyi, 1954; Fowell, 1955; Fowell and Moorse, 1960). Suitable media include 5% (w/v) glucose-nutrient broth-1% (w/v) yeast extract; 16% (w/v) malt extract-wort; and synthetic media such as Lodder and Kreger-van Rij's (1952) medium containing at least 5% (w/v) glucose.

In general, presporulation cultures should be incubated at 30° for 24–48 h but, for slow-growing yeasts such as wine yeasts, a longer period such as 72 h may be necessary (Fowell and Moorse, 1960). Periodical shaking of cultures to remove excess carbon dioxide is advised.

2. *Sporulation Treatment*

Most hybridization studies have been carried out on *Saccharomyces cerevisiae* and closely related species such as *Sacch. carlsbergensis*, *Sacch. chevalieri*, *Sacch. chodati*, *Sacch. diastaticus* and *Sacch. italicus* (e.g. Winge and Laustsen, 1939a; Lindegren, 1949). In the following account, therefore, priority is given to methods suitable for use with *Saccharomyces* spp.

Sporulation media were originally devised by taxonomists who, in their efforts to induce yeasts to form spores, experimented with a great variety of materials including natural products such as vegetables, fruits, malt extract, and inert substrates, e.g. gypsum, clay (see Phaff and Mrak, 1949; Roberts, 1950; Lodder and Kreger-van Rij, 1952; Reiff et al., 1960). Many of the older types of media are unsuitable for use in genetical investigations and industrial breeding projects for reasons such as difficulties involved in their preparation, e.g. gypsum blocks (Engel, 1872), inability to support good sporulation especially production of four-spored asci, e.g. McKelvey's (1926) carrot-calcium sulphate agar, and, in particular, the long time necessary for spores to appear on media such as carrot wedges (Reess, 1869), Gorodkowa agar (Gorodkowa, 1908) and V-8 agar, a mixture of vegetable juices (Wickerham et al., 1946) which contain sugars and other nutrients in sufficient amounts to support appreciable growth thereby causing spore formation to be delayed.

Increased knowledge of the factors which influence sporulation (Section III.C, p. 318), especially the effect of various compounds in different concentrations, has led to the development of synthetic media, such as sodium acetate-agar (Fowell, 1952), which are discussed in more detail below. Such media are particularly suitable for obtaining spores rapidly and in large numbers for most strains of Saccharomyces. Difficulty is, however, sometimes experienced with certain yeasts, especially those in other genera such as Hansenula and Debaryomyces, and in order to obtain good sporulation it is often necessary to experiment with other media including the classical types mentioned above. It should be noted that there are also certain yeasts which have been used for genetical research, such as Zygosaccharomyces spp. (Wickerham, 1955) and Schizosaccharomyces pombe (Leupold, 1950), in which malt-agar and other media containing nutrients capable of supporting growth are particularly suitable for sporulation, as this process is immediately preceded by conjugation; as previously indicated, a high concentration of energy source, such as sugar, and also nitrogenous compounds are required for conjugation.

Gypsum blocks, which are still preferred by some workers, were used to obtain spores for hybridization in the pioneer investigations of the Winge school (Winge and Laustsen, 1938, 1939a). They are prepared by mixing four parts of gypsum powder ($CaSO_4 \cdot \frac{1}{2}H_2O$) with one and a half parts of water. After hardening in conical forms of desired size, the blocks are placed in glass dishes and sterilized by heating at 120° for 2 h (Roberts, 1950). The surface of the block is inoculated with a dense suspension of yeast in water or wort, after which water is poured into the dish to saturate the block. The block is then incubated at 25°, and spores usually appear quite rapidly, 2–3 days often being sufficient for maximum sporulation in Saccharomyces and other genera. Many yeasts sporulate better on gypsum blocks than on other media such as carrot, malt-agar and Gorodkowa agar (Phaff and Mrak, 1949). Gypsum blocks, unfortunately, suffer from certain disadvantages. If too little water is used in their preparation, they become brittle and develop cracks after sterilization; if too much water is present, they fail to set properly and readily become infected with bacteria. When used repeatedly, they often become infected with spore-formers since they cannot withstand the strong heating necessary for absolute sterility.

In view of the above disadvantages, Hartelius and Ditlevsen (1951, 1953) devised cement blocks, consisting of cement and kieselguhr, capable of withstanding adequate sterilization. They claimed that several yeasts, including *Sacch. cerevisiae* (American Yeast Foam), sporulate better, and produce more four-spored asci, on cement than on gypsum. Cement blocks, however, have not proved popular, probably because they are tedious to prepare.

In America, the Lindegren school (Lindegren and Lindegren, 1944a; Lindegren, 1949) originally used gypsum slants in tubes, as first suggested by Graham and Hastings (1941). The slants are much easier to prepare than blocks, and there is less danger of infection while they are being prepared. Gypsum (100 g) is mixed to a thick cream with 100 g water, and poured into a test tube to a depth of 1 in. After slanting, the gypsum is hardened in an incubator at 50° for 24 h, the tube then being plugged and autoclaved. A dense suspension of yeast in water is deposited on the upper end of the slant, and about 3 ml sterile water, containing enough acetic acid to adjust the pH to 4, is pipetted over the lower end of the slant which is incubated for 1–2 days at 25°. The success achieved with this method was probably due in large measure to the introduction of acetate ions.

Agar-containing media have largely replaced gypsum and cement blocks in modern hybridization studies because of ease in preparation and handling. The superiority of media containing acetate for sporulation of *Saccharomyces* was first clearly established by Adams (1949) who, following the investigations of Stantial (1935) and Elder (1932, 1933, 1937) which indicated that a medium containing both a sugar and acetate gives higher yields of asci than either alone, recommended a sporulation medium consisting of 0·14% (w/v) anhydrous sodium acetate + 0·04% (w/v) glucose in distilled water or solidified by addition of 2% (w/v) agar. Some workers prefer to use liquid sporulation media, and satisfactory sporulation can be obtained if provision is made for adequate aeration, e.g. by shaking in rocker tubes (Stantial, 1935; Adams and Miller, 1954; Kirsop, 1954) or bubbling through air (Pazonyi, 1954; Fowell, 1967). It should be noted, however, that sporulation on agar-containing media supplemented with acetate is independent of cell population density over a wide range of values whereas, in liquid media, sporulation depends rather critically on cell density (Fowell, 1967).

Other types of acetate-containing media have been introduced in recent years. The value of acetate in stimulating the production of large numbers of four-spored asci was first reported by Fowell (1952) who introduced sodium acetate-agar as a sporulation medium. This consists of 0·3–0·5% (w/v) anhydrous sodium acetate + 1·5% (w/v) agar, pH 6·5–7·0 (sterilized for 20 min at 122°). The amount of acetate in Adams's acetate medium was found to be suboptimum for certain yeasts, and the inclusion of glucose is now known to be detrimental to sporulation (Sando, 1956; Fowell, 1967). Maximum sporulation on sodium acetate-agar is attained in 2–4 days, and in several hybrids up to 50% of the cells produce four-spored asci.

Kleyn (1954) devised a complex acetate medium of the following composition:

Difco Bacto-tryptose	0·25 g
Glucose	0·062 g
NaCl	0·062 g
Sodium acetate.3H$_2$O	0·5 g
Difco Bacto-agar	2·0 g
Distilled water	100 ml
(pH after sterilization, 6·9–7·1)	

Comparison of sporulation of a baker's yeast on Kleyn's medium and Fowell's medium revealed no differences either in total numbers of asci or numbers of four-spored asci (Fowell, 1967).

Addition of potassium ions to acetate-containing media improves the sporulation of yeasts (McClary et al., 1959), particularly of those which sporulate poorly on sodium acetate-agar (Fowell, 1967). The medium of McClary et al. (1959) has the composition:

Potassium acetate	1 g
Yeast extract	0·25 g
Glucose	0·1 g
Distilled water	100 ml

Aliquots (10 ml) of this liquid medium are inoculated with a washed yeast suspension to give a density of about 3×10^6 cells/ml, a concentration regarded as within the optimum range for sporulation (Miller et al., 1955). The medium is preferably used as a solid medium by the inclusion of 3% (w/v) agar (Lindegren, 1962).

The following procedure for obtaining spores is based on investigations of the factors controlling sporulation (Fowell and Moorse, 1960; Fowell, 1967): (1) Ten ml of a suitable liquid presporulation medium, containing 5–10% (w/v) glucose, is inoculated with yeast, and incubated at 30°, with occasional shaking only, for 24–48 h. (2) Yeast cells are separated by centrifugation, washed at least twice with distilled water (not saline), and then resuspended in 0·5 ml distilled water. (3) A 4 mm loop, which holds about 0·01 ml of the resultant aqueous suspension, is used to inoculate a slant composed of 10 ml of the following medium:

Sodium acetate (anhydrous)	0·5 g
KCl	1·0 g
Agar	1·5 g
Distilled water	100 ml

(4) After inoculation, the slant is incubated at 25° for 3–5 days. Experiments have shown that addition of glucose and also yeast extract to potassium-containing media can be detrimental to sporulation. The degree of sporulation on the above medium is comparable, for most yeasts, to that obtained by buffering at the optimum pH, usually 8–10.

Although potassium acetate-containing media have given good results with most strains of *Saccharomyces* tested, we have found certain yeasts, particularly some strains of *Sacch. carlsbergensis*, which form few if any spores. If such yeasts are considered sufficiently important, e.g. for production of hybrids for possible use in industry, two courses of action are possible. First, other sporulation media should be tested, e.g. nutrient agar, V-8 agar (Wickerham *et al.*, 1946), chalk-agar (Galloway, 1954), gypsum slants wetted with dilute (1° Balling) wort (Klöcker, 1924), 0·5% (w/v) K_2HPO_4 + 1·8% (w/v) mannitol (Saito, 1923), or dilute acetic acid, pH 4 (Lindegren and Lindegren, 1944a). Kirsop (1957) found that cell-free extracts of fresh yeast cells or of acetone-dried yeasts are extremely active in inducing the production of large numbers of spores. Crude extracts in dilutions as low as 2–3 ppm were found to stimulate spore formation. It is suggested that inoculation of an agar medium, containing cell-free extract prepared from a yeast with good powers of sporulation, may provide a method of stimulating spore formation in a yeast with little or no sporulating ability, although the feasibility of this procedure has yet to be demonstrated.

As already mentioned, many yeasts gradually lose the capacity to sporulate when subcultured over a long period; and, when such a decline has been observed, or is suspected to have occurred, methods of reviving sporulating ability should be considered. Repeated transfer of cells from a presporulation to a sporulation medium, and back again to the presporulation medium, appears to be an excellent method of improving sporulation (Fuchs, 1935; Bedford, 1942). Culturing yeasts on sterile garden soil improves sporulation by promoting selective growth of cell types with good spore formation (Hartelius and Ditlevsen, 1956); a similar selection probably operates in the previous method. Tsukahara and Yamada (1953) have claimed that the sporulating ability of saké yeasts can be restored by serial transfer on raw koji extract sterilized by filtration instead of heat. According to these workers, enzymes and other thermolabile substances in koji extract are necessary for sporulation. Finally, isolation of single cells, followed by sporulation tests on single-cell cultures, has proved useful, especially as a method of increasing the production of four-spored asci (Fowell and Moorse, 1960).

B. ISOLATION OF SPORES

After a yeast has been induced to sporulate, the next step in hybridization is to remove asci (preferably four-spored) from the non-sporulated cells; then the spores should be separated out from the asci.

1. *Dissection of Asci*

The classical method of isolating spores involves the use of a micromanipulator, the asci being ruptured and spores dissected out by means of glass microneedles. The process is carried out on the lower surface of a cover-slip inside a moist chamber. If the asci are allowed to dry out, they become hard and brittle, and almost impossible to break without damage to the spores; moisture is necessary also for movement and transfer of spores.

In the older types of micromanipulator, three-way movement of needles is effected mechanically by separate controls. A single lever, however, is used to move a needle in any desired plane in pneumatic models, e.g. the de Fonbrune and Singer models, and in the hydraulic model of Chas. W. Cook & Sons Ltd. Two glass needles are often used to dissect asci (Winge and Laustsen, 1937) but one alone is perfectly satisfactory. A microforge is often used to obtain needles of the required form (Thaysen and Morris, 1947; de Fonbrune, 1949), but they can, with practice, be made by hand using a dissecting microscope, a pilot burner and a hypodermic needle connected to a gas line (Lindegren, 1949; Fowell, 1955). A flat-tipped needle is useful for rupturing asci, and for moving and picking up spores (Lindegren, 1949), but some workers prefer a round tip.

The dissection of an ascus must be performed with considerable precision and care as one or more of the spores may be damaged. Usually it is necessary to apply momentary direct pressure to rupture the ascus, but if possible the ascus should be gently rolled about until it collapses.

Moist chambers suitable for the dissection of asci have been described by Winge (1935), Lindegren (1949) and Fowell (1955). In the Carlsberg Laboratory, the spores from a single ascus are moved into separate droplets of wort on the lower surface of a cover-slip forming the roof of a glass moist chamber, and allowed to develop into colonies a few cells of which are then transferred to a Freudenreich flask containing wort (Winge and Laustsen, 1937); alternatively, spores from different yeasts may be paired, to produce hybrids, in wort droplets (see Section V.C, p. 370). A similar method of isolating spores is used by Lindegren (1949) except that they are transferred to drops of agar, the resultant colonies being subsequently streaked on agar slants. Another method, due to Cox and Bevan (1961), involves the transfer of asci to cellophane slips which are incubated in contact with nutrient agar to soften their walls; the spores are then readily dissected out and allowed to germinate *in situ.*

Large numbers of spores are required for industrial breeding projects, mainly because spore viability is not very good in many of the yeasts employed. We devised a procedure (Fowell, 1955) in which spores are isolated on one cover-slip and then transferred to marked positions on a malt-agar film on the lower side of a second cover-slip. The cover slips measure $\frac{7}{8} \times \frac{7}{8}$ in. The agar film, which can accommodate up to 100 spores, is incubated overnight at 30° and then examined for spore germination. The film is afterwards cut up with a sterilized microspatula into small squares, corresponding to the Indian ink spots that indicate the positions of the spores; every agar square bearing a colony of at least 10 cells is transferred to a tube of malt wort. The tubes of wort are incubated at 30° and actively fermenting cultures are obtained in 2–14 days.

2. Enzymic Dissolution of Asci

Dissection of asci is a tedious and somewhat lengthy procedure, and there is always the risk of damage to some of the spores. Fortunately, it is now possible to obtain and separate out large numbers of spores by the use of an enzyme preparation that will dissolve ascal walls, without the viability of the spores being affected.

Wright and Lederberg (1957) first conceived the idea of dissolving ascal walls enzymically, using an extract obtained from *Bacillus polymyxa.* This method is now used as a routine procedure by Magni and von Borstel (1962) who have developed a strain with increased lytic activity; application of a dried enzyme extract to a sporulating yeast culture effects complete removal of ascal walls after 24–36 h at 28°, without any impairment of spore viability even after 5 days treatment. A number of other micro-organisms, including bacteria and moulds, are capable of dissolving the cell walls of yeasts (e.g. Kaisha and Kaisha, 1962; Tanaka and Phaff, 1965), and it is possible that enzyme extracts of some of these could be used for the same purpose.

Johnston and Mortimer (1959) first used a digestive enzyme extract, prepared from the crops of snails, to dissolve ascal walls and release spores. The Roman Snail (*Helix pomatia*) is the best source as it is larger and easier to dissect than other species. Recently, commercial extracts of snail digestive juice have become available (supplied by Endo Lab. Inc., New York and L'Industrie Biologique Francaise S.A., Gennevilliers, Paris), and their use greatly facilitates the task of isolating large numbers of spores. It is the practice in our laboratories to add 0·01 ml of the French enzyme preparation to 1 ml of a dense suspension of sporulated cells, and leave at room temperature overnight; use of undiluted extract has been found to impair spore viability.

Irrespective of the enzyme treatment used to dissolve ascal walls, separation of the spores is still necessary as they usually adhere firmly

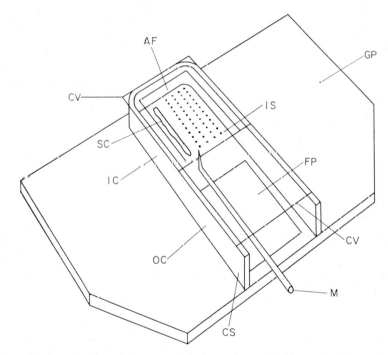

FIG. 12. Moist chamber for spore isolation. A sporulating culture, previously treated with snail enzyme extract to dissolve ascal walls, is streaked along one side of an agar film (e.g. malt agar) on the lower surface of a cover slip. Spores are separated out with a microneedle and spaced in rows of four; their positions are indicated, as shown, by Indian ink spots on the upper surface of the cover slip. AF indicates agar film; CS, copper strip, ¼ in; CV, cover slip; FP, moistened filter paper on floor of outer chamber; GP, glass plate; IC, inner chamber; IS, Indian ink spots, marking the positions of spores; M, microneedle inserted in moist chamber; OC, outer chamber; SC, sporulating culture streaked on agar film.

together. If tetrads of spores are required, some micromanipulation will be necessary. If the agar film method is used, the enzyme-treated suspension should be streaked along one side of the agar film on which the tetrads of spores are to be spaced out in rows (Fig. 12). The spores may be allowed to germinate *in situ*, and each resultant colony transferred on a small square of agar to a tube of nutrient medium (Fowell, 1955); alternatively, the complete film may be lifted with a sterile spatula and placed, spores uppermost, on the surface of a nutrient-agar plate containing glucose (Johnston and Mortimer, 1959), the resultant colonies being picked off for transfer. If there is no need for tetrad analysis, spores may be isolated by one of the methods described in the next section.

3. *Methods for Yeasts with Low Spore-Viability*

After spores are isolated, they must be induced to germinate and develop into cultures to provide mater (usually haploid) strains suitable for use in hybridization. A variable number of spores, depending on the yeast used, start to germinate but some spores, after producing a few cells often abnormal in shape and size, cease further growth probably because of lack of balance in their genetical constitution. It is necessary, therefore, to distinguish between germination, and spore viability which refers to the proportion of spores capable of unlimited growth, i.e. of producing definite cultures.

Little attention has been devoted to the physiology of spore germination in yeasts. Nagashima (1959) found that spores of a wine yeast failed to germinate on plain (well washed) agar. Addition of glucose or diluted tomato juice resulted in good germination, and 1% (w/v) glucose was optimal. Ammonium sulphate (0·05%, w/v) improved germination considerably, but higher concentrations were inhibitory. Inclusion of salts and vitamins appeared to be without effect. A medium consisting of 1% (w/v) glucose, 0·05% (w/v) $(NH_4)_2SO_4$ and 1·5% (w/v) agar was therefore recommended for the germination of spores; the pH value after autoclaving was 6·5 which appeared to be optimal. Palleroni (1961), however, found that the rate of germination was highest between pH 5 and 8, and glucose alone was necessary for good germination.

In experiments with a brewer's yeast, we found that no spores were capable of germinating on Nagashima's medium unless it was supplemented with casamino acids, and percentage germination was increased by addition of salts; casamino acids and salts could be replaced by yeast extract. Glucose (1%, w/v) proved to be optimal as Nagashima found. On a medium consisting of 1% (w/v) glucose, 1% (w/v) yeast extract and 2% (w/v) agar, 18% of the spores germinated and spore viability was 6%. Corresponding figures for 16% malt extract + 3%

(w/v) agar were not very different, namely 27% and 8·5% respectively. As a firm clear agar is required for spore isolation, we use 4% (w/v) malt extract + 2% (w/v) agar as routine procedure. In view of these observations, it is advisable to use a complete medium to ensure sustained development of all spores capable of producing definite cultures, since some of these are likely to exhibit nutritional deficiencies.

Many yeasts of industrial importance, especially bottom-fermenting brewer's yeasts (*Saccharomyces carlsbergensis*; Winge, 1944), sporulate very poorly if at all. Poor sporulation is unfortunately associated with low spore-viability. Thus the average sporulation of 11 strains of *Sacch. carlsbergensis* on sodium acetate-agar, studied by Emeis (1958b), was only 4% while spore viability was 0·07–2·0%. Top-fermenting brewer's yeasts (*Sacch. cerevisiae*) show a wider range of sporulating capacity (0–70%) on sodium acetate-agar. Thorne (1951) found that 12 out of 20 such yeasts from English breweries could sporulate on gypsum, and an average of 12% of the spores could germinate, but no figures were quoted for spore viability. The average spore viability of 33 top yeasts studied by Johnston (1965a) was 7·1% and only 25% of the viable spores yielded haploid cultures.

In order to obtain mater cultures from yeasts with low spore-viability, it is obviously necessary to isolate large numbers of spores, and the use of normal methods of micromanipulation for this purpose is likely to be time-consuming, tedious and also unsuccessful. There are, fortunately, various ways of circumventing the disadvantages of low spore-viability and these are discussed below.

a. Genetical devices. Inbreeding of yeasts usually results in degeneration of characteristics due to increased homozygosity. Spore viability is normally impaired, but Winge and Laustsen (1940) have reported improvement in this feature, as a result of inbreeding, in *Saccharomyces validus*. A similar effect, previously ascribed to inbreeding (Fowell, 1956b), is believed to be attributable to the segregation of diploid spores by a triploid yeast, the spore viability of which was less than 1%. The spore viability of diploid spore cultures was up to ten times that of the parent yeast. Following isolation of spores from two of these spore cultures, several haploids of both mating types were obtained with little difficulty.

A method with presents an opportunity of obtaining haploid cultures from a yeast, even if it is incapable of forming any spores, is the induction of mitotic haploidization. Spontaneous mitotic haploidization is a rare occurrence in the mould *Aspergillus*, but it is greatly accelerated by treatment with *p*-fluorophenylalanine (FPA; Lhoas, 1961). Emeis (1966) showed that the application of 0·3% (w/v) FPA to a gradient plate inoculated with *Saccharomyces* results in the production of mater cultures which are aneuploids and not true haploids, but this is of no practical significance in an industrial breeding project as they are satisfactory for use in hybridization. Genuine haploids can, however, be obtained with the same treatment from *Schizosaccharomyces* (Gutz, 1966). Following the application of this method to a poorly sporulating strain of *Sacch. carlsbergensis*, we were only able to isolate a single weak mater culture; but further trials seem warranted.

b. Isolation of intact asci: selective heat treatment. When spore viability is low, and the chances of fusion between spores inside an ascus are remote, isolation of intact asci by micromanipulation provides a quick method of obtaining at least a few haploid cultures. A much more effective method, however, of isolating intact asci, and in much larger numbers, is selective heat treatment which, in my opinion, is the best method of obtaining haploid cultures from yeasts of low spore-viability.

Selective heat treatment was originally devised by Wickerham and Burton (1954) as a method of isolating haploid cultures in the genus *Hansenula*. It is based on the fact that spores are slightly more resistant to heat than vegetative cells. The heat resistance of a vegetative culture in a nutrient medium is first determined by heating at 58–60° in a water bath. At intervals up to 20 min, a loopful of yeast suspension is streaked on an agar medium to ascertain the time required to kill the vegetative cells. Then a suspension of a sporulating culture is heated, under similar conditions, and agar plates are streaked at intervals of 1, 2, 3, 4, 5 and 10 min in excess of the time required to kill the vegetative cells. The plates are incubated, and a number of the smaller colonies are picked off from a plate bearing well-spaced colonies; the larger colonies are avoided as they may have arisen from vegetative cells which survived heating.

We have applied this method with considerable success to strains of *Saccharomyces*. Thermal resistance of vegetative cells is determined in malt-extract wort, and experiments have shown that a temperature of 54–60°, the exact temperature depending on the strain of yeast, is necessary to kill the majority of the cells in 5 min. The use of an aqueous suspension of sporulating cells is preferred, and in general heat treatment for 10–15 min at 59° has proved suitable; prolonged heating results in destruction of most of the spores. It should be noted that polyploid and aneuploid strains of *Saccharomyces* segregate spores of variable ploidy; streaking of heat-treated sporulating cultures on malt agar results in the development of large, usually diploid, colonies as well as small haploid colonies.

Microscopal examination has shown that ascal walls are not affected by heat treatment, i.e. intact asci and not spores are separated by plating. Large numbers of haploid cultures can be obtained by application of this method to yeasts with low spore-viability, so the chances of spores fusing inside the asci would appear to be small. Selective heat treatment can even be applied to yeasts with moderate or good spore viability; experiments have shown that, although many spores fuse inside the asci, surprisingly large numbers of haploid cultures can be obtained in this simple manner, without the use of any micromanipulation.

c. *Mass spore isolation*. German workers (Emeis, 1958a; Emeis and Gutz, 1958) introduced a novel technique of spore isolation which involves mechanical rupture of asci followed by uptake of spores in paraffin oil. A sporulating culture is mixed with water and powdered glass in a grinder containing a central piston rod, and this is rotated by an electric motor at about 100 rev/min for 10 min. This treatment destroys most of the cells, and liberates spores a few of which are unavoidably damaged. The resultant suspension is shaken with paraffin oil, the latter then being separated by centrifuging. The spores, because of their lipophilic properties, are taken up in the oil, while undamaged cells and cell debris remain in the aqueous phase. The oil suspension is streaked on a suitable agar medium in order to obtain single-spore colonies (usually haploids).

The spores have an unfortunate tendency to clump together after liberation by grinding, but the above workers found that most of the spores could be separated by mixing the oil layer with 15% (w/v) gelatin and grinding again without glass powder. It was admitted, however, that even after this treatment some colonies produced on plating were mixed. We have found that, with some yeasts, a considerable number of vegetative cells are taken up with spores in the oil. According to Gutz (1958), however, shaking of the oil suspension with repeated additions of water can effect complete elimination of vegetative cells. In spite of the above disadvantages, the method of Emeis and Gutz has been successfully used to hybridize brewer's yeasts (Emeis, 1959; Enebo et al., 1960; Johnston, 1963, 1965b).

We have used a method which combines application of snail enzyme extract to a sporulating culture with paraffin oil treatment. Addition of paraffin oil has proved effective in separating most of the spores in this method. Following enzymic dissolution of asci, sonication of spore groups can effect complete separation of spores (Magni, 1963; Gilmore, 1967).

It would appear that Enebo et al. (1960), like ourselves, have had difficulty in obtaining complete separation of spores and vegetative cells by the method of Emeis and Gutz, as they followed this with selective heat treatment to kill off the vegetative cells. We have had considerable success in obtaining haploid cultures from all types of yeast by combining enzymic dissolution of asci with selective heat treatment (Fowell, 1966). Paraffin oil is used to separate spores of yeasts showing moderate-to-good spore viability, but when spore viability is poor (and there is little chance of more than one spore per ascus being viable) oil treatment can be omitted. In fact, selective heat treatment alone is sufficient in such cases. Application of the enzyme-oil-selective heat treatment to a triploid baker's yeast with 50% spore viability, followed by plating on malt agar, resulted in the production of colonies of which 77% were mater cultures compared with 57% obtained with selective heat treatment alone.

Recently, Resnick *et al.* (1967) have shown that spores and vegetative cells have different electrophoretic mobilities, and by the use of stable-flow free boundary Staflo electrophoresis it is possible to obtain an essentially pure (99·04%) aqueous suspension of spores. In view of the difficulty we have experienced in separating spores of certain yeasts from vegetative cells by the oil method, this technique would appear to be superior.

It should be emphasized that, although mass spore isolation does not permit of tetrad analysis, it is very useful for study of different aspects of genetics such as recombination, linkage and gene conversion in yeasts of good spore viability.

C. METHODS OF HYBRIDIZATION

The three principal methods of hybridization are shown diagrammatically in Fig. 13. Variations of these methods, together with a further possible technique—transformation—are discussed below.

1. *Pairing of Spores*

The classical method of pairing spores in droplets of wort is due to Winge and Laustsen (1938). It is still widely used, although many workers prefer to pair spores on an agar film, e.g. malt agar. Roberts (1950) recommends incubation overnight for 16–18 h at 15° to permit zygote formation and observations of the first buds the following morning; the use of a low temperature accords with the observation that maximum frequency of conjugation occurs at 20° (Iguti and Ônobu, 1964).

The chief advantage of this method is that it is possible to distinguish under the microscope between zygotes formed by fusion of spores (hybrid formation) and those sometimes produced by fusion between haploid cells budded off from the same spore or by fusion between a spore and one of its own cells (self-diploidization). For yeasts such as *Saccharomyces chevalieri*, whose spores self-diploidize on or soon after germination, this is virtually the only method available for hybridization. The main disadvantage is the low rate of fusion between paired spores, about 1 in 15 (Roberts, 1950); fusion depends on the spores being viable, of opposite mating type and in the same physiological state. Another disadvantage, especially in industrial breeding projects, is that the spores can only be used once as they disappear in the process of fusion.

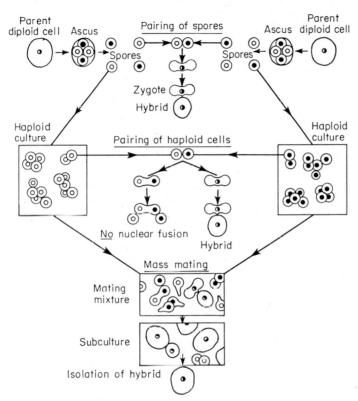

FIG. 13. The three principal methods of yeast hybridization. The black and white nuclei represent the two mating types. Nuclei of the diploid and hybrid cells are shown half black, half white, as they are heterozygous for mating type. Production of a hybrid depends on nuclear fusion in a zygote. Such a fusion often fails to occur when haploid cells fuse and, on rare occasions, when spores fuse; in such instances, zygotes bud off haploid cells. Mass mating provides an opportunity for successful nuclear fusion in some of the zygotes formed in a mating mixture (mixture of haploid cells of opposite mating types). Vigorous subculturing causes hybrid cells to outgrow haploid cells, thus facilitating their subsequent isolation, e.g. by micromanipulation. Reproduced from the *Journal of Applied Bacteriology* with permission.

2. *Pairing of Haploid Cells*

This procedure was first used by Chen (1950). Zygote formation, as in the previous method, can be observed microscopically, and there is the additional advantage that cells from the same haploid clone, i.e.

13

of the same genotype, can be used for several matings. Self-diploidization of germinating spores, when it occurs, can be delayed by transfer to sodium acetate-agar on which microcolonies of haploid cells are produced; this facilitates hybridization since the cells can then be paired with haploid cells of other yeasts (Palleroni, 1961).

Pairing of haploid cells often fails to yield hybrids for two reasons. First, the mating reaction may be so weak that no fusions occur even when several cells are paired. Secondly, fusion of haploid cells to form zygotes is not always accompanied by nuclear fusion (Fowell, 1951); in fact, as many as 50% of zygotes formed in this manner may bud off haploid cells instead of hybrid cells. Winge and Roberts (1952) recorded a single instance of similar failure of nuclear fusion in a zygote formed by fusion of spores. Diploid hybrid cells budded off from a zygote can usually be distinguished from haploid cells by their larger size and more oval shape. However, some haploid cells may self-diploidize and the diploid cells so produced, though smaller than hybrid cells, are easily mistaken for them. These observations emphasize the importance of isolating single cells budded off from zygotes, and the need for establishing hybridity by transfer of single-cell cultures to a sporulation medium, followed by genetical analysis of spores.

3. *Mass Mating*

Hybridization by mass mating was introduced by Lindegren and Lindegren (1943a). Their method involves mixing together, in a nutrient medium, large numbers of haploid cells of opposite mating type. The resultant mating mixture is left overnight at 16° and then examined for the presence of zygotes. If a mating is positive, the mixture of zygotes and cells is transferred to a presporulation medium and later to a sporulation medium. Four-spored asci are then isolated for genetical analysis of the spores (Lindegren, 1945, 1949). The method was devised primarily therefore for genetical studies, but it was subsequently recommended for the hybridization of commercial yeasts (Lindegren and Lindegren, 1943d; Lindegren, 1944).

Lindegren's technique has been criticized by Winge and Roberts (1948) on the grounds that some or all of the diploid cells in a mating mixture may arise by self-diploidization. Many yeasts, however, show little or no trace of self-diploidization, and Lindegren (1949) has to a large extent justified the use of his technique. First, he uses stocks which have been selected for types in which self-diploidization is rare, and the hybrids are heterozygous for several alleles so that they can be readily distinguished by genetical analysis.

Palleroni (1949) suggested plating the mating mixture on an agar plate, hybrid colonies being distinguished by their larger size; hybridity of colonies was confirmed by genetical analysis. This method ensures the production of pure hybrid cultures, but in my experience it is often unsuccessful, because differences in the size of hybrid and haploid colonies may be slight.

In genetical studies, hybrids may be conveniently isolated by prototrophic selection. This method, first used by Pomper and Burkholder (1949), involves crossing haploids with complementary nutritional deficiences, and plating the mating mixture on a minimal agar medium which only permits development of hybrid cells. Differences in pigment production and in carbon assimilation have also been utilized in the production and isolation of hybrids of *Saccharomyces* × *Zygosaccharomyces* (Wickerham, 1955; Wickerham and Burton, 1956a, b).

For industrial purposes, it is essential to obtain pure hybrid cultures because of the need for consistency of performance, e.g. as baker's yeasts. The application of the prototroph-recovery technique is regarded as impracticable, and also undesirable as the mutagenic treatment required for the induction of nutritional deficiencies in commercial yeasts may impair desirable characteristics such as dough-raising capacity and storage ability. The production of pure hybrids, free from contamination by haploid and other cells, can be assured by the following modification of Lindegren's technique (Fowell, 1951). After the mating mixture has been prepared and incubated overnight, it is subcultured vigorously 2–3 times in order to encourage the hybrid cells to outgrow the haploid cells. A number of large oval cells are then isolated by micromanipulation, and transferred to tubes of a medium such as malt-extract wort. The hybridity of the resultant cultures is checked by microscopal examination, and can be confirmed if necessary by subsequent genetical analysis. This modification of Lindegren's technique has proved suitable for the hybridization of all kinds of yeasts, including those which sporulate poorly, produce few viable spores, or whose haploid cultures have weak mating activity. Considerable success has been achieved in the production of improved types of industrial yeasts.

The hybridization of some yeasts by the above methods has sometimes proved difficult or impossible, but further improvements in technique have been rendered possible as the result of recent research on the physiology of conjugation (see Section IV.D, p. 355).

Jakob's (1962) three-stage technique has been slightly modified and used in conjunction with Fowell's (1951) modification of Lindegren's technique for the hybridization of yeasts (Fowell, 1966). In this method, it is not necessary to use haploids with complementary nutritional deficiencies as in the original technique of Jakob. Details of the procedure, which is based on the use of 16% (w/v) malt-extract wort, are shown in Fig. 14. Haploids of opposite mating type are mixed together in wort and aerated by shaking for 2 h. The cells are then compacted by centrifugation and left unshaken for 30 min; during this period, the biochemical reactions which precede conjugation are probably initiated. In the third stage, the cells are resuspended in fresh wort and aerated by shaking for 3 h in order to promote zygote formation. The mating mixture is then subcultured 2–3 times, and hybrid cells isolated by micromanipulation. According to Jakob, about two and a half times as many zygotes are produced by her technique than by the classical method of Lindegren. Our experiments, however, indicate that Jakob's technique may

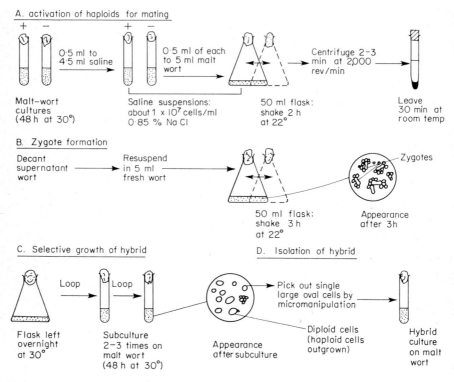

Fig. 14. Jakob's (1962) hybridization technique adapted for production of industrially useful hybrids.

only be suitable for use with haploids showing agglutinative properties. In the final aeration stage, the cells which are originally in close contact tend to be dispersed and sometimes very few zygotes are formed. A high degree of zygote formation has been observed in a centrifugate which was left standing for 4 h instead of being resuspended in wort and aerated; where difficulty is experienced in obtaining hybrids, therefore, it is advisable to omit the final aeration stage.

Haefner (1965) has devised a method of obtaining zygotes which is not only simpler than Jakob's technique but promises to be a more effective method of hybridization, particularly with certain yeasts. In this method, 0·2 ml of a liquid haploid culture, containing 10^7 cells, is spread on a plate of medium containing 2% (w/v) glucose, 1% (w/v) yeast extract, 0·5% (w/v) peptone and 2% (w/v) agar. The plate is then incubated at 30° for 24–36 h. The growth obtained in this way, for haploids of opposite mating type on separate plates, is transferred by replica plating to a common site on a fresh plate of the same medium. After this plate has been incubated for about 3 h at 30°, numerous zy-

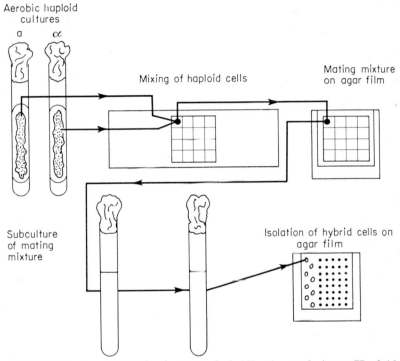

FIG. 15. Modification of Haefner's (1965) hybridization technique. Haploids of opposite mating type are incubated on agar slants for 24 h at 30°. Dense inocula of the two haploids are mixed in a water droplet on a glass slide, and the suspension spotted on an agar film (conjugation medium) on the lower surface of a cover slip (roof of the moist chamber). After incubation for 3–4 h at 20–25°, the mating mixture containing zygotes is subcultured 2–3 times, and hybrid cells are isolated, e.g. by micromanipulation on an agar film. A medium containing about 10% (w/v) sugar (e.g. 16%, w/v, malt agar) should be used for growth and conjugation of haploids. The grid marked on the slide and cover slip is to enable several matings of different haploids to be accommodated on the same agar film. Similarly, hybrid cells from several matings can be isolated on a common agar film. Indian ink spots on the cover-slip bearing the film mark the position of the isolated cells.

gotes can be observed under the low power of a microscope. Additional numbers of zygotes appear if the plate is afterwards placed at 4° for 2 h. It is claimed that, on the average, about 25% of the cells in a mating mixture form zygotes.

I have simplified Haefner's method to enable several hybrids to be prepared simultaneously with a minimum of materials (Fig. 15). Haploid cells are grown on slants of 16% (w/v) malt-extract agar for 24 h at 30°. A microspatula is used to transfer a small bead of cells, from haploid growths of opposite mating type, to a droplet of water inside one of the squares, marked out for the required number of matings, on a slide kept under moist conditions in a Petri dish. The microspatula is used to mix the cells together in the water and a small droplet, containing a dense suspension of cells, is then deposited in a marked position on a 16% malt-extract agar film in a moist chamber. The matings are left for 3–4 h at room temperature and then examined microscopically for zygote formation; after this, they are subcultured in the usual manner and hybrid cells isolated by micromanipulation.

4. *Transformation*

Oppenoorth (1961, 1962) has claimed that it is possible to change fermentation and other characteristics of one yeast by growth in a medium containing DNA from a donor yeast. Efforts to repeat this work (Harris and Thompson, 1960; Laskowski and Lochmann, 1961) have not been successful, and Oppenoorth's results are open to other interpretations. If the validity of this method could be established, it could prove useful for hybridization of yeasts which cannot sporulate or the spores of which are non-viable. Recent attempts at transformation have been briefly discussed by Mortimer and Hawthorne (1966).

References

Abdel-Wahab, M. F., Miller, J. J., Gabriel, O. and Hoffmann-Ostenhof, O. (1961). *Mh. Chem.* **92**, 22–30.
Adams, A. M. (1949). *Can. J. Res.* C. **27**, 179–189.
Adams, A. M. and Miller, J. J. (1954). *Can. J. Bot.* **32**, 320–334.
Ahmad, M. (1948). Ph.D. Thesis: Cambridge University.
Ahmad, M. (1952). *Nature, Lond.* **170**, 546–547.
Ahmad, M. (1953). *Ann. Bot.* **17**, 329–342.
Ahmad, M. (1965). *In* "Incompatibility in Fungi" (K. Esser and J. R. Raper, eds.), pp. 13–23. Springer-Verlag, New York.
Baltatu, G. (1939). *Zentbl. Bakt. ParasitKde (Abt II)* **101**, 196–225.
Barker, B. T. P. (1901). *Proc. R. Soc.* **194**, 467–485.
Barker, B. T. P. (1902). *J. Inst. Brew.* **8**, 26–76.
Bautz, E. (1955). *Z. Naturf.* **10b**, 313–316.
Beavens, E. A. (1940). Ph.D. Thesis: Cornell University.

Bedford, C. L. (1941). Ph.D. Thesis: California University.
Bedford, C. L. (1942). *Mycologia* **34**, 628–649.
Beijerinck, M. W. (1897). *Zentbl. Bakt. ParasitKde (Abt II)* **3**, 449–455.
Bettelheim, K. A. and Gay, J. L. (1963). *J. appl. Bact.* **26**, 224–231.
Brandt, K. (1941). *Protoplasma* **36**, 77–119.
Bright, T. B., Dixon, P. A. and Whymper, J. W. T. (1949). *Nature, Lond.* **164**, 544.
Brock, T. D. (1958a). *J. Bact.* **75**, 697–701.
Brock, T. D. (1958b). *J. Bact.* **76**, 334–335.
Brock, T. D. (1959a). *Science, N.Y.* **129**, 960.
Brock, T. D. (1959b). *J. Bact.* **78**, 59–68.
Brock, T. D. (1961). *J. gen. Microbiol.* **26**, 487–497.
Brock, T. D. (1965a). *J. Bact.* **90**, 1019–1025.
Brock, T. D. (1965b). *Proc. natn. Acad. Sci. U.S.A.* **54**, 1104–1112.
Catcheside, D. G. (1951). "The Genetics of Micro-organisms", 223 pp. Pitman, London.
Chen, S. Y. (1950). *C. r. hebd. Séanc. Acad. Sci., Paris* **230**, 1897–1899.
Clamp, J. R. and Hough, L. (1965). *Biochem. J.* **94**, 17–24.
Conti, S. F. and Brock, T. D. (1965). *J. Bact.* **90**, 524–533.
Conti, S. F. and Naylor, H. B. (1960). *J. Bact.* **79**, 331–340.
Cox, B. S. and Bevan, E. A. (1961). *Trans. Br. mycol. Soc.* **44**, 239–242.
Crandall, M. A. (1968). Ph.D. Thesis: Indiana University.
Crandall, M. A. and Brock, T. D. (1968a). *Science, N.Y.* **161**, 473–475.
Crandall, M. A. and Brock, T. D. (1968b). *Nature, Lond.* **219**, 533–534.
Crandall, M. A. and Brock, T. D. (1968c). *Bact. Rev.* **32**, 139–163.
Croes, A. F. (1967a). *Planta* **76**, 209–226.
Croes, A. F. (1967b). *Planta* **76**, 227–237.
Deysson, G. and Lau, N. T. (1963). *Annls pharm. fr.* **21**, 275–285.
Duggan, P. F. (1964). *Biochim. biophys. Acta* **88**, 223–224.
Eaton, N. R. (1960). *Archs Biochem. Biophys.* **88**, 17–25.
Eaton, N. R. and Klein, H. P. (1954). *J. Bact.* **68**, 110–116.
Elder, M. L. (1932). Ph.D. Thesis: Toronto University.
Elder, M. L. (1933). M.A. Thesis: Toronto University.
Elder, M. L. (1937). Ph.D. Thesis: Toronto University.
Emeis, C. C. (1958a). *Naturwissenschaften* **45**, 441.
Emeis, C. C. (1958b). *Brauerei wiss. Beil* **11**, 160–163.
Emeis, C. C. (1959). *Brauerei wiss. Beil.* **12**, 135–138.
Emeis, C. C. (1966). *Z. Naturf.* **21b**, 816–817.
Emeis, C. C. and Gutz, H. (1958). *Z. Naturf.* **13b**, 647–650.
Emeis, C. C. and Windisch, S. (1960). *Z. Naturf.* **15b**, 702–706.
Enebo, L., Johnsson, E., Nordström, K. and Möller, A. (1960). *Svensk BryggTidskr* **75**, 273–286.
Engel, L. (1872). D.Sc. Thesis: University of Paris.
Ephrussi, B. and Hottinguer, H. (1951). *Cold Spring Harb. Symp. quant. Biol.* **16**, 75–85.
Fonbrune, P., de (1949). "Technique de Micromanipulation", 203 pp. Masson, Paris.
Fowell, R. R. (1951). *J. Inst. Brew.* **57**, 180–195.
Fowell, R. R. (1952). *Nature, Lond.* **170**, 578.
Fowell, R. R. (1955). *J. appl. Bact.* **18**, 149–160.
Fowell, R. R. (1956a). *C. r. Trav. Lab. Carlsberg. Sér. physiol.* **26**, 117–138.

Fowell, R. R. (1956b). *Trans. Br. mycol. Soc.* **39**, 388–389.

Fowell, R. R. (1966). *Process Biochem.* **1** (1), 25–28.

Fowell, R. R. (1967). *J. appl. Bact.* **30**, 450–474.

Fowell, R. R. and Moorse, M. E. (1960). *J. appl. Bact.* **23**, 53–68.

Fuchs, J. (1935). *Wschr. Brau.* **52**, 165–166.

Galloway, L. D. (1954). *Lab. Pract.* **3**, 116.

Ganesan, A. T., Holter, H. and Roberts, C. (1958). *C. r. Trav. Lab. Carlsberg. Sér physiol.* **31**, 1–6.

Gilmore, R. A. (1967). *Genetics, Princeton* **56**, 641–658.

Goodman, J. and Rothstein, A. (1957). *J. gen. Physiol.* **40**, 915–923.

Gorodkowa, A. A. (1908). *Izv. imp. S.-Peterb. bot. Sada* **8**, 163–170.

Graham, V. R. and Hastings, E. G. (1941). *Can. J. Res.* C. **19**, 251–256.

Gronlund, A. F. and Campbell, J. J. R. (1961). *J. Bact.* **81**, 721–724.

Guilliermond, A. (1901). *C. r. hebd. Séanc. Acad. Sci., Paris* **133**, 1252.

Guilliermond, A. (1910). *Zentbl. Bakt. ParasitKde (Abt II)* **26**, 577–589.

Guilliermond, A. (1911a). *C. r. Séanc. Soc. Biol.* **70**, 442–444.

Guilliermond, A. (1911b). *C. r. hebd. Séanc. Acad. Sci., Paris* **152**, 448–450.

Guilliermond, A. (1918). *Bull. Soc. mycol. Fr.* **34**, 111–122.

Guilliermond, A. (1920a). *Bull. Soc. mycol. Fr.* **36**, 203–210.

Guilliermond, A. (1920b). "The Yeasts". (translated and revised by F. W. Tanner), 424 pp. Wiley, New York.

Guilliermond, A. (1928). "Titres et Travaux Scientifiques (1900–1928)". Laval.

Gutz, H. (1958). *Brauerei wiss. Beil.* **11**, 149–155.

Gutz, H. (1966). *J. Bact.* **92**, 1567–1568.

Gutz, H. (1967). *Science, N.Y.* **158**, 796–798.

Haefner, K. (1965). *Z. allg. Mikrobiol.* **5**, 77.

Hagedorn, H. (1964). *Protoplasma* **58**, 250–268.

Hansen, E. C. (1883). *C. r. Trav. Lab. Carlsberg* **2**, 13–47.

Hansen, E. C. (1889). *Zentbl. Bakt. ParasitKde* **5**, 632–640.

Hansen, E. C. (1891). *C. r. Trav. Lab. Carlsberg* **3**, 44–66.

Hansen, E. C. (1899). *Zentbl. Bakt. ParasitKde (Abt II)* **5**, 1–6.

Hansen, E. C. (1900). *C. r. Trav. Lab. Carlsberg* **5**, 1–38.

Hansen, E. C. (1902). *C. r. Trav. Lab. Carlsberg* **5**, 68–107.

Hansen, E. C. (1907). *Zentbl. Bakt. ParasitKde (Abt I)* **45**, 466–480.

Harris, G. and Thompson, C. C. (1960). *Nature, Lond.* **188**, 1212–1213.

Hartelius, V. and Ditlevsen, E. (1951). *Nature, Lond.* **168**, 385.

Hartelius, V. and Ditlevsen, E. (1953). *C. r. Trav. Lab. Carlsberg. Sér physiol.* **25**, 213–239.

Hartelius, V. and Ditlevsen, E. (1956). *C. r. Trav. Lab. Carlsberg. Sér. physiol.* **25**, 369–381.

Hashimoto, T., Conti, S. F. and Naylor, H. B. (1958). *J. Bact.* **76**, 406–416.

Hashimoto, T., Conti, S. F. and Naylor, H. B. (1959). *J. Bact.* **77**, 344–354.

Hashimoto, T., Gerhardt, P., Conti, S. F. and Naylor, H. B. (1960). *J. biophys. biochem. Cytol.* **7**, 305–310.

Hawthorne, D. C. (1963). *Genetics, Princeton* **48**, 1727–1729.

Herman, A. and Griffin, P. (1967). *Genetics, Princeton* **56**, 564.

Herman, A. and Roman, H. (1966). *Genetics, Princeton* **53**, 727–740.

Herman, A., Wickerman, L. J. and Griffin, P. (1966). *Genetics, Princeton* **54**, 339.

Hoffmeister, C. (1900). *Sber. dt. naturw.-med. Ver. Böhm. "Lotos"* **20**, 251–262.

Hunt, D. E. and Carpenter, P. L. (1963). *J. Bact.* **86**, 845–847.

Iguti, S. and Ônobu, T. (1964). *Bot. Mag., Tokyo* **77**, 181–190.

Ingram, M. (1955). "An Introduction to the Biology of Yeasts", 273 pp. Pitman, London.

Jakob, H. (1962). *C. r. hebd. Séanc. Acad. Sci., Paris* **254**, 3909–3911.

Johnston, J. R. (1963). *European Brewery Conv., Proc. Congr. 9th.*, Brussels, pp. 412–421.

Johnston, J. R. (1965a). *J. Inst. Brew.* **71**, 130–135.

Johnston, J. R. (1965b). *J. Inst. Brew.* **71**, 135–137.

Johnston, J. R. and Mortimer, R. K. (1959). *J. Bact.* **78**, 292.

Kaisha, K. Y. S. K. and Kaisha, M. S., Japan, Ltd. (1962). French Patent No. 1,333,739.

Kamisaka, S., Masuda, Y. and Yanagishima, N. (1967b). *Physiologia Pl.* **20**, 98–105.

Kamisaka, S., Masuda, Y. and Yanagishima, N. (1967c). *Pl. Cell Physiol., Tokyo* **8**, 121–127.

Kamisaka, S., Yanagishima, N. and Masuda, Y. (1967a). *Physiologia Pl.* **20**, 90–97.

Kirsop, B. H. (1954). *J. Inst. Brew.* **60**, 393–399.

Kirsop, B. H. (1957). Ph.D. Thesis: London University.

Kleyn, J. G. (1954). *Wallerstein Labs Commun.* **17**, 91–104.

Klöcker, A. (1924). "Die Gärungsorganismen". 3rd. ed. Urban and Schwarzenberg, Berlin.

Koninklijke Nederlandsche Gist– en Spiritusfabriek N. V. (1962). British Patent No. 989,247.

Kornberg, H. L. (1965). *In* "Function and Structure in Micro-organisms" (M. R. Pollock and M. H. Richmond, eds.), pp. 8–31. University Press, Cambridge.

Kornberg, H. L. and Elsden, S. R. (1961). *Adv. Enzymol.* **23**, 401–470.

Kruis, K. and Šatava, J. (1918). Cited in Winge, Ö. (1935). *C. r. Trav. Lab. Carlsberg. Sér. physiol.* **21**, Postscript, pp. 110–111.

Kudriavzev, V. I. (1954). "Yeast Systematics". Academy of Sciences, U.S.S.R., Moscow.

Kufferath, H. (1929). *Annls Soc. Zymol.* **1**, 214.

Kufferath, H. (1930). *Annls Soc. Zymol.* **2**, 33.

Langeron, M. and Luteraan, P. J. (1947). *Annls Parasit. hum. comp.* **22**, 254–275.

Laskowski, W. and Lochmann, E. R. (1961). *Naturwissenschaften* **48**, 225.

Leupold, U. (1950). *C. r. Trav. Lab. Carlsberg. Sér. physiol.* **24**, 381–480.

Leupold, U. (1956). *C. r. Trav. Lab. Carlsberg. Sér. physiol.* **26**, 221–251.

Leupold, U. (1958). *Cold Spring Harb. Symp. quant. Biol.* **23**, 161–170.

Levi, J. D. (1956). *Nature, Lond.* **177**, 753–754.

Lewis, M. J. and Phaff, H. J. (1964). *J. Bact.* **87**, 1389–1396.

Lhoas, P. (1961). *Nature, Lond.* **190**, 744.

Lindegren, C. C. (1944). *Wallerstein Labs Commun.* **7**, 153–168.

Lindegren, C. C. (1945). *Bact. Rev.* **9**, 111–170.

Lindegren, C. C. (1949). "The Yeast Cell, its Genetics and Cytology", 365 pp. Educational Publishers, St. Louis.

Lindegren, C. C. (1958). *Proc. Am. Soc. Brew. Chem.*, pp. 87–91.

Lindegren, C. C. (1962). "Yeast Genetics—1962", 21 pp. Southern Illinois University, Carbondale, Illinois.

Lindegren, C. C. and Hamilton, E. (1944). *Bot. Gaz.* **105**, 316–321.

Lindgren, C. C. and Lindegren, G. (1943a). *Proc. natn. Acad. Sci. U.S.A.* **29**, 306–308.

Lindegren, C. C. and Lindegren, G. (1943b). *Genetics, Princeton* **28**, 81.

Lindegren, C. C. and Lindegren, G. (1943c). *Ann. Mo. bot. Gdn* **30**, 453–469.

Lindegren, C. C. and Lindegren, G. (1943d). *J. Bact.* **46**, 405–419.

13*

Lindegren, C. C. and Lindegren, G. (1944a). *Bot. Gaz.* **105**, 304–316.

Lindegren, C. C. and Lindegren, G. (1944b). *Ann. Mo. bot. Gdn* **31**, 203–216.

Lindegren, C. C. and Lindegren, G. (1951). *J. gen. Microbiol.* **5**, 885–893.

Lindegren, C. C. and Lindegren, G. (1953). *Genetics, Princeton* **38**, 73–78.

Lindegren, C. C. and Lindegren, G. (1954). *Cytologia* **19**, 45–47.

Lindner, P. (1896). *Zenbtl. Bakt. ParasitKde (Abt II)* **2**, 537–539.

Lodder, J. and Kreger-van Rij. N. J. W. (1952). "The Yeasts: a Taxonomic Study", 713 pp. North Holland Publishing Co., Amsterdam.

Magasanik, B. (1961). *Cold Spring Harb. Symp. quant. Biol.* **26**, 249–256.

Magni, G. E. (1963). *Proc. natn. Acad. Sci. U.S.A.* **50**, 975–980.

Magni, G. E. and von Borstel, R. C. (1962). *Genetics, Princeton* **47**, 1097–1108.

Maneval, W. E. (1924). *Bot. Gaz.* **78**, 122–123.

Marquardt, H. (1963). *Arch. Mikrobiol.* **46**, 308–320.

McClary, D. O. and Nulty, W. L. (1957). *Bact. Proc.* 54.

McClary, D. O., Nulty, W. L. and Miller G. R. (1959). *J. Bact.* **78**, 362–368.

McClary, D. O., Williams, M. A. and Lindegren, C. C. (1957b). *J. Bact.* **73**, 754–757.

McClary, D. O., Williams, M. A., Lindegren, C. C. and Ogur, M. (1957a). *J. Bact.* **73**, 360–364.

McKelvey, C. E. (1926). *J. Bact.* **11**, 98–99.

Miller, G. R., McClary, D. O. and Bowers, W. D. (1963). *J. Bact.* **85**, 725–731.

Miller, J. J. (1957). *Can. J. Microbiol.* **3**, 81–90.

Miller, J. J. (1959). *Wallerstein Labs Commun.* **22**, 267–283.

Miller, J. J. (1963a). *Can. J. Microbiol.* **9**, 259–277.

Miller, J. J. (1963b). *Nature, Lond.* **198**, 214–215.

Miller, J. J., Calvin, J. and Tremaine, J. H. (1955). *Canad. J. Microbiol.* **1**, 560–573.

Miller, J. J., Gabriel, O., Scheiber, E. and Hoffmann-Ostenhof, O. (1957b). *Mh. Chem.* **88**, 417–420.

Miller, J. J. and Halpern, C. (1956). *Can. J. Microbiol.* **2**, 519–537.

Miller, J. J. and Hoffmann-Ostenhof, O. (1964). *Z. allg. Mikrobiol.* **4**, 273–294.

Miller, J. J., Hoffmann-Ostenhof, O., Scheiber, E. and Gabriel, O. (1959). *Can. J. Microbiol.* **5**, 153–159.

Miller, J. J., Scheiber, E., Gabriel, O. and Hoffmann-Ostenhof, O. (1957a). *Mh. Chem.* **88**, 271–274.

Morris, E. O. (1958). *In* "The Chemistry and Biology of Yeasts" (A. H. Cook, ed.), pp. 251–321. Academic Press, New York.

Mortimer, R. K. and Hawthorne, D. C. (1966). *A. Rev. Microbiol.* **20**, 151–168.

Motoc, D. and Dimitriu, C. (1966). *Inds aliment. agric.* **17**, 5–12.

Mrak, E. M., Phaff, H. J. and Douglas, H. C. (1942). *Science, N.Y.* **96**, 432.

Mundkur, B. (1961a). *Expl Cell Res.* **25**, 1–23.

Mundkur, B. (1961b). *Expl Cell Res.* **25**, 24–40.

Nadson, G. A. and Konokotina, A. G. (1911). *Izv. imp. S.-Peterb. bot. Sada* **11**, 117–143.

Nagashima, T. (1959). *Ecol. Rev., Sendai* **15**, 75–78.

Nagel, L. (1946). *Ann. Mo. bot. Gdn* **33**, 249–289.

Nägeli, C., von and Loew, O. (1880). *J. prakt. Chem.* **21**, 97–114.

Nickerson, W. J. and Thimann, K. V. (1941). *Am. J. Bot.* **28**, 617–621.

Nickerson, W. J. and Thimann, K. V. (1943). *Am. J. Bot.* **30**, 94–101.

Ochmann, W. (1932). *Zentbl. Bakt. ParasitKde (Abt II)* **86**, 458–465.

Oehlkers, F. (1923). *Ber. dt. bot. Ges.* **41**, 31–32.

Oeser, H. (1962). *Arch. Mikrobiol.* **44**, 47–74.

Okuda, S. (1959). *Ecol. Rev., Sendai* **15**, 25–30.

Okuda, S. (1961). *Pl. Cell Physiol., Tokyo* **2**, 371–381.
Olenov, J. M. (1936). *Arch. Mikrobiol.* **7**, 264–285.
Oppenoorth, W. F. F. (1956). *Nature, Lond.* **178**, 992–993.
Oppenoorth, W. F. F. (1957). *European Brewery Conv., Proc. Congr. 6th., Copenhagen*, 222–240.
Oppenoorth, W. F. F. (1957). *European Brewery Conv., Proc. Congr. 7th., Rome*, 180–207.
Oppenoorth, W. F. F. (1961). *European Brewery Conv., Proc. Congr. 8th., Vienna*, 172–204.
Oppenoorth, W. F. F. (1962). *Nature, Lond.* **193**, 706.
Palleroni, N. J. (1949). *Revta Fac. Agron. Vet. Univ. B. Aires* **12**, 302–317.
Palleroni, N. J. (1961). *Phyton, B. Aires* **16**, 117–128.
Pazonyi, B. (1954). *Acta microbiol. hung.* **1**, 49–70.
Pazonyi, B. and Márkus, L. (1955). *Agrokém. Talajt.* **4**, 225–234.
Phaff, H. J. and Mrak, E. M. (1948). *Wallerstein Labs Commun.* **11**, 261–279.
Phaff, H. J. and Mrak, E. M. (1949). *Wallerstein Labs Commun.* **12**, 29–44.
Pomper, S. and Burkholder, P. R. (1949). *Proc. natn. Acad. Sci. U.S.A.* **35**, 456–464.
Pomper, S., Daniels, K. M. and McKee, D. W. (1954). *Genetics, Princeton* **39**, 343–355.
Pontefract, R. D. and Miller, J. J. (1962). *Can. J. Microbiol.* **8**, 573–584.
Purvis, J. E. and Warwick, G. R. (1908). *Proc. Camb. phil. Soc. biol. Sci.* **14**, 30–40.
Race, R. R. (1960). *J. Am. med. Ass.* **174**, 1181–1187.
Ramirez, C. and Miller, J. J. (1963). *Nature, Lond.* **197**, 722–723.
Ramirez, C. and Miller, J. J. (1964). *Can. J. Microbiol.* **10**, 623–631.
Raper, J. R. (1954). *In* "Sex in Microorganisms" (D. H. Wenrich, I. F. Lewis and J. R. Raper, eds.), pp. 42–81. American Association for the Advancement of Science, Washington.
Reess, M. (1869). *Bot. Ztg.* **27**, 105–118.
Reess, M. (1870). "Botanische Untersuchungen über die Alkohol-gärungspilze". Felix, Leipzig.
Reiff, F., Kautzmann, R., Lüers, H. and Lindemann, M. (1960). "Die Hefen. I. Die Hefen in der Wissenschaft", 1024 pp. H. Carl, Nürnberg.
Renaud, J. (1937). *C. r. hebd. Séanc. Acad. Sci., Paris* **204**, 1274–1276.
Renaud, J. (1938). *C. r. hebd. Séanc. Acad. Sci., Paris* **206** 1397–1399.
Renaud, J. (1946). *Revue gén. Bot.* **629**, 193–211; **630**, 241–274; **631**, 289–314.
Resnick, M. A., Tippetts, R. D. and Mortimer, R. K. (1967). *Science, N.Y.* **158**, 803–804.
Richards, M. (1965). *J. Inst. Brew.* **71**, 459–460.
Roberts, C. (1950). *Meth. med. Res.* **3**, 37–50.
Roberts, C. and Walt, J. P., van der (1959). *C. r. Trav. Lab. Carlsberg. Sér. physiol.* **31**, 129–148.
Roman, H., Hawthorne, D. C. and Douglas, H. C. (1951). *Proc. natn. Acad. Sci. U.S.A.* **37**, 79–84.
Roman, H. and Sands, S. M. (1953). *Proc. natn. Acad. Sci. U.S.A.* **39**, 171–179.
Saito, K. (1916). *J. Coll. Sci. imp. Univ. Tokyo* **39**, 1–73.
Saito, K. (1923). *Bot. Mag., Tokyo* **37**, 63–66.
Sando, N. (1956). *Sci. Rep. Tôhoku Univ. Ser. IV (Biol.)* **22**, 99–113.
Sando, N. (1959). *Sci. Rep. Tôhoku. Ser. IV (Biol.)* **25**, 263–266.
Sando, N. (1960a). *Sci. Rep. Tôhoku Univ. Ser. IV (Biol.)* **26**, 139–152.
Sando, N. (1960b). *Sci. Rep. Tôhoku Univ. Ser. IV (Biol.)* **26**, 153–156.
Sando, N. (1960c). *Sci. Rep. Tôhoku Univ. Ser. IV (Biol.)* **26**, 157–161.
Santa María, J. (1957a). *J. Bact.* **74**, 692–693.

Santa María, J. (1957b). *Boln Inst. nac. Invest. agron., Madr.* **17**, 231–238.

Santa María, J. (1958). *Nature, Lond.* **181**, 1740.

Santa María, J. (1959). *Ann. Inst. nac. Invest. agron.* **8**, 737–750.

Šatava, J. (1918). Cited in Winge, Ö. (1935). *C. r. Trav. Lab. Carlsberg. Sér. physiol.* **21**. Postscript, pp. 110–111.

Satava, J. (1934). *Congr. int. tech. chim. Sucr. Distill.* 3. Paris.

Scheiber, E., Gabriel, O., Hoffmann-Ostenhof, O. and Miller, J. J. (1957). *Mh. Chem.* **88**, 414–417.

Schionning, H. (1895). *C. r. Trav. Lab. Carlsberg* **4**, 30–35.

Schopfer, W. H., Wustenfeld, D. and Turian, G. (1963). *Arch. Mikrobiol.* **45**, 304–313.

Schumacher, J. (1926). *Zentbl. Bakt. ParasitKde (Abt II)* **98**, 67–81.

Schwann, T. (1839). "Mikroskopische Untersuchungen über die Übereinstimmung in der Struktur und dem Wachstum der Tiere und Pflanzen". Berlin.

Seynes, J., de (1868). *C. r. hebd. Séanc. Acad. Sci., Paris* **67**, 105–109.

Shevchenko, L. A., Birusova, V. I. and Meissel, M. N. (1966). *Mikrobiologiya* **35**, 85–91.

Stantial, H. (1928). *Trans. R. Soc. Can. III* **22**, 257–261.

Stantial, H. (1935). *Trans. R. Soc. Can. III* **29**, 175–188.

Steinberg, A. G. and Giles, B. D. (1959). *Am. J. hum. Genet.* **11**, 380–384.

Stelling-Dekker, N. M. (1931). "Die Hefesammlung des Centraalbureau voor Schimmelcultures. I. Die Sporogenen Hefen", 547 pp. North Holland Publishing Co., Amsterdam.

Stoppani, A. O. M., Conches, L., Favelukes, S. L. S., de and Sacerdote, F. L. (1958). *Biochem. J.* **70**, 438–455.

Stovall, W. K. and Bulotz, A. (1932a). *J. infect Dis.* **50**, 73–88.

Stovall, W. K. and Bulotz, A. (1932b). *Am. J. publ. Hlth* **22**, 493–501.

Svihla, G., Dainko, J. L. and Schlenk, F. (1964). *J. Bact.* **88**, 449–456.

Takahashi, T. (1958). *Genetics, Princeton* **43**, 705–714.

Takahashi, T. (1961). *Seiken Ziho* **12**, 11–20.

Takahashi, T. (1964). *Bull. Brew. Sci.* **10**, 11–22.

Takahashi, T. and Ikeda ,Y. (1959). *Genetics, Princeton* **44**, 375–382.

Takahashi, T., Saito, H. and Ikeda, Y. (1958). *Genetics, Princeton* **43**, 249–260.

Takano, I and Oshima, Y. (1967). *Genetics, Princeton* **57**, 875–885.

Tamaki, H. (1965). *J. gen. Microbiol.* **41**, 93–98.

Tanaka, H. and Phaff, H. J. (1965). *J. Bact.* **89**, 1570–1580.

Taylor, N. W. (1964a). *J. Bact.* **87**, 863–866.

Taylor, N. W. (1964b). *J. Bact.* **88**, 929–936.

Taylor, N. W. (1965). *Archs Biochem. Biophys.* **111**, 181–186.

Taylor, N. W. and Orton, W. L. (1967). *Archs Biochem. Biophys.* **120**, 602–608.

Taylor, N. W., Orton, W. L. and Babcock, G. E. (1968). *Archs Biochem. Biophys.* **123**, 265–270.

Taylor, N. W. and Tobin, R. (1966). *Archs Biochem. Biophys.* **115**, 271–276.

Thaysen, A. C. and Morris, A. R. (1947). *J. gen. Microbiol.* **1**, 221–231.

Thorne, R. S. W. (1951). *C. r. Trav. Lab. Carlsberg. Sér. physiol.* **25**, 101–140.

Todd, R. L. and Hermann, W. W. (1936). *J. Bact.* **32**, 89–103.

Tremaine, J. H. and Miller, J. J. (1954). *Bot. Gaz.* **115**, 311–322.

Tremaine, J. H. and Miller, J. J. (1956). *Mycopath. Mycol. appl.* **7**, 241–250.

Tsukahara, T. and Yamada, M. (1953). *J. Soc. Brew. Japan* **48**, No. 8.

Wagner, F. (1928). *Zentbl. Bakt. ParasitKde (Abt II)* **75**, 4–24.

Walt, J. P., van der (1956). *Antonie van Leeuwenhoek* **22**, 265–272.

Welten, H. (1914). *Mikrokosmos* **8**, 3–5; 41–43.

Wickerham, L. J. (1951). "Taxonomy of Yeasts", 56 pp. U.S. Dept. Agric. Tech. Bull. No. 1029, Washington.

Wickerham, L. J. (1955). *Nature, Lond.* **176**, 22.

Wickerham, L. J. (1956). *C. r. Trav. Lab. Carlsberg. Sér. physiol.* **26**, 423–443.

Wickerham, L. J. (1958). *Science, N.Y.* **128**, 1504–1505.

Wickerham, L. J. (1960). U.S. Patent No. 2,960,445.

Wickerham, L. J. and Burton, K. A. (1954). *J. Bact.* **67**, 303–308.

Wickerham, L. J. and Burton, K. A. (1956a). *J. Bact.* **71**, 290–295.

Wickerham, L. J. and Burton, K. A. (1956b). *J. Bact.* **71**, 296–302.

Wickerham, L. J. and Burton, K. A. (1962). *Bact. Rev.* **26**, 382–397.

Wickerham, L. J., Flickinger, M. H. and Burton, K. A. (1946). *J. Bact.* **52**, 611–612.

Widra, A. and DeLamater, E. D. (1955). *Am. J. Bot.* **42**, 423–435.

Windisch, S. (1938). *Arch. Mikrobiol.* **9**, 551–554.

Winge, Ö. (1935). *C. r. Trav. Lab. Carlsberg. Sér. physiol.* **21**, 77–109.

Winge, Ö. (1944). *C. r. Trav. Lab. Carlsberg. Sér. physiol.* **24**, 79–95.

Winge, Ö. and Laustsen, O. (1937). *C. r. Trav. Lab. Carlsberg. Sér. physiol.* **22**, 99–117.

Winge, Ö. and Laustsen, O. (1938). *C. r. Trav. Lab. Carlsberg. Sér. physiol.* **22**, 235–245.

Winge, Ö. and Laustsen, O. (1939a). *C. r. Trav. Lab. Carlsberg. Sér. physiol.* **22**, 337–353.

Winge, Ö and Laustsen, O. (1939b). *C. r. Trav. Lab. Carlsberg. Sér. physiol.* **22**, 357–371.

Winge, Ö and Laustsen, O. (1940). *C. r. Trav. Lab. Carlsberg. Sér. physiol.* **23**, 17–39.

Winge, Ö. and Roberts, C. (1948). *C. r. Trav. Lab. Carlsberg. Sér. physiol.* **24**, 263–315.

Winge, Ö. and Roberts, C. (1949). *C. r. Trav. Lab. Carlsberg. Sér. physiol.* **24**, 341–346.

Winge, Ö. and Roberts, C. (1950a). *Nature, Lond.* **165**, 157–158.

Winge, Ö. and Roberts, C. (1950b). *C. r. Trav. Lab. Carlsberg. Sér. physiol.* **25**, 35–83.

Winge, Ö. and Roberts, C. (1952). *C. r. Trav. Lab. Carlsberg. Sér. physiol.* **25**, 141–171.

Winge, Ö. and Roberts, C. (1954). *C. r. Trav. Lab. Carlsberg. Sér. physiol.* **25**, 285–329.

Wright, R. E. and Lederberg, J. (1957). *Proc. natn. Acad. Sci. U.S.A.* **43**, 919–937.

Yanagishima, N. and Shimoda, C. (1967). *Pl. Cell. Physiol., Tokyo* **8**, 109–119.

Zetlin, S. (1914). *Biedermanns Zbl. AgrikChem.* **43**, 499–501.

Chapter 8

Yeast Genetics

ROBERT K. MORTIMER AND DONALD C. HAWTHORNE

*Donner Laboratory, University of California, Berkeley,
California 94720, U.S.A., and Department of Genetics,
University of Washington, Seattle, Washington 98105, U.S.A.*

I. Introduction

A prelude to genetic studies with yeast was the demonstration of haploid and diploid phases in the life cycle of *Saccharomyces ellipsoideus* Hansen by Winge (1935). This observation was followed by the first demonstration of Mendelian segregation in another species of yeast, *Saccharomycodes ludwigii* (Winge and Laustsen, 1939). In the ensuing three decades, interest in yeast genetics has increased greatly and, at present, the importance of yeast as an organism for genetic studies is widely recognized.

The yeasts are of intermediate biological complexity relative to bacteria and the higher fungi. Their study provides a test of the generality of the many concepts of molecular genetics established from studies on bacteria. Yeasts present many advantages for genetic studies, including rapid growth, clonability, and ease of handling and storage. They are adaptable to replica plating, micromanipulation, and an array of biochemical procedures.

The genetics of yeast has been the subject of a book by Lindegren (1949), and a chapter in a treatise on yeasts by Winge and Roberts (1958). A recent survey of the yeast genetics literature also has appeared (Mortimer and Hawthorne, 1966a). In this chapter, we wish to present a synthesis, based on representative genetic studies of yeast, to illustrate the current knowledge in this field and to demonstrate the potentialities of yeast for investigation of genetic problems.

The gene symbols used in this article are described in the report of the Carbondale Yeast Genetics Conference (1963), in our earlier article (Mortimer and Hawthorne, 1966b), or as they appear in the text. The symbols refer to the following phenotypes: a/α or h^+/h^- – mating type; nutritional requirements: ad – adenine, ar – arginine, hi – histidine, is – isoleucine or isoleucine plus valine, le – leucine, ly – lysine, met – methionine, ol – oleic acid, pha – phenylalanine, pn – pantothenate, py – pyridoxine, ser – serine, th – thiamine, thr – threonine plus methionine, tr – tryptophan, ty – tyrosine, tyrosine plus phenylalanine, or tyrosine plus phenylalanine plus tryptophan (also referred to as $arom$, Table V, p. 421), and ur – uracil; resistance genes: ac – actidione (cycloheximide), can – canavanine, CU – copper, ROC – roccal; carbohydrate fermentation: ga – galactose, MA – maltose, MEL – melibiose, MG – α-methylglucoside, MZ – melezitose, SU – sucrose; miscellaneous phenotypes: p – genetic petite, ρ – cytoplasmic petite, cy – cytochrome c, S – super-suppressor. Upper and lower case letters denote the dominant and recessive forms of the genes, respectively. Subscripts identify particular loci (first number) and alleles (second number).

II. Life Cycles

A. *SACCHAROMYCES*

Saccharomyces cerevisiae is diploid in the vegetative phase (Fig. 1). Meiosis precedes sporulation and normally four haploid spores are found in the ascus. Unless isolated individually, the spores will fuse with their complementary neighbours to restore the diploid state (Winge and Laustsen, 1937). In heterothallic strains, single-spore clones will remain haploid for many generations. By mixing four such haploid cultures derived from a single ascus in all possible pairwise combinations, Lindegren and Lindegren (1943) demonstrated that a single pair of alleles, a and α (see footnote for list of abbreviations), controlled the mating response. Mixing cultures of like mating-type gave no reaction, whereas mixing a and α cultures gave distorted premating cells and, within a few hours, zygotes. The mating-type alleles a and α and the

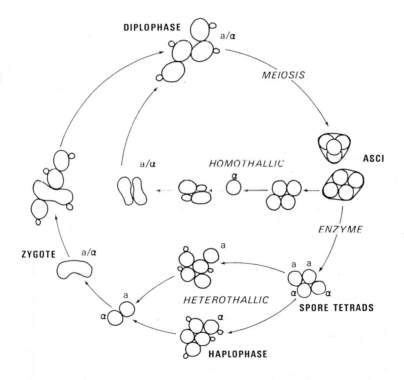

FIG. 1. Life cycle of *Saccharomyces* species. The diagram illustrates the alternation of haploid and diploid phases and the distinctions between heterothallic and homothallic species.

physiology of the mating process are common to several *Saccharomyces* species: *Sacch. bayanus, Sacch. carlsbergensis, Sacch. chevalieri, Sacch. chodati, Sacch. diastaticus, Sacch. italicus,* and *Sacch. oviformis.*

Even the homothallic strains, in which the haploid spores give rise to clones containing mostly diploid cells capable of sporulation, carry the same a and α mating-type alleles. This was demonstrated with the analysis of hybrids obtained by pairing spores of heterothallic *Sacch. cerevisiae* and homothallic *Sacch. chevalieri* (Winge and Roberts, 1949). The hybrids produced asci in which two spore clones remained haploid and displayed a mating specificity and two spore clones became diploid after a few generations and showed no further mating response. Haploid segregants of both a and α mating-type were recovered from each hybrid. The apparent lack of mating specificity in two of the spore clones is due to the segregation of a gene, D, for diploidization or homothallism. Although gene D is seemingly epistatic to the mating-type alleles, a and α, it in reality acts by causing their mutation. Under the influence of gene D, either a or α mutates to the complementary allele at rates attaining one switch in mating-type per two cell divisions (Oeser, 1962; Hawthorne, 1963a, b). The haploid cells fuse as soon as the mutant allele is expressed, and the zygotes give rise to stable diploids which are heterozygous for mating-type. The heterozygous condition of the mating-type alleles blocks any further action of the gene D. This phenomenon may be in the nature of a regulatory mechanism since there is evidence that the mating-type locus is a complex with cistrons for both a and α products (Hawthorne, 1963a; Takahashi, 1964).

Additional genes for homothallism have been detected in strains of *Sacch. cerevisiae* (Takahashi *et al.*, 1958; Takahashi, 1958). In one system, complementary genes were implicated by tetrad ratios of 0:4, 1:3, and 2:2 for homothallism versus heterothallism. Another case of complementary genes was described for *Sacch. oviformis* (Takano and Oshima, 1967). By itself, a gene designated HO_α could bring about the change of α to a, but the change from a to α required the presence of a modifier gene, HM, as well.

Another distinct interbreeding system in *Saccharomyces* is composed of *Sacch. lactis, Sacch. fragilis, Zygosacch. ashbyi,* and *Zygosacch. dobzhanskii* (Wickerham and Burton, 1956). Studies of the mating-type specificity and homothallism in *Sacch. lactis* demonstrated a close parallel to the above interpretation of the mating-type alleles and homothallism genes (Herman and Roman, 1966). In *Sacch. lactis,* the mating-type alleles a and α (these do not necessarily correspond to the alleles with the same designations in *Sacch. cerevisiae*) are subject to the influence of the respective mutator genes H_a and H_α; i.e. H_a effects the mutation of a to α while H_α does the converse. H_α is nine centimorgans from the mating-type locus while H_a is not linked to either mating-type or H_α. The centimorgan is the standard genetic map-unit. The length of a genetic interval in centimorgans is equal to one hundred times the average number of crossovers in that interval per chromatid (see p. 404).

Although the life cycle of sporogenous yeast normally alternates between haploid and diploid states, yeast cells of higher ploidy have been derived. In genetic analysis of diploid strains, occasional asci have been recovered that appear to have arisen from meiosis of a tetraploid cell (Roman *et al.*, 1951; Lindegren and Lindegren, 1951b; Leupold and Hottinguer, 1954). The spores derived from these asci have been shown to be diploid and their genotypes are those expected for meiotic products of a tetraploid cell.

Depending on the segregation of the mating-type alleles, these diploids may be maters (a/a and α/α) incapable of sporulation, or non-maters (a/α) which are able to sporulate. The diploids homozygous for mating-type have been used as parents in crosses to construct triploid ($a/a \times \alpha$; $a \times \alpha/\alpha$) or tetraploid strains ($a/a \times \alpha/\alpha$). Occasionally, diploids homozygous for mating-type arise spontaneously in haploid

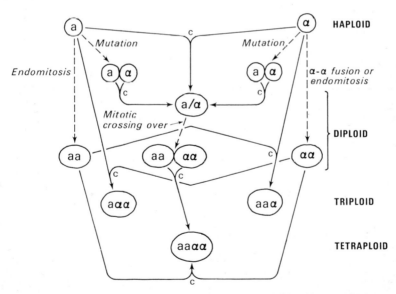

FIG. 2. Development of polyploid yeast cultures. Haploid cells may diploidize by mutation of the mating-type locus, by endomitosis, or by α-α fusion. Diploids homozygous for mating type can be used as parents for triploid and tetraploid crosses. Also, a/α diploids are proposed to yield tetraploids by occurrence of mitotic crossing-over between the mating-type locus and its centromere. The mating reaction (copulation) is indicated by c.

strains and these also have been used as parents in construction of polyploid lines (Roman and Sands, 1953; Mortimer, 1958; Laskowski, 1960). However, it is not essential to have diploids homozygous for mating-type in order to develop polyploid hybrids. Pomper *et al.* (1954) were able to effect hybridization of cells heterozygous for the mating-type locus ($a/\alpha \times a$, $a/\alpha \times \alpha$, $a/\alpha \times a/\alpha$). Parents that carried complementary nutritional requirements were incubated together and the mixture was plated on a medium on which only the hybrid would grow. By such procedures, rare matings were selected. A similar approach has been used to construct tetraploid series for gene dosage studies (Nelson and Douglas, 1963; Slonimski *et al.*, 1965). In addition, pentaploid and hexaploid cultures have been synthesized through selection of rare matings between triploids that bore complementary requirements (Mortimer, 1958; Laskowski, 1962). Some of the possible sequences of events leading to polyploid yeast are summarized in Fig. 2.

The average DNA content per cell was shown by Ogur *et al.* (1952) to increase in proportion to the ploidy in a haploid to tetraploid series. In addition, cell volume and dry weight, soluble protein, and RNA per cell, all have an approximate linear relationship to genome number. Genetic segregation in polyploids is discussed in Section IV, p. 405.

B. *SCHIZOSACCHAROMYCES*

The haplophase predominates in the life cycle of *Schizosaccharomyces pombe*. Malt-agar is used for the induction of the mating reaction, and the diploid zygote then normally undergoes meiosis to yield four haploid spores. Leupold (1950) reported the mating response to be under the control of three alleles, h^+ and h^- which are the heterothallic determinants, and h^{90} which confers homothallism. The complexity of the mating-type locus was revealed by crosses of $h^+ \times h^-$ which gave rare (0·4%) recombinants with the phenotype of the h^{90} allele. This result can be interpreted in the context of the observations already described (p. 388) with the *Saccharomyces* species to indicate that the heterothallic alleles are defective in one cistron or the other in the complex locus while the homothallic allele is free of defects. One only needs to postulate the presence of a diploidization gene, corresponding to the *D* gene of *Sacch. chevalieri*, which can switch the activity of the cistrons in the homothallic or 'wild type' gene complex but which is unable to effect a switch in a complex with a defective cistron.

Diploid cell lines of constitution $h^{90} h^+$ or $h^{90} h^-$ can be isolated on a selective medium and maintained vegetatively, enabling complementation studies or synthesis of tetraploids (Leupold, 1956a; Leupold and Gutz, 1963). The tetraploid zygotes give diploid spores, and from the appropriate crosses one can obtain $h^+ h^+$ or $h^- h^-$ segregants. These diploids strains, homozygous for the heterothallic alleles, were used to elucidate the role of the mating-type locus in the control of sporulation (Gutz, 1967). Diploids of constitution $h^{90} h^{90}$, $h^{90} h^+$, $h^{90} h^-$, and $h^+ h^-$ are able to undergo meiosis and form haploid spores directly when placed on malt-agar. The $h^+ h^+$ and $h^- h^-$ diploids do not sporulate directly, but when mixed give zygotic asci of two types. One type has four large diploid spores, the other has from five to eight smaller haploid spores. Genetic analysis indicated that the diploid spores were the products of a tetraploid meiosis, while the haploid spores came from two separate meioses of unfused parent diploid nuclei. Thus the presence of the h^+ and h^- gene products in the cytoplasm is sufficient for the initiation of meiosis; inclusion of the two alleles in the same nucleus is not essential.

III. Methodology

A. CULTURE CONDITIONS

Some of the media used in genetic studies with yeast are outlined below.

A non-synthetic complete medium, containing yeast extract, peptone, dextrose (YEPD) (and agar for solid medium) is used for maintenance of stocks and as a general purpose growth medium. A convenient recipe is 1%, 2%, 2% and 2%, respectively of the above constituents.

Synthetic media, containing trace elements, salts, a nitrogen source, biotin, and a carbon source, constitute a minimal medium suitable for yeast. The medium described by Wickerham (1946) and available commercially (Difco) is generally used. To the minimal medium are added combinations of amino acids, purines, pyrimidines, and vitamins to constitute synthetic complete and synthetic omission media.

The ability of a yeast to ferment a sugar can be assayed on the basis of gas formation. However, it is more convenient to score fermentation traits on solid medium in Petri plates. Non-synthetic complete medium is used in which dextrose has been replaced by the appropriate sugar. The medium is adjusted to neutrality or a slightly basic pH value and an indicator such as bromo-thymol blue is added. Acid production associated with fermentation results in yellow coloration of the agar.

To determine if a strain is respiratory-sufficient, the non-synthetic medium, modified to include a non-fermentable carbon source such as glycerol or lactate in place of dextrose, is used (Ogur and St John, 1956). Genic and cytoplasmic petite strains are unable to grow on this medium.

Media and general procedures used to induce sporulation of yeasts are described in Chapter 7 (p. 303) of this volume. The most commonly used sporulation medium is based on the one described by Fowell (1952.)

B. MUTAGENESIS

1. *Induction of Mutants*

Ultraviolet radiation, nitrous acid, ethylmethanesulphonate, and 1-methyl-1-nitro-nitrosoguanidine are among the mutagenic agents commonly used for inducing mutations in yeast. Protocols for the use of these agents are found, for example, in papers by Loprieno and Clarke (1965), Lindegren *et al.* (1965), Nasim and Auerbach (1967), and Guglielminetti *et al.* (1967). Also, most protocols used with bacteria are suitable for yeast with perhaps only slight modifications as to the time of treatment. With the chemical mutagens, care should be taken to see that the reaction mixture is adequately buffered so that the mutagen will be acting at the pH where its specificity is most pronounced. If the plating medium is not the standard maintenance medium (YEPD or its equivalent) then a post-treatment in YEPD broth for about one generation (90 min) is recommended for the isolation of forward mutations, i.e. mutations to auxotrophy or resistance to antimetabolites. A detailed study of post-treatment factors affecting mutation fixation has been presented by Zimmermann *et al.* (1966a).

2. RECOVERY OF MUTANTS

a. Total isolation of biochemical mutants. A haploid strain which has been mutagenized is plated on non-synthetic complete medium at dilutions to yield approximately 200 colonies per plate. Survival levels in the 10–50% range are optimum with regard to a balance between the frequency of induced mutations and other coincident changes such as petite induction or chromosome structural alterations. After 2–3 days' incubation, the resultant colonies are replica-plated to a synthetic medium lacking all nutrients for which mutants are sought. Any replicas failing to grow are classified as potential mutants. These are isolated from the original plate and retested on a series of synthetic media containing single nutrients or sets of nutrients in addition to those present in the minimal medium. From the pattern of growth on these various media, the specific requirement associated with the mutant can be deduced. The frequency of amino-acid and purine and pyrimidine mutants recovered following treatment with ultraviolet radiation or ethylmethanesulphonate is in the order of 1% (Lingens and Oltmanns, 1964; Lindegren *et al.*, 1965). For 1-methyl-3-nitro-1-nitrosoguanidine, a more potent mutagen, auxotroph frequencies in the order of 10% have been reported (Megnet, 1965a; Lingens and Oltmanns, 1966).

b. Isolation of mutants by colony morphology. Reaume and Tatum (1949) described an adenine-requiring mutant of *Saccharomyces cerevisiae* that was associated with red or pink colony colour. A second unlinked adenine-requiring red mutant was recovered by Ephrussi *et al.* (1949). Similar red mutants have been isolated in *Schizosaccharomyces pombe* (Leupold, 1955). This distinct colony morphology has provided a convenient selective system for isolation of a large number of mutants at these two genes (Inge-Vechtomov and Kozhin, 1964; Nashed and Jabbur, 1966; Loprieno, 1966; Raypulis and Kozhin, 1966; Woods and Bevan, 1967). The pigment accumulated by adenine-requiring red mutants has been shown to be polyribosylaminoimidazole to which are attached a number of different amino acids (Smirnov *et al.*, 1967). Roman (1956a) observed that red strains revert to white at a relatively high rate. The white revertants were found to be still adenine-dependent and genetic analyses revealed that they contained the original red mutation and in addition a mutation at one of several other adenine-white genes. Under aerobic conditions, the white revertants possess a selective advantage relative to the red parental cells. This provides a convenient selective procedure for isolation of spontaneous mutants in the adenine pathway. Induced forward mutation at the adenine-white loci also can be detected as white colonies or sectors in a strain that produces red colonies (Nasim and Clarke, 1965; Loprieno, 1966).

The ly_9 mutation in *Sacch. cerevisiae* confers a yellow colony colour, although no direct isolation of this mutation by colour has been reported (Bhattacharjee and Lindegren, 1964). On copper-containing media, copper-resistant methionine-dependent strains develop a brown pigment (Seno, 1963; R. K. Mortimer, unpublished observations). Again, there has been no application of this observation for the selection of mutants in the methionine pathway.

Petite mutations, which are of both genic and cytoplasmic origin, are readily detected as colonial variants. Colony diameter is greatly decreased and the white colour reflects the absence of cytochromes (Ephrussi, 1953). Petite mutants are unstained while wild-type colonies develop a deep red colour if overlayed with agar containing tetrazolium (Ogur *et al.*, 1957). However, one can select for petite mutants in a predominantly wild-type population by plating on a medium containing 2 mM-cobalt sulphate; the petite cells will form colonies while the respiratory-sufficient cells are inhibited (Horn and Wilkie, 1966a).

A class of smooth-colony mutants is obtained after prolonged culture of cells on lactate or ethanol. These mutants arise by mutation at a minimum of seven loci (Galzy, 1964; Galzy and Bizeau, 1965, 1966). Conversely, recessive rough-colony mutations have been identified at three genetic loci (Zakharov and Inge-Vechtomov, 1966).

Horn and Wilkie (1966b) have developed a technique that allows detection of most nutritional mutants on the basis of colony colour. The selection medium contains limiting concentrations of amino acids and bases and in addition the dye Magdala Red (Phloxin B; Squibb). On this medium, auxotrophs develop as red colonies whereas prototrophic colonies are white or light pink. This dye also stains the petite or respiration-deficient colonies (Nagai, 1963).

c. *Enrichment of nutritional mutants.* The frequency of mutant cells in a sample can be increased by selective killing of non-mutants. A number of fungicides, including nystatin, have been used and result in enrichments up to 100 × (Moat *et al.*, 1959; Snow, 1966; Strömnaes and Mortimer, 1968). The cells are incubated in a nitrogen-deficient minimal medium, treated with a mutagen, incubated in minimal medium containing nitrogen for 3–4 h, and then treated with the antibiotic. At this stage, prototrophic cells are in the exponential phase while auxotrophs are still not dividing. Nystatin acts selectively on growing cells apparently by binding to sterols in the cell membrane (Lampen, 1966). The treated cells are then plated on complete medium and the resultant clones are later tested for auxotrophy. With this approach, it is possible to enrich for specific classes of mutants by choice of the appropriate minimal medium.

A procedure similar to the above has been described for *Schizosacch. pombe* in which 2-deoxyglucose is used as the selective agent (Megnet, 1965a). This compound appears to inhibit cell-wall synthesis, leading to lysis of growing cells (Megnet, 1965b). The frequency of auxotrophs following nitrosoguanidine treatment was increased from 8% to 46% by this procedure and spontaneous mutants were enriched 1000-fold.

C. HYBRIDIZATION

The first hybridizations of *Saccharomyces* species were achieved by pairing spores from the parental stocks with a micromanipulator and following the mating visually to observe the formation of a zygote that yielded diploid buds (Winge and Laustsen, 1938). This procedure is still used to make crosses with spores bearing genes for lethality or diploidization (Hawthorne, 1963a). A haploid vegetative cell from a heterothallic stock can be used as the other parent in these crosses, but it is necessary to pre-incubate the spore for an hour or two to allow the onset of germination before it is paired to the vegetative cell.

For most crosses, one ordinarily chooses heterothallic haploid parents with complementary nutritional requirements. A mixture is made on a complete medium, incubated to allow mating, and a sample is then plated on a minimal medium to select the prototrophic diploid clones (Pomper and Burkholder, 1949). If the parents do not possess complementary requirements, it is possible to isolate a zygote from the mass-mating mixture by micromanipulation.

D. SPORULATION

Sporulation of the diploid hybrids is an essential step in most genetic analyses with yeast. Both genetic and environmental factors influence this process. The environmental conditions are discussed in detail in Chapter 7 (p. 311) of this volume. Before considering the genetic factors, it should be noted that the process of sporulation can be divided into two parts: (i) meiosis and the formation of four haploid nuclei; and (ii) the steps involved in incorporating the meiotic products in mature spores.

Heterozygosity for the mating-type alleles (a/α) is a necessary condition for the sporulation of diploid or polyploid cells. Diploids homozygous for mating-type (a/a or α/α) do not sporulate, but instead exhibit a mating response consistent with the mating-type allele present (Roman et al., 1955). The a/α diploids must be respiratory-sufficient to sporulate. Homozygosity at any of the genetic petite loci blocks sporulation as does any cytoplasmically inherited change that confers the petite phenotype (Ephrussi, 1953). It is likely that other genes are involved in the sporulation process; however, little work has been directed at delineating these factors.

Four spores are expected as the products of meiosis of a single nucleus. Most Saccharomyces species and Schizosaccharomyces pombe produce asci containing four spores. Nevertheless, asci both with fewer and with more than four spores are encountered. Two- and three-spored asci are common in sporulated samples of most Saccharomyces hybrids. Generally these are not selected for genetic analysis but, in random-spore analysis, they can contribute significantly to the sample. Takahashi (1962) and Takahashi and Akamatsu (1963) analysed segregations in two- and three-spored asci, and the ratios observed were those expected if the spores were random products of meiosis. In addition, it was shown that the spores from such asci were haploid. These results indicate that meiosis is normal, but spore formation is completed for only some of the resultant nuclei in such asci. They are not consistent with the interpretation that more than one nucleus is incorporated into one or two of the spores.

Asci with more than four spores are encountered much less frequently. Winge and Roberts (1954) concluded that such asci are a consequence of a post-meiotic mitosis. To account for asci in which five, six or seven spores were found, it was proposed that spore formation occurred for only some of the nuclei or that some spore pairs had fused. However, Lindegren and Lindegren (1953a) have shown that asci with more than four spores include spores of more than four distinct geno-types. Their results indicate that these asci contain the products of two separate meioses, presumably in adjoined cells.

E. RANDOM-SPORE ANALYSIS

Random spore samples can be used to establish linkage or to detect rare recombination events. In *Schizosaccharomyces pombe*, spores are released from the asci following incubation at low temperatures (Leupold, 1957). In *Saccharomyces* strains, however, the spores are not freely released and, if the asci are placed on nutrient medium, conjuga-tion of germinated spores occurs. Nevertheless, it is possible to obtain spore suspensions of these yeasts that are suitable for genetic analysis. Emeis and Gutz (1958) were able to liberate the spores from the asci by grinding a sporulated sample with glass beads. The ascus wall also can be digested enzymically. Enzymes that will act on the ascus wall are obtained from *Aerobacillus polymyxa* (Wright and Lederberg, 1957) or from terrestrial snails such as *Helix aspersa* or *Helix pomatia* (John-ston and Mortimer, 1959). These enzyme treatments leave the spore tetrads intact and as suitable material for tetrad analysis (see Section III.F. 1, p. 397). For random-spore analysis, the spore groups left after enzyme treatment can be disrupted by sonication (Magni, 1963), which results in a suspension containing single spores, unsporulated diploid cells, a few spore doubles, and cell debris. If this suspension is plated on a non-selective medium, colonies will develop from both spores and diploid cells.

If the diploid is heterozygous for recessive markers that affect colony morphology, colonies developing from spores can be identified. For example, Magni (1963) used a red-adenine marker, and Zakharov and Inge-Vechtomov (1964) used both a red-adenine gene and a marker causing rough colony development to identify spore colonies in random-spore analyses. If recessive resistance genes, for example for canavanine resistance, are included in the diploid, spores carrying the resistance marker can be selected (Sherman and Roman, 1963).

Additional modifications of the above random-spore procedures in-clude steps to eliminate selectively diploid cells by physical or chemical means. Following release from the asci, spores can be concentrated in a paraffin oil phase because of their lipophilic surface (Emeis, 1958). Also, selective inactivation of diploid cells by exposure to alcohol has been reported (Leupold, 1957; Zakharov and Inge-Vechtomov, 1964). Electrophoretic separation of spores and diploid cells also has been shown to be feasible (Resnick *et al.*, 1967).

F. ISOLATION OF SPORE TETRADS

1. *Dissection of Yeast Asci*

Winge and Laustsen (1937) first described a technique for the isolation and cultivation of spores from yeast asci. The ascus wall was ruptured by pressure applied with a microneedle, and the released spores were moved with the needle to droplets of nutrient medium. The operation, which required considerable patience and skill, was performed on a cover glass inverted over a dissection chamber. These authors noted that, "With some training, a person with the proper knack is able to isolate 70 spores a day". A much simpler method for isolation of spore tetrads developed from the observation that the ascus

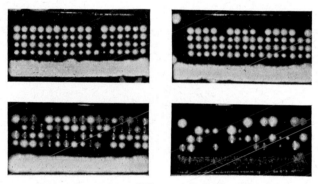

FIG. 3. Spore colonies from four different diploids. The horizontal streak is growth from the inoculum of enzyme-treated sample. Tetrads from individual asci are arranged vertically. The diploids used for the upper dissections were each heterozygous for six genes. The spores in the lower left slab were isolated from a diploid that was segregating nine genes including ad_2, which is present in the dark (red) spore clones. The diploid used in the lower right dissection was heterozygous for 18 genes, including ad_2 (red), and in addition was segregating a radiation-induced lethal.

wall could be digested enzymically without disturbing the spore tetrads (Wright and Lederberg, 1957; Johnston and Mortimer, 1959). The technique in general use proceeds as follows: an enzyme-treated sample of sporulated yeasts is streaked along one edge of a thin nutrient-agar slab located on a cover glass; the glass is inverted over a micro-chamber and individual tetrads are selected with a microneedle, separated, and the four spores positioned in a column across the slab. Ten to twenty tetrads can easily be located on a single slab which is then removed from the cover glass and placed, spores upwards, on the surface of a YEPD plate for incubation. The results of typical dissections are illustrated in Fig. 3.

2. Viability of Spores

TABLE I. *Fractional Viability of Spore Tetrads from Diploid and Tetraploid* Saccharomyces cerevisiae *Strains*

Strain	Degree of Ploidy	Fractional Viability					Percent Viable Spores	Reference
		4/4	3/4	2/4	1/4	0/4		
Miscellaneous hybrids	2	380	54	15	1	0	95·3	R. K. Mortimer, unpublished data
Z 28	2	272	35	7	0	0	96·1	S. Fogel, unpublished data
Z 34	2	702	121	35	12	3	93·2	S. Fogel, unpublished data
Miscellaneous hybrids	2	219	21	4	0	0	97·0	S. Fogel, unpublished data
C 814	2	550	103	33	19	3	91·6	Takahashi (1966)
X 1221	4	43	4	2	0	0	96·9	R. K. Mortimer, unpublished data
X 621	4	55	6	2	1	0	94·9	R. K. Mortimer, unpublished data

Diploid strains of *Saccharomyces* used in genetic analyses usually yield spores with high viability (> 90%). The same is obtained for tetraploid hybrids. In contrast, triploids characteristically have very low spore viability (< 10%), presumably because of imbalance occasioned by the aneuploid nature of most of the spores. In Table 1 are summarized data on the fractional viability of spore tetrads observed in a number of studies with *Saccharomyces* strains. The distributions for diploid and tetraploid hybrids can be explained by random-spore death at a frequency of approximately 5%. The basis of random-spore death is not clearly established. It is our experience that excessive treatment with enzyme or prolonged storage on sporulation medium increases the frequency of inviable spores. Spore inviability may also reflect some defect in spore development.

As stated above, triploid hybrids exhibit low spore viability, and this is characteristic of triploid organisms which in general are infertile. That approximately 10% of the spores from triploids are viable suggests that yeast cells are relatively insensitive to genic imbalance. With at least 16 chromosomes in *Sacch. cerevisiae* (1n), it is unlikely that either a haploid or diploid spore would be produced.

Other causes of spore inviability are recessive lethals, inversions, and translocations. A diploid heterozygous for a recessive lethal should yield asci with no more than two viable spores; however, there are mutations, such as is_3 and le_3, which can be considered as 'semi-lethals' in that they are associated with low (70%) spore viability. Some combinations of genes are lethal in the haploid spore; for example p_9 (op_1) in combination with either p_1, p_2, p_3 or glt_2 results in lethality (Beck *et al.*, 1968). Inversion and translocation heterozygotes should produce asci with four (4:0), two (2:2), or no (0:4) viable spores. The relative frequencies of these classes depend on the size and location of the structural alternations (Emerson, 1963; Esser and Kuenen, 1965).

G. SCORING GENETIC TRAITS

Most of the characters ordinarily followed in a cross can be scored on agar plates, and thus replica-plating procedures can be used. The spore clones are streaked onto a master plate of YEPD medium in a pattern that accommodates eight to ten tetrads per plate. From this master plate, direct replica platings are made to the test media for scoring the monogenic characters.

To test those characters under the control of two genes in the cross, it is necessary to replica-plate the master to a YEPD plate and cross stamp the tetrads with a and α tester stocks carrying only one of the genes. These secondary master plates are incubated overnight to permit matings of the spore segregants and testers and then are replica-plated to the appropriate diagnostic media. With these procedures, one can usually score the mating-type of the haploids as well as the segregation of the individual genes of a complementary pair (see Section III.I, p. 401).

H. MITOTIC RECOMBINATION

1. *Intergenic*

The infrequent occurrence of spontaneous mitotic recombination events leading to homozygosity of a marker (less than $1/10^5$ divisions) makes it necessary to use a selective system for the detection of the recombinants. If the diploid is heterozygous for a recessive gene that affects colony morphology (James, 1955; Roman, 1956a) or confers resistance to an antimetabolite such as actidione or canavanine (Srb, 1956), spontaneous recombinants can be detected. Treatment with X-rays, ultraviolet radiation or chemical mutagens increases the frequency of mitotic segregants to several percent of the survivors, and selection on the basis of pigment formation or by direct replica-plating is feasible. Even with these treatments, multiple exchanges on a chromosome arm are rare. Thus, detection of concomitant segregation of markers in mitotically dividing cells is a sensitive method for assigning genes to the same chromosome arm (Mortimer and Hawthorne, 1966b).

2. *Intragenic*

Diploids that carry in repulsion two non-complementing mutant alleles at a locus $(a_1 + / + a_2)$ will in most cases revert to normal phenotype at a rate much higher than is observed for either of the homoallelic diploids $(a_1 + /a_1 +$ or $+ a_2/ + a_2)$. The reversion of heteroallelic diploids is known to be the consequence of primarily non-reciprocal events (Roman, 1956b). If heteroallelic diploids are exposed to ultraviolet radiation (Roman and Jacob, 1957), X-rays (Mortimer, 1959; Manney and Mortimer, 1964), or a variety of chemical mutagens (Zimmerman and Schwaier, 1967), the frequency of reversion is increased greatly. With X-rays, the frequency of induced revertants is a linear function of X-ray dose (in rads). The slope of this induction curve varies considerably for different allele pairs. It has been found that the slopes serve as a consistent measure of the separation of the respective alleles (Manney and Mortimer, 1964; Manney, 1964). The fine structure map unit based on this method is defined as: 1 unit = 1 revertant/10^8 survivors·rad.

I. MULTIPLE GENE CONTROL

Diploid yeast hybrids may be heterozygous for more than one gene controlling a particular phenotype. This is revealed by asci with segregation ratios other than the 2:2 monogenic ratio for the parental phenotypes amongst the four spores. The relative frequencies with which different segregation ratios are obtained are determined by the type of multiple gene control and by the linkage relationships of the genes involved. The set of genes responsible for a particular function may be related in either a complementary or polymeric fashion. All genes in a complementary set must be functional for the cell to be functional, whereas any member of a set of polymeric genes is adequate for function. In some situations, a particular phenotype is under both complementary and polymeric gene control. These two genetic systems and the manner in which the segregation of individual genes within multigenic sets are scored are discussed below.

1. *Complementary Genes*

Most biosynthetic pathways involve a number of steps, each catalysed by a distinct enzyme which, in turn, is coded for by a single gene.

All the enzymes in the sequence must be functional for the synthesis of the end product of the pathway. For example, in *Saccharomyces* strains there are five steps in the synthesis of tryptophan from chorismic acid (DeMoss, 1965; Doy and Cooper, 1966). The loss of any one of the five enzyme activities by mutation of the corresponding gene results in a requirement for tryptophan.

A hybrid heterozygous for one of the five tryptophan genes will produce asci that contain two tryptophan-independent and two tryptophan-dependent spores. If a diploid is heterozygous for two of the genes, $1+:3-$ and $0+:4-$ ratios are also observed. The relative frequencies of the three classes of asci will depend upon the linkage relationships of the genes involved (see Sections IV.A and IV.B, p. 403 and 405).

The distribution of segregation ratios exhibited by the asci yields information only on the number and linkage of heterozygous genes controlling a particular pathway. To determine the segregation of each of the heterozygous genes uniquely, additional tests are needed. These tests include additional functional characterizations as well as intergenic complementation tests. For example, a *Saccharomyces* hybrid heterozygous for the first and last genes in the tryptophan pathway, tr_2 (anthranilate synthetase) and tr_5 (tryptophan synthetase), will produce four types of spores: A. $TR_2 \, TR_5$, B. $tr_2 \, TR_5$, C. $TR_2 \, tr_5$ and D. $tr_2 \, tr_5$. On synthetic medium lacking tryptophan, only spore A will grow. If this medium is supplemented with anthranilic acid, spore B will also grow. Thus, on this medium, the segregation of TR_5/tr_5 can be scored uniquely. However, the genotype with respect to TR_2/tr_2 can be determined only for spores A and B. The complete genotype of all four spores can be deduced only for asci that segregate $2:2$ or $0:4$.

TABLE II. *Tests to Determine the Genotype of Haploid Segregants of* Saccharomyces cerevisiae *Diploids Heterozygous for Two Genes Controlling a Particular Phenotype*

Genotype of Strain	Growth Response of Strain on Tryptophan-less medium		Growth response on Tryptophan-less Medium of Strains Mated with	
	Lacking anthranilic acid	Supplemented with anthranilic acid	$tr_2 \, TR_5$	$TR_2 \, tr_5$
$TR_2 \, TR_5$	+	+	+	+
$tr_2 \, TR_5$	−	+	−	+
$TR_2 \, tr_5$	−	−	+	−
$tr_2 \, tr_5$	−	−	−	−

The above procedure also requires the availability of utilizable intermediates of the pathway under consideration. Alternatively, the genotype of the spores can be determined by mating each spore with haploid cultures that carry single blocks. In the above example, each spore is cross-streaked with four cultures, $a\ tr_2\ TR_5$, $\alpha\ tr_2\ TR_5$, $a\ TR_2$ tr_5 and $\alpha\ TR_2\ tr_5$. After a period of incubation, to allow mating and growth, the cross-streaked plate is replica-plated to tryptophan-less medium. Growth occurs only if there has been mating and if the spore contains the dominant allele to complement the recessive gene in the tester.

The above tests are summarized in Table II. The tests illustrated in this table are generally applicable to multigenic situations involving most biosynthetic pathways.

2. Polymeric Genes

Some cellular functions in yeast are under a different type of genetic control in which any one of a set of genes is adequate for function. The genes controlling fermentation of certain sugars in Saccharomyces spp. are of this type. Presence of the dominant allele of any of at least seven genes confers ability to ferment 'the sugar maltose. Similarly, there are at least six polymeric genes that split sucrose and raffinose (Section VI.A.2, p. 423). Fermentation of the sugar α-methylglucoside is under the control of both complementary and polymeric genes (Hawthorne, 1958). The super-suppressors in both Saccharomyces cerevisiae and Schizosaccharomyces pombe also represent polymeric gene series. Within a class of suppressors, there are as many as eight genes, any one of which will restore function to a set of nonsense mutants (Hawthorne and Mortimer, 1963; Inge-Vechtomov, 1965; Barben, 1966; Gilmore, 1967). Also, there are a number of examples in which resistance to various metals or antifungal agents can arise by mutation at more than one gene.

If a diploid hybrid is heterozygous for two polymeric genes controlling a particular phenotype, asci that segregate 4:0, 2:2 and 3:1 will result. The relative frequencies of these ascal classes depends on the linkage relationships of the two genes (see Section IV.A and IV.B, p. 403 and 405). For random segregation, they should occur in relative frequencies of 1:1:4. For example, Winge (1952) found 37 (4:0), 36 (2:2) and 154 (3:1) asci from diploids heterozygous for two maltose genes, which is close to expectation. If more than two polymeric genes are heterozygous, there are relatively more 4:0 and 3:1 asci. With three independently segregating polymeric genes, one expects 19 (4:0):1 (2:2):16(3:1) asci. Winge found 47 asci from triply heterozygous diploids distributed 30:1:16 for these classes.

To score the segregation of individual genes in situations in which more than one polymeric gene is segregating is much more difficult than in the case with complementary genes. If the genes differ quantitatively in their action, for example in the rate of fermentation of a sugar or in the efficiency of suppression of a mutant phenotype, it is possible to deduce the genotype of some of the spores. Also, in some cases it is possible to distinguish polymeric genes qualitatively.

Unless the above procedures can be employed, spore genotypes can be determined only by secondary crosses and analysis for recombination. For example, from the cross $A_1 a_2 \times a_1 A_2$, where either A_1 or A_2 confers function, four spore-genotypes are possible: $A_1 A_2$, $A_1 a_2$, $a_1 A_2$ and $a_1 a_2$. The genotype of the fourth spore is easily deduced because this spore is non-functional. The genotypes of the other spore cultures can be determined by mating each of them with two tester cultures, one of each parental genotype. The resultant diploids will produce spores expressing the non-functional phenotype, hence genotypically $a_1 a_2$, only if the culture being tested does not carry the dominant gene present in the tester. This is the basis of the genetic procedures used to identify various polymeric gene series.

IV. Linkage Studies

The development of genetic maps is basic to many genetic investigations. Studies of recombination, both inter- and intragenic, as well as investigations of gene action and regulation are intimately concerned with gene location. As an ascomycete, yeast can be studied genetically by tetrad analysis. This approach provides certain information not readily obtainable by other means. The type of genetic control of a trait (complementary or polymeric) can readily be discerned. In addition, linkage between genes and between the gene and the centromere of the chromosome can be deduced from tetrad analysis data. The procedures for detecting linkage and for development of chromosome maps in yeast are described in the following sections.

A. GENE-GENE LINKAGE

1. *Tetrad Analysis*

The four spores in a yeast ascus represent the products of a single meiosis, and analysis of these tetrads provides information on recombination and linkage of genes in heterozygous state. A diploid heterozygous at two loci, $A/a\ B/b$, can produce three types of asci: parental ditype (PD) $AB\ AB\ ab\ ab$; non-parental ditype (NPD) $Ab\ Ab\ aB\ aB$; and tetratype (T) $AB\ Ab\ aB\ ab$. The relative frequencies

14

of these three ascal classes is determined by the positions of the two genes relative to each other.

If the genes are located on the same chromosome, NPD and T asci will result only from crossing over in the interval between A and B. Absence of crossing over in this interval is associated with a PD ascus. A single crossover between A and B leads to a T ascus. A second crossover can result in any of the three ascal classes depending on the number of chromatids involved in the two exchanges, PD, T, or NPD, respectively, for two-, three- and four-strand double crossovers. The relative frequencies of these double-crossover classes expected on random involvement of strands is $1:2:1$. Additional crossovers yield all three ascal classes. If A and B are widely separated on the chromosome such that many exchanges may occur, the PD:NPD:T ratio approaches $1:1:4$. two genes are considered to be linked if the ratio of PD:NPD asci is significantly greater than one. If relatively few NPD asci are found, the distance x, in centimorgans, between two genes can be calculated from the equation:

$$x = \frac{100(T + 6NPD)}{2(PD + NPD + T)}$$

This equation (Perkins, 1949) is derived on the assumption that the sample includes only tetrads in which 0, 1, or 2 crossovers have occurred in the interval between the two genes, and also that the NPD asci represent one-fourth of the double crossovers.

If A and B are located on different chromosomes, segregation of the two genes results from random disjunction of the respective centromeres at the first meiotic division as well as from exchanges between each of the genes and its centromere. If one or both genes exchange freely with their centromeres, the PD:NPD:T ratio should also be $1:1:4$. Thus, this ratio reflects random segregation of two genes whether on the same or different chromosomes. If both A and B are near their centromeres, tetratype asci will occur only if there has been a crossover between at least one of the genes and its centromere. This situation is indicated by a PD:NPD:T ratio of $1:1:< 4$. The frequency of tetratype asci is related to the second division segregation frequencies, w and y, of the two genes by the relation:

$$T = w + y - \frac{3}{2}wy \quad \text{(Perkins, 1949)}$$

2. Random-Spore Analysis

The alternation of haploid and diploid phases in the life cycle of yeast affords the opportunity of assaying directly the genotype of the products of meiosis. A random sample of spores from a doubly heterozygous diploid (A/a B/b) is expected to include spores of four genotypes, i.e. parental (AB and ab) and recombinant (Ab and aB). If A and B segregate independently, these four classes are expected in equal frequencies. Linkage is indicated if the fraction of recombinant spores is less than half. The separation of A and B equals the percentage of recombinant spores. In contrast to tetrad analysis, however, random-spore analysis affords no estimate of the extent of multiple crossing in a genetic interval, and hence distances determined by this method are reliable only for relatively short intervals ($<$ 30 centimorgans). Procedures for obtaining random-spore samples are described in Section II.E (p. 396).

B. GENE-CENTROMERE LINKAGE

In the absence of adequate cytological information, which is the situation that exists in yeast, it is possible to locate a gene relative to its centromere only by tetrad analysis. Three approaches based on the analysis of spore tetrads are available. These involve use of tetraploids, linear asci and unordered asci in which known centromere-linked genes are segregating.

1. *Tetraploids*

In both *Saccharomyces cerevisiae* and *Schizosaccharomyces pombe*, tetraploid hybrids are relatively easy to construct (see Section II, p. 389). A tetraploid carrying gene A in duplex state ($A/A/a/a$) is expected to form asci that segregate 4:0, 3:1, and 2:2 for the phenotypic characters determined by the dominant and recessive alleles, respectively. The genotypes of the spores in a 4:0 ascus all are A/a. In a 3:1 ascus, they are A/A, 2 A/a, and a/a, while in a 2:2 ascus two of the spores are A/A and two are a/a. For a gene that recombines freely with its centromere, the expected relative frequencies of 4:0, 3:1, and 2:2 asci (for bivalent pairing) are 4:4:1. The corresponding ratios for a gene located at a centromere are 2:0:1. Intermediate ratios are obtained for genes showing partial centromere linkage. Thus, the frequency of 3:1 asci can be used to determine the percent second division segregation of a gene (Roman *et al.*, 1955; Leupold, 1956b).

2. *Linear Asci*

In *Schizosaccharomyces pombe*, the asci are characteristically linear. The spores are arranged linearly in the classical Neurospora pattern in which the pairs at each end are sister spores. Centromere-linked genes are arranged linearly in the ascus in the pattern A A a a (Leupold, 1950). In *Saccharomycodes ludwigii*, Winge and Laustsen (1939) found that the spore pairs at each end were non-sister.

In *Saccharomyces* spp., linear asci are formed only in certain hybrids and under particular sporulation conditions. The vast majority of diploids form oval or tetrahedral asci in which the spore configuration carries no information. Magni (1961) found that opposite spores were sister spores in approximately one-third of a sample of oval asci analysed. This is the frequency expected for random location of the spores within such asci. In the linear asci, however, Hawthorne (1955) has demonstrated both cytologically and genetically that there is an alternation of non-sister nuclei. For the gene tr_1, known from other studies to be closely linked to its centromere, 95% of the asci had the linear configuration $TR\ tr\ TR\ tr$. Of the asci not showing this array, many more had either of the linear arrangements $TR\ tr\ tr\ TR$ or $tr\ TR\ TR\ tr$ than had the $TR\ TR\ tr\ tr$ array. This was interpreted as evidence of nuclear slippage which converts the alternating array to an array in which the spores at opposite ends are sister spores. The extent of nuclear slippage was found to be less than 5%. Second-division segregation frequency of a gene can be approximated as twice the frequency of asci with the $A\ A\ a\ a$ pattern because slippage of the alternating array apparently does not yield this arrangement.

3. *Unordered Asci*

Usually it is not convenient to determine the second-division segregation frequency of a gene either from tetraploid segregations or from analysis of linear asci. Most hybrids are diploid and form only oval or tetrahedral asci. However, as discussed in Section IV.A.1 (p. 404), centromere linkage of a gene can be detected if known centromere-linked genes are included in the cross. The ascus-type ratios, PD:NPD:T, of the gene relative to a centromere-linked gene will show a deficiency of tetratype asci (1:1:<4) if the unknown gene also is linked to its centromere. The second-division segregation frequency of the unknown gene can be calculated using the formula presented in Section IV.A.1 (p. 404) if the corresponding frequency is known for the other gene. If a centromere-linked gene such as tr_1 that recombines infrequently with its centromere (<1%) is included in the cross, it is possible to determine with reasonable accuracy the second-division segregation frequency of all other segregating genes. The frequency of tetratype asci between a gene and such a centromere marker-gene corresponds closely to the second-division segregation frequency of that gene.

For some hybrids, a number of known centromere-linked genes may be segregating. When this is the case, it is possible to determine unequivocally the first division spore array as that pattern that is consistent with first-division segregation of the majority of the centromere-

TABLE III. *Segregation in Asci from a Hybrid of* Saccharomyces cerevisiae *(X1914) Heterozygous for Centromere-Linked Genes*

Ascus	Spore	a/α	tr_1	ur_3	tr_5	hi_6	met_{14}	p_8	ar_8	ly_9	ad_2	First-Division Spore Array
1	A	a	+	−	+	−	−	+	−	−	−	●
	B	a	+	−	+	−	−	+	+	+	−	●
	C	α	−	+	−	+	+	−	+	−	+	○
	D	α	−	+	−	+	+	−	−	+	+	○
2	A	a	−	+	+	+	−	+	+	−	+	●
	B	a	−	+	+	−	−	+	−	−	−	●
	C	α	+	−	−	+	+	−	+	+	−	○
	D	α	+	−	−	−	+	−	−	+	+	○
3	A	a	−	−	−	−	−	−	−	−	−	●
	B	α	+	+	+	+	+	+	+	−	+	○
	C	a	−	−	+	−	−	−	−	+	+	●
	D	α	+	+	−	+	+	+	+	−	−	○
4	A	a	−	+	−	−	+	−	−	+	+	●
	B	α	+	−	+	+	−	+	−	−	−	○
	C	α	+	−	+	−	−	+	+	−	+	○
	D	a	−	+	−	+	+	−	+	+	−	●
5	A	α	−	−	+	+	+	−	−	+	−	●
	B	a	+	+	−	−	−	+	−	+	−	○
	C	α	−	−	+	−	+	−	+	−	+	●
	D	a	+	+	−	+	−	+	+	−	+	○
6	A	α	−	+	+	−	+	+	+	−	−	●
	B	a	−	+	−	−	+	+	−	−	+	●
	C	α	+	−	+	+	−	−	+	+	+	○
	D	a	+	−	+	−	+	−	−	+	+	○
7	A	α	+	−	+	−	+	+	+	+	−	●
	B	α	+	−	+	+	+	+	−	−	+	●
	C	a	−	+	−	−	−	−	−	−	−	○
	D	a	−	+	−	+	−	−	+	+	+	○
8	A	α	−	−	−	+	+	+	−	+	+	●
	B	a	+	+	+	−	−	−	+	+	+	○
	C	a	+	+	+	−	−	−	+	−	−	○
	D	α	−	−	−	+	+	+	−	+	−	◐
9	A	a	−	+	+	+	+	+	−	+	+	●
	B	α	+	−	−	+	−	−	−	−	−	○
	C	α	+	−	+	−	−	−	+	−	−	○
	D	a	−	+	+	−	+	+	+	+	+	●
10	A	a	+	−	+	−	+	+	−	−	+	●
	B	a	+	−	+	−	+	+	−	−	+	●
	C	α	−	+	−	+	−	−	+	+	−	○
	D	α	−	+	−	+	−	−	+	+	−	○
11	A	a	+	+	+	+	−	+	−	+	−	●
	B	α	−	−	−	−	+	−	+	+	+	○
	C	a	+	+	+	+	−	+	−	−	+	●
	D	α	−	−	−	−	+	−	+	−	−	○
12	A	α	−	+	+	+	−	+	−	−	+	●
	B	α	+	−	−	−	+	−	+	+	+	○
	C	a	+	−	−	−	+	−	+	+	−	○
	D	a	−	+	+	+	−	+	−	−	−	●
1st Division Segregation		10	12	12	10	7	12	12	4	6	3	
2nd Division Segregation		2	0	0	2	4	0	0	7	6	9	

TABLE IV. *Second Division Segregation Frequency of Miscellaneous Genes in Saccharomyces cerevisiae as Determined by Three Methods*

| | TETRAPLOID ASCI | | | | LINEAR ASCI | | | | UNORDERED ASCI | | |
| | Segregation ratios | | | % Second division segregation | Linear configuration | | | % Second division segregation | Division Segregation | | % Second division segregation |
Gene	4:0	3:1	2:2		++-- --++	+--+ -++-	+-+- -+-+		First	Second	
tr_1	42	5	23	5·6	9	67	439	3·5	2112	20	0·94
le_1	35	3	15	4·6	—	—	—	—	1518	78	4·9
ar_4	29	16	8	30	—	—	—	—	935	189	16·8
α	91	81	17	50	141	156	261	50·4	1463	1015	41·0
met_1	28	35	11	67	29	23	22	78	197	497	71·6
ur_1	9	8	4		24	22	25	66	263	516	66·2
ga_2	53	57	14	67	20	30	24	54	145	424	74·5

Percentages of second-division segregation for tetraploid asci were determined from values in Table 10 of Roman *et al.* (1955). Percentages in linear asci are quoted as twice the percentage of asci with the $+ + - -$, $- - + +$ configuration. In unordered asci, the percentages are the tetratype frequency of the gene relative to the deduced first-division spore array.

linked genes (Hawthorne and Mortimer, 1960). Second-division segregation of a gene corresponds to a tetratype ascus for this gene relative to the first-division spore array. This approach is illustrated in Table III in which the data are presented for a sample of asci from a diploid segregating for seven centromere-linked and three non-centromere-linked genes.

The three methods for determining second-division segregation frequency can be compared from the data presented in Table IV. In this table are relevant data for a group of genes studied in tetraploids, linear asci and unordered asci.

C. INTERFERENCE

Interference, as it pertains to genetic recombination, is defined as a deviation from random expectation in either the number (chiasma interference) or types (chromatid interference) of multiple crossovers. The degree of chiasma interference can be evaluated from the coincidence coefficient (C) which is defined as the observed number of coincident exchanges in two adjacent intervals divided by the number of double crossovers expected on the basis of the observed frequencies of exchanges in the two intervals. Values of C less than unity reflect positive chiasma interference while those greater than unity indicate negative chiasma interference. The double crossovers can be classified with regard to the number of strands involved in the two exchanges, i.e. two, three or four. If there is random involvement of chromatids in the two exchanges, the three classes of double crossovers are expected in relative frequencies of $1:2:1$. Departures from these ratios reflect chromatid interference.

In yeast, both types of interference have been observed. For crossovers in adjacent intervals along a chromosome arm, positive chiasma interference is usually found. In contrast, exchanges in regions located on opposite sides of a centromere appear to occur independently (Hawthorne and Mortimer, 1960; Desborough and Shult, 1962; Fogel and Hurst, 1967). In the absence of chiasma interference, second-division segregation frequencies higher than $\frac{2}{3}$ are not expected. The observation that a number of genes have second-division segregation frequencies as high as 79% is additional evidence that there is strong chiasma interference in regions adjacent to the centromere. For genes located within approximately 40 centimorgans of the centromere, chiasma interference can lead to a population of asci most of which have zero or one exchange in the gene-centromere interval (Hawthorne and Mortimer, 1960; Mortimer and Hawthorne, 1966b). No examples of negative chiasma interference in yeast have been reported.

Chromatid interference has been observed on only a few occasions.

With only a few exceptions, the distribution of double crossover asci does not depart significantly from 1 : 2 : 1 with respect to involvement of two, three and four strands (Hawthorne and Mortimer, 1960; Desborough and Shult, 1962). Fogel and Hurst (1967), in an analysis of more than 4000 asci for exchanges on the right arm of chromosome V of *Saccharomyces*, found 69 two-strand, 142 three-strand, and 57 four-strand double crossovers, which clearly indicates random involvement of strands in adjacent crossovers.

D. PREFERENTIAL SEGREGATION

Basic to the principle of independent assortment of genes is the assumption of random distribution of centromeres to the poles at the first meiotic division. In tetrad analysis, random disjunction of centromeres is indicated by PD:NPD ratios for genes linked to different centromeres that do not deviate significantly from unity. This is the usual result obtained for assortment of pairs of centromere markers on different chromosomes in *Saccharomyces* spp. (Hawthorne and Mortimer, 1960; Lindegren *et al.*, 1962; Hwang *et al.*, 1964; Mortimer and Hawthorne, 1966b). However, Shult and Lindegren (1956) suggested that random assortment does not always occur and that preferential segregation of certain centromeres or 'sites of affinity' could result in PD:NPD ratios different from unity. In a later analysis of 1487 tetrad distributions, Shult *et al.* (1962) detected 25 independent ratios that departed significantly from unity in the direction of an excess of NPD asci. At this level of significance (1%), only 7·5 such ratios were expected by chance. This excess of deviant ratios was considered as support for the hypothesis of preferential segregation. A dramatic example of preferential segregation of two fragments, one with the centromere and the other with the terminal region of the left arm of Chromosome VII, has recently been described by Shult *et al.* (1967). In certain hybrids, the gene le_1, which has consistently mapped near the centromere of this chromosome, was reported to segregate independently of other genes near the centromere of Chromosome VII. It was proposed that le_1 is located on a fragment that normally segregates preferentially with the centromere of Chromosome VII, and that in these exceptional hybrids this hypothesized affinity does not exist.

Unfortunately, these isolated cases of affinity have simply been tabulated. One would like to see the test crosses and predictions that would distinguish between affinity and other explanations such as translocations or mitotic recombination early in the growth of the diploid culture (Emerson, 1963).

E. MITOTIC SEGREGATION

The principal mechanisms by which the genotype of mitotically dividing diploid yeast cells can be altered are mitotic crossing over, gene conversion, and non-disjunction. These events occur at considerably higher frequencies than mutation, and are probably the main causes of instability of diploid cultures.

1. *Mitotic Crossing Over*

The conventional picture of mitotic crossing over assumes partial or complete pairing of homologous chromosomes, reciprocal exchanges between non-sister chromatids, and mitotic disjunction of centromeres as illustrated in Fig. 4. Mitotic crossing over is expressed phenotypically in the products of the first division of the cells. In half of these divisions, all heterozygous genes located distal to the point of exchange are made homozygous and the genotypes of the two daughter cells are reciprocal.

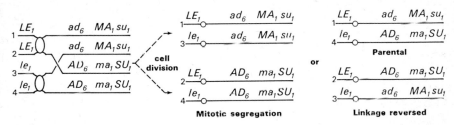

FIG. 4. Mitotic crossing over in diploids. Pairing of homologous chromosomes, exchange, and mitotic assortment of centromeres leads to mitotic descendants that are homozygous for characters formerly in heterozygous condition. The linkage group represented is a portion of Chromosome VII of *Saccharomyces* studied by Roman (1956a).

In the other half of the divisions, heterozygosity is maintained. However, in one of the two daughter cells, coupling of genes surrounding the point of exchange is reversed.

Homozygosity of recessive genes that were formerly heterozygous will lead to an alteration of phenotype, and the clone developing from a cell in which segregation due to mitotic crossing over has occurred will be sectored. Stern (1936) first proposed mitotic crossing over as an explanation of the 'twin spots' observed in somatic tissue of Drosophila. This phenomenon was later described in *Aspergillus spp.* (Pontecorvo *et al.*, 1953). Similarly, James and Lee-Whiting (1955) and Roman (1956a) explained the appearance of sectored colonies and homozygous mitotic segregants in diploid yeast as a consequence of

14*

mitotic crossing over. The genotypes of these mitotic segregants were shown to be consistent with those expected from mitotic crossing over. In addition, linked markers were found to sector concomitantly.

The frequency of mitotic crossing over of a gene is expected to increase with the distance of that gene from its centromere. Roman (1956a) observed the frequency of spontaneous mitotic segregation of ad_3 in *Saccharomyces* to be approximately ten times that of ad_6. Meiotic analysis has established that ad_6 is linked to the centromere of Chromosome VII, and that ad_3 is located at a considerable distance distal to ad_6. The frequencies of X ray-induced mitotic sectoring of different genes also are approximately linearly related to the distance of the genes from their centromeres (Mortimer, 1959; Haefner, 1966; Nakai and Mortimer, 1967).

In addition to X rays, ultraviolet radiation (Hurst and Fogel, 1964; Wilkie and Lewis, 1963) and a variety of chemical mutagens (Zimmermann *et al.*, 1966b) have been shown to induce mitotic crossing over in yeast.

2. *Mitotic Gene Conversion*

Diploid yeasts that carry in repulsion different mutant alleles of the same gene ($a_1 + / + a_2$) revert to wild type at a frequency much higher than observed for diploids homoallelic for either allele ($a_1 + /a_1 +$, or $+ a_2 / + a_2$; Roman, 1956b; 1958b). This effect occurs during mitotic divisions and can approach frequencies of 1 per 10^5 divisions. The frequency of reversion of heteroallelic diploids can be increased greatly by ultraviolet radiation (Roman and Jacob, 1957), X rays (Mortimer, 1959), and chemical mutagens (Zimmermann and Schwaier, 1967).

Reversion of heteroallelic diploids appears to be primarily non-reciprocal. Roman (1958a) analysed 88 revertants of a diploid heteroallelic at an adenine gene and found all of them to be heterozygous for one or the other of the input alleles. If conversion is a reciprocal event, many of these revertants should have been heterozygous for a chromosome bearing both alleles. Hurst and Fogel (1964) recovered five of these double mutants in a sample of 150 revertants of diploids heteroallelic at the hi_1 locus; the remainder were heterozygous for one or other of the alleles. Despite being mainly non-reciprocal, heteroallelic reversion is frequently associated with recombination of outside markers.

The sensitivity to X ray-induced mitotic reversion of heteroallelic diploids has been shown to depend on the allele pair (Manney and Mortimer, 1964). The slope of the plot of the number of revertants per survivor versus the dose serves as an estimate of the separation of the alleles, and consistent fine-structure maps of genes can be constructed using this method. A number of genes in *Saccharomyces* have been mapped in this fashion, including tr_5 (Manney, 1964), hi_4 (Fink, 1966), le_1 (Nakai and Mortimer, 1967) and ad_8 (Esposito, 1968).

As already discussed, sectoring of clones formed by diploid heterozygotes can be a consequence of mitotic crossing over. This is a reciprocal event and involves concomitant sectoring of linked markers. However, some sectored colonies are recovered in which the genotypes in opposite sectors are not reciprocal. In these colonies, one sector is genotypically like the parental cell (A/a) while the opposite sector is homozygous (a/a or A/A). Unlike mitotic crossing over, these events are almost always restricted to a single locus. Because the event is non-reciprocal and localized, it is akin to gene conversion. The frequency of such sectored colonies can be greatly increased by radiations (James, 1955; Yamasaki *et al.*, 1964; Nakai and Mortimer, 1967).

3. *Mitotic Non-disjunction*

In mitotic division, each daughter cell receives a complete set of chromosomes as a consequence of equal division and separation to the poles of all the chromosomes in the parent cell. Occasionally, however, one of the divided chromosomes fails to separate and both chromatids are included in one of the daughter nuclei. If the parent cell was diploid ($2n$), one daughter cell will be trisomic ($2n + 1$) and the other monosomic ($2n - 1$). Any recessive genes on the monosomic chromosome will be expressed, thus providing a powerful approach for localization of genes to linkage groups. In such monosomic cells, coincident expression of two recessive traits formerly in heterozygous condition is evidence that the two genes are on the same chromosome. This approach has been used successfully for development of linkage maps in *Aspergillus*. In various filamentous fungi, *p*-fluorophenylalanine has been shown to induce non-disjunction and, through a sequence of such events in a given pedigree, eventually to produce haploids. Emeis (1966) has examined the effect of this compound on diploid *Saccharomyces cerevisiae*. He observed expression of markers formerly in heterozygous state coincident with a decrease in chromosome number to various states of aneuploidy. Similarly, Strömnaes (1968) has found that *p*-fluorophenylalanine causes non-disjunction but that the chromosome number remains near diploid, presumably through compensating non-disjunctional events. In addition to non-disjunction, this compound was found to cause mitotic crossing over and translocations. Gutz (1966) similarly found mitotic segregants induced by *p*-fluorophenylalanine. These segregants were mainly aneuploid. Examples of spontaneous aneuploidy in yeast have been reported by Hawthorne and Mortimer (1960) and Cox and Bevan (1962).

F. CHROMOSOME STRUCTURAL CHANGES

Some yeast hybrids contain chromosome structural alterations that include translocations, inversions and deletions. A translocation results from breakage of two chromosomes followed by exchange and rejoining of the segments. Inversions involve single chromosomes and are characterized by a segment with a gene order reversed from that of the normal chromosome. Loss of a segment of a chromosome results in a deletion. These changes, in heterozygous condition, cause characteristic patterns of spore inviability in the tetrads. They also may be detected by anomolous linkage relationships of certain gene combinations. It is probable that one of the causes of low spore-viability often encountered with natural strains of yeast is heterozygosity for such alterations.

1. *Translocations*

A translocation heterozygote is expected to yield asci of three types with regard to the ratio of viable : non-viable spores, i.e. 4 : 0, 2 : 2, and 0 : 4 (Emerson, 1963; Esser and Kuenen, 1965). The 2 : 2 asci arise as a consequence of exchange in one of the centromere-breakpoint regions. The surviving spores are expected to be non-sister; that is, a centromere-linked gene, A/a, should segregate 1 A : 1 a in these asci. Also, only one of the two spores in such asci is expected to carry the translocation. In the absence of exchange in either region, random disjunction of centromeres will lead to asci with either four or no viable spores. Translocations in yeast have been studied by M. S. Esposito (personal communication), McKey (1967) and Strömnaes (1968).

2. *Inversions*

As with translocation heterozygotes, diploids heterozygous for an inversion should exhibit decreased spore viability. With no exchange in the inverted region, all four spores in the tetrad should be viable. A single exchange should result in two non-viable spores whereas a four-strand double crossover in the inversion should lead to a tetrad with no viable spores. The fraction of 0 : 4 asci is expected to be low except for long inverted regions. Also, if there has been exchange between the centromere and the inversion, the two viable spores in 2 : 2 asci may be sister spores. However, certain types of inversions will exhibit spore viability patterns similar to those expected of translocation heterozygotes and, in this case, it is possible to distinguish them only by linkage studies. An example of an inversion in a *Saccharomyces* hybrid has been described by Hwang and Lindegren (1966). The order of a group of linked genes on the right arm of Chromosome V was reversed in this hybrid in comparison with the order established in a number of other hybrids. No data were presented on spore viability, although it is probable that, in the development of this diploid, the inverted region was made homozygous. Homozygosity for either translocations or inversions is not expected to result in decreased spore viability.

3. Deletions

Deletion of a chromosome segment should be expressed as a recessive lethal, and all asci from a diploid heterozygous for a deletion should have only two viable spores. In the course of an experiment designed to select mutations of the mating-type locus in *Saccharomyces*, Hawthorne (1963a) obtained a diploid heterozygous for a deletion of a chromosome segment extending from the mating-type locus distally for approximately 30 centimorgans. This deletion was expressed as a lethal mutation closely linked to both a and MA_2 which normally maps 35 centimorgans from a. Moreover, a hybrid formed by crossing a spore bearing the lethal mutation to a haploid cell carrying the marker thr_4, located 16 centimorgans distal to mating-type, proved to be threonine-dependent.

G. GENETIC MAPS

The first linkage results for *Saccharomyces* were presented by Lindegren (1949). Three independently segregating centromere-linked genes, ad_1, ga_1, and α, gave evidence for at least three chromosomes. An additional chromosome was later identified by a centromere-linked uracil gene (Lindegren and Lindegren, 1951a). At present, the minimum haploid chromosome number in *Saccharomyces*, as determined by genetic mapping, is sixteen (Hawthorne and Mortimer, 1960, 1968; Lindegren *et al.*, 1962; Hwang *et al.*, 1963, 1964; Mortimer and Hawthorne, 1966b). Approximately 100 genes have been located on the genetic maps of this yeast. A genetic map of *Saccharomyces* is presented in Fig. 5 (Hawthorne and Mortimer, 1968).

The development of linkage maps for *Schizosaccharomyces pombe* is less advanced. Two chromosomes have so far been identified (Leupold 1958; Gutz, 1966; Ali, 1967). In *Saccharomyces lactis*, also, the genes so far mapped are located on two chromosomes. The distribution of histidine genes in this yeast, compared with that in *Sacch. cerevisiae*, suggests that *Sacch. lactis* has fewer chromosomes (Tingle *et al.*, 1968).

The lengths of intervals represented on the genetic maps have mostly been determined at temperatures for growth of 25–30°. However, yeast will sporulate over a fairly wide temperature range. Johnston and Mortimer (1967) studied recombination in a number of intervals for sporulation in the range 12–30°. Except for one region, in which recombination was decreased at lower temperatures, no pronounced

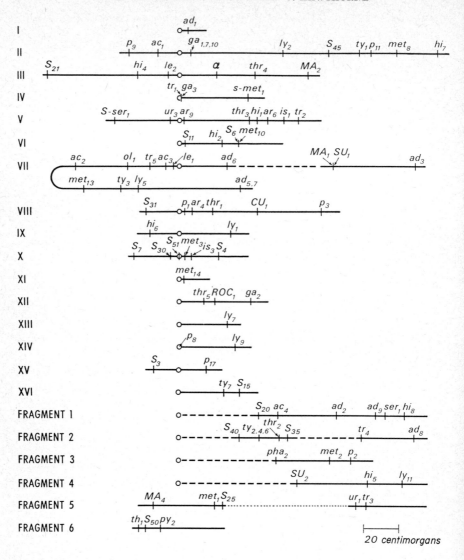

Fig. 5. Genetic map of *Saccharomyces cerevisiae*. Sixteen chromosomes, each identified by one or more centromere-linked genes, and six fragments that may be new chromosomes or segments of chromosomes I–XVI, are represented. The gene symbols used are explained on p. 386. This map, which appeared in Hawthorne and Mortimer (1968), is reproduced by permission of the editors of *Genetics*.

effect of temperature on crossing over was observed. A similar effect of decreased temperatures on crossing over in this region has been reported by Shult *et al.* (1967).

V. Intragenic Recombination

In the preceding sections, recombination between genes on the same or on different chromosomes was discussed. A diploid heterozygous for a single pair of alleles, A/a, was assumed to produce asci that contained two A and two a spores. Multiply heterozygous diploids were assumed to segregate in the same fashion (2:2) for every gene in heterozygous condition. These are basic assumptions in the conventional picture of genetic segregation. However, asci with 4:0, 3:1, 1:3 and 0:4 ratios of spore phenotypes also occur in meiotic tetrads of yeast and other ascomycetes.

Exceptional ratios can arise when a given phenotype is under the control of more than one heterozygous gene. Examples of such genetic systems are the complementary genes controlling various biosynthetic pathways, the polymeric gene sets that control the fermentation of sugars such as sucrose and maltose, and suppressors (Sections III.I, p. 400, VI, p. 423, and VIII, p. 442). Also, polyploidy and polysomy result in segregation ratios different than those expected of diploids. A mitotic segregation preceding meiosis may convert a heterozygous to a homozygous condition which will be expressed as 4:0 or 0:4 segregation ratios.

Even when such explanations have been excluded, asci are recovered that segregate irregularly for particular genes. This phenomenon of irregular segregation, called gene conversion by Winkler (1930), was described first in *Saccharomyces* by Lindegren (1955). A diploid heterozygous at the AD_1 locus yielded an ascus that contained one adenine-independent (white) and three adenine-dependent (pink) spores. In this ascus, other heterozygous genes segregated 2:2. Subsequent studies with yeast and other fungi have established gene conversion as a general phenomenon. This topic has been extensively reviewed by Roman (1963), Holliday (1964), Whitehouse and Hastings (1965) and Emerson (1967). Some of the general features of gene conversion and other intragenic events will be discussed briefly in this section.

Gene conversion appears to occur for any gene in heterozygous condition. Of 34 loci for which data were presented by Roman (1963), irregular segregations were observed for 25. Too few tetrads were studied at the other genes to attach significance to the fact that no irregular segregations were observed for them. The frequencies of irregular segregation varied widely for different heterozygous combinations, from 3 in 990 for TR_1/tr_1 to 13 in 84 for AD_5/ad_5. The irregular ratios were almost entirely of the 3:1 and 1:3 types and these occurred in approximately equal frequencies; in 11,209 segregations at these 34 loci, there were 117 3:1 and 104 1:3 ratios. In addition, there were two each of $4:0$ and $0:4$ segregations. Nine genes studied by Takahashi (1964, 1966) all exhibited irregular segregations ($3:1$ and $1:3$) ascribable to gene conversion at an average frequency of about 1%. Treatment of the diploids with ultraviolet radiation before sporulation increased this frequency considerably, particularly for the mating-type locus.

More information about gene conversion has been obtained from analyses of diploids that carry in transconfiguration two mutant alleles at a locus ($a_1 + / + a_2$). As expected, most asci from such diploids contain two $a_1 +$ and two $+ a_2$ spores. However, exceptional asci are encountered that arise either by conversion of one or both alleles or by reciprocal recombination between the alleles (Roman, 1963; Fogel and Hurst, 1967; Fogel and Mortimer, 1968). The normal and main types of asci recoverable from such heteroallelic diploids are presented in Fig. 6.

(a)	(b)	(c)	(d)	(e)	(f)	(g)	(h)
$a_1 +$	$a_1 +$	$a_1 +$	$a_1 +$	$a_1 +$	$a_1 +$	$a_1 +$	$a_1 +$
$a_1 +$	$+ +$	$a_1 +$	$a_1 +$	$a_1\ a_2$	$+ a_2$	$a_1 +$	$a_1\ a_2$
$+ a_2$	$+ a_2$	$a_1\ a_2$	$+ +$	$+ a_2$	$+ a_2$	$a_1 +$	$+ +$
$+ a_2$	$+ a_2$	$+ a_2$	$+ a_2$	$+ a_2$	$+ a_2$	$+ a_2$	$+ a_2$
Normal	Conversion of $+/a_1$		Conversion of $+/a_2$		Symmetrical conversion of $+/a_1$ and $+/a_2$		Reciprocal recombination

Fig. 6. Main Types of Asci Recoverable from Heteroallelic Diploids

In three of the exceptional ascal types (b, d and h in Fig. 6), one of the spores contains neither mutant allele and is therefore wild-type. In many experiments in which intragenic events are studied, these wild-type spores, because they can be selected, are used to signal the event. It is clear, however, that the asci in which such prototrophic spores occur represent only a fraction of the exceptional asci. Also, these spores can arise from both reciprocal and non-reciprocal events within the gene. Nevertheless, one of the established procedures for obtaining fine-structure maps of genes is based on determining the frequencies of such wild-type spores from diploids heteroallelic for different combinations of alleles.

Fogel and Hurst (1967) have developed a technique for detecting and isolating spore tetrads that include one prototrophic spore. These authors analysed 1081 such asci from a diploid that was heteroallelic at the hi_1 locus. Of these asci, 101 arose by reciprocal recombination between the alleles (class h in Fig. 6) and the remainder had a $3:1$ conversion of one of the alleles (classes b and d). Conversion of the proximal allele occurred in 847 asci, while conversion of the distal allele was found in 133 asci. The conversion asci had a high frequency of exchange in the corresponding adjacent intervals while the reciprocal recombination events interfered with exchanges in adjoining regions.

Through analysis of unselected tetrads, the additional classes of exceptional asci represented in Fig. 6 can be recovered. However, because the irregular asci are relatively rare, this type of study is very laborious. It is necessary to analyse a large number of asci to obtain a reasonable sample of the different exceptional classes. In a sample of 201 unselected asci from a heteroallelic diploid analysed by Roman (1958a), five asci were identified that had segregated irregularly at the hetero-allelic locus. Three of these asci segregated 3 : 1 for one of the alleles, one segregated 1 : 3 for one allele, and one was the result of a reciprocal recombination between the alleles. In addition to asci in which only one allele at a heteroallelic locus segregates irregularly, asci can be recovered that have simultaneous conversion of both alleles. Such asci were recovered in a sample of unselected asci from *Saccharomyces* hybrids that were heteroallelic for different allele combinations at the ar_4 (argininosuccinase) gene (Fogel and Mortimer, 1968). The asci with conversion of both alleles are nearly entirely of the symmetrical types f and g shown in Fig. 6. Among 697 asci from a diploid heteroallelic for alleles *4* and *17* located at opposite ends of the ar_4 gene, 56 were irregular. Forty-six of these segregated irregularly (3 : 1 or 1 : 3) for only one of the alleles (classes b, c, d and e) and three converted simultaneously for both alleles (classes f and g). In addition, nine asci had a reciprocal recombination between alleles *4* and *17* (class h). A sample of 547 asci was obtained from another diploid bearing alleles *2* and *17* at the same locus. The separation of these alleles is only about one-tenth that of alleles *4* and *17*. Of 36 irregular asci in this sample, 28 had converted simultaneously for both alleles and only eight of the asci involved conversion of just one allele. No reciprocal recombinations between the alleles were observed. Thus, gene conversion appears to represent a process in which the genetic information on a section of one chromatid is replaced by the information on its homologue. The average length of the segment involved in conversion is sufficient usually to involve alleles *2* and *17* in a single event, whereas alleles *4* and *17* are too widely spaced to be included, except rarely, in the same conversion event. Independent estimates of the separations of these two pairs of alleles are approximately 100 and 1000 nucleotides respectively.

On the basis of the above studies, a rationale can be provided for genetic fine-structure mapping in which the frequency of wild-type spores is used as an estimate of the separation of alleles. These spores arise mostly by 3 : 1 conversion of one of the alleles, and the frequency of conversion of a given allele does not seem to depend on the presence of other alleles in heterozygous condition in the same gene. However, if the second allele is far removed, these 3:1 conversions will usually be associated with 2:2 segregation of the other allele, and a prototrophic spore will arise. For allele pairs located closer together, a greater percentage of the 3:1 segregations will be associated with symmetrical 1:3 segregations of the other allele, and these events will not result in a prototrophic spore. If the probability of simultaneous symmetrical conversion increases in some regular way with the proximity of alleles, a reasonably consistent map of the gene will result. Reciprocal recombination between the alleles, which also contribute in a minor way to the prototrophic spore sample, should decrease in frequency with a decrease in separation of alleles. Maps have been constructed for a number of genes in both *Saccharomyces cerevisiae* and *Schizosaccharomyces pombe* using meiotic mapping procedures (Leupold and Gutz, 1963; Gutz, 1966; Dorfman, 1964; Esposito, 1968).

Models of gene conversion have been presented by Holliday (1964) and Whitehouse and Hastings (1965). These models, which differ considerably in detail, are based on the formation of single-strand breaks in the region of heterozygosity, unwinding of the double helices, and formation of a heteroduplex region in which mismatched base pairs occur at the mutant site. These mismatched pairs are assumed to be corrected by a repair system that involves degradation and resynthesis of DNA in the heteroduplex region.

VI. Gene Action

A. GENE-ENZYME RELATIONSHIPS

Our understanding of the manner in which genes act to determine the phenotype of an organism has been greatly extended by the many advances in molecular genetics in recent years. Genes are responsible for determining the amino-acid sequences of proteins through the mechanisms of transcription and translation. The specificity of a protein and the manner in which it is folded is established by the primary amino-acid sequence which in turn is determined by the base sequence in the DNA of the gene encoding the particular protein. Proteins function as enzymes, as components of cell structures, or as regulatory macromolecules.

Proteins with enzymic activity serve diverse functions in the cell. Many catalyse various reactions in the biosynthesis of vitamins, amino acids, purines, pyrimidines and lipids. Fermentation of sugars is mediated by specific enzymes. Other enzymes function in the replication and repair of DNA, in the different steps of protein synthesis, and in the transport of compounds into and out of the cell. In this section,

TABLE V. *Gene-Enzyme Relationships in* Saccharomyces cerevisiae

Gene	Phenotype	Enzyme	Reference
ar_3	arginine-requiring	ornithine carbamoyl-transferase	Lacroute et al. (1965)
ar_4	arginine-requiring	argininosuccinase	Grenson et al. (1966)
ar_7	arginine-requiring	N-α-acetylornithine transacetylase	De Deken (1963)
cpa_1	see Section VI B (p. 431)	carbamoyl phosphate synthetase	Lacroute et al. (1965)
cpa_2	see Section VI B (p. 431)	carbamoyl phosphate synthetase	Lacroute et al. (1965)
glt_1	glutamate-requiring	aconitate hydratase	Ogur et al. (1964)
glt_2 (ly_8)	glutamate-requiring; lysine-requiring; genic petite	aconitate hydratase	Ogur et al. (1965)
hi_1	histidine-requiring	phosphoribosyl-ATP-pyrophosphorylase	Fink (1964)
hi_2	histidine-requiring	histidinol phosphate phosphatase	Fink (1964)
hi_3	histidine-requiring	imidazoleglycerol phosphate dehydrase	Fink (1964)
hi_{4A}	histidine-requiring	phosphoribosyl-AMP-hydrolase	Fink (1964)
hi_{4B}	histidine-requiring	phosphoribosyl-AMP-pyrophosphorylase	Fink (1964)
hi_{4C}	histidine-requiring	histidinol dehydrogenase	Fink (1964)
hi_5	histidine-requiring	imidazole acetolphosphate transaminase	Fink (1964)
is_1 (V)	isoleucine-requiring	threonine deaminase	Kakar and Wagner (1964); de Robichon-Szulmajster and Magee (1968)
is_2 (IV)	isoleucine- and valine-requiring	acetohydroxy acid synthetase	Magee and de Robichon-Szulmajster (1968)
is_3 (III)	isoleucine- and valine-requiring	dihydroxy acid dehydrase	Kakar and Wagner (1964)
is_4 (II)	isoleucine- and valine-requiring	reducto-isomerase	Kakar and Wagner (1964)
is_5 (I)	isoleucine- and valine-requiring	reducto-isomerase	Kakar and Wagner (1964)
ly_1	lysine-requiring	saccharopine dehydrogenase	Jones and Broquist (1965)
ly_9	lysine-requiring	saccharopine reductase	Jones and Broquist (1965)
thr_1	threonine-requiring	homoserine kinase	de Robichon-Szulmajster (1967)
thr_2	threonine- and methionine-requiring	aspartate semialdehyde dehydrogenase	de Robichon-Szulmajster et al. (1966)
thr_3	threonine- and methionine-requiring	aspartokinase	de Robichon-Szulmajster et al. (1966)
thr_4	threonine-requiring	threonine synthetase	de Robichon-Szulmajster (1967)
thr_5	threonine- and methionine-requiring	aspartate aminotransferase	de Robichon-Szulmajster et al. (1966)

TABLE V. (*Cont.*)

Gene	Phenotype	Enzyme	Reference
thr_6	threonine- and methionine-requiring	homoserine dehydrogenase	de Robichon-Szulmajster *et al.* (1966)
tr_1	tryptophan-requiring	N-(5′phosphoribosyl) anthranilate isomerase	DeMoss (1965); Doy and Cooper (1966)
tr_2	tryptophan-requiring	anthranilate synthetase	DeMoss (1965); Doy and Cooper (1966)
tr_3	tryptophan-requiring	anthranilate synthetase; indole-glycerol phosphate synthetase	DeMoss (1965); Doy and Cooper (1966)
tr_4	tryptophan-requiring	phosphoribosyl transferase	DeMoss (1965); Doy and Cooper (1966)
tr_5	tryptophan-requiring	tryptophan synthetase	Manney (1964); Doy and Cooper (1966)
$arom_{1A}$	tyrosine-, phenylalanine-, tryptophan-, and *p*-aminobenzoate-requiring	3-enolpyruvylshikimate 5-phosphate synthetase	de Leeuw (1967)
$arom_{1B}$	tyrosine-, phenylalanine-, tryptophan- and *p*-aminobenzoate requiring	shikimate kinase	de Leeuw (1967)
$arom_{1C}$	tyrosine-, phenylalanine-, tryptophan- and *p*-aminobenzoate-requiring	5-dehydroquinate synthetase	de Leeuw (1967)
$arom_{1D}$	tyrosine-, phenylalanine-, tryptophan- and *p*-aminobenzoate-requiring	5-dehydroshikimate reductase	de Leeuw (1967)
$arom_2$	tyrosine-, phenylalanine-, tryptophan- and *p*-aminobenzoate-requiring	chorismate synthetase	de Leeuw (1967)
ur_1	uracil-requiring	dihydro-orotate dehydrogenase	Lacroute (1964a)
ur_2	uracil-requiring	aspartate carbamoyltransferase	Lacroute (1964a)
ur_3	uracil-requiring	orotidine 5-phosphate decarboxylase	Lacroute (1964a)
ur_4	uracil-requiring	dihydro-orotase	Lacroute (1964a)
cpu	See Section VI B (p. 431)	carbamoyl phosphate synthetase	Lacroute *et al.* (1965)
ga_1	galactose non-fermenter	galactokinase	de Robichon-Szulmajster (1958)
ga_2	galactose non-fermenter	galactose permease	Douglas and Condie (1954)
ga_5	galactose non-fermenter	phosphoglucomutase	Douglas (1961)
ga_7	galactose non-fermenter	galactose transferase	Douglas and Hawthorne (1964)
ga_{10}	galactose non-fermenter	galactose epimerase	Douglas and Hawthorne (1964)
cy_1	cytochrome *c*-deficient	cytochrome *c*	Sherman *et al.* (1966)

we will discuss the gene-enzyme relationships for those functions that have been established in yeast.

1. *Biosynthetic Pathways*

Mutation of a gene that encodes a particular enzyme may result in the loss of activity associated with that enzyme. If this enzyme catalyses one of the steps of a biosynthetic pathway, for example biosynthesis of an amino acid, the mutation will be expressed as a requirement for that amino acid. Mutation of other genes associated with the same pathway would also impose this requirement on the cell. Study of mutants blocked at different steps in various biosynthetic pathways has been of great value in establishing the intermediates and enzymes of these pathways. With relatively few exceptions, the reactions involved in synthesis of amino acids and pyrimidines determined from work on other organisms, e.g. *Neurospora crassa* and *Escherichia coli*, apply to yeast. Many of the gene-enzyme relationships associated with synthesis of these compounds in yeast have been determined; they are summarized in Table V. The genetic control of some of these pathways is discussed in more detail in the section on regulation (Section VI.B, p. 430). Mutants of yeast that have a requirement for unsaturated fatty acids have been isolated. These mutants presumably are associated with a defect in one of the steps in biosynthesis of lipids (Resnick and Mortimer, 1966).

2. *Fermentation*

Genes conferring the ability to ferment the sugars galactose, sucrose, maltose and melibiose were among the first markers to be used for genetic studies in yeast (Winge and Roberts, 1948; Lindegren, 1949). Many of the early analyses were made on hybrids between species with different fermentation characteristics: e.g. *Sacch. cerevisiae* × *Sacch. carlsbergensis*, *Sacch. cerevisiae* × *Sacch. chodati*, *Sacch. cerevisiae* × *Sacch. bayanus*, *Sacch. cerevisiae* × *Sacch. chevalieri*, and *Sacch. cerevisiae* × *Sacch. microellipsoideus*. Winge and Roberts (1948, 1950, 1952) found four dominant genes for maltose fermentation; because any one of these genes would suffice, this was termed a polymeric gene system. Some species have several genes; *Sacch. cerevisiae*, for example, had MA_1, MA_2 and MA_3. Species with only a single maltose gene, such as *Sacch. italicus* and *Sacch. diastaticus*, generally were found to carry MA_1 (Gilliland, 1954). An analogous situation is seen with sucrose fermentation. Any one of three genes, R_1, R_2 or R_3, confers the ability to ferment sucrose or one-third of raffinose (Gilliland, 1949; Winge and Roberts, 1952). For sucrose fermentation, the ubiquitous gene is R_2 or SU_2. The symbol SU has been adopted to accommodate several additional genes, SU_4, SU_5 and SU_6, which enable the yeast to ferment sucrose but not raffinose (Carbondale Yeast Genetics Conference, 1963). If a species has several maltose genes, it generally has no more than one sucrose gene, and *vice versa*. One observation that helps to account for this situation is the finding that two pairs of maltose and sucrose genes (MA_1-su_1 and MA_3-su_3) are so closely linked that they can be treated as alleles, and in nature they are generally found in repulsion (Winge and Roberts, 1952). Only non-crossover (PD) tetrads were recovered from 108 asci of MA_1 su_1 × ma_1 SU_1 crosses, and 124 asci of MA_3 su_3 × ma_3 SU_3 crosses (Mortimer and Hawthorne, 1966b). However, several instances of gene conversion brought the two dominant genes into coupling, thus indicating they occupy separate sites on the chromosome. Also, Winge and Roberts (1952) observed one tetratype ascus in a sample of 64 asci from a hybrid in which MA_1 and SU_1 were in repulsion. This demonstrates that these are tightly linked but distinct genes.

A number of additional genes for maltose fermentation have since been reported (Gilliland, 1954; Winge and Roberts, 1956; Terui *et al.*, 1959; Oeser and Windisch, 1964; Oshima, 1967). As the number of loci involved increases, the allelism tests to identify these loci become progressively more tedious (Section III.I, p. 402). Tetrad analysis is essential for a reliable identification of these dominant genes, particularly because material from new sources is involved. This greatly increases the hazards of polysomy which could go undetected if the test stocks are not well marked with recessive genes. Polysomy could simulate multiple-gene segregation.

Baffling tetrad ratios were encountered with the study of α-methylglucoside fermentation (Lindegren, 1953) and were proposed as examples of gene conversion. Some of the segregation ratios, however, can be explained with a combination of complementary and polymeric genes that control fermentation ability (Hawthorne, 1958; Takahashi and Ikeda, 1959). At least five genes are involved; MG_4 in combination with MA_1, or MG_2 in combination with either MG_1 or MG_3, will enable the strain to ferment α-methylglucoside. These five genes are unlinked and are not centromere-linked. For example, if a cross is MA_1 MG_1 MG_2 mg_3 mg_4 (+) × ma_1 mg_1 mg_2 MG_3 MG_4 (−), the expected distribution of tetrad ratios is 19 (4:0), 352 (3:1), 703 (2:2), 216 (1:3), and 6 (0:4).

Another locus concerned with α-methylglucoside, as well as turanose, maltose, sucrose and melezitose fermentations, namely the gene MZ, has been the subject of a number of studies by Lindegren and co-workers (Lindegren and Lindegren, 1953b; Lindegren *et al.*, 1957; Ouchi and Lindegren, 1963; Hwang *et al.*, 1964). They report multiple alleles at this locus whose activities can be ranked in a progressive step-

wise pattern starting with the fermentation of turanose, then turanose + maltose, turanose + maltose + sucrose, turanose + maltose +
sucrose + α-methylglucoside and finally turanose + maltose + sucrose + α-methylglucoside + melezitose. The differences in specificity
apparently lie with some aspect of the mechanism for inducing this
adaptive enzyme. Strains with alleles which allow the initiation of
fermentation on only turanose or maltose are, after being adapted on
these sugars, able to ferment sucrose, α-methylglucoside, and melezitose.

From a yeast which was responsible for superattenuation in beer
wort, *Sacch. diastaticus*, Gilliland (1954) detected a gene controlling the
fermentation of starch or dextrin. Lindegren and Lindegren (1956)
report that two genes are involved: DX controls the synthesis of a
dextrinase capable of splitting dextrin and glycogen but not starch or
amylopectin, and ST controls the synthesis of an amylase which acts
upon the starch and amylopectin.

The rapid fermentation of galactose can be prevented by a mutation
at any one of seven loci. The first block in the pathway is introduced by
ga_2 which causes the loss of a permease (Douglas and Condie, 1954).
Next are two genes, ga_3 (Winge and Roberts, 1948; Spiegelman et al.,
1950) and ga_4 (Douglas and Hawthorne, 1964), imposing blocks in the
adaption system. Mutations in the structural genes in the Leloir pathway are ga_1 (galactokinase), ga_7 (galactotransferase) and ga_{10} (galacto-
epimerase). Mutants defective in the major isozyme of phosphogluco-
mutase are designated ga_5 (Tsoi and Douglas, 1965). Regulatory
mutants conferring constitutive synthesis of the enzymes catalysing
reactions involved in galactose utilization will be considered in the
section on regulation (Section VI.B, p. 434).

3. Miscellaneous Genetic Loci

a. Permeases. Certain amino-acid analogues are inhibitory to the
growth of yeast, presumably because of incorporation of these compounds into protein. Mutants of *Sacch. cerevisiae* have been isolated that
are resistant to certain of these antimetabolites. Possible mechanisms
by which a mutational change can render a cell resistant to an amino-
acid analogue are: (i) by alteration of an enzyme so that it can detoxify
the analogue; (ii) by modification of the specific activating enzyme or
transfer-RNA to restrict incorporation of the analogue during translation; (iii) by a change in the regulation of synthesis of the normal
amino acid so that its endogenous concentration is increased to an
effectively competitive level relative to the analogue for incorporation
into proteins; or (iv) by loss of a permeation system that normally
promotes entrance of the amino acid and its analogue into the cell.

Most of the analogue-resistant mutants so far described in yeast appear to involve permeases.

Sorsoli et al. (1964) recovered mutants of *Sacch. cerevisiae* that grew in the presence of the methionine analogue ethionine. One of these mutants, which was recessive, also conferred resistance to *p*-fluorophenylalanine. This strain incorporated ethionine at a much lower rate than sensitive strains. Also, methionine and a random mixture of labelled amino acids were found to enter the cell much less readily. Surdin et al. (1965) also found an ethionine-resistant mutant to have a ten-fold lower capacity than a sensitive strain for accumulating amino acids. These authors concluded that the mutation was in a gene controlling synthesis of a general amino-acid permease. Further studies of this strain have demonstrated that three genes were involved in determining resistance or sensitivity to ethionine (Cherest and de Robichon-Szulmajster, 1966; de Robichon-Szulmajster and Cherest, 1966). The amino-acid permease gene, *AAP*, governs the capacity to accumulate amino acids and their analogues. Mutant strains, *aap*, have decreased accumulating capacity and, in addition, are resistant to a number of analogues including ethionine. Another semidominant mutation, eth_1^r, confers specific resistance to this analogue and a third mutation, eth_2^s, suppresses the resistance conferred by *aap* or eth_1^r.

Mutants resistant to canavanine have been shown to be deficient in a permease specific for arginine and its analogues (Wiame et al., 1962; Grenson et al., 1966). The $cana_1$ (arg-p_1) mutation is recessive, and its dominant allele shows a gene-dosage effect for uptake of arginine when studied in heterozygous and homozygous condition. Grenson (1966) also demonstrated that lysine uptake is facilitated by the arginine permease. However, a second mutation, *lys-p*, specifically blocks accumulation of lysine. Gits and Grenson (1967) found that methionine incorporation was under the control of two systems, the general amino-acid permease system and a specific methionine-permease system which has a much higher affinity for methionine.

Lacroute and Slonimski (1964) described four unlinked mutations that confer resistance to the pyrimidine analogue 5-fluorouracil. Two of the mutations block uptake of uracil and its analogue and, at the same time, cause a leakage of uracil from the cell. A third mutation to resistance was at one of the structural genes (ur_2) in uracil biosynthesis, and apparently involved a modification of aspartate carbamoyl-transferase, the enzyme under end-product regulation, leading to insensitivity to the analogue.

b. Resistance mutations. Micro-organisms are capable of developing, through mutation, resistance to an array of toxic compounds. Study of the mechanism of toxicity and the manner in which resistance is conferred by mutation often has provided insight to fundamental cellular processes. In the preceding section, amino acid- and pyrimidine-analogue resistant mutants that had lost specific permease activities were described. Mutants of yeast resistant to a number of other toxic compounds have been reported and, in some cases, the mechanism of resistance has been determined.

The antibiotic cycloheximide (actidione) is toxic to certain yeasts at very low concentrations (< 1 μg/ml). This drug inhibits synthesis of both DNA and protein. Siegal and Sisler (1965) have found that *Saccharomyces fragilis*, which is resistant to cycloheximide, possesses ribosomes insensitive to this drug. Rao and Grollman (1967) have identified this resistance with the 60S ribosome subunit. In *Saccharomyces cerevisiae*, both dominant and recessive mutations that permit growth in the presence of the drug have been identified at eight genetic loci (Wilkie and Lee, 1965; Mortimer and Hawthorne, 1966b). An extract of one of the recessive mutants was shown to function in an *in vitro* protein synthesis assay in the presence of cycloheximide, and the component of this extract that permitted synthesis was the ribosomes (Cooper *et al.*, 1967). This strain and three other strains with semi-dominant resistance genes were shown to be freely permeable to the drug.

The polyene antifungal agent, nystatin, has been employed in mutant-enrichment procedures because of its specific toxicity towards dividing cells (Moat *et al.*, 1959; Snow, 1966; Strömnaes and Mortimer, 1968). Lampen (1966) has proposed that nystatin combines with membrane sterols to cause a deleterious change in permeability. Mutants of yeast resistant to this drug have been isolated by Ahmed and Woods (1967). The mutants were recessive and located at three genetic sites. Two dominant modifying genes also were identified. It was proposed that the mutations to resistance occurred at genes determining the composition of the cell membrane.

2-Deoxyglucose is toxic to growing cells of *Schizosaccharomyces pombe*, possibly because of interference in the synthesis of structural polysaccharides. The sites of 2-deoxyglucose-induced lysis were shown to correspond with the regions of growth of the glucan layers in the cell wall (Johnson, 1968). A mutant that was resistant to this compound was found to be partially deficient in hexokinase, indicating that an enzymic product of 2-deoxyglucose was the toxic substance (Megnet, 1965b). Similarly, mutants of this yeast that were resistant to allyl alcohol have been found to be partially deficient in alcohol dehydrogenase as a consequence of mutations in the structural gene of this enzyme (Megnet, 1965c, 1967).

A number of antibacterial antibiotics inhibit respiration of yeast, and sensitivity of these drugs can be assayed on a medium in which the carbon source is non-fermentable. Mutants resistant to erythromycin were shown to be of either genic or cytoplasmic origin. The alteration in the cytoplasmic mutants was identified with the mitochondria because resistance was lost coincident with a change to the petite phenotype (Wilkie *et al.*, 1967; Thomas and Wilkie, 1968a). Recombinants have been recovered from mitotic descendants of hybrids between strains cytoplasmically resistant to different antibiotics (Thomas and Wilkie, 1968b). Further discussion of these mitochondrial mutants is presented in Section IX (p. 451).

Copper ions are inhibitory to yeast at relatively high concentrations. Mutations to copper resistance have served as genetic markers in linkage studies (Hawthorne and Mortimer, 1960; Lindegren *et al.*, 1962). Other mutations have been described which confer resistance to cadmium (Middlekauf *et al.*, 1956) and lithium (Laskowski, 1956).

c. *Morphological mutants*. Winge and Laustsen (1937) demonstrated genetic control of colony morphology in a homothallic *Saccharomyces* species. The colonies developing from the four spores of a single ascus possessed unique morphologies which were transmitted through meiosis. These traits, which were expressed in giant colonies on gelatin agar, appeared to be under complex genetic control. Single-gene control of cell shape was observed in *Saccharomycodes ludwigii* (Winge and Laustsen, 1938). A number of additional genetic loci in yeast that affect colony size, colour or texture have since been identified. Many of these are described in Section III.B.1.b (p. 393).

d. *Mutations affecting response to radiation*. The response of yeast cells to radiation depends both on the ploidy of the cells and on the stage in the division cycle at which the cells are irradiated (for a review of radiobiological studies on yeast see James and Werner, 1965). In addition, genetic control of radiation sensitivity has been demonstrated by the recovery of mutations affecting radiation sensitivity. Moustacchi (1965) observed that a relatively high percentage (10–80%) of survivors of exposure to ^{32}P decay, X-rays or ultraviolet radiation were resistant to ionizing radiation. Resistance was found to be recessive and appeared to be due to mutation at any of a number of genetic loci. Unless an unusually large number of genes are proposed, however, it is difficult to explain the very high mutation frequency.

Nakai and Matsumoto (1967) have reported mutations in yeast that cause increased sensitivity to radiation. Two mutants were associated with increased sensitivity of cells to ultraviolet radiation, and mutation at a third locus rendered the cells more susceptible to X ray-inactivation. The X ray-sensitive mutation had the effect of sensitizing a resistant fraction of the cell population. The resistant cells had been shown previously to be budding cells. In addition, when homozygous, this mutation eliminated the characteristic shoulder on diploid survival curves. Snow (1967) identified at least six loci in *Saccharomyces cerevisiae* that control sensitivity of cells to ultraviolet radiation. The increases in sensitivity associated with the different mutations varied considerably. These mutations also increased the susceptibility of cells to nitrous acid inactivation, but the relative sensitivities of the mutants differed from that found for ultraviolet radiation. The mutant most sensitive to nitrous acid was least sensitive to ultraviolet radiation. Laskowski *et al.* (1968) have discovered a mutation that renders cells sensitive to both ultraviolet and X-radiation. This mutant appears to be different from any of those discussed above. Radiation-sensitive mutants have also been recovered in *Schizosaccharomyces pombe* (Haefner and Howrey, 1967).

The lack of a thymineless mutant in yeast, which precludes specific labelling of the DNA, makes it difficult to establish the mechanism by which cells are made more sensitive by the mutation. It is likely, however, that some of the mechanisms for repair of radiation damage established in bacteria, such as pyrimidine dimer removal, also exist in yeast. The sensitive mutants presumably could be deficient in these repair steps.

e. Temperature mutants. Mutations in genes that control steps in the biosynthesis of amino acids, purine and pyrimidine bases, and vitamins can be compensated for because these components can be taken into the cell from its surroundings. However, many genetically-controlled cellular functions, when lost by mutation, cannot be replaced by additions to the growth medium. For example, activating enzymes, polymerases, ribosomal proteins, and structural proteins are essential cellular components loss of which cannot be replaced. Thus, mutation in genes controlling synthesis of such constituents would be lethal. However, in some cases, the mutant defect may be rectified by a shift in incubation temperature. If the mutation is of the base-pair substitution type, it is possible that the conformational defect introduced in the protein by the associated amino-acid replacement can be altered by incubation at a different temperature. Hartwell (1967) has isolated about 400 such conditional lethal mutants in *Saccharomyces cerevisiae*. These mutants fail to grow on enriched medium at 36° but are able to grow on the same medium at 23°. It is estimated that such non-supplementable mutations occur in about 90% of the genome of yeast. The functions controlled by most of these genes are still unknown, and study of these temperature mutants will provide an approach to this important segment of the genome. Some of the temperature mutants isolated by Hartwell have been identified with genes controlling synthesis of particular activating enzymes (Hartwell and McLaughlin, 1968).

B. REGULATION

Cells must synthesize a great variety of molecules, some in large quantities and others only in small amounts. The total resources of cells clearly are limited and, to achieve optimum synthesis of all the components needed for division, they must possess regulatory mechanisms. Optimum concentrations within the cell of the end-products of various biosynthetic pathways, or of the enzymes required for the fermentation of particular sugars, for example, are controlled in two different ways. The *rate of synthesis* of an enzyme or enzymes can be controlled by a repressor molecule which binds to the operator region of the corresponding gene or operon that codes for enzymes that catalyse reactions on a pathway. The operator region serves as a switch and determines whether messenger RNA (m-RNA) is synthesized. For some genes, the repressor is active (and no m-RNA is made) unless combined with an inducer such as the substrate on which the enzyme acts. Enzyme will then be synthesized only if the cells have been exposed to the substrate. Other genes are regulated by a repressor molecule that is inactive unless it is combined with specific small molecules. For example, the rate of synthesis of the enzyme that catalyses the first step in a biosynthetic pathway is often repressed by a combination of the end-product of that pathway with the specific repressor molecule. Thus, overproduction of a particular end-product can be avoided. The *activity* of enzymes already synthesized also can be controlled by interaction with small molecules. Frequently, the first enzyme in a biosynthetic pathway is subject to such feedback inhibition by the end-product of the pathway. This inhibition appears to be the result of a reversible coupling of the enzyme and the end-product of the pathway which interferes with the reaction between the enzyme and its substrate. Many of the details of regulatory mechanisms have developed from investigations with bacteria. However, it appears from studies on other organisms, including yeast, that these mechanisms are quite general.

Two approaches have been followed in the search for genes involved in the regulation of biosynthetic pathways in yeast. Mutants resistant to analogues of amino acids or purine- and pyrimidine bases have been sought with the idea that, among the possible changes conferring resistance, there could be over-production of normal metabolites due to a repressor or operator mutation, or there could be a mutation in a structural gene for an enzyme in the pathway which abolishes its sensitivity to feedback inhibition. Also, mutants sensitive to normal metabolites have been examined because of the possibility that these might arise in branched or interacting pathways and be indicative of feedback inhibition (Meuris *et al.*, 1967).

These approaches can be illustrated by a consideration of the pathways leading to synthesis of uracil and arginine in *Sacch. cerevisiae* (Fig. 7). An intermediate common to both pathways, carbamoyl phosphate, is synthesized from glutamine, ATP and bicarbonate by two parallel enzyme systems. One enzyme system is controlled by a single locus, *cpu*, that codes for an enzyme which is highly sensitive to feedback inhibition by uridine 5'-triphosphate. In the second system, the same reaction is controlled by two genes, *cpa₁* and *cpa₂*, which together are repressed by arginine (Lacroute *et al.*, 1965). Either pathway can be

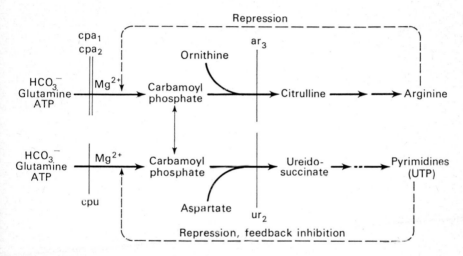

FIG. 7. Regulation of the arginine and pyrimidine pathways in *Saccharomyces cerevisiae*. For details see text. Reproduced by permission of F. Lacroute.

lost without imposing a requirement for uracil or arginine. This indicates that the carbamoyl phosphate synthesized by the two reactions forms a common pool which is available for subsequent steps in either the arginine or pyrimidine pathways. However, with either a *cpa₁* or *cpa₂* mutation, addition of uracil to the medium imposes a requirement for arginine. Arginine added to the minimal medium retards growth of the *cpu* mutant but does not result in an absolute requirement for uracil. Haploids with both a *cpa* and a *cpu* mutation require both uracil and arginine.

The gene *cpu* is closely linked or allelic with *ur₂*, the gene for the enzyme aspartate carbamoyltransferase that controls the next step in the uracil pathway. Mutants in which both enzyme activities have been lost can be obtained in a single step ; of 20 mutants identified as ur_2, 10 were shown to have lost the *cpu* activity (carbamoyl phosphate synthetase) as well (Lacroute, 1966). The double mutants can be explained in terms of a single protein with two active sites, or as polar mutants in the first cistron of a polycistronic messenger. Even if there are two peptides encoded by adjacent cistrons, they probably are complexed in a heteromultimer. This assertion arises from the following observations. Both enzymic activities are subject to feedback inhibition by uridine 5′-triphosphate. One of the genes conferring resistance to 5-fluorouracil, FU_2^r, is allelic with *cpu* and ur_2 ; and independent isolates of FU_2^r have different phenotypes with respect to the two enzyme activities. For example, with FU_{2-5}^r the carbamoyl phosphate synthetase activity is lost, and the aspartate carbamoyltransferase is no longer sensitive to feedback inhibition. With another mutant, FU_{2-2}^r, both enzyme activities are retained but neither is sensitive to feedback inhibition.

By using combinations of multiple mutants in the analysis of the enzyme levels for the remaining steps in the uracil pathway, Lacroute (1968) was able to show that the genes involved were not regulated by end-product repression, but rather were induced by the intermediary metabolites. The product of the reaction catalysed by aspartate carbamoyltransferase, ureidosuccinic acid, is the inducer for the synthesis of the next enzyme, dihydro-orotase, while the product of this reaction, dihydro-orotic acid, is the inducer for the synthesis of the three enzymes that catalyse subsequent reactions, namely dihydro-orotate dehydrogenase, orotidine 5′-phosphate pyrophosphorylase and orotidine 5′-phosphate decarboxylase.

In the arginine pathway, the enzyme that catalyses the coupling of carbamoyl phosphate and ornithine, ornithine carbamoyltransferase, is subject to both feedback inhibition by arginine and to the control of a regulator gene (Bechet *et al.*, 1962, 1965). Bechet and Wiame (1965) give the following evidence when suggesting that the product of the regulator gene is a protein which binds or inactivates the carbamoyltransferase. The addition of arginine to preparations of cells that were rendered permeable initiates a drop in the carbamoyltransferase activity, but a comparable loss of activity is not seen when arginine is simply added to cell extracts. The addition of actidione to stop protein synthesis halts the drop in carbamoyltransferase activity. The enzyme activity can be restored in the permeable cells that have been exposed to arginine by heating them for five minutes at 60°. Permeable cells of the non-repressible regulatory mutant do not show a drop in ornithine carbamoyltransferase activity when exposed to arginine.

The investigation of mutants sensitive to tyrosine, *ty-i*, and phenylalanine, *phe-i*, revealed that *Saccharomyces cerevisiae* has two isofunctional enzymes that catalyse the first step in the aromatic pathway, the condensation of erythrose 4-phosphate with phosphoenolpyruvic acid (3-deoxy-D-arabinoheptulosonic acid 7-phosphate (DAHP) synthetase). One enzyme is sensitive to feedback inhibition by tyrosine, and the other is inhibited by phenylalanine (Meuris, 1967). Presumably the phenylalanine-sensitive activity is lost in the *ty-i* mutants, while the tyrosine-sensitive activity is missing in *phe-i* mutants. From crosses of *ty-i* × *phe-i*, the double-mutant segregants *ty-i phe-i* were obtained and, as expected, were found to require all of the products of the aromatic pathway, namely tyrosine, phenylalanine, tryptophan and *p*-aminobenzoic acid, for growth. That two DAHP synthetase activities are present in *Sacch. cerevisiae* has been demonstrated directly by Lingens *et al.* (1966). The activity present in crude extracts was sensitive to inhibition by both tyrosine and phenylalanine. However, this activity could be separated into tyrosine-sensitive and phenylalanine-sensitive fractions. While the DAHP synthetase activities are subject to endproduct inhibition, no repression of this activity by growth in the presence of these end-products could be demonstrated (Lingens *et al.*, 1967).

Regulation of other steps in the aromatic pathway has also been examined (Doy and Cooper, 1966; Lingens *et al.*, 1966, 1967). The first enzyme on the tryptophan branch, anthranilate synthetase, was found to be sensitive to feedback inhibition by tryptophan. In addition, synthesis of this enzyme was repressed 3–4-fold when the tryptophan concentration in the medium was increased from 10^{-4} M to 10^{-3} M. The first enzyme on the tyrosine-phenylalanine branch, chorismate mutase, is subject to feedback inhibition by tyrosine and activation by tryptophan. In addition, synthesis of this enzyme is induced by tryptophan. A summary of the regulation of the entire aromatic pathway in *Sacch. cerevisiae* has been presented by Lingens *et al.* (1967).

Both repression by exogenous methionine and feedback inhibition by S-adenosylmethionine are operative with the first enzyme for methionine biosynthesis, homoserine O-transacetylase (de Robichon-Szulmajster and Cherest, 1967). On the isoleucine pathway, the first enzyme, threonine deaminase, is sensitive to feedback inhibition by isoleucine (de Robichon-Szulmajster and Magee, 1968). The second enzyme, acetohydroxyacid synthetase, which is shared with the valine pathway, is inhibited by valine alone; isoleucine has little or no effect on this enzyme (Magee and de Robichon-Szulmajster, 1968a, b).

More complicated systems than those illustrated in the above examples have been found for the regulation of the catabolic pathways with inducible enzymes. For example, the regulation and linkage relationships of the genes involved in galactose utilization have several aspects of the operon model. The genes ga_1, ga_7 and ga_{10}, for the first three enzymes in the Leloir pathway, namely galactokinase, transferase and epimerase, are closely linked (Douglas and Hawthorne, 1964). Synthesis of these enzymes is induced simultaneously when the wild-type yeast is exposed to galactose (de Robichon-Szulmajster, 1968). There are four regulatory genes affecting this process: ga_3, i, C and ga_4. The precise role of GA_3 is not clear; but it is known that the ga_3 mutation inhibits only the initiation of synthesis of the galactose enzymes; once enzyme molecules have been formed, there is no longer an impediment to adaptation (Spiegelman et al., 1950). The ga_3 block can be by-passed by a recessive mutation, i^-, which causes the constitutive synthesis of these enzymes (Douglas and Pelroy, 1963). Constitutive synthesis is also achieved by a much rarer dominant mutant, C (Douglas and Hawthorne, 1966). The C mutant has the properties expected of the O^c type operator mutations; however, it is not linked to the structural gene complex $ga_{1,7,10}$. Instead, C is closely linked to the ga_4 locus. The ga_4 mutation causes the loss of all three enzyme activities. The C mutant must be coupled with the wild-type allele, GA_4, for constitutive synthesis of the galactose enzymes. Diploids with the genotype $C\ GA_4/c\ ga_4$ are constitutive, while $C\ ga_4/c\ GA_4$ diploids are inducible. By analogy with the lactose system in Escherichia coli, the i locus codes for a repressor. The C locus is the site of repressor recognition, and it in turn controls the expression of the adjacent cistron, GA_4. The product of GA_4 is a protein because both complementing and suppressible alleles at this locus have been found. Moreover, GA_4 has a positive role in the expression of the enzymes coded by the three structural genes GA_1, GA_7 and GA_{10}.

A sequential induction of the respiratory enzymes in yeast is suggested by the following observations. In fully grown aerobic cultures, two cytochrome c species, iso-1 and iso-2, are synthesized in a ratio of $10:1$. However, if the kinetics of their synthesis during adaption to aerobic growth is followed, it is seen that the minor species, iso-2 cytochrome c, appears first. It is proposed that the polypeptide chain of iso-2 cytochrome c, without the haem group, may function as a repressor of the synthesis of iso-1 cytochrome c. Addition of the haem group to the polypeptide to form the iso-2 molecule relieves this repression and permits the synthesis of iso-1 cytochrome c. There is no evidence for a reciprocal repression by the iso-1 peptide; mutations at the cy_1 locus which causes the loss of the iso-1 peptide do not result in a significant increase in iso-2 cytochrome c (Slonimski et al., 1965).

Five recessive modifying genes, cy_2, cy_3, cy_4, $c\ y_5$ and cy_6, decrease the levels of both iso-1 and iso-2 cytochrome c (Sherman, 1964). The mutant alleles cy_{3-5} and cy_{3-6} markedly increase the sensitivity of cytochrome a synthesis to glucose repression, whereas other alleles at this locus, cy_{3-1} and cy_{3-3}, had no effect (Reilly and Sherman, 1965). A correlation was seen between a very low level of cytochrome c and the extra sensitivity to glucose repression of the cytochrome a synthesis.

From the analyses of specific activities of different enzymes during the cell cycle in synchronized cultures, Tauro *et al.* (1968) argue that the genome is only periodically available for transcription, i.e. there is one period of enzyme synthesis per structural gene per generation, and all mechanisms for regulation are superimposed upon this restriction. The various enzymes assayed showed a discrete and characteristic period of increment. The genes for the enzymes involved had been mapped, so it was possible to relate the sequence of increments in enzyme activity with the linkage relationships of the corresponding genes. The timing of enzyme synthesis was not correlated with centromere distances. However, synthesis of the enzymes coded by four genes on Chromosome V followed a temporal sequence corresponding to their order on the chromosome: the synthesis of orotidine 5-phosphate decarboxylase (ur_3), aspartokinase (thr_3), phosphoribosyl-ATP-pyrophosphorylase (hi_1), and threonine deaminase (is_1; Table V, p. 421), proceeded in that order. With these results, the authors propose that the chromosomes are transcribed undirectionally and, presumably, this timing is correlated with their replication.

VII. Mutant Characterization

The metabolic blocks imposed by different mutations are dealt with in Section VI.A (p. 420) dealing with gene-enzyme relationships. Mutants also can be characterized by various properties of the mutant alleles such as revertibility with different mutagens, susceptibility to phenotypic reversion by culture conditions, response to supersuppressors, and intracistronic complementation response. From these characterizations, deductions can be made about the nature of the lesion in the DNA of the gene, i.e. whether there has been a base substitution that results in a missense or nonsense codon, a single base addition or deletion that causes a reading-frame shift, or a gross aberration.

An intimation as to the nature of the mutational defect is provided by the spontaneous reversion rates of the mutant. *A priori*, one expects higher reversion rates with missense and nonsense mutants than with the reading-frame mutants; while no revertants are likely to be obtained with a gross aberration such as a deletion. However, there are only a few reports of reversion studies in yeast in which the defect could be ascertained by other means. Among the 10 adenine-1 alleles (*Schizo-*

15

saccharomyces pombe) studied by Clarke (1965), at least five respond to external suppressors. These suppressors are presumably allele-specific super-suppressors and thus the suppressible mutants are nonsense mutants. The five suppressible mutants had reversion rates in the range 10^{-8}–10^{-7}.

Base-addition or -deletion mutations are more difficult to define experimentally. However, observations consistent with the expected behaviour of such mutants have been reported by Magni and von Borstel (1962) and Magni (1963). These authors proposed that the most likely way a base could be deleted or added to restore the original reading frame would be through pairing errors and unequal exchange in crossing-over. Thus, with these mutants, reversion rates during meiosis would be expected to be higher than during mitosis. Two mutants which fit this pattern are thr_{4-1} and hi_{1-1}. In the first case, the thr_4/thr_4 diploids had mitotic and meiotic rates of $3/10^{10}$ and $100/10^{10}$ reversions per cell per generation, respectively.

The importance of chromosome pairing and, presumably, crossing over was tested by using a diploid heterozygous for a deletion covering the thr_4 region. The reversion rates for this diploid were approximately $3/10^{10}$ for the mitotic division and $4/10^{10}$ for the meiotic division. For the diploid homozygous for hi_{1-1}, the reversion rates were $7/10^{10}$ in mitosis and $200/10^{10}$ in meiosis. This diploid was heterozygous for the genes thr_3 and ar_6 which flank the hi_1 locus. Of 163 revertant spore clones, 71% were recombinant for the outside markers, while only 15% of a sample of 202 non-revertant spores were recombinant. No suppressors have been obtained with the hi_{1-1} and thr_{4-1} alleles and, moreover, these mutants are not 'leaky'. These observations also are consistent with the properties expected of reading frame-shift mutants.

The above 'meiotic effect' was seen with two nonsense mutants, ly_{1-1} and hi_{5-2} (Magni *et al.*, 1966). The mitotic reversion rates for the homozygous diploids were about $4/10^8$, and the meiotic rates were about $5/10^7$. However, this tenfold enhancement in the reversion rates associated with meiosis was shown to be due to an increase in the number of suppressor mutations and not to any significant change in the number of back mutants. For example, for ly_{1-1}, the back mutation rates were $5/10^9$ in mitosis and $7/10^9$ in meiosis, while the suppressor mutation rates were $3/10^8$ for mitosis and $5/10^7$ for meiosis. It was proposed that these suppressors arose by base additions or deletions resulting from unequal crossing over in genes coding for transfer-RNA molecules. Although these rates are approximately ten times those seen for the reversion of the frame-shift mutants documented above, it should be kept in mind that ten or more suppressor loci may be involved in the reversion of these two nonsense mutants.

The most precise description of the mutational lesion is possible with the nonsense mutants that are suppressed by the super-suppressors. The nonsense codon can be identified as UAG (amber) or UAA (ochre) by crossing the mutant to stocks with the wild-type allele which also carry known amber- or ochre-specific suppressors. If tetrad analysis reveals $4+:0-$ and $3+:1-$ ratios for the phenotype of the gene in question, the mutant allele is considered to be suppressible. The necessity to make crosses and perform tetrad analyses to test the suppressibility of new mutants can be circumvented by isolating the mutants in a parental stock containing sets of at least two amber and two ochre mutants chosen so that one can select, with the appropriate omission medium, super-suppressors in a given class.

The recognition of missense mutants is often facilitated by their response to different culture conditions. Two factors that can be varied over a range sufficient to be of use without undue adverse effects on the growth of the wild-type yeast are temperature and osmotic pressure. Nutritional requirements and fermentative abilities are customarily assayed at 30°; however, it has been our experience that most of these tests can be conducted at temperatures from 18° to 35°. The osmotic pressure of the test media can be raised by adding any of a number of relatively innocuous solutes such as sorbitol, diethylene glycol, glycerol, sodium chloride and potassium chloride. The least toxicity is seen with potassium chloride; the wild-type yeast will still grow with 2 M-KCl in the synthetic media. The minimal medium with 2% glucose and an assortment of mineral salts gives an osmotic pressure equivalent to approximately 0.1 M-KCl.

Three temperatures, 18°, 25° and 33°, and media with 0, 0.5 M and 1.0 M-KCl were used to screen for temperature-sensitive and osmotic remedial alleles among 231 mutants at 43 loci (Hawthorne and Friis, 1964). Thirty-six mutants were classified as osmotic-remedial with the criterion that they had shown a greater growth rate on media supplemented with potassium chloride for at least one of the three test temperatures. Twenty of these mutants could be classified as temperature-sensitive since they were able to grow on media free from potassium chloride at certain temperatures, and there were three additional temperature-sensitive mutants which were unaffected or inhibited by the potassium chloride.

The interdependence of temperature and osmotic pressure was manifested by a shift in the optimum osmotic pressure for the fastest growth rate with a change in temperature. The commonest interaction was an increased osmotic pressure optimum with an increase in the incubation temperature. Plots of the osmotic pressure optima revealed peaks of only 0.6 M–0.8 M-KCl in breadth. Some mutants, regarded as 'leaky' under the customary assay conditions, gave no growth under the restrictive conditions.

The osmotic-remedial property shows a negative correlation with suppressibility. Twenty-three suppressible mutants were screened and none responded to the increased osmotic pressure. Conversely, 16 osmotic-remedial mutants were found to be unaffected by a super-suppressor. However, modifier genes have been found for an osmotic-remedial allele, ga_{7-1}, in the galactose pathway (Bassel and Douglas, 1968). A partial restoration of enzyme activity, galactose 1-phosphate uridylyltransferase, is obtained in the presence of the modifiers without the addition of solutes to the galactose-containing medium.

A positive correlation is seen for osmotic-remediability and interallelic complementation. For those genes where interallelic complementation has been demonstrated, the osmotic-remedial alleles are found among the complementing alleles. For example, a sample of 27 ad_7 alleles that contained nine osmotic-remedial mutants, had 16 alleles which showed complementation at 30°. Seven osmotic-remedial alleles were in this class, and the remaining two were classified among the complementers when the tests were conducted at 21°. Also, osmotic-remedial alleles are found more frequently for those genes with interallelic complementation than for those genes where there is no complementation. For example, Nashed and Jabbur (1966) found 34 of 194 strains with different ad_2 alleles, but none of 28 strains with different ad_1 alleles, to respond to $1 \cdot 0$ M-KCl at 30°. No complementation has been seen in extensive tests with ad_1 alleles (Woods and Bevan, 1966).

From the above observations, it is argued that mutants sensitive to osmotic pressure and temperature are missense mutants that form complete polypeptides which, as a consequence of the associated single amino-acid substitution, are more sensitive to these physical parameters. It is proposed that, under the normal culture conditions, the peptides would be folded abnormally while, under the permissive conditions, an active protein could be formed. It is not implied, however, that an equilibrium need exist between the two configurations. It is conceivable that correct folding can only be achieved during growth of the nascent polypeptide chain.

Information pertinent to mutant characterization is given by two complementation phenomena: interallelic (intracistronic) complementation and intercistronic complementation among the genes of an operon that yields a polycistronic messenger.

Interallelic complementation is restricted to those genes that code for enzymes that are composed of two or more polypeptide chains. The haploid-mutant parents each form defective polypeptides, either of which, if aggregated to form the homomultimer, would be inactive;

however, in the hetero-allelic diploid, the two species of defective peptides are combined in some of the polymers and, in complementing
combinations, this heteromultimer is catalytically active. Presumably
the defective monomers in some way are mutually compensating to
yield the complementation reaction. Chances for complementation are
the greatest if both parents contribute a complete polypeptide molecule, i.e. if they are missense mutants which have only a single amino-
acid change in the polypeptide. Only rarely are nonsense (suppressible) mutants found to participate in interallelic complementation
(Manney, 1964; Inge-Vechtomov and Simarov, 1967; Barben, 1966).
It is even less likely that frame-shift mutants would be complementers
because, instead of chain termination at the mutant site, there would be
a continuation of the peptide chain with a completely different sequence
of amino acids until a nonsense triplet was read.

The results of the tests of alleles for complementation can be por-
trayed in a complementation map. In the construction of such a map,
complementing allele combinations are represented as non-overlapping
lines, and non-complementing combinations as overlapping lines. An
attempt is made to portray the complementation responses of the alleles
as continuous lines. This usually results in a complex map whenever a
large number of alleles are involved. For example, a circular map with a
linear spur was obtained in two studies of complementation at the ad_7
locus (Costello and Bevan, 1964; Dorfman, 1964). A map in the form of
two circles connected by a line was constructed for the ad_6 (red adenine-
less) gene in *Schizosaccharomyces pombe* (Leupold and Gutz, 1963). A
similar map has been obtained by Woods and Bevan (1966) for the
equivalent gene (ad_2) in *Saccharomyces cerevisiae*. However, a linear map
was derived from another study of complementation with ad_2 alleles
(Soidla *et al.*, 1967). Linear maps have also been presented for the ad_8
gene in *Schizosacch. pombe* (Megnet and Giles, 1964), and the tr_5 gene
(Manney, 1964), and the hi_{4A} and hi_{4C} cistrons (Fink, 1966) in *Sacch.
cerevisiae*.

Fewer complementing alleles were examined in the last three studies,
but the alleles were mapped by random-spore analysis (ad_8) or by X
ray-induced mitotic recombination (tr_5 and hi_4). With only a few exceptions, there was colinearity of the sequence of alleles on the fine structure map and the position of the corresponding subunits on the
complementation map. These exceptions to colinearity may presage
more complex complementation maps which might be derived with the
use of additional alleles and which would bring the subunits and mutant
sites into juxtaposition. A case in point is the study of the ad_6 locus
(in *Schizosacch. pombe*) by Leupold (1961) and Leupold and Gutz (1963).
The fine-structure map of this gene could be wrapped around the two

circles of the complementation map to achieve a good correspondence of sites and subunits. While it is probable that the complexities of the complementation map reflect in some measure the folding of the peptides and their association in the polymer, the map should not be considered as an actual representation of the protein configuration (Fincham, 1966).

Comparison of the complementation and fine-structure maps can reveal the direction of translation of the messenger-RNA if nonsense mutants are included. This was done by Manney (1964) in his analysis of the tr_5 locus. Two nonsense mutants were found among the 19 complementing alleles. By X-ray mapping, one was placed at one end and the other in the centre of the locus. There was no complementation between these two nonsense mutants. The missense mutants they complemented always mapped to the right of the nonsense mutant involved. The nonsense mutant at the left end was presumed to form a nearly complete peptide chain because it was able to complement alleles located throughout the remainder of the map. The centrally located nonsense mutant, on this argument, would form only a peptide fragment which could still make an active polymer with the peptide from mutants to its right. Thus, it was concluded that translation proceeded from right to left with respect to the fine-structure map. A similar polarity of complementation of nonsense mutants has been observed in other genes in yeast (Mortimer and Manney, 1964; Barben, 1966; Fink, 1966).

The complementation responses of conditional mutants have been examined under both restrictive and permissive conditions. As anticipated, the presence of a super-suppressor frequently enabled nonsense mutants of ad_2 to participate in interallelic complementation. However, there were cases of negative complementation, i.e. nonsense mutants, which respond to the suppressor in either the haploid or homo-allelic diploid, and were not suppressed when combined with a missense mutant in a hetero-allelic diploid (Inge-Vechtomov et al., 1966; Inge-Vechtomov and Simarov, 1967). Osmotic-remedial and temperature-sensitive mutants were used in another study involving negative complementation at the ad_2 locus (Nashed et al., 1967). Leaky mutants were excluded from the collection of 208 mutants used. Nevertheless, there were 19 mutants for which growth was restored under rather rigid conditions: 11 for $1 \cdot 0$ M-KCl at $25°$, three for $1 \cdot 0$ M-KCl at $35°$, and five in the absence of KCl at $35°$. Out of 1625 combinations involving these conditional mutants, there were 367 that showed negative complementation, i.e. there was no growth for the hybrid under the condition that was permissive for the conditional parent. The mutants that responded at $35°$ to the presence or absence of $1 \cdot 0$ M-KCl were the ones that were most sensitive when paired with a non-remedial allele; about a third of such pairings gave negative complementation.

In the above cases of negative complementation, it can be assumed that the hybrid protein has no activity under conditions that are permissive for the remedial parent. But this raises the question why there are no or insufficient homologous polymers coded from the remedial allele to show activity. The possibility that the enzyme is composed of a large number of monomers, thus greatly reducing the chance of attaining a homologous polymer, is discounted by Nashed et al. (1967). They argue that, with a large polymer, there would be relatively few mutant combinations that would show complementation, and the opposite situation is observed for ad_2. Instead, they argue that there is a preferential aggregation of monomers from different allelic origin to give primarily heterologous multimers.

Some complex loci are known in yeasts, including $ad_{5,7}$, hi_{4ABC} and ty_2 ($arom_{1ABCDE}$). These loci are characterized by: (i) complementation between mutants in different cistrons that is comparable to intergenic complementation between unlinked loci; and (ii) the occurrence of a class of mutants which fail to complement mutants in two or more of the cistrons. The non-complementing mutants are interpreted as polar mutants, either addition-deletion or nonsense mutants, located in the first cistron and preventing translation of the subsequent cistron. For example, the ad_5 and ad_7 alleles were located in two separate regions, and the non-complementing $ad_{5,7}$ alleles mapped among the ad_5 sites (Dorfman, 1964). From this evidence, it was concluded that there was a polycistronic messenger and that ad_5 was the first cistron to be translated.

A study of complementation and fine structure at the hi_4 locus was accompanied by assays of the corresponding enzymes (Fink, 1966). Three cistrons at this locus control the synthesis of phosphoribosyl-AMP 1,6 cyclohydrolase (hi_{4A}), phosphoribosyl-AMP pyrophosphohydrolase (hi_{4B}), and histidinol dehydrogenase (hi_{4C}; Table V, p. 421), and map in the order A, B, C. The two classes of polar mutants obtained—'ABC⁻' and 'BC⁻'—indicate that translation of hi_{4A} occurs first. The ABC⁻ mutants mapped in the A region, and the BC⁻ mutants in the B region. The enzyme assays are compatible with the complementation tests except for the BC⁻ mutants. The BC⁻ haploids had no activity for the A enzyme, yet they were able to complement A⁻ mutants in the hetero-allelic diploid. A multi-enzyme aggregate was postulated to explain this observation. It was proposed that the A enzyme must be complexed with the B and C proteins to have activity.

The term 'intercistronic complementation' has been used in a different sense by Oshima (1967) to designate the interaction of defective peptides from genes that form isozymes of α-glucosidases. Inter-allelic complementation was demonstrated between two ma_1 alleles; moreover, one of these ma_1 alleles in combination with mutant alleles at either of two unlinked maltose genes enabled haploids to ferment maltose. Presumably there is enough similarity between the peptides coded by the different maltose genes to permit this heterologous aggregation to restore enzyme activity.

VIII. Suppression

As discussed in the preceding section, forward mutations may result either from substitution of a base pair or from addition or deletion of base pairs in the DNA. Reverse mutation may involve exact restoration of the DNA to its original state. However, reversion very often is the result of a change at a genetic site different from that of the original mutation. In such cases, the second mutation is known as a suppressor.

The second mutation may occur in the same gene but at a different site from that of the original mutation. For example, frameshift mutations may be reversed by a nearby base-pair addition or deletion that restores the correct reading-frame. This is the mechanism proposed for reversion of meiotic-effect mutants in yeast (Magni and von Borstel, 1962; Magni, 1963). It is also possible that a missense mutation, with a non-functional amino-acid replacement in the gene product, may be reversed by a second-site mutation resulting in the substitution of another amino acid elsewhere in the protein. Examples of this type of intralocus suppression have been described for bacteria, but none has been reported for yeast.

The wild-type phenotype is frequently restored by a mutation in a gene other than the one with the original defect. These external suppressors act either by opening up a new pathway that by-passes the block imposed by the first mutation (indirect suppression) or by modifying one of the steps in the transfer of information from the DNA to protein such that the defect is not translated into the gene product (informational suppression).

Examples of indirect suppression are seen in the locus-specific suppressors obtained for four of the five mutants in the isoleucine-valine pathway. The isoleucine requirement imposed by is_1 can either be completely alleviated by one suppressor or can be satisfied by threonine when another suppressor is present (Kakar, 1963). Each of the iso-leucine-valine mutants, is_2, is_3 and is_5, has a suppressor which eliminates only one of the requirements (Kakar et al., 1964). For is_2 and is_3, the suppressed 'half-revertant' stocks still require isoleucine. The half-revertants of is_5 are valine-dependent. This pattern of suppression suggests that, at one time, *Saccharomyces* had separate but parallel pathways for the synthesis of isoleucine and valine. When these pathways were combined into one, it was, for example, the acetohydroxyacid synthetase (is_2) of the isoleucine pathway which was modified to accept both substrates. Thus, the suppressor of is_2 which alleviates the valine requirement would be a vestige of the valine-specific pathway.

Other locus-specific suppressors have been found for ad_3 (Zimmermann and Schwaier, 1963) and ser_1 (Mortimer and Hawthorne, 1966b; R. K. Mortimer, unpublished observations). Although only a single ser_1 allele has been examined, the suppressor is presumed to be providing an alternate pathway since it acted upon no other phenotype and was found to occur naturally in several strains of diverse origin.

The most thoroughly investigated suppressors in yeast, the super-suppressors, are analogous to the informational suppressors in *Escherichia coli* which permit the translation of the nonsense codons, UAA, UAG and UGA (Garen, 1968). The super-suppressors are characterized as allele-specific and locus-non-specific. The suppressible alleles occur in genes distributed throughout the linkage maps. In the original study, 50 mutant alleles located at 37 loci were tested against the suppressors S_1 and S_2 (Hawthorne and Mortimer, 1963). Fourteen of the alleles were suppressible, and they occurred in 10 genes causing adenine, uracil, arginine, histidine, leucine, lysine, tryptophan, and tyrosine requirements as well as a respiratory deficiency. Similar frequencies of suppressible alleles were found when many alleles at specific loci were tested. In *Sacch. cerevisiae*, 13 of 32 tr_5 alleles were suppressed by a super-suppressor, S_{11} (Manney, 1964); five of 27 ad_7 mutants were found to be suppressible (Hawthorne and Friis, 1964), as were 13 of 39 hi_4 mutants (Fink, 1966), 15 of 96 ad_2 mutants (Inge-Vechtomov and Simarov, 1967), eight of 20 le_1 mutants (Nakai and Mortimer, 1967), and 10 of 43 ar_4 mutants (Mortimer and Gilmore, 1968). In *Schizosacch. pombe*, 32 suppressible alleles have been identified from a sample of 138 mutants at three different adenine loci (Barben, 1966).

As discussed in the preceding section on mutant characterization, the suppressible alleles have the properties expected with nonsense mutations, i.e. mutations that lead to a termination of the polypeptide synthesis at the mutant codon. For some alleles, one can demonstrate a polarized pattern of complementation in either intragenic or intercistronic complementation tests (Manney, 1964; Fink, 1966). Moreover, Manney (1968) has demonstrated directly the formation of peptide fragments of tryptophan synthetase in strains carrying suppressible alleles of tr_5, the structural gene for this enzyme.

Yeasts would appear to have many more suppressor genes than *Escherichia coli*. In *Schizosacch. pombe*, Barben (1966) has found seven

15*

suppressor genes with four distinct patterns of suppression. A conservative estimate of the number of super-suppressors in *Sacch. cerevisiae* would be about 30; however, the task of making allelism tests to identify the suppressors isolated by different investigators has only just commenced. Inge-Vechtomov (1964, 1965) has found six dominant, four semidominant, and two recessive super-suppressors for ad_2 alleles. At least 20 suppressors, on the basis of both phenotypic differences and linkage relationships, have been isolated by Gilmore (1967). The linkage relationships of 22 suppressors were examined by Hawthorne and Mortimer (1968), and 16 were placed on the linkage maps of *Saccharomyces* (see Fig. 5, p. 416).

As a first step towards identifying the super-suppressors, Gilmore and Mortimer (1966) and Gilmore (1967) have classified the suppressors by their action spectra against five suppressible alleles. Eight distinct phenotypic classes were obtained with the examination of 83 suppressor isolates. Most classes contained several suppressors mapping at different loci. For example, the first class contains suppressors mapping at seven different loci, while the third class has five different suppressor genes. Ten phenotypic classes are seen in a classification scheme based upon 18 suppressible alleles (Hawthorne and Mortimer, 1968). An abridged version of this scheme, which uses the 15 most widely circulated alleles, is given in Table VI.

Two different nonsense codons are represented by the suppressible alleles used in the latter classification scheme. If we accept the universality of the genetic code, we can deduce that the two nonsense codons are UAA and UAG from the observation that there are suppressors specific for alleles in one or the other classes of nonsense and a third group of suppressors that act upon alleles in both classes. Suppressors that act upon both types of nonsense mutants are expected only with the ochre (UAA) and amber (UAG) codons under the 'wobble hypothesis' (Crick, 1966). No single suppressor should act upon both UAA and UGA or UAG and UGA mutants.

Reversion studies with chemical mutagens are consistent with the above interpretation. Magni and Puglisi (1966) found that none of the three suppressible alleles hi_{5-2}, ly_{1-1}, and ar_{4-17} gave back mutants with hydroxylamine-induced mutagenesis. This was expected since hydroxylamine induces primarily $GC \rightarrow AT$ transitions. The DNA base-pair triplets that code the nonsense codons are:

$$\frac{ATT}{TAA} \text{ (ochre)}, \quad \frac{ATC}{TAG} \text{ (amber)} \quad \text{and} \quad \frac{ACT}{TGA} \text{ (UGA)}$$

The $GC \rightarrow AT$ transition will convert an amber or UGA mutant to an ochre mutant, whereas ochre mutants should be unaffected because

TABLE VI. *Classification of Super-Suppressors in* Saccharomyces cerevisiae *Based on their Action Spectra*

Suppressible Alleles	Ochre-Specific Suppressors					Ochre-Amber Suppressors			Amber Suppressors	
	I	II	III	IV	V	VI	VII	VIII	IX	X
	S_1–S_8	S_{11}	S_{15}	S_{20}–S_{25}	S_{30}–S_{31}	S_{35}	S_{40}	S_{45}	S_{50}–S_{51}	S_{60}
tr_{5-2}	+	+	−	−	−	−	−	−	−	−
ur_{4-1}	+	+	−	−	−	−	−	−	−	−
ad_{2-1}	+	±	±	−	−	±	−	−	−	−
ty_{1-1}	±	±	+	±	−	−	−	−	−	−
ly_{1-1}	+	+	+	±	−	−	−	−	−	−
ly_{2-1}	+	+	+	+	−	−	−	−	−	−
hi_{5-2}	+	+	+	+	−	−	−	−	−	−
ad_{6-3}	+	+	+	+	−	±	−	−	−	−
met_{4-1}	+	+	+	+	−	±	±	−	−	−
hi_{4-1}	+	+	+	+	−	±	±	±	−	−
le_{2-1}	+	+	+	+	±	+	+	+	−	−
is_{1-1}	+	+	+	+	±	+	+	+	−	−
ty_{7-1}	−	−	−	−	−	±	+	+	+	+
ty_{6-1}	−	−	−	−	−	+	−	±	+	+
tr_{1-1}	−	−	−	−	−	−	+	±	+	−

In the scoring of the nutritional requirements, + indicates good growth from replica prints in two days; ± visible growth by 3–5 days; − indicates no signs of growth by five days.

they contain no GC pair. The hi_{5-2}, ly_{1-1} and ar_{4-17} alleles all have the same nonsense codon, the one recognized by the original class of super-suppressors. Hydroxylamine and ethylmethanesulphonate, which also favour the GC → AT transition, were used in an experiment to decide which of the two nonsense codons contains the GC pair (D. C. Hawthorne, unpublished observations). The experimental design was to treat strains bearing one class of nonsense mutants and the super-suppressor for the other class. Only strains in which the second nonsense codon was paired with the original suppressors responded to the mutagens. The revertants were due to a mutation to an allele susceptible to the suppressor. Thus, the second nonsense codon contains the GC pair and the original class of super-suppressors are ochre-specific suppressors.

In deciding the nature of the suppressor mutation, we again can draw upon the findings in *Escherichia coli* where the suppressors have been demonstrated to be genes coding for transfer-RNAs (Garen, 1968). If one assumes that the mutational event is occurring in the triplet coding for the anticodon of the t-RNA, then we can expect specific mutagens to act upon only certain species of t-RNA to give particular suppressor mutations. Treatment with ethylmethanesulphonate does indeed enhance the recovery of several classes of both amber- and ochre-specific suppressors as much as twenty-fold (D. C. Hawthorne, in preparation). In contrast, Magni and Puglisi (1966) found that hydroxylamine was only slightly effective for inducing the ochre suppressor mutation, whereas the frameshift mutagen ICR-170, which is an acridine mustard, induced suppressors efficiently and, at the same time, was without effect at the suppressible allele. The fact that a base insertion or deletion should result in an active product indicates that the suppressor gene is not translated into protein, and is thus consistent with the concept that these genes are coding for transfer-RNAs.

Granted that the suppressor genes are coding for transfer-RNAs, then one expects that they should yield an active product that should be manifested in diploid as well as haploid cells. In other words, the super-suppressors should behave as dominant genes. Many suppressors are dominant or at least semidominant, but a few appear to be recessive. This feature of dominance or recessiveness depends as much upon the allele being suppressed as upon the suppressor. For example, S_{11} suppresses mutations such as hi_{5-2} or ly_{1-1} dominantly, but is able to suppress ad_{2-1} only when homozygous. Whether a suppressor is dominant or recessive most likely reflects both the level of suppression attained (percentage wild-type activity) and the threshold of the particular enzyme activity needed for growth. A 'low-level suppressor' is likely to be recessive in the suppression of most alleles.

The designation 'low-level suppressor' is applied to those ochre-specific suppressors with a relatively limited action spectra, the Class IV and Class V suppressors. We now interpret the differences in action spectra for the various classes of ochre-specific suppressors to be the result of quantitative differences in suppressor efficiency rather than qualitative differences, i.e. different amino-acid substitutions. Support for this argument comes from the observation that three different low-level suppressors in the same haploid clone will together suppress alleles which would not be suppressed by one of the suppressors acting alone. The low-level suppressors acting in concert can mimic the action spectra of the Class I suppressors (R. K. Mortimer, unpublished observations). It might be noted that the presence of two different Class I suppressors in the same haploid is very deleterious, and results in a drastic decrease in growth rate on the complete medium (Gilmore, 1967).

The eight Class I suppressors, S_1–S_8, all substitute tyrosine in the peptide chain when an ochre codon is read in the messenger-RNA. Gilmore et al. (1968) have reached this conclusion from an examination of the peptides in the digests of cytochrome c isolated from strains having a suppressible allele of the structural gene, cyt_{1-2}, combined with the various suppressors. Thus, we can argue that the drastic effects of two Class I suppressors in the same haploid are not due to a deficiency in the normal level of tyrosyl-t-RNA because the ochre-specific suppressor t-RNA would recognize the tyrosine codons, UAU and UAC, as well as the ochre codon, UAA. Therefore, we conclude that it is the excessive translation of the normal chain-terminating codons or 'periods' in the message which is deleterious.

Cells apparently have means of counteracting the effects of super-suppressors. A cytoplasmically inherited 'suppressor', ψ, of an ochre-specific suppressor has been described by Cox (1965). A genic modifier of suppressors acting upon ad_{2-1} has been reported by Inge-Vechtomov (1967). The term 'anti-suppressors' has been proposed by Hawthorne (1967) for genic modifiers which greatly restrict the action spectra of amber suppressors. The investigation of anti-suppressors is being pursued as a genetic approach to the study of the processes involved in messenger translation and protein synthesis.

IX. Cytoplasmic Inheritance

A. RESPIRATION-DEFICIENT MUTANTS

Respiration deficiency in *Saccharomyces cerevisiae*, i.e. the inability to grow on non-fermentable substrates, may result from either a nuclear gene mutation or the loss of a cytoplasmic factor (ρ^-). In either case, the phenotype on the usual growth media is a small white colony, and the term *petite* is generally used for these mutants. The genic or segregational petites when crossed to *grandes* (respiratory-sufficient haploids) give hybrids whose asci contain two petite and two grande segregants (Chen *et al.*, 1950). The mutations result in the loss of one or more of the respiratory enzymes (Sherman and Slonimski, 1964). All cytoplasmic petites examined have the following features in common: (i) the mutational event is irreversible; (ii) there is a simultaneous loss of several enzymes associated with the mitochondria, principally cytochrome c oxidase and cytochromes a, a_5, and b; and (iii) the mode of inheritance is non-Mendelian.

Evidence for a cytoplasmic factor in the inheritance of respiratory sufficiency has been documented by Ephrussi (1953). The essential observation was that crosses of petite with grande haploids yielded grande diploids which, when sporulated, gave only grande progeny. The involvement of the cytoplasm in the transmission of a hereditary factor was corroborated by the experiments of Wright and Lederberg (1957). By crossing petites with grandes derived from stocks which formed zygotic cells in which nuclear fusion was often delayed, they were able to recover haploid progeny that carried the recessive nuclear markers from the petite parent, but which were respiratory sufficient.

The above observations were made with 'neutral' petites which are actually a minority among the spontaneously arising vegetative petites. Generally, a mass mating of a petite with a grande strain will give a high proportion of petite diploid clones. The explanation proposed is that the normal cytoplasmic factor is supplanted or suppressed by a defective factor from the 'suppressive' petite (Ephrussi et al., 1955). The suppressive factor is not immediately expressed in the zygote, and functioning and synthesis of respiratory enzymes continues for a time after copulation. The zygote can exist in a 'premutational' state during which it can, subject to culture conditions, give rise to either grande or petite buds (Ephrussi et al., 1966). Therefore, it is suggested that the suppressive factor does not interfere with the expression of the normal factor but with its replication.

Given petite isolates show characteristic degrees of suppressiveness which can range from zero for neutral petites to nearly 100 for the highly suppressive petites (Sherman and Ephrussi, 1962). Strains with intermediate values have been investigated to see if they were composed of a mixture of neutral and highly suppressive petites, or were homogeneous with an intermediate degree of suppressiveness reflecting the probability that the particular defective factor would prevail in descendants from petites by grande zygotes. The latter interpretation was favoured by sub-cloning experiments which showed a strong mother-daughter correlation (Ephrussi and Grandchamp, 1965).

Vegetative petites occur spontaneously at rates much higher than those associated with gene mutations. Most haploid cultures contain 1–5% petite cells. Massive induction of petites occurs in cultures treated with a variety of chemicals, including acriflavine, caffeine, copper sulphate, manganese chloride and tetrazolium chloride (Nagai et al., 1961). To obtain petites by the use of these chemicals, it is necessary to provide conditions that will permit growth of the culture. Acriflavine is used extensively because, with concentrations that cause negligible killing, one can obtain essentially 100% petite induction of newly formed buds (Marcovich, 1953). Growth at elevated temperatures also results in petite buds (Sherman, 1959; Ogur et al., 1960), and can be used in place of the acriflavine treatment to distinguish between the mother cells and progeny (James, 1961). Mother cells as well as the buds become petite when cultures are treated with 5-fluorouracil (Moustacchi and Marcovich, 1963; Lacroute, 1963) or ethidium bromide (Slonimski et al., 1968). Ethidium bromide is highly effective in causing petite induction without any lethality. Even non-growing cells can be converted to petites with this agent.

Irradiation with ultraviolet light is another means of inducing vegetative petites. Experiments with monochromatic light have implicated a nucleic acid as the target (Raut and Simpson, 1955; Wilkie, 1963). Reports that induction by ultraviolet radiation is sensitive to photoreactivation support this conclusion (Sarachek, 1958; Pittman et al., 1959). A comparison of the ultraviolet dosage curves for the induction of petites in anaerobically and aerobically grown haploid cultures indicates that a single hit suffices in the former, while multiple events are required with the latter (Wilkie, 1963). Mitochondrial DNA is proposed as the likely target; a single template is postulated for the anaerobic cells, while the aerobic cells would contain many copies.

Because the respiratory enzymes that are missing in the vegetative petites are associated with mitochondria, it was natural to equate this organelle with the cytoplasmic factor (Ephrussi, 1953; Wilkie, 1964). However, loss of the cytoplasmic factor in petites does not mean loss of the mitochondria; they are found in petite strains, albeit deformed, as revealed by electron microscopy (Yotsuyanagi, 1962).

Mitochondria-rich fractions can be prepared from petite as well as from grande cells in sufficient quantity for chemical analyses and enzyme assays (Mahler et al., 1964). A finding which may prove to be of particular significance in the elucidation of the 'petite mutation' came from the comparison of mitochondrial DNA from grande, neutral petite, and highly suppressive petite strains (Mounolou et al., 1966). Mitochondrial DNA (m-DNA) isolated from the three strains differed in density: values obtained were $1\cdot687$ g/cm^3 for the grande, $1\cdot683$ g/cm^3 for the neutral petite, and $1\cdot695$ g/cm^3 for the suppressive petite. Among the explanations to be considered are that: (i) petite induction results in a rearrangement in the m-DNA molecule, a deletion or duplication, which changes the GC:AT ratio; or (ii) under the conditions of petite induction, there is selection of a minority species from a population of m-DNA molecules.

Cytochemical investigations of petites, 'near-petites', and wild-type strains for cytochrome oxidase and succinate dehydrogenase activities have shown that different mitochondrial populations can persist in the same cell (Avers et al., 1965a, b). Of particular interest were the near-petites, isolated after acriflavine treatment and scored as petite by the tetrazolium assay (Ogur et al., 1957), but which were still respiratory sufficient. These variants had decreased numbers (90–40%) of cytochrome oxidase-positive mitochondria and generally more (116–150%) succinate dehydrogenase-positive mitochondria. The wild-type parental stock, a diploid, had an average of 45 cytochrome oxidase-positive and

30 succinate dehydrogenase-positive mitochondria per cell. The classical petite strains have no cytochrome oxidase-positive mitochondria, and the average number of succinate dehydrogenase-positive mitochondria was the same as that of wild-type.

Since the first report of a nuclear gene which gave the petite phenotype (Chen *et al.*, 1950), at least 17 different segregational petites have been investigated (Raut, 1953; Sherman, 1963; Mackler *et al.*, 1965; Hawthorne and Mortimer, 1968). The petite mutants are found with a rather high incidence, 1–3% of the survivors, after a mutagenesis regimen with nitrous acid, ethylmethanesulphonate, or ultraviolet radiation that gives about 50% kill. A systematic study of petite mutants by complementation tests for allelism may be ambiguous because both parents may be lacking the cytoplasmic factor, ρ, and thus the zygotes would still be ρ^- and incapable of respiration. This could incorrectly indicate allelism between petite mutations in different genes. Haploids with the petite genes p_3 or p_{12} do not maintain ρ^+ and will not complement neutral vegetative petites. With other mutants, for example, p_2, only a fraction of the cells in a fresh isolate are found to carry ρ^+, and the factor is rapidly lost with continued culturing (Sherman and Ephrussi, 1962). It is not known, in the case of p_3 or p_{12}, whether the primary function of these genes is the maintenace of the ρ^+ factor, or if the loss of ρ^+ is a secondary effect of the particular genetic block in the respiratory system. However, segregation of a gene conferring a propensity for the vegetative petite mutation has been noted. The mutant strains contained 12–75% petite cells, while the wild-type segregants had the usual 1–5% petites (Ephrussi and Hottinguer, 1951). Negrotti and Wilkie (1968) have described a recessive gene, *gi* (glucose induction), that results in the loss of ρ^+ under conditions of glucose repression or anoxia. The ρ^+ factor is retained if the stock is maintained on media containing glycerol, galactose or maltose; ρ^+ is lost when the strain is grown on glucose or on maltose under anoxia. With the customary protocol of isolating and maintaining mutants on media containing glucose, this gene could easily be mistaken for the usual segregational petite.

For mutants which retain ρ^+, a relationship between the cytoplasmic factor and given petite genes can be demonstrated. The onset of respiration has been studied in synchronized populations of zygotes from crosses of genic petites, p_1, p_5 and p_7, with a neutral vegetative petite (Jakob, 1965). Each cross had a characteristic lag, 0·5 h for $p_5 \times \rho^-$, 4·8 h for $p_7 \times \rho^-$, and 9·7 h for $p_1 \times \rho^-$. When two genic petites are crossed, the lag observed is that of the parent having the shortest lag; i.e. the $p_5 \times p_7$ zygotes started to respire in 0·5 h. However, when a $p_5\rho^-$ stock was used in the cross, $p_5\rho^- \times p_7\rho^+$, the lag period was 4·8 h. It is proposed that, with ρ^+ present in a mutant haploid, development of the mitochondria will proceed until the stage of the particular genetic block. This will vary with the different genes; the ρ^+ mutant that is the furthest along in the maturation of the mitochondria will show the shortest lag.

The respiration-deficient phenotype is seen with diverse genetic defects that may not be directly related to mitochondrial maturation. A mutant with a block in protoporphyrin synthesis gave a petite phenotype when cultured on synthetic media (Yčas and Starr, 1953). Certain alleles of cy_3, a regulatory gene for cytochrome c synthesis, exhibit the petite phenotype (Reilly and Sherman, 1965). Mutants glt_1 and glt_2, which lack aconitate hydratase activity, have pleiotropic phenotypes; besides a requirement for glutamate, both have the petite morphology and, in addition, glt_2 (formerly designated ly_8) causes a lysine requirement. Most glt_2 segregants were found to be ρ^-. Only a few clones had a small fraction (0·01–0·03%) ρ^+ cells (Ogur et al., 1965). The glt_1 mutant will respire on lactate medium and will utilize lactate for growth in a medium with 0·2% glucose (Bowers et al., 1967). Another mutant with the petite phenotype but capable of respiration is p_9 or op_1; the latter designation is for oxidative phosphorylation, the physiological reaction which is actually defective (Kováč et al., 1967).

B. MITOCHONDRIAL DRUG RESISTANCE

A new dimension to studies of extranuclear inheritance in yeast has stemmed from the discovery that antibiotics such as chloramphenicol and erythromycin, which inhibit protein synthesis in bacteria, prevent synthesis of respiratory enzymes in yeast (Clark-Walker and Linnane, 1966; Huang et al., 1966). In the presence of these drugs, the wild-type sensitive strains can grow on glucose or other sugars but not on non-fermentable substrates. Thus, one is able to select for resistant mutants on a medium containing glycerol together with chloramphenicol or one of the other drugs (Wilkie et al., 1967).

Both cytoplasmic and nuclear mutations can confer resistance to erythromycin (Thomas and Wilkie, 1968a; Linnane et al., 1968). The nuclear genes are dominant and show Mendelian inheritance with two resistant and two sensitive segregants in asci from resistant × sensitive crosses. The cytoplasmic mutations were located on the ρ factor, described above, by the following experiments. Zygotes from a cross of resistant × sensitive grande haploids form clones in which the percentage of resistant diploid cells varies from 0% to 100%. If the cross is repeated using a ρ^- isolate of the resistant clone, none of the zygotes gives rise to clones containing resistant cells. Conversely, if the sensitive stock is made the ρ^- parent, then the zygotes give pure clones of resistant cells. Tetrad analysis of asci from the resistant diploids gave only 4 : 0 ratios for resistance versus sensitivity.

Only limited instances of cross-resistance were seen when mutants selected on one drug were tested for resistance to a variety of other drugs. Mutants selected on chloramphenicol were generally resistant to tetracycline, but chloramphenicol resistance and erythromycin resistance were achieved independently (Wilkie *et al.*, 1967). With the isolation of mutants having specific resistance characteristics, it is possible to look for evidence of recombination between differently marked ρ factors.

Thomas and Wilkie (1968b) have made crosses with three different cytoplasmic mutants for resistance to erythromycin, spiramycin, and paromycin. Their protocol called for the crosses to be made under anaerobic conditions. The rationale was that parents grown anaerobically and mated anaerobically would not develop mitochondria enclosing the DNA and, moreover, replication of the m-DNA would be minimal; thus, recombination processes would be facilitated. The zygotes were plated on minimal medium and incubated aerobically until the prototrophic diploid colonies were large enough to pick and test on the glycerol-drug plates. This regimen resulted in clones that nearly always contained only one type of mitochondria; either one or the other parental types or a recombinant type. Recombinant types from erythromycin × paromycin crosses were frequent. Generally from 25% to 40% of the clones were recombinant; but resistances to erythromycin and spiramycin were closely linked, less than 5% of the clones being recombinant. Multiple resistance was demonstrated to be due to recombinant ρ factors, and not mixed populations of mitochondria, by testing the clones on media containing combinations of the drugs. When crosses were carried out aerobically, the zygotes gave clones of mixed parental mitochondrial types with very few recombinants.

C. KILLER CHARACTER

Bevan and Makower (1963) reported observations on three yeast cell types with respect to the killer characters; these were described as killers, sensitives and neutrals. Sensitive cells are killed when mixed and grown together with killer cells or when placed in the cell-free filtrate from a 48-h liquid culture of a killer stock. The neutral cells are resistant to the toxic agent of the killer cells, and in turn show no toxicity towards the sensitive cells. The toxic agent of killer cells has been characterized as a protein.

The killer phenotype is determined by a cytoplasmic factor 'k' and a nuclear gene M controlling the maintenance of 'k'. Another cytoplasmic factor, 'n', confers the neutral phenotype, and again M is needed for its maintenance. Cells without a cytoplasmic factor are sensitive; they can be either M or m with respect to the nuclear gene. Crosses of killers × sensitives give diploid killers. These hybrids give tetrad ratios of 4:0 (killer:sensitive) if the sensitive parent also carried M; if not, there are two killer clones per ascus, the segregants carrying M. If the microcolonies are sampled within the first few divisions after the germination of the spore by pairing cells to a sensitive M tester lacking the cytoplasmic factor, the presence of the 'k' factor can be demonstrated in the m segregants because the diploids obtained are again killers heterozygous for M. The same pattern of inheritance is seen with the neutral character in crosses of neutrals with sensitives. Crosses of killers × neutrals give 'weak killers'. These hybrids yield asci with abnormal tetrad ratios and, even with vegetative propagation, they will give variant clones of strong killer or neutral cells (Somers and Bevan, 1969; Bevan and Somers, 1969).

References

Ahmed, K. A. and Woods, R. A. (1967). *Genet. Res.* **9**, 179 193.

Ali, A. M. M. (1967). *Can. J. Genet. Cytol.* **9**, 473–481.

Avers, C. J., Pfeffer, C. R. and Rancourt, M. W. (1965a). *J. Bact.* **90**, 481–494.

Avers, C. J., Rancourt, M. W. and Lin, F. H. (1965b). *Proc. natn. Acad. Sci. U.S.A.* **54**, 527–535.

Barben, H. (1966). *Genetica* **37**, 109–148.

Bassel, J. and Douglas, H. C. (1968). *J. Bact.* **95**, 1103–1110.

Bechet, J. and Wiame, J. M. (1965). *Biochem. biophys. Res. Commun.* **21**, 226–234.

Bechet, J., Wiame, J. M. and De Deken-Grenson, M. (1962). *Archs int. Physiol. Biochim.* **70**, 564–565.

Bechet, J., Wiame, J. M. and Grenson, M. (1965). *Archs int. Physiol. Biochim.* **73**, 137–139.

Beck, J. C., Mattoon, J. R., Hawthorne, D. C. and Sherman, F. (1968). *Proc. natn. Acad. Sci. U.S.A.* **60**, 186–193.

Bevan, E. A. and Makower, M. (1963). *Proc. 11th Intern. Congress Genetics* **1**, 203 (abstract).

Bevan, E. A. and Somers, J. M. (1969). *Heredity, Lond.* in press.

Bhattacharjee, J. K. and Lindegren, G. (1964). *Biochem. biophys. Res. Commun.* **17**, 554–558.

Bowers, W. D., McClary, D. O. and Ogur, M. (1967). *J. Bact.* **94**, 482–484.

"Carbondale Yeast Genetics Conference" (1963). *Microbial Genetics Bull.* **19**, suppl.

Chen, S. Y., Ephrussi, B. and Hottinguer, H. (1950). *Heredity, Lond.* **4**, 337–351.

Cherest, H. and de Robichon-Szulmajster, H. (1966). *Genetics, Princeton* **54**, 981–991.

Clarke, C. H. (1965). *Genet. Res.* **6**, 433–441.

Clark-Walker, G. D. and Linnane, A. W. (1966). *Biochem. biophys. Res. Commun.* **25**, 8–13.

Cooper, D., Banthorpe, D. U. and Wilkie, D. (1967). *J. molec. Biol.* **26**, 347–350.

Costello, W. P. and Bevan, E. A. (1964). *Genetics, Princeton* **50**, 1219–1230.

454 ROBERT K. MORTIMER AND DONALD C. HAWTHORNE

ROBERT K. MORTIMER AND DONALD C. HAWTHORNE

454 ROBERT K. MORTIMER AND DONALD C. HAWTHORNE

Cox, B. S. (1965). *Heredity, Lond.* **20**, 505–521.

Cox, B. S. and Bevan, E. A. (1962). *New Phytol.* **61**, 342–355.

Crick, F. H. C. (1966). *J. molec. Biol.* **19**, 548–555.

De Deken, R. H. (1963). *Biochim. biophys. Acta* **78**, 606–616.

de Leeuw, A. (1967). *Genetics, Princeton* **56**, 554 (abstract).

De Moss, J. A. (1965). *Biochem. biophys. Res. Commun.* **18**, 850–857.

de Robichon-Szulmajster, H. (1958). *Science, N. Y.* **127**, 28–29.

de Robichon-Szulmajster, H. (1967). *Bull. Soc. Chim. biol.* **49**, 1431–1462.

de Robichon-Szulmajster, H. and Cherest, H. (1966). *Genetics, Princeton* **54**, 993–1006.

de Robichon-Szulmajster, H. and Cherest, H. (1967). *Biochem. biophys. Res. Commun.* **28**, 256–262.

de Robichon-Szulmajster, H. and Magee, P. T. (1968). *Eur. J. Biochem.* **3**, 492–501.

de Robichon-Szulmajster, H., Surdin, Y. and Mortimer, R. K. (1966). *Genetics, Princeton*, **53**, 609–619.

Desborough, S. and Shult, E. E. (1962). *Genetica* **33**, 69–78.

Dorfman, B. (1964). *Genetics, Princeton* **50**, 1231–1243.

Douglas, H. C. (1961). *Biochim. biophys. Acta* **52**, 209–211.

Douglas, H. C. and Condie, F. (1954). *J. Bact.* **68**, 662–670.

Douglas, H. C. and Hawthorne, D. C. (1964). *Genetics, Princeton* **49**, 837–844.

Douglas, H. C. and Hawthorne, D. C. (1966). *Genetics, Princeton* **54**, 911–916.

Douglas, H. C. and Pelroy, G. (1963). *Biochim. biophys. Acta* **68**, 155–156.

Doy, C. H. and Cooper, J. M. (1966). *Biochim. biophys. Acta* **127**, 302–316.

Emeis, C. C. (1958). *Die Brauerei* **11**, 160–163.

Emeis, C. C. (1966). *Z. Naturf.* **21**, 816–817.

Emeis, C. C. and Gutz, H. (1958). *Z. Naturf.* **13b**, 647–650.

Emerson, S. (1963). *In* "Methodology in Basic Genetics" (W. J. Burdette, ed.), pp. 167–208. Holden-Day, San Francisco.

Emerson, S. (1967). *A. Rev. Genet.* **1**, 201–220.

Ephrussi, B. (1953). "Nucleo-Cytoplasmic Relations in Microorganisms". Clarendon Press, Oxford, England.

Ephrussi, B. and Grandchamp, S. (1965). *Heredity, Lond.* **20**, 1–7.

Ephrussi, B. and Hottinguer, H. (1951). *Cold Spring Harbor Symp. quant. Biol.* **16**, 75–84.

Ephrussi, B., Hottinguer, H. and Tavlitski, J. (1949). *Ann. Inst. Pasteur* **76**, 419–450.

Ephrussi, B., Hottinguer, H. de M. and Roman, H. (1955). *Proc. natn. Acad. Sci. U.S.A.* **41**, 1065–1071.

Ephrussi, B., Jakob, H. and Grandchamp, S. (1966). *Genetics, Princeton* **54**, 1–29.

Esposito, M. S. (1968). *Genetics, Princeton* **58**, 507–527.

Esser, K. and Kuenen, R. (1965). "Genetik der Pilze". Springer-Verlag, New York.

Fincham, J. R. S. (1966). "Genetic Complementation". W. A. Benjamin, Inc., New York.

Fink, G. R. (1964). *Science, N. Y.* **146**, 525–527.

Fink, G. R. (1966). *Genetics, Princeton* **53**, 445–459.

Fogel, S. and Hurst, D. D. (1967). *Genetics, Princeton* **57**, 455–481.

Fogel, S. and Mortimer, R. K. (1969). *Proc. natn. Acad. Sci. U.S.A.* in press.

Fowell, R. R. (1952). *Nature, Lond.* **170**, 578.

Galzy, P. (1964). *Heredity, Lond.* **19**, 731–733.

Galzy, P. and Bizeau, C. (1965). *Heredity, Lond.* **20**, 31–36.

Galzy, P. and Bizeau, C. (1966). *Ann. Technol. agric.* **15**, 289–294.

Garen, A. (1968). *Science, N. Y.* **160**, 149–159.

Gilliland, R. B. (1949). *C. r. Trav. Lab. Carlsberg* **24**, 347–356.

Gilliland, R. B. (1954). *Nature, Lond.* **173**, 409.

Gilmore, R. A. (1967). *Genetics, Princeton* **56**, 641–658.

Gilmore, R. A. and Mortimer, R. K. (1966). *J. molec. Biol.* **20**, 307–311.

Gilmore, R. A., Stewart, J. W. and Sherman, F. (1968). *Biochim. biophys. Acta* **161**, 270–272.

Gits, J. J. and Grenson, M. (1967). *Biochim. biophys. Acta* **135**, 507–516.

Grenson, M. (1966). *Biochim. biophys. Acta* **127**, 339–346.

Grenson, M., Mousset, M., Wiame, J. M. and Bechet, J. (1966). *Biochim. biophys. Acta* **127**, 325–338.

Guglielminetti, R., Bonatti, S., Loprieno, N. and Abbondandolo, A. (1967). *Mutation Res.* **4**, 441–447.

Gutz, H. (1966). *J. Bact.* **92**, 1567–1568.

Gutz, H. (1967). *Science, N. Y.* **158**, 796–798.

Haefner, K. (1966). *Z. VererbLehre* **98**, 82–90.

Haefner, K. and Howrey, L. (1967). *Mutation Res.* **4**, 219–221.

Hartwell, L. H. (1967). *J. Bact.* **93**, 1662–1670.

Hartwell, L. H. and McLaughlin, C. S. (1968). *Proc. natn. Acad. Sci. U. S. A.* **59**, 422–428.

Hawthorne, D. C. (1955). *Genetics, Princeton* **40**, 511–518.

Hawthorne, D. C. (1958). *Heredity, Lond.* **12**, 273–284.

Hawthorne, D. C. (1963a). *Genetics, Princeton* **48**, 1727–1729.

Hawthorne, D. C. (1963b). *Proc. 11th Intern. Congr. Genet.* **1**, 34–35 (abstract).

Hawthorne, D. C. (1967). *Genetics, Princeton* **56**, 563.

Hawthorne, D. C. and Friis, J. (1964). *Genetics, Princeton* **50**, 829–839.

Hawthorne, D. C. and Mortimer, R. K. (1960). *Genetics, Princeton* **45**, 1085–1110.

Hawthorne, D. C. and Mortimer, R. K. (1963). *Genetics, Princeton* **48**, 617–620.

Hawthorne, D. C. and Mortimer, R. K. (1968). *Genetics, Princeton* **60**, in press.

Herman, A. and Roman, H. (1966). *Genetics, Princeton* **53**, 727–740.

Holliday, R. (1964). *Genet. Res.* **5**, 282–304.

Horn, P. and Wilkie, D. (1966a). *Heredity, Lond.* **21**, 625–635.

Horn, P. and Wilkie, D. (1966b). *J. Bact.* **91**, 1388.

Huang, M., Biggs, D. R., Clark-Walker, G. D. and Linnane, A. W. (1966). *Biochim. biophys. Acta.* **114**, 434–436.

Hurst, D. D. and Fogel, S. (1964). *Genetics, Princeton* **50**, 435–458.

Hwang, D. S., Lindegren, C. C., Bhattacharjee, J. K. and Roshanmanesh, A. (1964b). *Can. J. Genet. Cytol.* **6**, 414–418.

Hwang, Y. L. and Lindegren, G. (1966). *Can. J. Genet. Cytol.* **8**, 677–694.

Hwang, Y. L., Lindegren, G. and Lindegren, C. C. (1963). *Can. J. Genet. Cytol.* **5**, 290–298.

Hwang, Y. L., Lindegren, G. and Lindegren, C. C. (1964). *Can. J. Genet. Cytol.* **6**, 373–380.

Inge-Vechtomov, S. G. (1964). *Vest. Leningrad Univ.* **9**, 112–117.

Inge-Vechtomov, S. G. (1965). *Genetika* **2**, 22–26.

Inge-Vechtomov, S. G. (1967). *Genetika* **9**, 176–178.

Inge-Vechtomov, S. G. and Kozhin, S. A. (1964). *Issledovania po Genetike* **2**, 77–85.

Inge-Vechtomov, S. G. and Simarov, B. V. (1967). *Issledovania po Genetike* **3**, 127–148.

Inge-Vechtomov, S. G., Simarov, B. V., Soidla, T. R. and Kozhin, S. A. (1966). *Z. VererbLehre* **98**, 375–384.

Jakob, H. (1965). *Genetics, Princeton* **52**, 75–98.

James, A. P. (1955). *Genetics, Princeton* **40**, 204–213.

James, A. P. (1961). *Can. J. Genet. Cytol.* **3**, 128–134.

James, A. P. and Lee-Whiting, B. (1955). *Genetics, Princeton* **40**, 826–831.

James, A. P. and Werner, M. M. (1965). *Radiat. Bot.* **5**, 359–382.

Johnson, B. F. (1968). *J. Bact.* **95**, 1169–1172.

Johnston, J. R. and Mortimer, R. K. (1959). *J. Bact.* **78**, 292.

Johnston, J. R. and Mortimer, R. K. (1967). *Heredity, Lond.* **22**, 297–303.

Jones, E. E. and Broquist, H. P. (1965). *J. biol. Chem.* **240**, 2531–2536.

Kakar, S. N. (1963). *Genetics, Princeton* **48**, 967–979.

Kakar, S. N. and Wagner, R. P. (1964). *Genetics, Princeton* **49**, 213–222.

Kakar, S. N., Zimmerman, F. and Wagner, R. P. (1964). *Mutation Res.* **1**, 381–386.

Kováč, L., Lachowicz, T. M. and Slonimski, P. P. (1967). *Science, N. Y.* **158**, 1564–1567.

Lacroute, F. (1963). *C. r. hebd. Séanc. Acad. Sci., Paris* **257**, 4213–4216.

Lacroute, F. (1964). *C. r. hebd. Séanc. Acad. Sci., Paris* **258**, 2884–2886.

Lacroute, F. (1966). Thesis: L'Université de Paris.

Lacroute, F. (1968). *J. Bact.* **95**, 824–832.

Lacroute, F. and Slonimski, P. P. (1964). *C. r. hebd. Séanc. Acad. Sci., Paris* **235**, 2172–2174.

Lacroute, F., Piérard, A., Grenson, M. and Wiame, J. M. (1965). *J. gen. Microbiol.* **40**, 127–142.

Lampen, J. O. (1966). *Symp. Soc. gen. Microbiol.* **16**, 111–130.

Laskowski, W. (1956). *Genetics, Princeton* **41**, 98–106.

Laskowski, W. (1960). *Z. Naturf.* **15b**, 495–506.

Laskowski, W. (1962). *Z. Naturf.* **17b**, 93–108.

Laskowski, W., Lochmann, E. R., Jannsen, S. and Fink, E. (1968). *Biophysik* **4**, 233–242.

Leupold, U. (1950). *C. r. Trav. Lab. Carlsberg* **24**, 381–480.

Leupold, U. (1955). *Arch. Julius Klaus-Stift. VererbForsch.* **30**, 506–516.

Leupold, U. (1956a). *C. r. Trav. Lab. Carlsberg* **26**, 221–251.

Leupold, U. (1956b). *J. Genet.* **54**, 411–426.

Leupold, U. (1957). *Allg. Pathol. Bakteriol.* **20**, 535–544.

Leupold, U. (1958). *Cold Spring Harb. Symp. quant. Biol.* **23**, 161–170.

Leupold, U. (1961). *Arch. Julius Klaus-Stift. VererbForsch.* **36**, 89–117.

Leupold, U. and Hottinguer, H. (1954). *Heredity, Lond.* **8**, 243–258.

Leupold, U. and Gutz, H. (1963). *In* "Genetics Today", Proc. XIth Intern. Cong. Genet., Vol. 1, pp. 31–35.

Lindegren, C. C. (1949). "The Yeast Cell. Its Genetics and Cytology". Educational Publishers, St. Louis.

Lindegren, C. C. (1953). *J. Genet.* **51**, 625–637.

Lindegren, C. C. (1955). *Science, N. Y.* **121**, 605–607.

Lindegren, C. C. and Lindegren, G. (1943). *Proc. natn. Acad. Sci. U. S. A.* **29**, 306–308.

Lindegren, C. C. and Lindegren, G. (1951a). *Indian Phytopathol.* **4**, 11–20.

Lindegren, C. C. and Lindegren, G. (1951b). *J. gen. Microbiol.* **5**, 885–893.

Lindegren, C. C. and Lindegren, G. (1953a). *Genetics, Princeton* **38**, 73–78.

Lindegren, C. C. and Lindegren, G. (1953b). *Genetica* **26**, 430–444.

Lindegren, C. C. and Lindegren, G. (1956). *J. gen. Microbiol.* **15**, 19–28.

Lindegren, C. C., Pittman, D. D. and Ranganathan, B. (1957). *Proc. Intern. Genet. Symp., Japan. Cytologia Suppl.* 42–50.

Lindegren, C. C., Lindegren, G., Shult, E. and Hwang, Y. L. (1962). *Nature, Lond.* **194**, 260–265.

Lindegren, G., Hwang, Y. L., Oshima, Y. and Lindegren, C. C. (1965). *Can. J. Genet. Cytol.* **7**, 491–499.

Lingens, F. and Oltmanns, O. (1964). *Z. Naturf.* **19b**, 1058–1065.

Lingens, F. and Oltmanns, O. (1966). *Z. Naturf.* **21b**, 660–663.

Lingens, F., Goebel, W. and Uesseler, H. (1966). *Biochem. Z.* **346**, 357–367.

Lingens, F., Goebel, W. and Uesseler, H. (1967). *Eur. J. Biochem.* **1**, 363–374.

Linnane, A. W., Lamb, A. J., Christodoulou, C. and Lukins, H. B. (1968). *Proc. natn. Acad. Sci. U. S. A.* **59**, 1288–1293.

Loprieno, N. (1966). *Mutation Res.* **3**, 486–493.

Loprieno, N. and Clarke, C. H. (1965). *Mutation Res.* **2**, 312–319.

Mackler, B., Douglas, H. C., Will, S., Hawthorne, D. C. and Mahler, H. R. (1965). *Biochemistry, N. Y.* **4**, 2016–2020.

Magee, P. T. and de Robichon-Szulmajster, H. (1968a). *Eur. J. Biochem.* **3**, 502–506.

Magee, P. T. and de Robichon-Szulmajster, H. (1968b). *Eur. J. Biochem.* **3**, 507–511.

Magni, G. E. (1961). *Atti. Assoc. Genetica Ital.* **6**, 47–50.

Magni, G. E. (1963). *Proc. natn. Acad. Sci. U. S. A.* **50**, 975–980.

Magni, G. E. and von Borstel, R. C. (1962). *Genetics, Princeton* **47**, 1097–1108.

Magni, G. E. and Puglisi, P. (1966). *Cold Spring. Harb. Symp. quant. Biol.* **31**, 699–704.

Magni, G. E., von Borstel, R. C. and Steinberg, C. M. (1966). *J. molec. Biol.* **16**, 568–570.

Mahler, B., Mackler, B., Grandchamp, S. and Slonimski, P. P. (1964). *Biochemistry, N. Y.* **3**, 668–676.

Manney, T. R. (1964). *Genetics, Princeton* **50**, 109–121.

Manney, T. R. (1968). *Genetics, Princeton* **60**, in press.

Manney, T. R. and Mortimer, R. K. (1964). *Science, N. Y.* **143**, 581–582.

Marcovich, H. (1953). *Ann. Inst. Pasteur* **85**, 199–216.

McKey, T. (1967). *U.S.A.E.C. Document UCRL 18066*, pp. 38–42.

Megnet, R. (1965a). *Mutation Res.* **2**, 328–331.

Megnet, R. (1965b). *J. Bact.* **90**, 1032–1035.

Megnet, R. (1965c). *Pathol. Microbiol.* **28**, 50–57.

Megnet, R. (1967). *Archs Biochem. Biophys.* **121**, 194–201.

Megnet, R. and Giles, N. H. (1964). *Genetics, Princeton* **50**, 967–971.

Meuris, P. (1967). *Bull. Soc. Chim. biol.* **49**, 1573–1578.

Meuris, P., Lacroute, F. and Slonimski, P. P. (1967). *Genetics, Princeton* **56**, 149–161.

Middlekauf, J. E., Hino, S., Yang, S. P., Lindegren, G. and Lindegren, C. C. (1956). *J. Bact.* **72**, 796–801.

Moat, A. G., Peters, N., Jr. and Srb, A. M. (1959). *J. Bact.* **77**, 673–677.

Mortimer, R. K. (1958). *Radiat. Res.* **9**, 312–326.

Mortimer, R. K. (1959). *Radiat. Res. Suppl.* **1**, 394–402.

Mortimer, R. K. and Gilmore, R. A. (1968). *Adv. biol. med. Phys.* **12**, 319–331.

Mortimer, R. K. and Hawthorne, D. C. (1966a). *A. Rev. Microbiol.* **20**, 151–168.
Mortimer, R. K. and Hawthorne, D. C. (1966b). *Genetics, Princeton* **53**, 165–173.
Mortimer, R. K. and Manney, T. R. (1964). *Genetics, Princeton* **50**, 270 (abstract).
Mounolou, J. C., Jakob, H. and Slonimski, P. P. (1966). *Biochem. biophys. Res. Commun.* **24**, 218–224.
Moustacchi, E. (1965). *Mutation Res.* **2**, 403–412.
Moustacchi, E. and Marcovich, H. (1963). *C. r. hebd. Séanc. Acad. Sci., Paris* **256**, 5646–5648.
Nagai, S. (1963). *J. Bact.* **86**, 299–302.
Nagai, S., Yanagashima, N. and Nagai, H. (1961). *Bact. Rev.* **25**, 404–426.
Nakai, S. and Matsumoto, S. (1967). *Mutation Res.* **4**, 129–136.
Nakai, S. and Mortimer, R. (1967). *Radiat. Res. Suppl.* **7**, 172–181.
Nashed, N. and Jabbur, G. (1966). *Z. VererbLehre* **98**, 106–110.
Nashed, N., Jabbur, G. and Zimmermann, F. K. (1967). *Molec. gen. Genet.* **99**, 69–75.
Nasim, A. and Auerbach, C. (1967). *Mutation Res.* **4**, 1–14.
Nasim, A. and Clarke, C. H. (1965). *Mutation Res.* **2**, 395–402.
Negrotti, T. and Wilkie, D., (1968). *Biochim. biophys. Acta* **153**, 341–349.
Nelson, N. M. and Douglas, H. C. (1963). *Genetics, Princeton* **48**, 1585–1591.
Oeser, H. (1962). *Arch. Mikrobiol.* **44**, 47–74.
Oeser, H. and Windisch, S. (1964). *Naturwissenschaften* **5**, 122.
Ogur, M. and St John, R. (1956). *J. Bact.* **72**, 500–504.
Ogur, M., Minckler, S., Lindegren, G. and Lindegren, C. C. (1952). *Archs Biochem. Biophys.* **40**, 175–184.
Ogur, M., St John, R. and Nagai, S. (1957). *Science, N. Y.* **125**, 928–929.
Ogur, M., Ogur, S. and St John, R. (1960). *Genetics, Princeton* **45**, 189–194.
Ogur, M., Coker, L. and Ogur, S. (1964). *Biochem. biophys. Res. Commun.* **14**, 193–197.
Ogur, M., Roshanmanesh, A. and Ogur, S. (1965). *Science, N. Y.* **147**, 1590.
Oshima, J. (1967). *J. ferment. Technol.* **45**, 550–565.
Ouchi, S. and Lindegren, C. C. (1963). *Can. J. Genet. Cytol.* **5**, 257–267.
Perkins, D. D. (1949). *Genetics, Princeton* **34**, 607–626.
Pittman, D., Ranganathan, B. and Wilson, F. (1959). *Expl Cell Res.* **17**, 368–377.
Pomper, S. and Burkholder, P. R. (1949). *Proc. natn. Acad. Sci. U. S. A.* **35**, 456–464.
Pomper, S., Daniels, K. M. and McKee, D. W. (1954). *Genetics, Princeton* **39**, 343–355.
Pontecorvo, G., Roper, J. A., Hemmons, L. M., MacDonald, K. D. and Bufton, A. W. J. (1953). *Adv. Genet.* **5**, 141–238.
Rao, S. S. and Grollman, A. P. (1967). *Biochem. biophys. Res. Commun.* **29**, 696–704.
Raut, C. (1953). *Expl Cell Res.* **4**, 295–305.
Raut, C. and Simpson, L. W. (1955). *Archs Biochem. Biophys.* **57**, 218–228.
Raypulis, E. P. and Kozhin, S. A. (1966). *Trans. Moscow Soc. Natural.* **22**, 135–139.
Reaume, S. E. and Tatum, E. L. (1949). *Archs Biochem.* **22**, 331–338.
Reilly, C. and Sherman, F. (1965). *Biochim. biophys. Acta* **95**, 640–651.
Resnick, M. A. and Mortimer, R. K. (1966). *J. Bact.* **92**, 597–600.
Resnick, M. A., Tippetts, R. D. and Mortimer, R. K. (1967). *Science, N. Y.* **158**, 803–804.
Roman, H. (1956a). *C. r. Trav. Lab. Carlsberg* **26**, 299–314.
Roman, H. (1956b). *Cold Spring Harb. Symp. quant. Biol.* **21**, 175–185.

Roman, H. (1958a), *Annls Genet.* **1**, 11–17.

Roman, H. (1958b). *Cold Spring Harb. Symp. quant. Biol.* **23**, 155–160.

Roman, H. (1963). *In* "Methodology in Basic Genetics" (W. J. Burdette, ed.), pp. 209–227. Holden-Day, San Francisco.

Roman, H. and Jacob, F. (1957). *C. r. hebd. Séanc. Acad. Sci., Paris* **245**, 1032–1034.

Roman, H. and Sands, S. M. (1953). *Proc. natn. Acad. Sci. U. S. A.* **39**, 171–179.

Roman, H., Hawthorne, D. C. and Douglas, H. C. (1951). *Proc. natn. Acad. Sci. U. S. A.* **37**, 79–84.

Roman, H., Phillips, M. M. and Sands, S. M. (1955). *Genetics, Princeton* **40**, 546–561.

Sarachek, A. (1958). *Cytologia* **23**, 143–158.

Seno, T. (1963). *Mem. Coll. Sci. Kyoto Univ.* **30**, 1–8.

Sherman, F. (1959). *J. cell. comp. Physiol.* **54**, 37–52.

Sherman, F. (1963). *Genetics, Princeton* **48**, 375–385.

Sherman, F. (1964). *Genetics, Princeton* **49**, 39–48.

Sherman, F. and Ephrussi, B. (1962). *Genetics, Princeton* **47**, 695–700.

Sherman, F. and Roman, H. (1963). *Genetics, Princeton* **48**, 255–261.

Sherman, F. and Slonimski, P. P. (1964). *Biochim. biophys. Acta* **90**, 1–15.

Sherman, F., Stewart, J. W., Margoliash, E., Parker, J. and Campbell, W. (1966). *Proc. natn. Acad. Sci. U. S. A.* **55**, 1498–1504.

Shult, E. E. and Lindegren, C. C. (1956). *Genetica* **28**, 165–176.

Shult, E. E., Desborough, S. and Lindegren, C. C. (1962). *Genet. Res.* **3**, 196–209.

Shult, E. E., Lindegren, G. and Lindegren, C. C. (1967). *Can. J. Genet. Cytol.* **9**, 723–759.

Siegal, M. R. and Sisler, H. D. (1965). *Biochim. biophys. Acta* **103**, 558–567.

Slonimski, P. P., Acher, R., Péré, G., Sels, A. and Somlo, M. (1965). *In* "Mécanismes de Régulation des Activités Cellulaires chez les Microorganismes", Centre Natl. Rech. Sci., No. 124, pp. 435–461. Paris, 1965.

Slonimski, P. P., Perrodin, G. and Croft, J. H. (1968). *Biochem. biophys. Res. Commun.* **30**, 232–239.

Smirnov, M. N., Smirnov, V. N., Budowsky, E. I., Inge-Vechtomov, S. G. and Serebriakov, N. G. (1967). *Molec. Biologia* **1**, 639–647.

Snow, R. (1966). *Nature, Lond.* **211**, 206–207.

Snow, R. (1967). *J. Bact.* **94**, 571–575.

Soidla, T. R. and Inge-Vechtomov, S. G. (1966). *Genetika* **9**, 141–150.

Soidla, T. R., Inge-Vechtomov, S. G. and Simarov, B. V. (1967). *Issledovania po Genetike* **3**, 148–164.

Somers, J. M. and Bevan, E. A. (1969). *Heredity, Lond.* in press.

Sorsoli, W. A., Spence, K. D. and Parks, L. W. (1964). *J. Bact.* **88**, 20–24.

Spiegelman, S., Sussman, R. R. and Pinska, E. (1950). *Proc. natn. Acad. Sci. U. S. A.* **36**, 591–606.

Srb, A. M. (1956). *C. r. Trav. Lab. Carlsberg* **26**, 363–380.

Stern, C. (1936). *Genetics, Princeton* **21**, 625–730.

Strömnaes, Ö. (1968). *Hereditas* **59**, 197–220.

Strömnaes, Ö. and Mortimer, R. K. (1968). *J. Bact.* **95**, 197–200.

Surdin, Y., Sly, W., Sire, J., Bordes, A. M. and de Robichon-Szulmajster, H. (1965). *Biochim. biophys. Acta* **107**, 546–566.

Takahashi, T. (1958). *Genetics, Princeton* **43**, 705–714.

Takahashi, T. (1962). *Bull. Brew. Sci.* **8**, 1–9.

Takahashi, T. (1964). *Bull. Brew. Sci.* **10**, 11–22.

Takahashi, T. (1966). *Bull. Brew. Sci.* **12**, 15–34.

Takahashi, T. and Akamatsu, K. (1963). *Seiken Zihô* **15**, 54–58.

Takahashi, T. and Ikeda, Y. (1959). *Z. VererbLehre* **90**, 66–73.
Takahashi, T., Saito, H. and Ikeda, Y. (1958). *Genetics, Princeton* **43**, 251–260.
Takano, I. and Oshima, Y. (1967). *Genetics, Princeton* **57**, 875–885.
Tauro, P., Halvorson, H. O. and Epstein, R. L. (1968). *Proc. natn. Acad. Sci. U. S. A.* **59**, 277–284.
Terui, G., Okada, H. and Oshima, Y. (1959). *Tech. Rep. Osaka Univ.* **9**, 237–259.
Thomas, D. Y. and Wilkie, D. (1968a). *Genet. Res.* **11**, 33–41.
Thomas, D. Y. and Wilkie, D. (1968b). *Biochem. biophys. Res. Commun.* **30**, 368–372.
Tingle, M., Herman, A. and Halvorson, H. O. (1968). *Genetics, Princeton* **58**, 361–371.
Tsoi, A. and Douglas, H. C. (1965). *Biochim. biophys. Acta* **92**, 513–520.
Whitehouse, H. L. K. and Hastings, P. J. (1965). *Genet. Res.* **6**, 27–92.
Wiame, J. M., Bechet, J., Mousset, M. and de Deken, M. (1962). *Archs int. Physiol. Biochim.* **70**, 766–767.
Wickerham, L. J. (1946). *J. Bact.* **52**, 293–301.
Wickerham, L. J. and Burton K. A. (1956). *J. Bact.* **71**, 290–295.
Wilkie, D. (1963). *J. molec. Biol.* **7**, 527–533.
Wilkie, D. (1964). "The Cytoplasm in Heredity". Methuen and Co., Ltd., London.
Wilkie, D. and Lee, B. K. (1965). *Genet. Res.* **6**, 130–138.
Wilkie, D. and Lewis, D. (1963). *Genetics, Princeton* **48**, 1701–1716.
Wilkie, D., Saunders, G. and Linnane, A. W. (1967). *Genet. Res.* **10**, 199–203.
Winge, Ö. (1935). *C. r. Trav. Lab. Carlsberg* **21**, 77–112.
Winge, Ö. (1952). *Heredity, Lond.* **6**, 263–269.
Winge, Ö. and Laustsen, O. (1937). *C. r. Trav. Lab. Carlsberg* **22**, 99–116.
Winge, Ö. and Laustsen, O. (1938). *C. r. Trav. Lab. Carlsberg* **22**, 235–244.
Winge, Ö. and Laustsen, O. (1939). *C. r. Trav. Lab. Carlsberg* **22**, 357–374.
Winge, Ö. and Roberts, C. (1948). *C. r. Trav. Lab. Carlsberg* **24**, 263–315.
Winge, Ö. and Roberts, C. (1949). *C. r. Trav. Lab. Carlsberg* **24**, 341–346.
Winge, Ö. and Roberts, C. (1950). *C. r. Trav. Lab. Carlsberg* **25**, 35–83.
Winge, Ö. and Roberts, C. (1952). *C. r. Trav. Lab. Carlsberg* **25**, 141–171.
Winge, Ö. and Roberts, C. (1954). *Heredity, Lond.* **8**, 295–304.
Winge, Ö. and Roberts, C. (1956). *Nature, Lond.* **177**, 383–384.
Winge, Ö. and Roberts, C. (1958). *In* "The Chemistry and Biology of Yeasts" (A. H. Cook, ed.), pp. 123–156. Academic Press, New York.
Winkler, H. (1930). "Die Konversion der Gene". Verlag Gustav Fischer, Jena.
Woods, R. A. and Bevan, E. A. (1967). *Heredity, Lond.* **21**, 121–130.
Yčas, M. and Starr, T. J. (1953). *J. Bact.* **65**, 83–88.
Wright, R. E. and Lederberg, J. (1957). *Proc. natn. Acad. Sci. U. S. A.* **43**, 919–923.
Yamasaki, T., Ito, T. and Matsudaira, Y. (1964). *Jap. J. Genet.* **39**, 147–150.
Yotsuyanagi, Y. (1962). *J. Ultrastruct. Res.* **7**, 141–158.
Zakharov, I. A. and Inge-Vechtomov, S. G. (1964). *Issledovania po Genetike* **2**, 134–139.
Zakharov, I. A., and Inge-Vechtomov, S. G. (1966). *Genetika* **8**, 112–118.
Zimmermann, F. K. and Schwaier, R. (1963). *Z. VererbLehre* **94**, 253–260.
Zimmermann, F. K. and Schwaier, R. (1967). *Molec. gen. Genet.* **100**, 63–76.
Zimmermann, F. K., Schwaier, R. and v. Laer, U. (1966a). *Z. VererbLehre* **98**, 152–166.
Zimmermann, F. K., Schwaier, R. and v. Laer, U. (1966b). *Z. VererbLehre* **98**, 230–246.

Chapter 9

Life Cycles in Yeasts

R. R. FOWELL

The Distillers Co. (Yeast) Ltd., Great Burgh, Epsom,
Surrey, England

I. Introduction

The occurrence in yeasts of an alternation of generations (haplophase and diplophase) was established by Kruis and Šatava (1918), Šatava (1918, 1934) and Winge (1935) (see Chapter 7, p. 339). Phaff and Mrak (1948) described six types of life cycle in yeasts. The first two types were exemplified by yeasts which are commonly distinguished as haploid and diploid respectively, i.e. yeasts of type 1 are haploid for most of their existence while, in type 2, the diploid condition is predominant. The criteria adopted for differentiation of the remaining types were less satisfactory. Differences in a heterogamous sexual process were used to distinguish between type 3 (*Debaryomyces*) and type 5 (*Nadsonia*). Type 4 was represented by *Saccharomycodes ludwigii* in which the ascospores conjugate on germination (haplophase restricted to the ascospores). Type 6 would not appear to be valid; it was based on the life cycle described by Windisch (1938, 1940) for *Candida pulcherrima* and *C. tropicalis* which were considered to be diploid in the vegetative state. In old cultures, thick-walled "pulcherrima" cells (chlamydospores) are produced and, according to Windisch, these form buds containing ascospores, and fusion of cells from budding ascospores was reported. However, other workers (Langeron and Luteraan, 1947; van der Walt, 1952; Lodder and Kreger-van Rij, 1952) have shown that the so-called ascospores cannot be distinguished from vegetative cells by

shape, refraction, staining and other properties, and they are probably
cells formed by budding within loose remnants of cell walls. Certainly
the available evidence indicates that species of *Candida* are haploid;
many strains are known to be imperfect forms of diploid genera such
as *Hansenula, Kluyveromyces* and *Pichia* (see Chapter 2, p. 69).

In a review of life cycles, sexuality and sexual mechanisms in fungi,

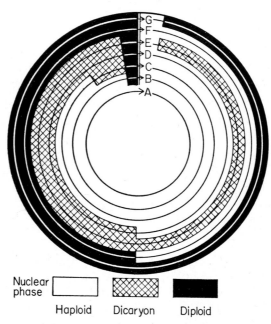

FIG. 1.   Schematic comparison of life cycles in fungi. In each cycle, changes in
nuclear phase are indicated progressing clockwise by changes in shading. The
double vertical line at the top of the diagram represents meiosis, and each of
the two narrow sectors adjacent to the line represents a single nuclear generation.
Reproduced with permission from Raper (1954). (Copyright 1954 by the American
Association for the Advancement of Science.)

Raper (1954) showed that there are seven basic types of life cycle
(Fig. 1) in which the temporal spacing of changes in nuclear phase is
considered to be of primary importance. The life cycles, which he
designated by the letters A to G, range from completely haploid to
completely diploid, and the intercalation of a unique nuclear association, the dicaryon, greatly increases the variability of the life cycles.
Thus types C, D and E are characterized by the insertion of a dicaryophase between haplophase and diplophase. The dicaryophase arises with
the fusion of sexual cells or organs unaccompanied by nuclear fusion.
Haploid nuclei become associated in one or more pairs (dicaryons)
and these divide mitotically to provide dicaryons for further cells.

Eventually, nuclear fusion occurs in an ascus or basidium where meiosis immediately ensues and haploid spores are formed; the diplophase is thus restricted to a single cell (ascus or basidium). The dicaryophase becomes increasingly dominant at the expense of the haplophase between the higher Ascomycetes (type C) and the Basidiomycetes (types D and E); in the latter group, smuts (type E) exhibit extreme reduction of the haplophase as the haploid basidiospores fuse to initiate the dicaryophase.

II. Types of Life Cycle

Although Raper (1954) quoted examples of yeasts among fungal representatives of the different types of life cycle, a detailed consideration of the applicability of his scheme to yeasts has not been previously undertaken. Most of the basic types of life cycle are found within the group of yeasts, but his scheme requires some modification to accommodate most of the common strains of *Saccharomyces cerevisiae* (Fig. 2).

Type A is represented by exclusively haploid yeasts such as *Torulopsis* (Fig. 2(a)), *Candida*, *Rhodotorula* and other members of the Cryptococcaceae in which the existence of a perfect or diploid stage is unknown (many strains in this family are now known to be imperfect forms of diploid sporogenous yeasts). In type B, the haplophase is still predominant; the diplophase is usually restricted to the zygote formed by the fusion of two haploid cells accompanied by nuclear fusion. The zygote functions as an ascus; its diploid nucleus undergoes meiosis which is followed by the formation of haploid spores. Yeasts of this type include species of *Coccidiascus*, *Debaryomyces*, *Endomyces*, *Endomycopsis*, *Lipomyces*, *Nematospora*, *Pichia*, *Schizosaccharomyces* (Fig. 2(b)), *Schwanniomyces*, *Wingea* and *Zygosaccharomyces*. The zygote frequently arises by a heterogamous process involving conjugation between a mother cell and its bud. In *Schwanniomyces* (Ferreira and Phaff, 1959; Fig. 2(c)) and *Wingea* (van der Walt, 1967), the diploid nucleus, formed by fusion of the nuclei of the two cells, undergoes meiosis in the bud (meiosis bud), but the mother cell functions as the ascus; in *Schwanniomyces*, this usually contains only a single spore. It is probable that similar cytological events are involved in heterogamous strains of *Debaryomyces*, *Pichia* and *Zygosaccharomyces*. Parthenogenetic asci, formed without preceding conjugation, occur in yeasts such as species of *Nematospora*, *Endomyces* and *Endomycopsis*; in *Schizosaccharomyces pombe*, polyploidy accounts for the production of such asci (Leupold, 1956).

The type B life cycle is also found in certain species of *Hansenula* which exist largely or entirely in a haploid state in nature (Wickerham,

1951; Wickerham and Burton, 1962). In *H. capsulata*, the diplophase is momentary and has not yet been detected in natural habitats: a haploid cell forms a bud, the two nuclei fuse to form a diploid nucleus in the mother cell, and then meiosis is followed by the formation of two hat-shaped spores. Other species, such as *H. silvicola*, *H. angusta* and *H. californica*, are capable of producing a few diploid cells; once diploid cells are formed, they tend to outgrow haploid cells because of their

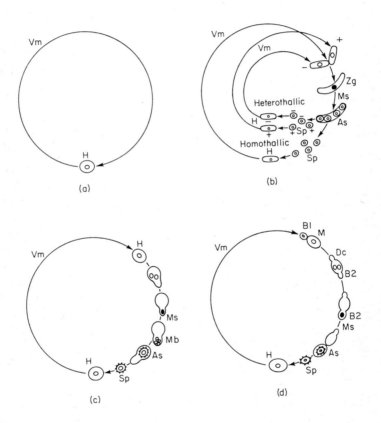

FIG 2. Life cycle in yeasts. (a) *Terulopsis* (type A). (b) *Shizosaccharomyces pembe* (type B). (c) *Schwanniomyces* (type B). (d) *Nadsonia* type C, showing short dicaryophase). (e) *Hansenula wingei* (type F). (f) *Saccharomyces cerevisiae* (type G, with transient dicaryophase). (g) *Saccharomyces chevalieri* and certain strains of *Saccharomyces cerevisiae* (type G; spores self-diploidize on or soon after germination). (h) *Saccharomycodes ludwigii* (type G; spores conjugate on germination). Haplophase is indicated by thin lines, diplophase by thick lines. Mating types are indicated as a, α (+,−). As, indicates an ascus; B1, first bud; B2, second bud; CSp, conjugating spores; D, diploid cell; H, haploid cell; Mb, meiosis bud; Ms, meiosis; Sp, spore; Vm, vegetative multiplication by budding; and Zg, zygote.

faster growth rate, especially when sporulation is prevented by cultivation on rich laboratory media.

As already mentioned, the next three types of life cycle found in fungi, namely C, D and E, are characterized by a dicaryophase. There have been few reports of dicaryosis in yeasts, but it may be more common than is generally supposed. The dicaryophase, when present, is nearly always of short duration; but though this is the distinguishing

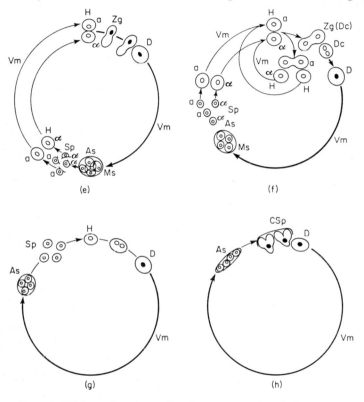

feature of type C life cycle, the only clear example of this type in yeasts is in *Nadsonia* (Fig. 2(d)). This yeast is haploid for most of its existence. Conjugation occurs between a mother cell and its bud, the nucleus of which becomes associated with the nucleus of the mother cell to form a dicaryon; the nuclei are carried with the protoplasmic contents into a second bud formed at the opposite end of the mother cell and nuclear fusion in this bud, which functions as an ascus, is followed by meiosis and the formation of one or more spores (Nadson and Konokotina, 1911, 1912, 1926).

The existence of a stable dicaryophase of long duration in yeasts is rare, and has only been reported for *Sporidiobolus* (Nyland, 1948, 1949;

Laffin and Cutter, 1959) and, more recently, *Rhodosporidium* (Banno, 1967). The latter genus was created for a new species, *Rhodosporidium toruloides*, consisting of a dicaryotic mycelium with clamp connections formed after conjugation of haploid cells of opposite mating-type from *Rhodotorula glutinis*. Nuclear fusion occurs in chlamydospores produced on the mycelium. When a chlamydospore germinates, meiosis occurs and a promycelium of 2–4 haploid cells is produced; these bud off cells to form the *Rhodotorula* phase. The life cycle (long haplophase, long dicaryophase and short diplophase) is a good example of type D. Although *Sporidiobolus* resembles *Rhodosporidium* in many respects, the chlamydospores are reported to bud off diploid cells (Laffin and Cutter, 1959). In the absence of definite information about the occurrence of meiosis and the haplophase, it is not possible to allocate the life cycle in Raper's (1954) scheme.

The stability of the dicaryotic condition in the above yeasts is to be expected, as the development of mycelium with clamp connections, in addition to other features, indicates their close relationship with the Basidiomycetes. Dicaryosis in the genus *Saccharomyces*, however, would appear to be a transient and inherently unstable condition. The existence of a dicaryophase in yeasts was first described by Guilliermond (1910) for the Johannisberg II strain of *Sacch. cerevisiae* and also *Sacch. paradoxus*. In certain zygotes, the nuclei failed to fuse and divided side by side, two nuclei remaining in the zygote and two passing into the first bud where nuclear fusion finally occurred. Renaud (1938) observed a delay in nuclear fusion to the third cell in a chain of cells proliferated from a zygote of a wine yeast (*Sacch. cerevisiae* var. *ellipsoideus*). Later, he reported considerable variation in frequency of dicaryotic zygotes among four wine yeasts (Renaud, 1946); in some instances, zygotes budded off haploid cells which later conjugated to form new zygotes. Fowell (1951) found that a high proportion of zygotes in a baker's yeast (*Sacch. cerevisiae*) were dicaryotic and budded off dicaryotic or haploid cells. Certain pairs of haploid cultures fail to produce diploid yeasts in spite of zygote formation, i.e. the dicaryophase breaks down permanently into haplophase. According to Lindegren and Lindegren (1954), all zygotes of *Saccharomyces* are dicaryons, nuclear fusion occurring in the buds proliferated in succession from their surface. The evidence adduced for dicaryosis in the above yeasts was entirely cytological; it remained for Wright and Lederberg (1957) to provide conclusive genetical proof of its occurrence by crossing double and triple mutants of Fowell's haploid strains and demonstrating complete absence of genic recombination.

In Raper's (1954) scheme of life cycles, dicaryosis is always associated with a short diplophase. In *Saccharomyces* spp., however, the diplophase is usually dominant and the haplophase of variable duration. Dicaryosis in this genus may be regarded as a degenerative condition associated with a decline in mating activity of haploid cells. It tends to result in prolongation of the haplophase, and the life cycle in typical members of this genus, e.g. many strains of *Sacch. cerevisiae* (Fig. 2(f)), is to be considered as type G, modified by the insertion of an unstable dicaryophase which often breaks down into haplophase (sometimes permanently): thus it tends to assume the features of a type F life cycle which has a long haplophase.

The type F life cycle, in which there are only two stages, haplophase and diplophase, of about equal duration, is well represented by *Hansenula wingei* (Wickerham and Burton, 1962; Fig. 2(e)); in this species, nuclear fusion occurs very early in the conjugation of haploid cells, before cell wall dissolution is complete (Conti and Brock, 1965), i.e. there is no dicaryophase.

Yeasts with type G life cycle are predominantly diploid, the haplophase being of short duration, the spores often conjugating at or soon after germination. Several species of *Hansenula* exist largely in the form of diploid cells, e.g. *H. saturnus*, *H. suaveolens* and *H. subpelliculosa* (Wickerham and Burton, 1962). It is interesting to note that vegetative cells of *Kluyveromyces africanus* (Roberts, 1960) are predominantly diploid, while the related species, *Kluyveromyces polysporus*, has a type B life cycle, the diplophase being restricted to the zygote (Roberts and van der Walt, 1959). In many yeasts, especially strains of *Sacch. cerevisiae*, haploid cells have a tendency to self-diploidize and, where this tendency is strong, the haplophase is considerably reduced. *Saccharomyces chevalieri* (Fig. 2(g)) possesses a gene D which causes self-diploidization of all spore cultures soon after their formation (Winge and Roberts, 1949). In extreme cases, the haplophase is represented only by the ascospores. The ascospores of *Saccharomycodes ludwigii* (Fig. 2(h)) conjugate in pairs inside the asci (Hansen, 1891). In some yeasts, the ascospores diploidize on germination. Winge and Laustsen (1937, 1940) showed that, in a strain of *Sacch. cerevisiae* var. *ellipsoideus*, the nucleus in certain spores divides into two, and these fuse when the spores germinate. The spores in some strains of *Sacch. cerevisiae*, e.g. Johannisberg II, all diploidize on or soon after germination, while other strains (including a wine yeast) show a 2:2 segregation of diploid and haploid spore cultures (R. R. Fowell, unpublished observations); it is probable that these strains are homozygous and heterozygous respectively for the gene D. It should be noted that polyploid yeasts also segregate spores of different ploidy but diploid spores are usually in the majority. Finally, many yeasts, including certain strains of brewer's yeasts, have apparently lost the ability to sporulate as the result of

16

prolonged cultivation or storage on artificial media, i.e. the haplophase is completely eliminated.

III. Evolutionary Significance of Life Cycles

In his review of sexual processes in fungi, Raper (1954) showed that there is a striking lack of correlation between various combinations of sexual features (life cycle, pattern of sexuality, sexual mechanism) and accepted phylogenetic groups. In particular, homothallism, often considered the most primitive pattern of sexuality, occurs in conjunction with all types of life cycle and in every major group of fungi. In yeasts, also, homothallism is associated with different life cycles, e.g. species of *Schizosaccharomyces*, *Debaryomyces* (type B), *Nadsonia* (type C) and *Saccharomyces chevalieri* (type G).

Nevertheless, as Raper (1954) pointed out, there is a loose correlation in fungi between each of the separate facets of sexuality and morphological specialization. In general, life cycles have assumed their most complex form in the evolution of the higher fungi. Haploidy predominates in the Phycomycetes, and dicaryosis, which is restricted in the Ascomycetes, assumes major importance in the Basidiomycetes. There are, however, many exceptions to this generalization, notably the restriction of haploid–diploid and diploid cycles to one group of aquatic Phycomycetes and the yeasts. To account for the random distribution of homothallism in fungi, Raper (1954) suggested that the different types of heterothallism found in this group probably evolved independently, at different levels, from a primitive homothallic stock; then, within each heterothallic group, homothallic species arose secondarily, probably as the result of chromosome changes. There is clear evidence for the existence of two types of homothallism in yeasts (see Chapter 7, p. 345); the homothallism of *Sacch. chevalieri* and related yeasts is superimposed on a heterothallic condition from which it is obviously derived.

Guilliermond (1931, 1937, 1940) maintained that haploid and diploid yeasts represent two distinct phylogenetic lines, with haploid yeasts evolving from a form similar to *Eremascus*, and diploid yeasts being derived from *Taphrina*. As indicated above, Raper (1954) showed that that the evolution of fungi has involved a general progression from haploidy towards diploidy. In agreement with this view, detailed studies of the yeast genera *Hansenula* and *Dekkeromyces* (Wickerham, 1951; Wickerham and Burton, 1956, 1962) have revealed that the most fundamental evolutionary trend in these yeasts has been an increase in the ratio of diploid to haploid cells. Moreover, the evolution of most diploid species has been accompanied by an increase in fermentative powers (ability to ferment more strongly and ferment more sugars), increasing ability to synthesize vitamins and produce esters, and progressive changes in morphological properties including the ability to produce mycelium and changes in colony appearance; primitive species form mucoid colonies which are replaced by mat or rugose colonies in more recently evolved species. At present, five phylogenetic branches are recognized in the genus *Hansenula* (Wickerham and Burton, 1962). The two main lines are a homothallic and a heterothallic chain of species respectively; these have evolved from primitive haploid forms living on coniferous trees, and later deciduous trees, to diploid species in habitats such as soil and fruit juices where free-living conditions encourage the development of a highly competitive flora. Two other lines, however, although showing some progression towards diploidy, have developed an increasing dependence on coniferous trees; this has been accompanied by some loss of physiological activity, especially a diminution in fermentative properties.

The above observations certainly support the view that diploidy has conferred a definite advantage in the evolution of yeasts. Barnett (1959) has suggested that the simultaneous existence of haplophase and diplophase among yeasts in natural habitats may be of survival value, and this applies in particular to yeasts such as *Sacch. rouxii* (Hjort, 1956) and *H. wingei* (Wickerham and Burton, 1962) that show rapid oscillations between the two phases. The haplophase permits of the rapid appearance and selection of mutants, and therefore of early adaptation to changes in the environment. The diplophase provides a stable phenotype because the expression of recessive genes is masked; but it also permits of favourable genetic variation and the opportunity of gene recombination and heterosis. Accumulation of unfavourable genes in this phase is counteracted by the opportunity for their elimination when diplophase is succeeded by haplophase. The ability of diploid cells to grow more rapidly than haploid cells assists in the development and extension of a diplophase. Under laboratory conditions, involving culture on rich media, mixtures of haploid and diploid cells eventually become exclusively diploid, especially when prevented from sporulating.

It may be doubted whether all haploid yeasts are primitive. *Torulopsis*, according to Šatava (1934), is a haploid yeast derived from *Saccharomyces*, a view shared by Winge and Laustsen (1937) and also Lindegren and Lindegren (1943). It may be noted that loss of mating activity among haploid cultures of *Saccharomyces* spp. maintained over long periods is not always due to diploidization (see Chapter 7, p. 341). Lindegren and Lindegren (1943) found that conjugation occurred between strains of *Torulopsis* spp. when they were properly paired, and Wickerham and Burton (1952) showed that certain asporogenous (haploid) yeasts in nature represent the mating types of known sporogenous yeasts; thus strains of *Torulopsis sphaerica*, when mixed in the right combinations, produced ascospores typical of the sporogenous yeast, *Zygosaccharomyces lactis*. Several other yeasts, classified with *Torulopsis* in the Cryptococcaceae, are now known to be the haploid forms of sporogenous genera (see Chapter 2, p. 65). It would be premature, however, to assume that all yeasts in this family are the haploid counterparts of such genera.

It is possible that, in nature, many yeasts exist for long periods or even indefinitely in the haploid state because conditions are rarely, if ever, favourable for conjugation. In heterothallic yeasts, conjugation depends on contact between haploid strains of opposite mating-type; this is not a limiting factor in homothallic yeasts in which, however, the value of ensured conjugation is apparently offset by the disadvantages of homozygosity, i.e. lack of genetic recombination. Physiological factors, also, may limit conjugation as this is influenced by degree of aeration, sugar concentration, salts and temperature (see Chapter 7, pp. 354–357).

The evolution of diploid yeasts has probably involved, first, the acquisition and enhancement of mating activity, assisted in some yeasts by an agglutinative mechanism (Wickerham, 1956) and, secondly, the stabilization of the diplophase. A high degree of mating activity promotes a rapid transition from haplophase to diplophase as soon as conditions become favourable for conjugation. The selective development of diploid cells, at the expense of haploid cells, must have depended initially on the creation of a stable nuclear condition under normal growth conditions. Thus, in *Schizosaccharomyces* spp., predominance of the haplophase is due to the instability of the zygotic diplophase. It would appear, therefore, that the evolution of a stable diplophase in yeasts was rendered possible by the development of a genetic mechanism which was only able to initiate meiosis and induce spore formation under conditions unfavourable to growth.

References

Banno, I. (1967). *J. gen. appl. Microbiol., Tokyo* **13**, 167–196.
Barnett, J. A. (1959). *Experientia* **15**, 99–100.
Conti, S. F. and Brock, T. D. (1965). *J. Bact.* **90**, 524–533.
Ferreira, J. D. and Phaff, H. J. (1959). *J. Bact.* **78**, 352–361.
Fowell, R. R. (1951). *J. Inst. Brew.* **57**, 180–195.
Guilliermond, A. (1910). *Zentbl. Bakt. ParasitKde (Abt II)* **26**, 577–589.
Guilliermond, A. (1931). *C. r. hebd. Séanc. Acad. Sci., Paris* **192**, 577–599.
Guilliermond, A. (1937). "La Sexualité, le Cycle de Développement, la Phylogénie et la Classification des Levures". Masson et Cie, Paris.
Guilliermond, A. (1940). *Bot. Rev.* **6**, 1–24.
Hansen, E. C. (1891). *C. r. Trav. Lab. Carlsberg* **3**, 44–66.
Hjort, A. (1956). *C. r. Trav. Lab. Carlsberg. Sér. physiol.* **26**, 161–179.

Kruis, K. and Šatava, J. (1918). Cited in Winge, Ö. (1935). *C. r. Trav. Lab. Carlsberg. Sér. physiol.* **21**, Postscript, pp. 110–111.

Laffin, R. J. and Cutter, V. M. (1959). *J. Elisha Mitchell scient. Soc.* **75**, 89–96, 97–100.

Langeron, M. and Luteraan, P. J. (1947). *Annls Parasit. hum. comp.* **22**, 254–275.

Leupold, U. (1956). *C. r. Trav. Lab. Carlsberg. Sér. physiol.* **26**, 221–251.

Lindegren, C. C. and Lindegren, G. (1943). *Proc. natn. Acad. Sci. U.S.A.* **29**, 306–308.

Lindegren, C. C. and Lindegren, G. (1954). *Cytologia* **19**, 45–47.

Lodder, J. and Kreger-van Rij, N. J. W. (1952). "The Yeasts: a Taxonomic Study", 713 pp. North Holland Publishing Co., Amsterdam.

Nadson, G. A. and Konokotina, A. G. (1911). *Izv. imp. S.-Peterb. bot. Sada* **11**, 117–143.

Nadson, G. A. and Konokotina, A. G. (1912). *Wschr. Brau.* **29**, 309–313, 332–336.

Nadson, G. A. and Konokotina, A. G. (1926). *Annls Sci. nat. (Bot.)* **8**, 165–182.

Nyland, G. (1948). *Mycologia* **40**, 478–481.

Nyland, G. (1949). *Mycologia* **41**, 686–701.

Phaff, H. J. and Mrak, E. M. (1948). *Wallerstein Labs Commun.* **11**, 261–279.

Raper, J. R. (1954). *In* "Sex in Microorganisms" (D. H. Wenrich, I. F. Lewis and J. R. Raper, eds.), pp. 42–81. American Association for the Advancement of Science Publ. No. 37, Washington.

Renaud, J. (1938). *C. r. hebd. Séanc. Acad. Sci., Paris* **206**, 1397–1399.

Renaud, J. (1946). *Revue gén. Bot.* **629**, 193–211; **630**, 241–274; **631**, 289–314.

Roberts, C. (1960). *C. r. Trav. Lab. Carlsberg. Sér. physiol.* **31**, 325–341.

Roberts, C. and Walt, J. P. van der (1959). *C. r. Trav. Lab. Carlsberg. Sér. physiol.* **31**, 129–148.

Šatava, J. (1918). Cited in Winge, O. (1935). *C. r. Trav. Lab. Carlsberg. Sér. physiol.* **21**, Postscript, pp. 110–111.

Šatava, J. (1934). *Congr. int. tech. chim. Sucr. Distill.* **3**. Paris.

Walt, J. P. van der (1952). Thesis: University of Delft.

Walt, J. P. van der (1967). *Antonie van Leeuwenhoek* **33**, 97–99.

Wickerham, L. J. (1951). "Taxonomy of Yeasts", 56 pp. U.S. Dept. Agric. Tech. Bull. No. *1029*, Washington.

Wickerham, L. J. (1956). *C. r. Trav. Lab. Carlsberg. Sér. physiol.* **26**, 423–443.

Wickerham, L. J. and Burton, K. A. (1952). *J. Bact.* **63**, 449–451.

Wickerham, L. J. and Burton, K. A. (1956). *J. Bact.* **71**, 290–295.

Wickerham, L. J. and Burton, K. A. (1962). *Bact. Rev.* **26**, 382–397.

Windisch, S. (1938). *Arch. Mikrobiol.* **9**, 551–554.

Windisch, S. (1940). *Arch. Mikrobiol.* **11**, 368.

Winge, Ö. (1935). *C. r. Trav. Lab. Carlsberg. Sér. physiol.* **21**, 77–109.

Winge, Ö. and Laustsen, O. (1937). *C. r. Trav. Lab. Carlsberg. Sér. physiol.* **22**, 99–117.

Winge, Ö. and Laustsen, O. (1940). *C. r. Trav. Lab. Carlsberg. Sér. physiol.* **23**, 17–39.

Winge, Ö. and Roberts, C. (1949). *C. r. Trav. Lab. Carlsberg. Sér. physiol.* **24**, 341–346.

Wright, R. E. and Lederberg, J. (1957). *Proc. natn. Acad. Sci. U.S.A.* **43**, 919–937.

Author Index

Italic numbers indicate pages on which a reference is listed

A

Abadie, F., 49, 51, *74*

Abbondandolo, A., 392, *455*

Abdel-Wahab, M. F., 334, *376*

Abrahams, I., 147, *180*

Acher, R., 390, 434, *459*

Acta, H. W., 159, *174*

Adams, A. M., 100, *102*, 311, 314, 315, 317, 318, 319, 320, 321, 323, 326, 327, 360, *376*

Adams, K. F., 187, 188, *216*

Adzet, J. M., 48, *74*

Afzelius, B. A., 250, *297*

Agar, H. D., 19, *73*, 237, 260, *297*

Ahearn, D. G., 55, 66, *73*, *75*, *76*, 101, 102, *104*, 156, *174*

Ahmad, K. U., 198, *215*

Ahmad, M., 198, *215*, 342, 343, 344, *376*

Ahmed, K. A., 427, *453*

Ainley, R., 139, *174*

Ainsworth, G. C., 128, 145, *174*

Aist, J. R., 276, *297*, *302*

Ajello, E., 150, *179*

Ajello, L., 145, 149, 150, 155, 165, *174*

Akamatsu, K., 395, *459*

Al-Doory, Y., 100, *102*

Ales, J. M., 188, *216*

Ali, A. M. M., 415, *453*

Almon, L., 171, *174*

Alper, R. E., 241, *297*

Anderson, K. W., 203, *215*

Anderson, T., 117, *174*

Andrieu, S., 35, *74*

Andriole, V. T., 140, 141, *174*

Andrus, C. F., 276, *297*

Arblaster, P. G., 114, 117, *181*

Arentzen, W. P., 147, *178*

Arisi, C., 157, *176*

Aristova, M. V., 218, *216*

Arpin, M., 90, 92, 93, 98, 100, *103*

Artargaveytia-Allende, R. C., 156, *174*

Aschner, M., 33, *73*, 151, *174*

Ashby, S. F., 204, *215*

Assis-Lopes, L., 129, *182*, 196, *218*

Atchinson, R. W., 148, *182*

Aubert, J. P., 147, 148, *176*

Auerbach, C., 392, *458*

Ausherman, R. J., 150, *179*

Austwick, P. K. C., 128, 129, 137, 141, 145, 157, *174*, *177*, *178*, 196, *216*

Avers, C. J., 250, 263, 264, 271, *297*, *298*, 449, *453*

Azare-Nyako, A., 214, *215*

B

Babcock, G. E., 351, *382*

Bab'eva, I. P., 54, *73*, *75*, 208, 209, 210, *215*

Bacon, J. S. D., 229, 230, *297*

Baillon, H., 164, *174*

Baker, J. M., 45, *73*, 97, 98, *102*

Baker, R. D., 143, 145, *174*, *177*

Bakerspigel, A., 279, 287, *300*

Balows, A., 150, *179*

Baltatu, G., 330, *376*

Bandoni, R. J., 34, 63, *76*

Banno, I., 30, 63, 65, 66, *73*, *74*, 465, *470*

Banthorpe, D. U., 427, *453*

Barben, H., 402, 439, 440, 443, *453*

Barbesier, J., 137, *174*

Barfatani, M., 71, *73*

Barker, B. T. P., 311, 317, 318, 319, 320, 339, 353, *376*

Barker, D. C., 289, *301*

Barker, E. R., 60, *75*

Barnett, J. A., 31, 32, *73*, *74*, 469, *470*

Barthe, J., 156, *182*

Bartley, W., 263, 264, 267, 269, *297*, *300*, *302*

478

Gilmore, R. A., 369, *378*, 402, 443, 444, 446, 447, *455, 457*
Giordano, A., 152, *180*
Girbardt, M., 283, 284, *298*
Gits, J. J., 426, *455*
Gitter, M., 129, *177*
Gloor, F., 140, *180*
Goebel, W., 433, *457*
Gold, W., 173, *177*
Gonzales, R. O., 156, *179*
Gonzales-Ochoa, A., 109, *177*
Goodchild, R. T., 121, *176*
Goodman, J., 327, *378*
Gordon, M. A., 148, 158, 160, 161, 162, 163, 170, 171, 172, *175, 177*
Gorin, P. A. J., 34, *76*
Gorodkowa, A. A., 313, 359, *378*
Goslings, W. R. O., 116, *176*
Goto, M., 158, *180*
Graciansky, P. de, 170, *177*
Graf, K., 137, *177*
Graham, V. R., 360, *378*
Grandchamp, S., 265, *299*, 448, 449, *454, 457*
Gray, J. E., 140, *178*
Green, D. E., 273, *298*
Greenberg, S. M., 156, 166, *179*
Gregory, P. H., 100, 102, *103*, 186, 187, 194, *216*
Gregson, A. E. W., 121, *177*
Grenson, M., 421, 422, 426, 431, *453, 455, 456*
Gresham, G. A., 117, 140, *177, 182*
Grieshaber, E., 249, *298*
Griffin, P., 348, 351, 352, *378*
Grimley, P. M., 156, *177*
Grinbergs, J., 97, 99, *103*
Grollman, A. P., 427, *458*
Gronlund, A. F., 336, *378*
Grose, E., 147, *177*
Gruby, M., 108, *177*
Guerrier, G., 156, *182*
Guglielminetti, R., 392, *455*
Guilliermond, A., 3, *4*, 37, 44, 45, *74*, 204, *216*, 248, 251, 253, 255, 258, 260, 279, *298*, 322, 339, 341, 353, *378*, 466, 468, *470*
Guillon, J-C., 128, *182*
Gunkel, W., 101, 102, *104*
Günther, T., 237, *298*
Gustafson, B. A., 96, *103*

Gutz, H., 304, 309, 345, 368, 369, *377*, *378*, 390, 391, 396, 413, 415, 420, 439, *454, 455, 456*
Guze, L. B., 138, *177*

H

Haddow, W. R., 206, *216*
Haefner, K., 357, 375, *378*, 412, 429, *455*
Hagedorn, H., 260, *298*, 307, 309, *378*
Hagihara, B., 263, 265, *300, 301*
Hajsig, M., 100, *103*, 140, 157, *177*
Haley, L. D., 117, 138, *177*
Halle, M. A., 171, *181*
Halpern, C., 311, 319, 320, 321, 336, *380*
Halvorson, H. O., 271, *299*, 415, 435, *460*
Hamilton, E., 315, 318, *379*
Hamilton, E. D., 187, 188, *216*
Hanabusa, J., 121, *179*
Hansemann, D., 144, *182*
Hansen, E. Chr., 37, *74*, 184, 197, 198, 206, 215, *216*, *298*, 307, 311, 312, 313, 316, 318, 319, 339, *378*, 467, *470*
Hansen, P., 139, *177*
Harley, J. L., 195, 196, *216*
Harper, R. A., 276, 291, 292, 293, 294, 296, *298*
Harris, G., 376, *378*
Harris, L. J., 113, 114, 115, *177*
Hartelius, V., 330, 360, 363, *378*
Harter, L. L., 276, 283, *297*
Hartwell, L. H., 429, 430, *455*
Harven, E. de, 283, *298*
Hasenclever, H. F., 138, 147, 148, 157, 171, *177*
Hasegawa, T., 66, *74*
Hashimoto, T., 19, *74*, 308, 309, 311, 334, *378*
Haskins, R. H., 72, *74*
Haslbrunner, E., 266, *301*
Hastings, E. G., 360, *378*
Hastings, P. J., 417, 420, *460*
Haugen, R. K., 143, 145, *174, 177*
Havelkova, M., 250, *298*
Havener, W. H., 139, *181*
Hawker, L. E., 279, *298*

17*

Subject Index

A

Abies concolor, yeasts in slime fluxes from, 90

Acetate-containing media for inducing yeast sporulation, 361

Acetate, stimulatory effect of on yeast sporulation, 320

Acetohydroxyacid synthetase in yeast, 433

Acid phosphatase activity of the yeast wall, 234

Acid production by yeasts, 33

Acriflavine, use of to induce petites, 448

Actidione as an anti-yeast antibiotic, 172

Actidione resistance of yeasts, 33, 427

Adenine-requiring yeast mutants, 393

Adenosine triphosphatase activity of yeast plasmalemma, 240

Age of culture, effect of on yeast sporulation, 316

Agglutination in *Hansenula wingei*, biochemical basis of, 348

Agglutination, sexual, in yeasts, 347

Agglutinative factors in *Hansenula wingei*, 350

Amber codons, mutations in with yeasts, 444

Ambrosia bark beetles, yeasts associated with, 97

Amino-acid content of yeast spores, 334

Amino acids, effect of on yeast sporulation, 323

Aminopeptidase activity of the yeast wall, 234

Ammonium ions, effect of on yeast sporulation, 325

Ammonium nitrogen, utilization of by yeasts, 80

Amphotericin B as an antibiotic active against yeasts, 172

Anascosporogenous yeasts, 65

Anastomosis in yeasts, 19

Aneuploid segregants in yeasts, 413

Aneuploidy in yeasts, 341

Angular cheilosis, 119

Animal pathogens, yeasts as, 107

Animals, incidence of *Candida albicans* in, 128

yeasts associated with, 93

Anions, effect of on yeast sporulation, 327

Antagonistic effects of yeasts, 80

Antarctic soils, yeasts in, 83

Anthranilate synthase in yeast, 433

Antibiotics active against yeast diseases, 172

Antibiotics, effect of on incidence of *Candida albicans*, 117

effects of on yeasts, 81

Antifungal agents, use of in treating yeast diseases, 173

Antigenicity of pathogenic yeasts, 170

Antigens of yeasts, 35

Anti-suppressors in yeasts, 447

Apiculate yeasts, budding in, 232

Apples, yeasts associated with, 92

Aquatic environments, yeasts in, 102

Arginine-requiring yeast, 421

Aromatic amino-acid pathway, genetics of in yeast, 433

Arthrospores, yeast, 7

Ascal, classes in yeasts, 404

Ascal spores, fusion of in yeasts, 339

Asci, dissection of, 363

yeast, nature of, 307

Ascomycete nucleus, nature of, 274

B

C